Springer Series in Materials Science

Volume 222

The Springer Series in Materials Science covers the complete spectrum of materials physics, including fundamental principles, physical properties, materials theory and design. Recognizing the increasing importance of materials science in future device technologies, the book titles in this series reflect the state-of-the-art in understanding and controlling the structure and properties of all important classes of materials.

More information about this series at http://www.springer.com/series/856

Claudia Felser · Atsufumi Hirohata
Editors

Heusler Alloys

Properties, Growth, Applications

Springer

Editors
Claudia Felser
Max-Planck-Institut für Chemische Physik fester Stoffe
Dresden
Germany

Atsufumi Hirohata
Department of Electronics
University of York
York
UK

ISSN 0933-033X ISSN 2196-2812 (electronic)
Springer Series in Materials Science
ISBN 978-3-319-21448-1 ISBN 978-3-319-21449-8 (eBook)
DOI 10.1007/978-3-319-21449-8

Library of Congress Control Number: 2015945328

Springer Cham Heidelberg New York Dordrecht London

Printed on acid-free paper

Springer International Publishing AG Switzerland is part of Springer Science+Business Media (www.springer.com)

Preface

Since the discovery of Cu_2MnSn by Friedrich Heusler in 1903, this type of ternary alloys has been intensively investigated due to its ferromagnetism without containing any ferromagnetic elements, such as Fe, Co, Ni and Gd. Such an alloy, known as Heusler alloy, has been demonstrated to form thermal antiphase boundaries, which can be utilised for shape-memory materials and magnetic refrigeration. The Heusler alloys have then attracted more attention from the spintronics community since the first prediction of the half-metallicity on a half-Heusler alloy, NiMnSb, by de Groot et al. in 1983. The Heusler alloys are therefore known to be one of the best half-metallic ferromagnets to achieve 100 % spin polarisation at room temperature due to their high Curie temperature and large intrinsic magnetic moments. In order to implement the Heusler alloys into spintronic devices, the half-metallicity needs to be maintained in their thin film form. This requires to overcome major challenges; the formation of fully-ordered phase, the avoidance of atomic disorder at surfaces and the interfaces and introduction of strong magnetic anisotropy.

These three challenges have been initially discussed from the theoretical point of view, followed by experimental analysis on the properties of the Heusler alloys in a bulk form. Based on the technical advancement in growth, these alloys have been grown in a thin-film form over the last decades. These films have recently patterned into a nanometre-scale junction for their device implementation. In such a nano-junction, the surface contributions dominate the overall properties and hence emphasise the above challenges even more than ever. It is therefore a good time to overview the development of the theory of the Heusler alloys as well as understanding and improvement of the properties of the alloys.

This book consists of seven parts from fundamental theory to device applications. The first part focuses on the theory of Heusler compounds, including two chapters by the pioneering research groups on half-metallic Heusler alloys; theory of Heusler and full-Heusler compounds by Galanakis and basics and perspectives of magnetic Heusler compounds by Felser. The properties of the alloys are described in the following part. Two techniques are used to characterise the alloys, i.e., spin-resolved photoemission (Aeschlimann) and nuclear magnetic resonance

(Wurmehl) in the first two chapters. Their properties in nanometric scale (Hütten) as well as their atomic interactions (Grin) are shown afterwards. The magnetic and electric properties of binary-form Heusler alloys (Coey) are then discussed. Their off-stoichiometric properties (Leighton) are also discussed at the end of the part.

The third part highlights three major applications of the Heusler alloys; spintronics (Hirohata), thermoelectrics (Balke) and thermodynamics (James). These chapters refer to the advantages of the usage of these alloys as well as the obstacles for their applications as related to the above challenges. Such applications rely on the two-dimensional form of the alloys, which is covered by the following part (Hono). The evaluation of the magnetic properties of these films is reviewed in Part V, including ultrafast optics (Hillebrands), Mössbauer spectroscopy (Mibu) and X-ray magnetic circular dichroism (Elmer). In the next part, further evaluation of these films in multilayered junctions is shown to study their giant magnetoresistive (Takanashi) and tunnelling magnetoresistive behaviours (Mitani). The influence of their interfacial disorder (Yamamoto) and exchange coupling (O'Grady) onto their behaviours are also explained in the latter half of this part. At the end of this book, new emerging applications of the Heusler alloys are discussed, especially topological insulators (Chadov).

On behalf of the authors, we sincerely wish this book serves as a comprehensive handbook of Heusler alloys in a timely manner and contributes to further development in this exciting research field. We are grateful to all the authors who spent their precious time and knowledge to realise this book. We also thank the tremendous technical support from Springer.

Dresden Claudia Felser
York Atsufumi Hirohata

Contents

Part I Theory of Heusler Compounds

Part II Properties

Part VI Device Applications

Contributors

Martin Aeschlimann Department of Physics and Research Center OPTIMAS, University of Kaiserslautern, Kaiserslautern, Germany

Alexander Auge Department of Physics, Thin Films and Physics of Nanostructures, Bielefeld University, Bielefeld, Germany

Benjamin Balke Institut für Anorganische und Analytische Chemie, Johannes Gutenberg-Universität, Mainz, Germany

D. Bende Max-Planck-Institut für Chemische Physik fester Stoffe, Dresden, Germany

Kanwal Preet Bhatti Department of Chemical Engineering and Materials Science, University of Minnesota, Minneapolis, MN, USA

Stanislav Chadov Max-Planck-Institute für Chemische Physik fester Stoffe, Dresden, Germany

Mirko Cinchetti Department of Physics and Research Center OPTIMAS, University of Kaiserslautern, Kaiserslautern, Germany

J.M.D. Coey Department of Physics, Trinity College Dublin, Dublin 2, Ireland

Sami El-Khatib Department of Physics, American University of Sharjah, Sharjah, United Arab Emirates; NIST Center for Neutron Research, National Institute of Standards and Technology, Gaithersburg, MD, USA

Hans-Joachim Elmers Johannes Gutenberg-Universität, Mainz, Germany

H. Endo Nihon University, Koriyama, Fukushima, Japan

G.H. Fecher Max-Planck-Institut für Chemische Physik fester Stoffe, Dresden, Germany

Claudia Felser Max-Planck-Institute für Chemische Physik fester Stoffe, Dresden, Germany

G. Vallejo Fernandez Department of Physics, University of York, York, UK

Roman Fetzer Department of Physics and Research Center OPTIMAS, University of Kaiserslautern, Kaiserslautern, Germany

Luke R. Fleet Department of Physics, University of York, Heslington, UK; Department of Materials, Imperial College London, London, UK

Iosif Galanakis Department of Materials Science, School of Natural Sciences, University of Patras, Patras, Greece

Yu. Grin Max-Planck-Institut für Chemische Physik fester Stoffe, Dresden, Germany

Kohei Hamaya Graduate School of Engineering Science, Osaka University, Toyonaka, Japan

Burkard Hillebrands Fachbereich Physik and Forschungszentrum OPTIMAS, Technische Universität Kaiserslautern, Kaiserslautern, Germany

Atsufumi Hirohata Department of Electronics, University of York, Heslington, UK; PRESTO, Japan Science and Technology Agency, Kawaguchi, Japan

Kazuhiro Hono Magnetic Materials Unit, National Institute for Materials Science, Tsukuba, Japan

Andreas Hütten Department of Physics, Thin Films and Physics of Nanostructures, Bielefeld University, Bielefeld, Germany

Richard D. James Department of Aerospace Engineering and Mechanics, University of Minnesota, Minneapolis, MN, USA

Julia Krez Institut für Anorganische und Analytische Chemie, Johannes Gutenberg-Universität, Mainz, Germany

Huseyin Kurt Department of Engineering Physics, Istanbul Medeniyet University, Istanbul, Turkey

Chris Leighton Department of Chemical Engineering and Materials Science, University of Minnesota, Minneapolis, MN, USA

Judith Meyer Department of Physics, Thin Films and Physics of Nanostructures, Bielefeld University, Bielefeld, Germany

Ko Mibu Graduate School of Engineering, Nagoya Institute of Technology, Nagoya, Japan

Seiji Mitani Magnetic Materials Unit, National Institute for Materials Science, Tsukuba, Japan

T. Nakayama Nagaoka University of Technology, Nagaoka, Niigata, Japan

K. O'Grady Department of Physics, University of York, York, UK

Stuart S.P. Parkin Almaden Research Center, IBM, San Jose, CA, USA; Max Planck Institute, Halle, Germany

Daniel P. Phelan Department of Chemical Engineering and Materials Science, University of Minnesota, Minneapolis, MN, USA

James Sagar Department of Physics, University of York, Heslington, UK; London Centre for Nanotechnology, University College London, London, UK

Yuya Sakuraba National Institute for Materials Science (NIMS), Tsukuba, Ibaraki, Japan

Thomas Sebastian Fachbereich Physik and Forschungszentrum OPTIMAS, Technische Universität Kaiserslautern, Kaiserslautern, Germany; Helmholtz-Zentrum Dresden-Rossendorf, Institute of Ion Beam Physics and Materials Research, Dresden, Germany

Yintao Song Department of Aerospace Engineering and Mechanics, University of Minnesota, Minneapolis, MN, USA

Vijay Srivastava Department of Aerospace Engineering and Mechanics, University of Minnesota, Minneapolis, MN, USA

Yukiko K. Takahashi Magnetic Materials Unit, National Institute for Materials Science, Tsukuba, Japan

Koki Takanashi Institute for Materials Research, Tohoku University, Sendai, Japan

Masaaki Tanaka Graduate School of Engineering, Nagoya Institute of Technology, Nagoya, Japan

Niclas Teichert Department of Physics, Thin Films and Physics of Nanostructures, Bielefeld University, Bielefeld, Germany

Tetsuya Uemura Division of Electronics for Informatics, Graduate School of Information Science and Technology, Hokkaido University, Sapporo, Japan

Changhai Wang Max Planck Institute for Chemical Physics of Solids, Dresden, Germany

F.R. Wagner Max-Planck-Institut für Chemische Physik fester Stoffe, Dresden, Germany

L. Wollmann Max-Planck-Institut für Chemische Physik fester Stoffe, Dresden, Germany

S. Wurmehl Leibniz Institute for Solid State and Materials Research Dresden IFW, Dresden, Germany; Institut für Festkörperphysik, Technische Universität Dresden, Dresden, Germany

M. Wójcik Institute of Physics, Polish Academy of Sciences, Warszawa, Poland

Masafumi Yamamoto Division of Electronics for Informatics, Graduate School of Information Science and Technology, Hokkaido University, Sapporo, Japan

Part I
Theory of Heusler compounds

Chapter 1
Theory of Heusler and Full-Heusler Compounds

Iosif Galanakis

Abstract Spintronics/magnetoelectronics brought at the center of scientific research the Heusler and full-Heusler compounds, since several among them have been shown to be half-metals. In this introductory chapter we present a study of the basic electronic and magnetic properties of both Heusler families; the so-called semi-Heusler alloys like NiMnSb and the full-Heusler alloys like Co_2MnGe (usual full-Heuslers), Mn_2CoAl (inverse full-Heuslers) and (CoFe)MnAl (LiMgPdSn-type full-Heuslers). First-principles calculations are employed to discuss the origin of the gap which is fundamental for the understanding of their electronic and magnetic properties. For half-metallic Heusler compounds the total spin magnetic moment M_t scales linearly with the number of the valence electrons Z_t in the unit cell. These simple rules connect directly the magnetic to the electronic properties opening the way to engineer new half-metallic alloys with "à la carte" magnetic properties such as the quaternary half-metals, the so-called half-metallic antiferromagnets, magnetic semiconductors or even the more exotic spin-gapless semiconductors. Finally, special topics like exchange constants, defects, vacancies, surfaces and interfaces are being discussed.

1.1 Introduction

Half-metallic magnets attracted the last two decades a lot of attention due to their possible applications in spintronics/magnetoelectronics [1, 2]. The addition of the spin degree of freedom to the conventional electronic devices based on semiconductors has several advantages like non-volatility, increased data processing speed, decreased electric power consumption and increased integration densities [3–5]. In the half-metallic materials the two spin bands show a completely different behavior. The majority spin band shows the typical metallic behavior, and the minority

I. Galanakis (✉)
Department of Materials Science, School of Natural Sciences, University of Patras, 26504 Patras, Greece
e-mail: galanakis@upatras.gr

C. Felser and A. Hirohata (eds.), *Heusler Alloys*,
Springer Series in Materials Science 222, DOI 10.1007/978-3-319-21449-8_1

spin band exhibits a semiconducting behavior with a gap at the Fermi level. The existence of the gap leads to 100 % spin-polarization at the Fermi level and thus a fully spinpolarised current should be feasible in these compounds maximizing the efficiency of magnetoelectronic devices [6, 7]. Bowen et al. have observed such a current in trilayers made up of half-metallic $La_{0.7}Sr_{0.3}MnO_3$ as electrodes and SrO_3 as a barrier [8].

The family of Heusler compounds, named after von Heusler [9], incorporates a huge number of magnetic members exhibiting diverse magnetic phenomena like itinerant and localized magnetism, antiferromagnetism, helimagnetism, Pauli paramagnetism or heavy-fermionic behavior [10–17]. The first Heusler alloys studied were crystallizing in the $L2_1$ structure which consists of 4 fcc sublattices (see Fig. 1.1). Afterwards, it was discovered that it is possible to leave one of the four sublattices unoccupied ($C1_b$ structure). The latter compounds are often called half- or semi-Heusler compounds or simply Heuslers, while the $L2_1$ compounds are known as full-Heusler compounds. The most-well known semi-Heusler compound is NiMnSb [18]. In 1983 de Groot and his collaborators [19] showed using first-principles elec-

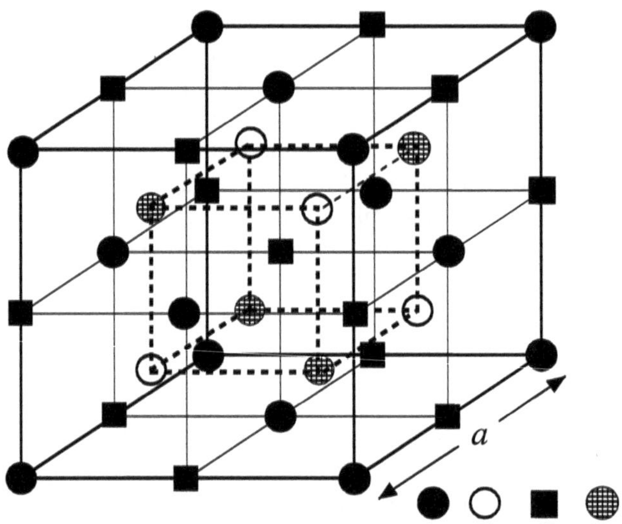

	A	B	C	D
Semi–Heuslers ($C1_b$) : XYZ	*X*	*Y*		*Z*
Full–Heuslers ($L2_1$) : X_2YZ	*X*	*Y*	*X*	*Z*
Quaternary Heuslers : (XX')YZ	*X*	*Y*	*X'*	*Z*
Inverse Heuslers (XA) : X_2YZ	*X*	*X*	*Y*	*Z*

Fig. 1.1 Schematic representation of the various structures of the semi- and full-Heusler compounds. In all cases the lattice is consisted of 4 interpenetrating f.c.c. lattices. Note also that if all atoms were identical, the lattice would be simply the b.c.c

tronic structure calculations that this compound is in reality half-metallic, i.e. the minority band is semiconducting with a gap at the Fermi level E_F, leading to 100 % spin polarization at E_F. Other known half-metallic materials except the semi- and full-Heusler alloys [20–24] are some oxides (e.g. CrO_2 and Fe_3O_4) [25], the manganites (e.g. $La_{0.7}Sr_{0.3}MnO_3$) [25], the double perovskites (e.g. Sr_2FeReO_6) [26], the pyrites (e.g. CoS_2) [27], the transition metal chalcogenides (e.g. CrSe) and pnictides (e.g. CrAs) in the zinc-blende or wurtzite structures [28–51], the europium chalcogenides (e.g. EuN) [52], the diluted magnetic semiconductors (e.g. Mn impurities in Si or GaAs) [53, 54] and the so-called d^0 ferromagnets like CaAs [55]. Although thin films of CrO_2 and $La_{0.7}Sr_{0.3}MnO_3$ have been verified to present practically 100 % spin-polarization at the Fermi level at low temperatures [25, 56], the Heusler alloys remain attractive for technical applications like spin-injection devices [57], spin-filters [58], tunnel junctions [59], or GMR devices [60, 61] due to their relatively high Curie temperature compared to these compounds [10].

NiMnSb is the half-metallic semi-Heusler compound which attracted a lot of experimental interest. Its half-metallic character in single crystals has been well-established by infrared absorption [62] and spin-polarized positron-annihilation [63, 64] experiments. Also high quality films of NiMnSb have been grown [65–73] but contrary to single crystals they are not half-metallic [25, 74–77]; a maximum value of 58 % for the spin-polarization of NiMnSb was obtained by Soulen et al. [25]. These polarization values are consistent with a small perpendicular magnetoresistance measured for NiMnSb in a spin-valve structure [78, 79], a superconducting tunnel junction [59] and a tunnel magnetoresistive junction [80]. The loss of half-metallicity in the NiMnSb thin films is due to the segregation of Sb and Mn atoms to the surface, which is far from being perfect [81–84].

NiMnSb has attracted also considerable attention among theoreticians and several first-principles calculations have confirmed its half-metallic character [85–90]. Larson et al. have shown that the $C1_b$ structure of NiMnSb is the most stable with respect to an interchange of the atoms [91]. Orgassa et al. showed that a few percent of disorder induce states within the gap but does not destroy the half-metallicity [92]. Surfaces and interfaces with semiconductors have been found not to be half-metallic [93–99] but Wijs and de Groot as well as Debernardi et al. proposed that at some interfaces it is possible to restore the half-metallic character of NiMnSb [100, 101]. Finally, Kübler calculated the Curie temperature of NiMnSb [102] which was in excellent agreement with the experimental value of 770 K [10]. Doping in similar semi-Heusler compounds can be employed to tune their magnetic properties [103, 104].

Full-Heusler compounds were the first to be synthesized [105–107]. In a pioneering paper Kübler et al. studied the mechanisms stabilizing the ferro- or the antiferromagnetism in these compounds [108]. Japanese research groups were the first to predict the existence of half-metallicity in the case of full-Heusler compounds using *ab-initio* electronic structure calculations: Ishida and collaborators studied the Co_2MnZ compounds, where Z stands for Si and Ge [109–111], and Fujii and collaborators studied the Fe_2MnZ compounds [112]. But Brown et al. [113] using polarized neutron diffraction measurements have shown that there is a finite very small spin-down density of states (DOS) at the Fermi level instead of an absolute gap in agreement with the *ab-initio* calculations of Kübler et al. for the Co_2MnAl and Co_2MnSn compounds [108]. Full-Heusler compounds became very popular for potential applications and several groups managed to grow Co_2MnGe and Co_2MnSi thin films [114–122]. Geiersbach and collaborators have grown (110) thin films of Co_2MnSi, Co_2MnGe and Co_2MnSn using a metallic seed on top of a MgO(001) substrate [123, 124] and studied also the transport properties of multilayers of these compounds with normal metals [125]. But as Picozzi et al. and Galanakis et al. have shown that the interfaces of such structures are not half-metallic [126–130]. Although, the cubic structure and ferromagnetism in these compounds is particular stable [131–133], it has been shown that defects and vacancies can lead to loss of the half-metallic character [134–138], although a small degree of disorder may lead to half-metallic ferrimagnetism instead of ferromagnetism [139, 140]. Finally, Kämmerer and collaborators managed to built magnetic tunnel junctions based on Co_2MnSi [141, 142]. Similar experiments have been undertaken by Inomata and collaborators using $Co_2Cr_{0.6}Fe_{0.4}Al$ as the magnetic electrode [143, 144].

In this chapter, we present a study of the basic electronic and magnetic properties of the half-metallic Heusler alloys. Analyzing the *ab-initio* results using the group-theory and simple models we explain the origin of the gap in both the semi- and full-Heusler alloys, which is fundamental for understanding their electronic and magnetic properties. For both families of compounds the total spin magnetic moment scales with the number of valence electron, thus opening the way to engineer new half-metallic Heusler alloys with the desired magnetic properties as the quaternary half-metallic ferromagnets, the so-called half-metallic antiferromagnets, the magnetic semiconductors including also the spin-gapless semiconductors.

1.2 Semi-Heusler Compounds

1.2.1 Band Structure of Semi-Heusler Compounds

We will start our discussion from the semi-Heusler compounds and more precisely from NiMnSb which is the most studied representative of the half-metallic semi-Heuslers [19]. Figure 1.2 shows the density of states (DOS) of NiMnSb in a non-spin-polarized calculation and in a calculation correctly including the spin-polarization. Given are the local contributions to the density of states (LDOS) on the Ni-site, the Mn-site and the Sb-site. The calculations have been performed using the Vosko, Wilk and Nusair parametrization [145] for the local density approximation (LDA) to the exchange-correlation potential [146, 147] to solve the Kohn-Sham equations within the full-potential screened Korringa-Kohn-Rostoker (FSKKR) method [148, 149].

In the non-magnetic case the DOS of NiMnSb has contributions from 4 different bands: Each Sb atom with the atomic configuration $5s^2 5p^3$ introduces a deep lying s band, which is located at about -12 eV and is not shown in the figure, and three

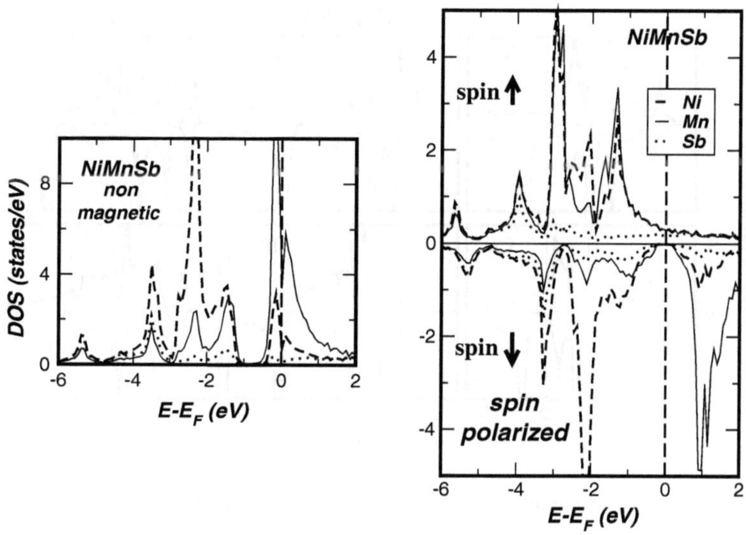

Fig. 1.2 Atom-resolved density of states (DOS) of NiMnSb for a non-magnetic (*left*) and ferromagnetic (*right*) calculation. The zero energy value corresponds to the Fermi level E_F

p-bands in the regions between -5.5 and -3 eV. These bands are separated by a deep minimum in the DOS from 5 Ni d bands between -3 and -1 eV, which themselves are separated by a sizeable band gap from the upper 5 d-bands of Mn. Since all atomic orbitals, i.e. the Ni d, the Mn d and the Sb sp orbitals hybridize with each other, all bands are hybrids between these states, being either of bonding or antibonding type. Thus the Ni d-bands contain a bonding Mn d admixture, while the higher Mn d-bands are antibonding hybrids with small Ni d-admixtures. Equally the Sb p-bands exhibit strong Ni d- and somewhat smaller Mn d-contributions. This configuration for NiMnSb is energetically not stable. The Fermi energy lies in the middle of an antibonding band and the Mn atom can gain considerable exchange energy by forming a magnetic moment. Therefore the spin-polarized results show a considerably different picture. In the majority (spin $\uparrow$) band the Mn d states are shifted to lower energies and form a common d band with the Ni d states, while in the minority band (spin $\downarrow$) the Mn states are shifted to higher energies and are unoccupied, so that a band gap at E_F is formed separating the occupied d bonding from the unoccupied d-type antibonding states. Thus NiMnSb is a half-metal, with a band gap at E_F in the minority band and a metallic DOS at E_F in the majority band. The total magnetic moment, located mostly at the Mn atom, can be easily estimated to be exactly 4 μ_B. Note that NiMnSb has 22 valence electrons per unit cell, 10 from Ni, 7 from Mn and 5 from Sb. Since, due to the gap at E_F, in the minority band exactly 9 bands are fully occupied (1 Sb-like s band, 3 Sb-like p bands and 5 Ni-like d bands) and accommodate 9 electrons per unit cell, the majority band contains $22 - 9 = 13$ electrons, resulting in a moment of 4 μ_B per unit cell. Notice that the

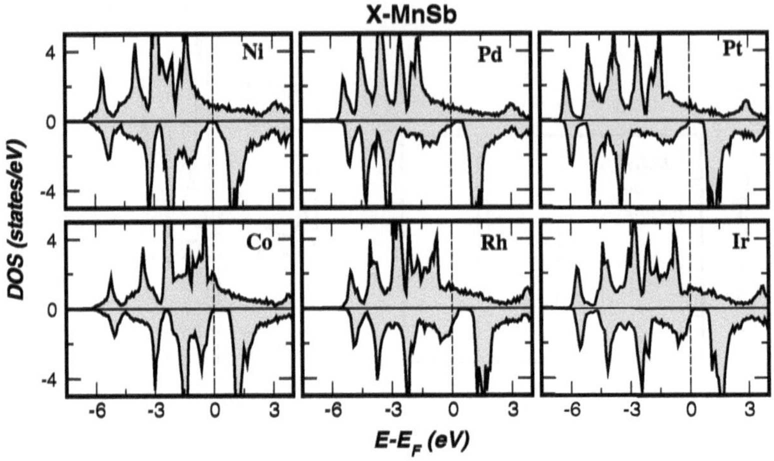

Fig. 1.3 DOS of XMnSb compounds for X = Ni, Pd, Pt and Co, Rh, Pd

compounds with 18 valence electrons like CoTiSb and FeVSn are well-known to be stable semiconductors [13–15].

As representative examples of the semi-Heuslers we will shortly discuss the electronic structure of the XMnSb compounds, with X being an element of the Co or Ni columns in the periodic table. These compounds are known experimentally to be ferromagnets with high Curie temperatures ranging between 500 K and 700 K for the Co, Ni, Pd and Pt compounds, while the Curie temperatures of the Ir and Rh compounds are around room temperature [10]. All six compounds (see Fig. 1.3) present a minority-spin band energy gap. As above Sb p states occupy the lowest part of the DOS shown in the figure, while the Sb s states are located ~12 eV below the Fermi level. For all compounds under study the Fermi level is within the gap with the exception of PdMnSb and IrMnSb where it is close to the band edge. The DOS of all compounds is characterized by the large exchange-splitting of the Mn d states which is around 3 eV. This leads to large localized spin moments at the Mn site, the existence of which has been verified also experimentally [150, 151]. The localization of the spin magnetic moment stems from the fact that although the 3d valence electrons of Mn are itinerant, the spin-down electrons are almost excluded from the Mn site. In Table 1.1 we present the calculated spin magnetic moments at the different sites for all the compounds under study.

The total spin magnetic moment in μ_B is just the difference between the number of spin-up occupied states and the spin-down occupied states. As explained above, the number of occupied spin-down states is given by the number of spin down bands, i.e. 9, so that the number of occupied spin-up states is $22 - 9 = 13$ for NiMnSb and the isovalent compounds with Pd and Pt, but $21 - 9 = 12$ for CoMnSb, RhMnSb and IrMnSb and $20 - 9 = 11$ for FeMnSb, provided that the Fermi level stays within the gap. Therefore one expects total moments of $4\mu_B$ for Ni-, Pd- and PtMnSb, 3

Table 1.1 Calculated spin magnetic moments in μ_B for the XMnSb compounds using the FSKKR method

$m^{spin}(\mu_B)$	X	Mn	Sb	Void	Total
NiMnSb	0.26	3.71	−0.06	0.05	3.96
PdMnSb	0.08	4.01	−0.11	0.04	4.02
PtMnSb	0.09	3.89	−0.08	0.04	3.94
CoMnSb	−0.13	3.18	−0.10	0.01	2.96
RhMnSb	−0.13	3.57	−0.14	0.00	3.29
IrMnSb	−0.19	3.33	−0.11	−0.00	3.02
FeMnSb	−0.70	2.72	−0.05	0.02	1.98

The total spin magnetic deviate from the ideal integer values due to the due to problems with the ℓ_{max} cutoff [259] (The experimental lattice constants [10] have been used)

μ_B for the compounds with Co, Rh and Ir and 2 μ_B for FeMnSb. In general, for a total number Z_t of valence electrons in the unit cell, the total moment M_t is given by $M_t = Z_t - 18$, since with 9 electron states occupied in the minority band, $Z_t - 18$ is just the number of uncompensated electron spins. The local moment per unit cell as given in Table 1.1 is close to 4 μ_B in the case of NiMnSb, PdMnSb and PtMnSb, which is in agreement with the half-metallic character (or nearly half-metallic character in the case of PdMnSb) observed in Fig. 1.3. We also find that the local moment of Mn is not far away from the total number of 4 μ_B although there are significant (positive) contributions from the X-atoms and a negative contribution from the Sb atom. For the half-metallic CoMnSb and IrMnSb compounds the total moment is about 3 μ_B. Also the local moment of Mn is reduced, but only by about 0.5 μ_B. The reduction of the total moment to 3 μ_B is therefore accompanied by negative Co and Ir spin moments. The hybridization between Co and Mn is considerably larger than between Ni and Mn being a consequence of the smaller electronegativity difference and the larger extend of the Co orbitals. Therefore the minority valence band of CoMnSb has a larger Mn admixture than the one of NiMnSb whereas the minority conduction band of CoMnSb has a larger Co admixture than the Ni admixture in the NiMnSb conduction band, while the populations of the majority bands are barely changed. As a consequence, the Mn moment is reduced by the increasing hybridization, while the Co moment becomes negative. The table also shows that further substitution of Fe for Co leads also to a half-metallic alloy with a total spin magnetic moment of 2 μ_B as has been already shown by de Groot et al. in [152].

1.2.2 Origin of the Gap

The inspection of the local DOS shown in Fig. 1.2 for the ferromagnet NiMnSb shows that the DOS close to the gap is dominated by d-states: in the valence band by bonding hybrids with large Ni or Co admixture and in the conduction band by

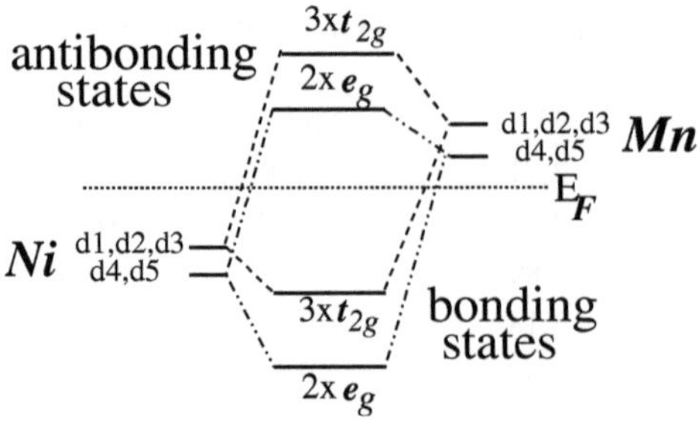

Fig. 1.4 Schematic illustration of the origin of the gap in the minority band in semi-Heusler alloys and in compound semiconductors: The energy levels of the energetically lower lying bonding hybrids are separated from the energy levels of the antibonding hybrids by a gap, such that only the bonding states are occupied. Due to legibility reasons, we use *d1*, *d2* and *d3* to denote the d_{xy}, d_{yx} and d_{zx} orbitals, respectively, and *d4*, *d5* for the d_{r^2}, $d_{x^2-y^2}$ orbitals

the antibonding hybrids with large Mn admixture. Thus the gap originates from the strong hybridization between the d states of the higher valent and the lower valent transition metal atoms. This is shown schematically in Fig. 1.4. Therefore the origin of the gap is somewhat similar to the gap in compound semiconductors like GaAs which is enforced by the hybridization of the lower lying As sp-states with the energetically higher Ga sp-states. Note that in the $C1_b$-structure the Ni and Mn sublattices form a zinc-blende structure, which is important for the formation of the gap. The difference with respect to GaAs is than only, that 5 d-orbitals, i.e. 3 t_{2g} and 2 e_g orbitals, are involved in the hybridization, instead of 4 sp^3-hybrids in the compound semiconductors.

The gap in the half-metallic $C1_b$ compounds is normally an indirect gap, with the maximum of the valence band at the Γ point and the minimum of the conduction band at the X-point. For NiMnSb we obtain a band gap of about 0.5 eV, which is in good agreement with the experiments of Kirillova and collaborators [62], who, analyzing their infrared spectra, estimated a gap width of ~0.4 eV. As seen already from Fig. 1.3 the gap of CoMnSb is considerable larger (~1 eV) and the Fermi level is located at the edge of the minority valence band.

As it is well-known, the local density approximation (LDA) and the generalized gradient approximation (GGA) strongly underestimate the values of the gaps in semiconductors, typically by a factor of two. However, very good values for these gaps are obtained in the so-called GW approximation of Hedin and Lundqvist [153], which describes potential in semiconductors very well. On the other hand the minority gap in the half-metallic systems might be better described by the LDA and GGA since in these system the screening is metallic.

1.2.3 Role of *sp*-Elements

While the *sp*-elements are not responsible for the existence of the minority gap, they are nevertheless very important for the physical properties of the Heusler alloys and the structural stability of the $C1_b$ structure, as we discuss in the following.

While an Sb atom has 5 valence electrons ($5s^2$, $5p^3$), in the NiMnSb compound each Sb atom introduces a deep lying *s*-band, at about -12 eV, and three *p*-bands below the center of the *d*-bands. These bands accommodate a total of 8 electrons per unit cell, so that formally Sb acts as a triple charged Sb^{3-} ion. Analogously, a Te-atom behaves in these compounds as a Te^{2-} ion and a Sn-atom as a Sn^{4-} ion. This does not mean, that locally such a large charge transfer exists. In fact, the *s*- and *p*-states strongly hybridize with the transition-metal *d*-states and the charge in these bands is delocalized and locally Sb even looses about one electron, if one counts the charge in the Wigner-Seitz cells. What counts is that the *s*- and *p*-bands accommodate 8 electrons per unit cell, thus effectively reducing the *d*-charge of the transition-metal atoms.

This is nicely illustrated by the existence of the semiconducting compounds CoTiSb and NiTiSn. Compared to CoTiSb, in NiTiSn the missing *p*-charge of the Sn atom is replaced by an increased *d* charge of the Ni atom, so that in both cases all 9 valence bands are occupied.

The *sp*-atom is very important for the structural stability of the Heusler alloys. For instance, it is difficult to imagine that the half-metallic NiMn and PtMn alloys with zinc-blende structure actually exist, since metallic alloys prefer highly coordinated structures like fcc, bcc, hcp etc. Therefore the *sp*-elements are decisive of the stability of the $C1_b$ compounds. A careful discussion of the bonding in these compounds has been recently published by Nanda and Dasgupta [154] using the crystal orbital Hamiltonian population (COHP) method. For the semiconductor FeVSb they find that while the largest contribution to the bonding arises from the V-*d* – Fe-*d* hybridization, contributions of similar size arise also from the Fe-*d* – Sb-*p* and the V-*d* – Sb-*p* hybridization. Similar results are also valid for the semiconductors like CoTiSb and NiTiSn and in particular for the half-metal NiMnSb. Since the majority *d*-band is completely filled, the major part of the bonding arises from the minority band, so that similar arguments as for the semiconductors apply.

Another property of the *sp*-elements is worthwhile to mention: substituting the Sb atom in NiMnSb by Sn, In or Te destroys the half-metallicity [20]. This is in contrast to the substitution of Ni by Co or Fe, which is documented in Table 1.1. The total moment of 4 μ_B for NiMnSb is reduced to 3 μ_B in CoMnSb and 2 μ_B in FeMnSb, thus preserving half-metallicity. In NiMnSn the total moment is reduced to 3.3 μ_B (instead of 3) and in NiMnTe the total moment increases only to 4.7 μ_B (instead of 5). Thus by changing the *sp*-element it is rather difficult to preserve the half-metallicity, since the density of states changes more like in a rigid band model [20].

1.2.4 Slater-Pauling Behavior

As discussed above the total moment of the half-metallic $C1_b$ Heusler alloys follows the simple rule: $M_t = Z_t - 18$, where Z_t is the total number of valence electrons. In short, the total number of electrons Z_t is given by the sum of the number of spin-up and spin-down electrons, while the total moment M_t is given by the difference

$$Z_t = N_\uparrow + N_\downarrow \quad , \quad M_t = N_\uparrow - N_\downarrow \quad \rightarrow \quad M_t = Z_t - 2N_\downarrow \tag{1.1}$$

Since 9 minority bands are fully occupied, we obtain the simple "rule of 18" for half-metallicity in $C1_b$ Heusler alloys

$$M_t = Z_t - 18 \tag{1.2}$$

the importance of which has been recently pointed out by Jung et al. [155] and Galanakis et al. [20]. It is a direct analogue to the well-known Slater-Pauling behavior of the binary transition metal alloys [156–158]. The difference with respect to these alloys is, that in the semi-Heusler alloys the minority population is fixed to 9, so that the screening is achieved by filling the majority band, while in the transition metal alloys the majority band is filled with 5 d-states and charge neutrality is achieved by filling the minority states. Therefore in the transition-metal alloys the total spin magnetic moment is given by $M_t = 10 - Z_t$. Similar rules with integer total moments are also valid for other half-metals, e.g. for the full-Heusler alloys like Co_2MnGe with $L2_1$ structure. For these alloys we will in Sect. 1.3.2 derive the "rule of 24": $M_t = Z_t - 24$, with arises from the fact that the minority band contains 12 electrons. For the half-metallic zinc-blende compounds like CrAs the rule is: $M_t = Z_t - 8$, since the minority As-like valence bands accommodate 4 electrons [28–30]. In all cases the moments are integer.

In Fig. 1.5 we have gathered the calculated total spin magnetic moments for the semi-Heusler alloys which we have plotted as a function of the total number of valence electrons. The dashed line represents the rule $M_t = Z_t - 18$ obeyed by these compounds. The total moment M_t is an integer quantity, assuming the values 0, 1, 2, 3, 4 and 5 if $Z_t \geq 18$. The value 0 corresponds to the semiconducting phase and the value 5 to the maximal moment when all 10 majority d-states are filled. Firstly we varied the valence of the lower-valent (i.e. magnetic) transition metal atom. Thus we substitute V, Cr and Fe for Mn in the NiMnSb and CoMnSb compounds using the experimental lattice constants of the two Mn compounds. For all these compounds we find that the total spin magnetic moment scales accurately with the total charge and that they all present the half-metallicity.

As a next test we have substituted Fe for Mn in CoMnSb and NiMnSb, but both CoFeSb and NiFeSb loose their half-metallic character. In the case of NiFeSb the majority d-states are already fully occupied as in NiMnSb, thus the additional electron

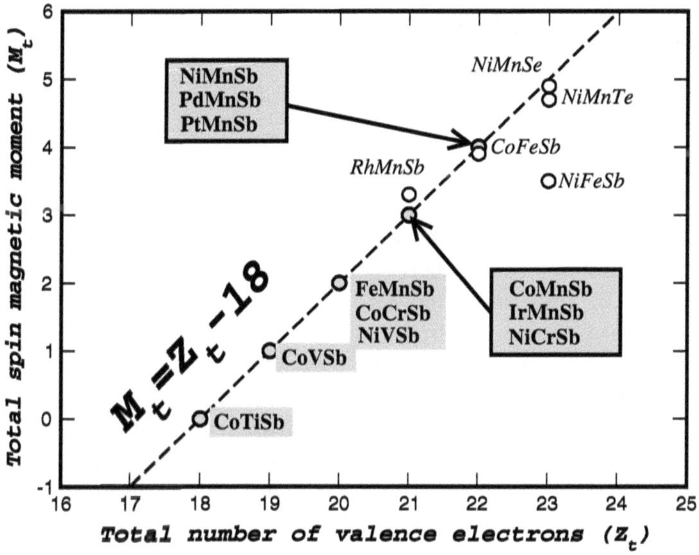

Fig. 1.5 Calculated total spin moments for all the studied semi Heusler alloys. The *dashed line* represents the Slater-Pauling behavior. With *hollow circles* we present the compounds deviating from the SP curve. Some experimental values for bulk systems near the SP curve from [10]: NiMnSb 3.85 μ_B, PdMnSb 3.95 μ_B, PtMnSb 4.14 μ_B and finally CoTiSb non-magnetic

has to be screened by the minority d-states, so that the Fermi level falls into the minority Fe states and the half-metallicity is lost; for half-metallicity a total moment of 5 μ_B would be required which is clearly not possible. For CoFeSb the situation is more delicate. This system has 22 valence electrons and if it would be a half-metal, it should have a total spin-moment of 4 μ_B as NiMnSb. In reality our calculations indicate that the Fermi level is slightly above the gap and the total spin-moment is slightly smaller than 4 μ_B. The Fe atom possesses a comparable spin-moment in both NiFeSb and CoFeSb compounds contrary to the behavior of the V, Cr and Mn atoms. Except NiFeSb other possible compounds with 23 valence electrons are NiMnTe and NiMnSe. We have calculated their magnetic properties using the lattice constant of NiMnSb. As shown in Fig. 1.5, NiMnSe almost makes the 5 μ_B (its total spin moment is 4.86 μ_B) and is nearly half-metallic, while its isovalent, NiMnTe, has a slightly smaller spin moment. NiMnSe and NiMnTe show big changes in the majority band compared to systems with 22 valence electrons as NiMnSb or NiMnAs, since antibonding p-d states, which are usually above E_F, are shifted below the Fermi level, thus increasing the total moment to nearly 5 μ_B.

1.3 Full Heusler Compounds

1.3.1 Electronic Structure

The second family of Heusler alloys, which we discuss, are the full-Heusler alloys. We consider in particular compounds containing Co and Mn, as these are the full-Heusler alloys that have attracted most of the attention. They are all strong ferromagnets with high Curie temperatures (above 600 K) and except the Co_2MnAl they show very little disorder [10]. They adopt the $L2_1$ structure shown in Fig. 1.1. Each Mn or sp atom has eight Co atoms as first neighbors, sitting in an octahedral symmetry position, while each Co has four Mn and four sp atoms as first neighbors and thus the symmetry of the crystal is reduced to the tetrahedral one. The Co atoms occupying the two different sublattices are chemically equivalent as the environment of the one sublattice is the same as the environment of the second one but rotated by 90°. The occupancy of two fcc sublattices by Co (or in general by X) atoms distinguishes the full-Heusler alloys with the $L2_1$ structure from the semi-Heusler compounds with the $C1_b$ structure, like e.g. CoMnSb, where only one sublattice is occupied by Co atoms and the other one is empty. Although in the $L2_1$ structure, the Co atoms are sitting on second neighbor positions, their interaction is important to explain the magnetic properties of these compounds as we will show in the next section. Finally, it should be noted that orbital magnetism in these compounds plays a minor role and thus can be neglected when discussing their magnetic properties [159].

In Fig. 1.6 we have gathered the spin-resolved density of states (DOS) for the Co_2MnAl, Co_2MnGa, Co_2MnSi and Co_2MnGe compounds calculated using the FSKKR. Firstly as shown by photoemission experiments by Brown et al. in the case of Co_2MnSn [160] and verified by our calculations, the valence band extends around 6 eV below the Fermi level and the spin-up DOS shows a large peak just below the

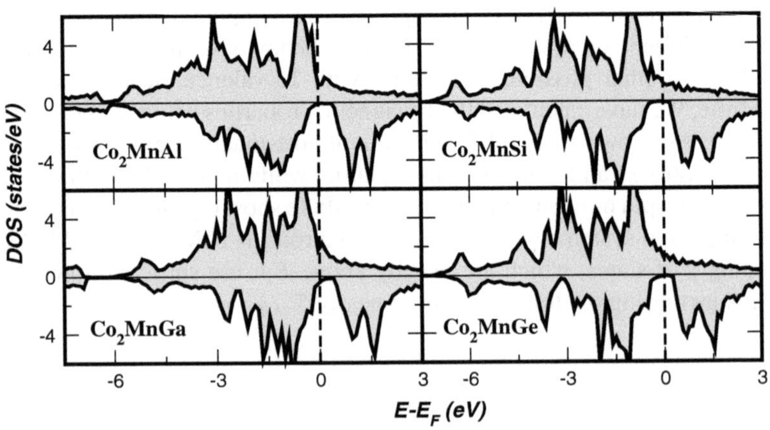

Fig. 1.6 Total DOS for the Co_2MnZ compounds with Z = Al, Si, Ge, Sn compounds

Table 1.2 Calculated spin magnetic moments in μ_B using the experimental lattice constants (see [10]) for the Co_2MnZ compounds, where Z stands for the *sp* atom

$m^{spin}(\mu_B)$	Co	Mn	Z	Total
Co_2MnAl	0.77	2.53	−0.10	3.97
Co_2MnGa	0.69	2.78	−0.09	4.06
Co_2MnSi	1.02	2.97	−0.07	4.94
Co_2MnGe	0.98	3.04	−0.06	4.94
Co_2MnSn	0.93	3.20	−0.08	4.98

Fermi level for these compounds. Although Ishida et al. [109, 110] predicted them to be half-metals with small spin-down gaps ranging from 0.1 to 0.3 eV depending on the material, our previous calculations showed a very small DOS at the Fermi level, in agreement with the ASW results of Kübler et al. [158] for Co_2MnAl and Co_2MnSn. However a recalculation of our KKR results with a higher ℓ-cut-off of $\ell_{max} = 4$ restores the gap and we obtain good agreement with the results of Picozzi et al. using the FLAPW method. The gap is an indirect gap, with the maximum of the valence band at Γ and the minimum of the conduction band at the X-point. In the case of the full-Heusler alloys each Mn atom has eight Co atoms as first neighbors instead of four as in CoMnSb and the hybridization effect is very important decreasing even further the Mn spin moment to less than 3 μ_B except in the case of Co_2MnSn where it is comparable to the CoMnSb compound. The Co atoms are ferromagnetically coupled to the Mn spin moments and they posses a spin moment that varies from $\sim$0.7 to 1.0 μ_B. The Co moment is large and positive and arises basically from two unoccupied Co bands in the minority conduction band, as explained below. Therefore both Co atoms together can have a moment of about 2 μ_B, if all majority Co states are occupied. This is basically the case for Co_2MnSi, Co_2MnGe and Co_2MnSn (see Table 1.2). In contrast to this the *sp* atom has a very small negative moment which is one order of magnitude smaller than the Co moment. The negative sign of the induced *sp* moment characterizes most of the studied full and semi Heusler alloys with very few exceptions. The compounds containing Al and Ga have 28 valence electrons and the ones containing Si, Ge and Sn 29 valence electrons. The first compounds have a total spin magnetic moment of 4 μ_B and the second ones of 5 μ_B which agree with the experimental deduced moments of these compounds [161, 162]. So it seems that the total spin magnetic moment, M_t, is related to the total number of valence electrons, Z_t, by the simple relation: $M_t = Z_t - 24$, while in the semi-Heusler alloys the total magnetic moment is given by the relation $M_t = Z_t - 18$. In the following section we will analyze the origin of this rule.

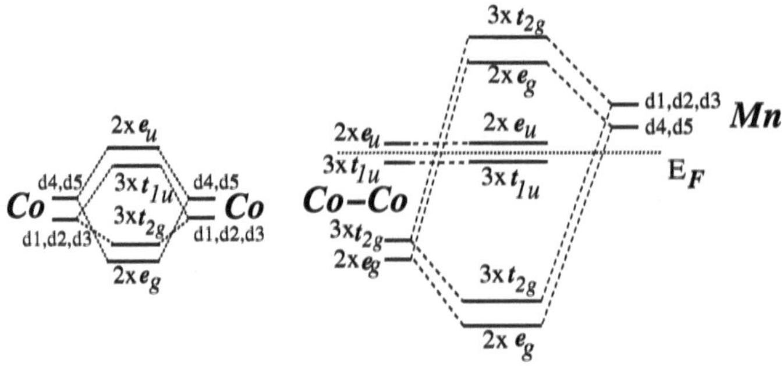

Fig. 1.7 Schematic illustration of the origin of the gap in the minority band in full-Heusler alloys. Due to legibility reasons, we use *d1*, *d2* and *d3* to denote the d_{xy}, d_{yx} and d_{zx} orbitals, respectively, and *d4*, *d5* for the d_{r^2}, $d_{x^2-y^2}$ orbitals

1.3.2 Origin of the Gap in Full-Heusler Alloys

Since, similar to the semi-Heusler alloys, the four sp-bands are located far below the Fermi level and thus are not relevant for the gap, we consider only the hybridization of the 15 d states of the Mn atom and the two Co atoms. For simplicity we consider only the d-states at the Γ point, which show the full structural symmetry. We will give here a qualitative picture, since a thorough group theoretical analysis has been given in [21]. Note that the Co atoms form a simple cubic lattice and that the Mn atoms (and the Ge atoms) occupy the body centered sites and have 8 Co atoms as nearest neighbors. Although the distance between the Co atoms is a second neighbors distance, the hybridization between these atoms is qualitatively very important. Therefore we start with the hybridization between these Co atoms which is qualitatively sketched in Fig. 1.7. The 5 d-orbitals are divided into the twofold degenerate d_{r^2}, $d_{x^2-y^2}$ and the threefold degenerate d_{xy}, d_{yx}, d_{zx} states. The e_g orbitals (t_{2g} orbitals) can only couple with the e_g orbitals (t_{2g} orbitals) of the other Co atom forming bonding hybrids, denoted by e_g (or t_{2g}) and antibonding orbitals, denoted by e_u (or t_{1u}). The coefficients in front of the orbitals give the degeneracy.

In a second step we consider the hybridization of these Co-Co orbitals with the Mn d-orbitals. As we show in the right-hand part of Fig. 1.7, the double degenerated e_g orbitals hybridize with the d_{r^2} and $d_{x^2-y^2}$ of the Mn that transform also with the same representation. They create a doubly degenerate bonding e_g state that is very low in energy and an antibonding one that is unoccupied and above the Fermi level. The $3 \times t_{2g}$ Co orbitals couple to the $d_{xy,yx,zx}$ of the Mn and create 6 new orbitals, 3 of which are bonding and are occupied and the other three are antibonding and high in energy. Finally the $2 \times e_u$ and $3 \times t_{1u}$ Co orbitals can not couple with any of the Mn d-orbitals since none of these is transforming with the u representations and they are orthogonal to the Co e_u and t_{1u} states. With respect to the Mn atoms these states are therefore non-bonding. The t_{1u} states are below the Fermi level and they are occupied while the e_u are just above the Fermi level. Thus in total 8 minority d-bands are filled and 7 are empty.

Therefore all 5 Co-Mn bonding bands are occupied and all 5 Co-Mn antibonding bands are empty, and the Fermi level falls in between the 5 non-bonding Co bands, such that the three t_{1u} bands are occupied and the two e_u bands are empty. The maximal moment of the full Heusler alloys is therefore 7 μ_B per unit cell, which is achieved, if all majority d-states are occupied.

1.3.3 Slater-Pauling Behavior of the Full-Heusler Alloys

Following the above discussion we will investigate the Slater-Pauling behavior and in Fig. 1.8 we have plotted the total spin magnetic moments for all the compounds under study as a function of the total number of valence electrons. The dashed line represents the half metallicity rule: $M_t = Z_t - 24$ of the full Heusler alloys. This rule arises from the fact that the minority band contains 12 electrons per unit cell: 4 are occupying the low lying s and p bands of the sp element and 8 the Co-like minority d bands ($2 \times e_g$, $3 \times t_{2g}$ and $3 \times t_{1u}$), as explained above (see Fig. 1.7). Since 7 minority bands are unoccupied, the largest possible moment is 7 μ_B and occurs when all majority d-states are occupied.

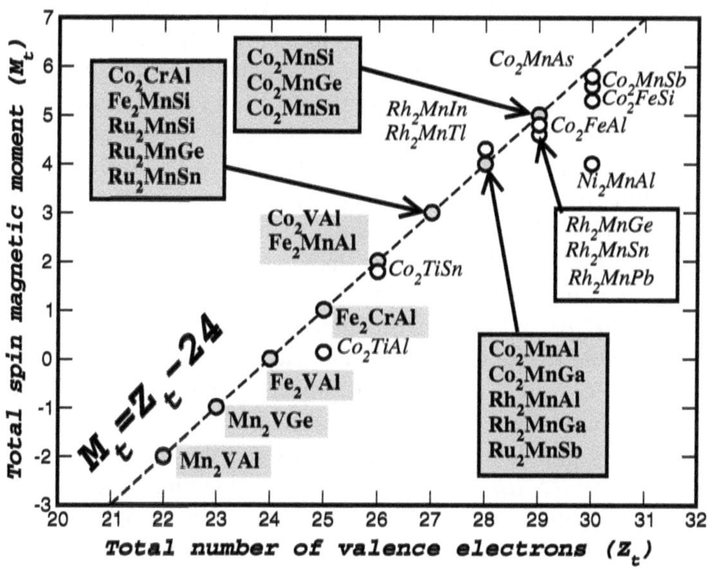

Fig. 1.8 Calculated total spin moments for all the studied full Heusler alloys. The *dashed line* represents the Slater-Pauling behavior. With *open circles* we present the compounds deviating from the SP curve. Some experimental values for bulk systems near the SP curve from [10]: Co_2MnAl 4.01 μ_B, Co_2MnSi 5.07 μ_B, Co_2MnGa 4.05 μ_B, Co_2MnGe 5.11 μ_B, Co_2MnSn 5.08 μ_B, Co_2FeSi 5.9 μ_B, Mn_2VAl -1.82 μ_B and finally Fe_2VAl non-magnetic

Overall we see that many of our results coincide with the Slater-Pauling curve. Some of the Rh compounds show small deviations which are more serious for the Co_2TiAl compound. We see that there is no compound with a total spin magnetic moment of 7 μ_B or even 6 μ_B. Moreover we found also examples of half-metallic materials with less than 24 electrons, Mn_2VGe with 23 valence electrons and Mn_2VAl with 22 valence electrons. Firstly, we have calculated the spin moments of the compounds Co_2YAl where Y = Ti, V, Cr, Mn and Fe. The compounds containing V, Cr and Mn show a similar behavior. As we substitute Cr for Mn, which has one valence electron less than Mn, we depopulate one Mn spin-up state and thus the spin moment of Cr is around 1 μ_B smaller than the Mn one while the Co moments are practically the same for both compounds. Substituting V for Cr has a larger effect since also the Co spin-up DOS changes slightly and the Co magnetic moment is increased by about 0.1 μ_B compared to the other two compounds and V possesses a small moment of 0.2 μ_B. This change in the behavior is due to the smaller hybridization between the Co atoms and the V ones as compared to the Cr and Mn atoms. Although all three Co_2VAl, Co_2CrAl and Co_2MnAl compounds are on the SP curve as can be seen in Fig. 1.8, this is not the case for the compounds containing Fe and Ti. If the substitution of Fe for Mn followed the same logic as the one of Cr for Mn then the Fe moment should be around 3.5 μ_B which is a very large moment for the Fe site. Therefore it is energetically more favorable for the system that also the Co moment is increased, as it was also the case for the other systems with 29 electrons like Co_2MnSi, but while the latter one makes it to 5 μ_B, Co_2FeAl reaches a value of 4.9 μ_B. In the case of Co_2TiAl, it is energetically more favorable to have a weak ferromagnet than an integer moment of 1 μ_B as it is very difficult to magnetize the Ti atom. Even in the case of the Co_2TiSn the calculated total spin magnetic moment of 1.78 μ_B (compared to the experimental value of 1.96 μ_B [163]) arises only from the Co atoms as was also shown experimentally by Pendl et al. [164], and the Ti atom is practically nonmagnetic and the latter compound fails to follow the SP curve.

As a second family of materials we have studied the compounds containing Fe. Fe_2VAl has in total 24 valence electrons and is a semi-metal, i.e. nonmagnetic with a very small DOS at the Fermi level, as it is already known experimentally [165–168]. All the studied Fe compounds follow the SP behavior as can be seen in Fig. 1.8. In the case of the Fe_2CrAl and Fe_2MnAl compounds the Cr and Mn atoms have spin moments comparable to the Co compounds and similar DOS. In order to follow the SP curve the Fe in Fe_2CrAl is practically nonmagnetic while in Fe_2MnAl it has a small negative moment. When we substitute Si for Al in Fe_2MnAl, the extra electron exclusively populates Fe spin-up states and the spin moment of each Fe atom is increased by 0.5 μ_B contrary to the corresponding Co compounds where also the Mn spin moment was considerably increased. Finally we calculated as a test Mn_2VAl and Mn_2VGe that have 22 and 23 valence electrons, respectively, to see if we can reproduce the SP behavior not only for compounds with more than 24, but also for compounds with less than 24 electrons. As we have already shown Fe_2VAl is nonmagnetic and Co_2VAl, which has two electrons more, has a spin moment of $2\mu_B$. Mn_2VAl has two valence electrons less than Fe_2VAl and its total spin moment is -2 μ_B and thus it follows the SP behavior [169]; negative total spin moment means that the "minority" band with the gap has more occupied states than the "majority" one.

As we have already mentioned the maximal moment of a full-Heusler alloy is seven μ_B, and should occur, when all 15 majority d states are occupied. Analogously for a semi-Heusler alloy the maximal moment is 5 μ_B. However this limit is difficult to achieve, since due to the hybridization of the d states with empty sp-states of the transition metal atoms (sites X and Y in Fig. 1.1), d-intensity is transferred into states high above E_F, which are very difficult to occupy. Although in the case of semi-Heusler alloys, we could identify systems with a moment of nearly 5 μ_B, the hybridization is much stronger in the full-Heusler alloys so that a total moment of 7 μ_B seems to be impossible. Therefore we restrict our search to possible systems with 6 μ_B, i.e. systems with 30 valence electrons, but as shown also in Fig. 1.8, none of them makes exactly the 6 μ_B. Co_2MnAs shows the largest spin moment: 5.8 μ_B. The basic reason, why moments of 6 μ_B are so difficult to achieve, is that as a result of the strong hybridization with the two Co atoms the Mn atom cannot have a much larger moment than 3 μ_B. While due to the empty e_u-states the two Co atoms have no problem to contribute a total of 2 μ_B, the Mn moment is hybridization limited. Recent calculations employing the Hubbard U have shown that Co_2FeSi is susceptible of being a half-metal with a 6 μ_B total spin magnetic moment [170]. Further calculations by Tsirogiannis and Galanakis, using the on-site Coulomb parameters from [171], have shown that the result depends strongly on the choice of the functional accounting for the double term in these hybrid methods [172].

1.4 Inverse Full-Heusler Compounds

Except the usual full-Heusler compounds studied in Sect. 1.3 there exist also the so-called inverse full-Heusler compounds. The latter compounds have also the chemical formula X_2YZ but in their case the valence of the X transition-metal atom is smaller than the valence of the Y transition metal atom. As a consequence, the inverse Heusler compounds crystallize in the so-called *XA* or X_α structure, where the sequence of the atoms is X-X-Y-Z and the prototype is Hg_2TiCu [173]. Several inverse Heuslers have been studied using first-principles electronic structure calculations in literature [173–183]. In all cases the *XA* structure is energetically preferred to the $L2_1$ structure of the usual full-Heusler compounds where the sequence of the atoms is X-Y-X-Z. The latter has been also confirmed by experiments on Mn_2CoGa and Mn_2CoSn films as well as Co doped Mn_3Ga samples [184–187], but experiments on Mn_2NiSb revealed that the actual arrangement of the atoms at the various sites can be influenced by the preparing method [188]. Inverse Heuslers are interesting for applications since they combine coherent growth on semiconductors with large Curie temperatures which can exceed the 1000 K as in the case of Cr_2CoGa [189].

In [190], extensive first-principles calculations have been presented on the inverse full-Heusler compounds having the chemical formula X_2YZ where (X = Sc, Ti, V, Cr or Mn), (Z = Al, Si or As) and the Y ranged from Ti to Zn. Several of these

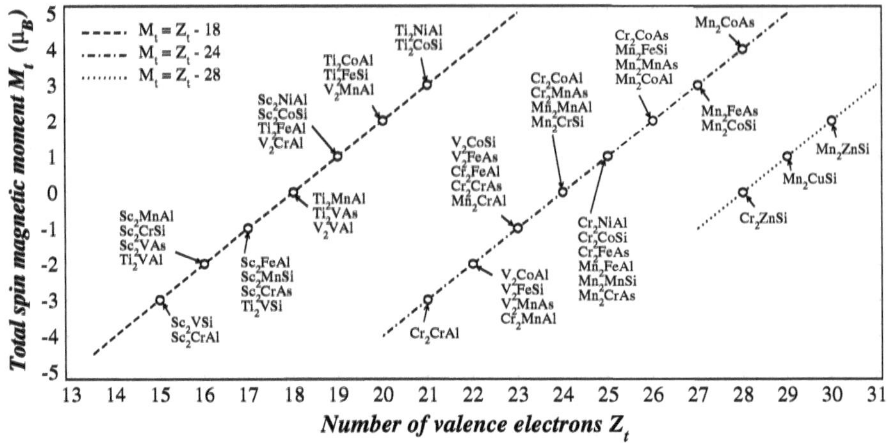

Fig. 1.9 Calculated total spin moments for several inverse Heusler compounds studied in [190]. The *dashed lines* represent the three variants of the Slater-Pauling behavior for these compounds

compounds have been identified to be half-metallic magnets. The appearance of half-metallicity is associated in all cases to a Slater-Pauling behavior of the total spin-magnetic moment. It was shown that when X is Sc or Ti, the total spin magnetic moment per formula unit (or unit cell) in μ_B follows the rule $M_t = Z_t - 18$ where Z_t is the total number of valence electrons in the unit cell. When X = Cr- or Mn, the variant followed by M_t is $M_t = Z_t - 24$, and when X = V the form of the Slater-Pauling rule was found to be material specific. The occurrence of these rules can be explained using simple hybridization arguments of the transition metal d-orbitals. In fact, when X is Cr or Mn the situation is similar to the usual Heusler compounds discussed above, but when X is Sc or Ti, the Fermi level in the minority-spin band structure is located below the non-bonding t_{1u} states discussed in Sect. 1.3 leading to the other variant of the Slater-Pauling rule. In Fig. 1.9 we present for several inverse Heusler compounds the calculated total spin-magnetic moment in μ_B per formula unit as a function of the total number of valence electrons. Finally, a third variant of the Slater-Pauling rule occurs, as shown in [190], when X is Cr or Mn and Y is Cu or Zn. In this case Cu or Zn d-states are completely occupied being in energy below the X d-states, and the half-metallic compounds follow a $M_t = Z_t - 28$ rule.

1.5 LiMgPdSn-Type Heusler Compounds

Except the usual and inverse full-Heusler compounds, another full-Heuslers family is the LiMgPdSn-type ones; also known as LiMgPdSb-type Heusler compounds [191]. These are quaternary compounds with the chemical formula (XX′)YZ where X, X′ and Y are transition metal atoms. The valence of X′ is lower than the valence of X atoms, and the valence of the Y element is lower than the valence of both X and X′. The sequence of the atoms along the fcc cube's diagonal is X-Y-X′-Z which is energetically the most stable [192]. A few LiMgPdSn-type half-metallic compounds have been studied [191, 193, 194]. Recent studied have been devoted to the study of a large series of such compounds [195, 196].

More precisely in [196], first-principles electronic structure calculations have been employed to study the electronic and magnetic properties of the 60 LiMgPdSn-type multifunctional quaternary Heusler compounds. It was shown that most of the compounds were half-metals obeying the same Slater-Pauling rule for full-Heusler compounds, $M_t = Z_t - 24$, with only few exceptions. The driving force behind the Slater-Paling rule was shown to be the same hybridization scheme as in full-Heusler compounds. In Table 1.3, the calculated total and atom-resolved spin magnetic moments, for the case where X is Co, from [196] are presented.

Table 1.3 Calculated equilibrium lattice constant and spin magnetic moments in μ_B for the LiMgPdSn-type Heusler compounds, where X is Co, obeying the $M_t = Z_t - 24$ Slater-Pauling rule using the FPLO method code [260, 261]

(XX′)YZ	a(Å)	m^X	$m^{X'}$	m^Y	m^{total}	Z_t
(CoCr)TiAl	5.95	−0.23	−2.15	0.28	−2.00	22
(CoV)TiSi	5.90	−0.26	−1.67	−0.12	−2.00	22
(CoCr)VAl	5.82	−0.34	−1.28	0.59	−1.00	23
(CoMn)TiAl	5.86	−0.23	−1.03	0.23	−0.98	23
(CoCr)TiSi	5.80	−0.21	−0.92	0.09	−1.00	23
(CoV)TiAs	5.97	−0.13	−0.90	−0.03	−1.00	23
(CoMn)CrAl	5.71	0.66	−1.09	1.50	1.00	25
(CoFe)VAl	5.73	0.61	0.51	−0.09	0.97	25
(CoMn)VSi	5.65	0.67	−0.05	0.38	0.97	25
(CoFe)TiSi	5.73	0.58	0.65	−0.19	1.00	25
(CoCr)VAs	5.80	0.72	−0.41	0.68	1.00	25
(CoMn)TiAs	5.83	0.62	0.54	−0.15	1.00	25
(CoMn)VAs	5.77	1.15	0.47	0.35	2.00	26
(CoFe)MnAl	5.68	0.73	−0.18	2.59	3.00	27
(CoFe)CrSi	5.61	1.00	0.33	1.79	3.00	27
(CoMn)CrAs	5.75	1.06	−0.44	2.38	3.00	27
(CoFe)VAs	5.78	1.05	1.19	0.75	3.00	27
(CoFe)MnSi	5.60	0.79	0.55	2.81	4.00	28
(CoFe)CrAs	5.75	0.91	0.80	2.34	4.00	28
(CoFe)MnAs	5.74	0.94	0.99	3.12	5.00	29

The total spin magnetic moment is given per formula unit (which coincides with the per unit cell value)

1.6 Disordered Quaternary Heusler Alloys

We proceed our study by examining the behavior of the so-called quaternary Heusler alloys [22, 197–199]. In the latter compounds, one of the four sites is occupied by two different kinds of neighboring elements like $Co_2[Cr_{1-x}Mn_x]Al$ where the Y site is occupied by Cr or Mn atoms There is even the possibility of growing quinternary Heusler compounds where also the D site is occupied by two species of *sp*-atoms [200]. To perform this study we used the KKR method within the coherent potential approximation (CPA) as implemented by H. Akai [54], which has been already used with success to study the magnetic semiconductors [54]. For all calculations we assumed that the lattice constant varies linearly with the concentration x which has been verified for several quaternary alloys [10, 11]. To our knowledge from the systems under study only $Co_2Cr_{0.6}Fe_{0.4}Al$ has been studied experimentally [201–205].

We calculated the total spin moment for several quaternary alloys taking into account several possible combinations of chemical elements and assuming in all cases a concentration increment of 0.1. The first possible case is when we have two different low-valent transition metal atoms at the Y site like $Co_2[Cr_{1-x}Mn_x]Al$. The total spin moment varies linearly between the 3 μ_B of Co_2CrAl and the 4 μ_B of Co_2MnAl. In the case of the $Co_2[Cr_{1-x}Fe_x]Al$ and $Co_2[Mn_{1-x}Fe_x]Al$ compounds and up to around $x = 0.6$ the total spin moment shows the SP behavior but for larger concentrations it slightly deviates to account for the non-integer moment value of Co_2FeAl. The second case is when one mixes the *sp* elements, but as we just mentioned these compounds also obey the rule for the total spin moments. The third and final case is to mix the higher valent transition metal atoms like in $[Fe_{1-x}Co_x]_2MnAl$ and $[Rh_{1-x}Co_x]_2MnAl$ alloys. In the first case the total spin moment varies linearly between the 2 and 4 μ_B of Fe_2MnAl and Co_2MnAl compounds, respectively. Rh is isoelectronic to Co and for the second family of compounds we find a constant integer value of 4 μ_B for all the concentrations. A special case is Mn_2VAl which has less than 24 electrons and the total spin moment is $-2\,\mu_B$. If now we mix Mn and Co, we get a family of compounds where the total spin moment varies linearly between the $-2\,\mu_B$ and the 2 μ_B and for $x = 0.5$ we get the case of a paramagnetic compound consisting of magnetic elements. Thus all the compounds obey the rule $M_t = Z_t - 24$, showing the Slater-Pauling behavior regardless of the origin of the extra charge.

As a rule of thumb we expect, that for two half-metallic alloys like XYZ and X'YZ (or XY'Z, XYZ'), which both lay on the Slater-Pauling curve, also the mixtures like $X_{1-x}X'_xYZ$ lay on the Slater Pauling curve, with an average moment of $< M_t > = (1-x)M_t^{XYZ} + xM_t^{X'YZ}$. However, if these intermediate structures are stable, is not guaranteed in particular if the parent compounds are not neighbors on the Slater-Pauling curve.

1.7 Half-Metallic Antiferromagnets

A special case for applications are the compounds made up from magnetic elements with exactly 18 (in the case of semi-Heuslers) or 24, in the case of full-Heuslers, valence electrons which should have a total zero spin magnetic moment in case of half-metallicity. These alloys should be of special interest for applications since they create no external stray fields and thus exhibit minimal energy losses [206]. In literature they are named either half-metallic fully-compensated ferrimagnets (HM-FCF) [207] or half-metallic antiferromagnets (HMAs) which was the initial term used by van Leuken and de Groot when studying the semi-Heusler compound CrMnSb in 1995 [208]. The HMA character of CrMnSb has been also confirmed by calculations made by Shaughnessy and collaborators [209]. Contrary to conventional antiferromagnets here the compensation of the spin magnetic moments stems from different magnetic sublattices, e.g. in CrMnSb Cr and Mn atoms have antiparallel spin magnetic moments of about the same magnitude [208]. However, in contrast to zero temperature limit in which the total magnetization vanishes, at finite temperatures spin fluctuations induce a net magnetization in HMAs leading to a ferrimagnetic state [210]. Except the semi-Heusler CrMnSb, also full-Heusler alloys with 24 valence electrons have been predicted to be HMAs including the Mn_3Ga [207, 211, 212], Cr_2MnZ (Z = P, As, Sb, Bi) alloys [173, 213, 214], the Co-doped Mn_2VZ (Z = Al, Si) half-metallic ferrimagnetic alloys [215] and the Cr-doped Co_2CrAl [216]. Heusler compounds are not the only family of alloys where the half-metallic anitferromagnetism has been predicted. Potential HMA candidates include also the double-perovskites [217–221], superlattice structures [222, 223], diluted magnetic semicondutors [224, 225] and even Fe-based superconductors [226].

A special case of interest would be full-Heusler compounds where X and Y are of the same chemical species. This may lead to an easier growth of these materials and thus to enhanced properties of devices based on them. Moreover in this case we would also avoid the degradation of the magnetic properties caused by impurities and atomic swaps when the X and Y are of different chemical species like in Co_2MnSi [227]. The cubic structure which results when X and Y are the same in full-Heusler is known as the $D0_3$ structure. We mentioned in the above paragraph that Mn_3Ga which has 24 valence electrons has been predicted to be a half-metallic antiferromagnet [207]. Several experiments have been lately devoted to the growth of high quality samples of Mn_3Ga alloy [212, 228, 229]. In [177], Li et al. have shown that also Mn_3Al which has 24 valence electrons is a HF-FCF and when Cr was substituted for Mn, Cr_3Al was found to have a total spin-magnetic moment of $-3\,\mu_B$ in accordance to the Slater-Pauling rule for half-metallic Heusler compounds. Doping of Mn_3Al with vanadium leads to the loss of half-metallicity [230]. In a recent publication [189] it was shown that Cr_3Se which has also 24 valence electrons is almost a half-metal with an almost zero total spin magnetic moment. The Cr atoms in the unit cell were antiferromagnetically coupled giving rise to ferrimagnetic behavior and the estimated Curie temperature was as high as $\sim$700 K. Thus this compound is attractive for spintronics applications. Moreover the entire $Cr_{2+x}Se$ family, with

$0 \leq x \leq 1$, shows the half-metallic antiferromagnetic behavior [231]. We should mention finally that one could envisage of growing a half-metallic antiferromagnet by combing alternate layers of half-metallic ferromagnetic and ferrimagnetic Heusler compounds as suggested in [232].

1.8 Magnetic Semiconductors

Magnetic semiconductors are compounds combining both the semiconducting behavior with the magnetic properties. Such compounds can offer novel functionalities to spintronic and magnetoelectronic devices, e.g. they can act as spin-filter materials. Such materials can find application in magnetic tunnel junctions (MTJ). In usual MTJs the magnetic electrodes are separated by an insulating barrier and ballistic transport is achieved through the tunnelling of the electrons via the barrier. The alternative is to use a spin-filter material as the barrier and have metallic electrodes. Then the probability for electrons tunnelling through the spin-filter barrier is different for the two spin-directions and the flow of a spin-polarized current can be achieved [233, 234]. Among Heusler compounds there are two families of compounds studied recently which are magnetic semiconductors and can be used as spin-filter materials: (i) (CoV)XAl with X being Ti, Zr or Hf which are ferromagnetic semiconductors [235], and (ii) (CrV)XAl with X being Ti, Zr or Hf which are fully-compensated ferrimagnetic semiconductors [236]. The latter also combine magnetic semiconducting behavior with zero magnetization leading to minimum energy losses in devices. In Fig. 1.10 we present the total density of states for (CoV)TiAl and (CrV)TiAl where the energy gaps in both spin-directions are present.

A special class are the so-called spin-gapless semiconductors. These materials combine the properties of half-metals and magnetic semiconductors, and they are actually magnetic semiconductors where there is an almost vanishing zero-width energy gap at the Fermi level in the majority-spin direction and a usual energy gap in the other spin-direction [237]. Spin-gapless semiconductors offer also novel functionalists due to their unique properties: (i) the mobility of carriers is considerably larger than in usual semiconductors, (ii) excited carriers include both electrons and holes which can be 100 % spin-polarized simultaneously, and (iii) a vanishing amount of energy is enough to excite majority-spin electrons from the valence to the conduction band.

Although gapless-semiconductors are well known in literature [237], it was not until 2008, that Wang proposed that the doping of $PbPdO_2$, a gapless semiconductor, with transition metal atoms would lead to a spin gapless semiconductor [238, 239]. Experimental confirmation was offered in 2014 by Kim and collaborators who studied pollycrystalline films of Mn and Co doped $PbPdO_2$ [240]. Among Heusler compounds several have been identified to be spin-gapless semiconductors [195, 196, 241, 242]. The main attention was given to Mn_2CoAl, an inverse full-Heusler compound, due to its successful growth in the form of films. In 2008 Liu and collaborators synthesized Mn_2CoAl using an arc-melting technique and found

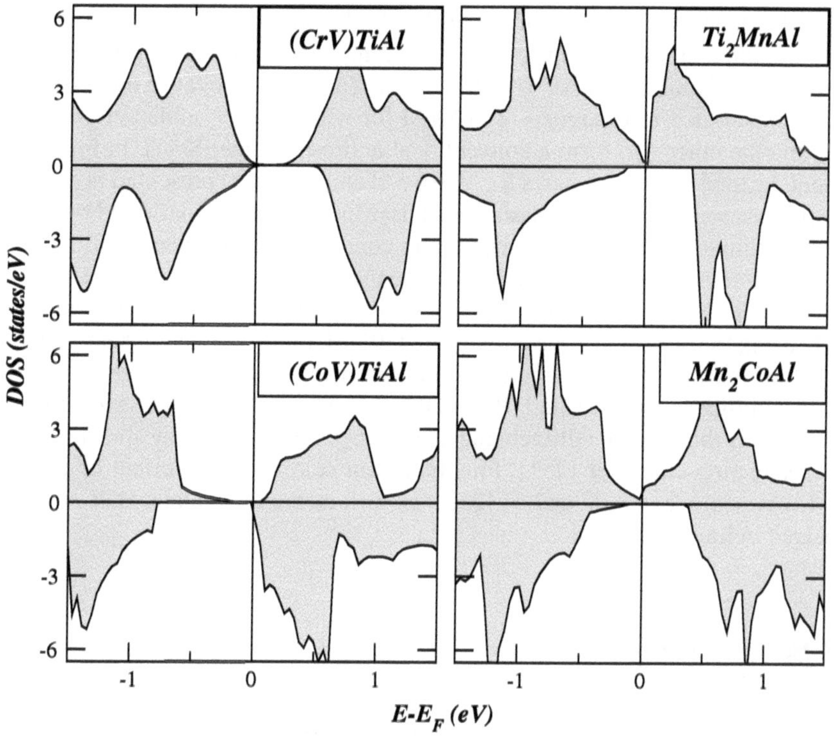

Fig. 1.10 Total DOS per formula unit around the Fermi level for four magnetic semiconductors. (CrV)TiAl is a fully-compensated ferrimagnetic semiconductor, (CoV)TiAl is a ferromagnetic semiconductor, and Ti_2MnAl and Mn_2CoAl show a spin-gapless semiconducting behavior

that it adopted the lattice structure of inverse full-Heuslers with a lattice constant of 5.8388 Å and a total spin magnetic moment of 1.95 μ_B per formula unit [174]. Electronic structure calculations yielded a ferrimagnetic state with a total spin magnetic moment of 1.95 μ_B per formula unit and an antiparallel coupling between the Mn nearest-neighboring atoms [174]. In 2011 Meinert and collaborators studied again theoretically this compound reproducing the calculated results of Liu et al. and also studied the exchange constants revealing that magnetic interactions in these compounds are short range direct interactions [243]. The breakthrough took place in 2013, when Ouardi et al. identified the spin-gapless behavior of Mn_2CoAl and have confirmed it experimentally in bulk-like pollycrystalline films [244]. They found an experimental lattice constant of 5.798 Å, a Curie temperature of 720 K and a total spin magnetic per formula unit of 2 μ_B at a temperature of 5 K [244]. Following this research work, Jamer and collaborators have grown thin films of 70 nm thickness on top of GaAs [245], but these films were found to deviate from the spin-gapless semiconducting behavior [246]. On the contrary films, grown on top of a thermally oxidized Si substrate, were found to be spin-gapless semiconductors with a Curie temperature of 550 K [247]. *Ab-initio* calculations of Skaftouros et al. identified among the inverse Heusler compounds another four potential SGS materials: Ti_2CoSi, Ti_2MnAl, Ti_2VAs and Cr_2ZnSi, with the latter three being also fully-compensated ferrimagnets, and V_3Al for which one V sublattice is not magnetic and the other two form a conventional antiferromagnet [241]. In Fig. 1.10 we present the total density of states for Mn_2CoAl and Ti_2MnAl presented in [241]. The spin-gapless semiconducting character of Ti_2MnAl was also confirmed by Jia et al. [248]. Wollman et al. [249] confirmed the conclusion of Meinert et al. that direct exchange interactions are responsible for the magnetic order in Mn_2CoAl studying a wide range of Mn_2-based Heusler compounds and predicted a Curie temperature of 740 K using the spherical approximation [250]. Skaftouros et al. discussed in detail the behavior of the total magnetic moment in inverse Heusler compounds including the spin-gapless materials [190]. Galanakis and collaborators have shown that defects keep the half-metallic character of Mn_2CoAl but destroy the spin-gapless semiconducting character [251]. Finally, recent studies on the effect of doping of Mn_2CoAl with Co, Cu, V and Ti [252], as well as the anomalous Hall effect have appeared in literature [253].

1.9 Special Topics

1.9.1 Exchange Constants and Curie Temperature

In [131] the exchange interactions are studied. In the case of semi-Heuslers, like NiMnSb and CoMnSb, the dominant interaction is the indirect interaction between the Mn atoms. Magnetic interactions are much more complex in the case of full-Heusler compounds, like Co_2MnSi and Co_2CrAl. Now there are three magnetic atoms in the unit cell and the ferromagnetic order is stabilized by the intersublattice interactions between the Mn(Cr) and Co atoms and between Co atoms belonging to different sublattices (see Fig. 1.1 for the structure) [131]. Contrary to the usual full-Heusler compounds, where both direct and indirect exchange interactions are present, the properties in the inverse Heusler compounds are dominated by short range interactions [175, 249].

To calculate the Curie temperature in magnetic materials usually either the mean-field (MFA) or the random-phase approximations (RPA) are employed. It has been shown in the case of Heusler compounds that RPA is more adequate to estimate the Curie temperature yielding values close to the experimental results [131]. The reason is that MFA is the numerical average over the magnons (spin-waves) while RPA is the harmonic average and thus in the latter the magnons with lower energy have a more significant impact on the Curie temperature which represents more accurately the experimental situation. In general Heusler compounds containing Mn atoms have Curie temperature much higher than the room temperature [10].

A cases of interest is the behavior of exchange constants upon doping. In [132], it was shown that in the case of the semi-Heusler compounds, NiMnSb and CoMnSb, the artificial shift of the Fermi level using a rigid band model results in a change in the relative strength of the Ruderman-Kittel-Kasuya-Yosida (RKKY)-like ferromagnetic and the superexchange antiferromagnetic interactions. The Curie temperature takes its maximum value when the Fermi level is located exactly at the middle of the minority-spin energy gap. This shift of the Fermi level in real systems can be achieved either by substituting Cu for Ni or Co in NiMnSb/CoMnSb or by substituting Sb for Sn in AuMnSn. In both cases for large concentrations of Cu or Sb atoms the superexchange antiferromagnetic interaction dominates over the RKKY-like ferromagnetic interaction and antiferromagnetism becomes the stable magnetic state [103, 104].

1.9.2 Defects and Vacancies

Spin-orbit interaction is mixing the two spin-channels and thus no real half-metal can exist. In reality, in Heusler compounds the effect of the spin-orbit coupling on the minority-spin band gap can be neglected [254]. Even when heavy elements are present like Sb in NiMnSb, the *p*-states of Sb are located low in energy and have vanishing weight near the Fermi level. Thus in bulk Heusler compounds only defects can affect their half-metallic character.

There are several derivatives of the initial ideal Heusler structures presented in Fig. 1.1 due either to atomic swaps or disorder [133]. In the structures derived from the occurrence of Co antisites in Co_2MnAl and Co_2MnSi compounds the half-metallicity is destroyed, while the Mn-Al(Si) swaps preserve the half-metallic character of the parent compounds [133]. In both Co_2MnAl and Co_2MnSi compounds when Fe or Cr is substituted for Mn the half-metallicity is preserved [134, 135]. In the case of Co_2MnSi, when a surplus of Mn or Si atoms is present, like in the $Co_2Mn_{1+x}Si_{1-x}$ compounds (with x ranging between -0.2 and 0.2), the half-metallicity is preserved contrary to the $Co_2Mn_{1+x}Al_{1-x}$ compounds where the half-metallicity is destroyed [134, 135]. A very interesting case of defects occurs when Cr or Mn atoms substitute Co atoms in Co_2CrZ or Co_2MnZ compounds with Z being a *sp*-elements. The Cr and Mn impurity atoms are antiferromagnetically coupled to the Co atoms as well as to the Cr and Mn atoms sitting at the perfect C sites, and thus a half-metallic ferrimagnetic state occurs [139, 140].

Vacancies are also a very common defect occurring during the growth of samples of Heusler compounds. In the case of Co_2CrZ or Co_2MnZ compounds when the vacancy occurs at sites occupied by Co atoms the half-metallicity is destroyed contrary to vacancies at the Cr(Mn) or Z sites which leave the half-metallic character of the parent compounds unaltered [136]. In [138] it was shown for several families of Heusler compounds that the occurrence of vacancies at the A or C sites (see Fig. 1.1) occupied by the higher-valent transition-metal atoms alters the local environment of the atoms destroying the half-metallic character of the compounds under study.

1.9.3 Surfaces and Interfaces

Surfaces of Heusler compounds have been found not be half-metallic. This is valid irrespectively of the orientation of the surface and it is valid for both the semi-Heusler compounds, as it was shown in the case of NiMnSb (111), (110) and (100) surfaces [93–95], and the full-Heuslers like Co_2MnSn [129]. The loss of half-metallicity is due to surface states pinned exactly at the Fermi level. A similar situation occurs in the case of interfaces with semiconductors [96, 97] where also interface states appear within the minority-spin energy gap. An interesting case occurs at interfaces of full-Heusler compounds containing Cr. The large enhancement of the Cr moment at the interface between a CrAl-terminated $Co_2CrAl(001)$ spacer and the InP(001) semiconductor weakens the effect of the interface states, resulting in the high value of the spin polarization at the Fermi level calculated in [128]. Finally, also the case of interfaces between the full-Heusler compounds and magnetic binary compounds has been studied. In cases where both magnetic spacers contain the same transition metal atoms, like the case of Co_2MnSi/CoPd multilayers, a high degree of spin polarization at the interface is present [130].

1.10 Summary and Outlook

In this chapter we have given an introduction into the electronic structure and the resulting magnetic properties of half-metallic Heusler alloys, which represent interesting hybrids between metallic ferromagnets and semiconductors. Many unusual features arise from the half-metallicity induced by the gap in the minority band, and therefore the understanding of the gap is of central importance.

For the semi-Heusler alloys like NiMnSb, crystallizing in the $C1_b$ structure, the gap arises from the hybridization between the d-wavefunctions of the lower-valent transition metal atom (e.g. Mn) with the d-wavefunctions of the higher-valent transition metal atom (e.g. Ni). Thus the d-d hybridization leads to 5 occupied bonding bands, which have a larger Ni and smaller Mn admixture. These states form the valence band, being separated by a band gap from the conduction band which is formed by five antibonding hybrids with a large Mn d- and a small Ni d-admixture. The role of the sp atoms like Sb is very different. Firstly they are important for the bonding, in particular for the stabilization of the $C1_b$ structure. Secondly the sp atom creates for each spin direction one s and three p bands in the energy region below the d states which by hybridization can accommodate also transition metal electrons, such that e.g. Sb formally acts like a Sb^{3-} and Sn as a Sn^{4-} anion. In this way the effective number of valence d-electrons can be changed by the valence of the sp elements.

Since the minority valence band consist of 9 bands, compounds with 18 valence electrons like CoTiSb have the same density of states for both spin directions and are semiconductors. More general, compounds with a total number of Z_t valence electrons per unit cell are ferromagnets and have an integer total spin moment of $M_t = Z_t - 18$, since $Z_t - 18$ is the number of uncompensated spins. For instance, NiMnSb has 22 valence electrons and therefore a total moment of exactly 4 μ_B. This relation is similar to the well known Slater-Pauling behavior observed for binary transition-metal alloys and allows to classify the half-metallic $C1_b$ Heusler alloys into classes with integer moments between 0 and 5 μ_B. The maximum moment of 5 μ_B is difficult to achieve, since it requires that all majority d-sates are occupied.

In the case of the full-Heusler alloys like Co_2MnGe, there are, in addition to the Co-Mn bonding and antibonding d-hybrids, also Co states which cannot hybridize with both the Mn and the Ge atoms and are exclusively localized at the two Co sublattices. Thus in addition to the 5 Co-Mn bonding and 5 Co-Mn antibonding bands, there exist 5 such "non-bonding" bands which are only splitted-up by the weaker Co-Co hybridization into 3 occupied d states of t_{1u} symmetry and 2 unoccupied e_u states, which are located just below and just above the Fermi level such that the indirect gap in these materials is smaller than in the semi-Heuslers. Due to the additional 3 occupied t_{1u} cobalt bands, the full-Heusler alloys have 12 occupied minority bands instead of 9 in the case of the semi-Heusler compounds and their relation for the total spin magnetic moment becomes $M_t = Z_t - 24$. Thus systems like Fe_2VAl with 24 valence electrons are semiconductors, Co_2VAl (26 valence electrons) has a total spin moment of 2 μ_B, Co_2CrAl 3 μ_B, Co_2MnAl 4 μ_B and finally Co_2MnSi which has 29 valence electrons has a total spin moment of 5 μ_B. The maximal total spin moment for these alloys is 7 μ_B, but as has been shown even the 6 μ_B are difficult to be achieved. Special cases of half-metallic full-Heuslers are the inverse Heuslers, the LiMgPdSn-type Heuslers, the disordered quaternary ones (where at one site there exist in a random way atoms of two different chemical species), the so-called half-metallic antiferromagnets which have 24 valence electrons in the unit cell and thus zero net magnetization and finally the magnetic semiconductors including the Heusler compounds showing spin-gapless semiconducting behavior.

The existence of the minority gap is central for any application of half-metals in spintronics, and thus it is of great importance to understand and control all mechanisms that can destroy the gap. The spin-orbit interaction couples the two spin-bands and induces states in the gap; however this effect is weak and the spinpolarization remains in most cases as high as ~99 % [254]. There are, secondly, excitation effects leading to states in the gap. In the simplest approach one can consider in the adiabatic approximation "static spin waves", which are superpositions of spin-up and spin-down states. At higher temperatures spinwaves excitations will smear out the gap [255, 256]. These excitations drive the system to the paramagnetic state above the Curie temperature. At low temperatures the interaction of the electrons with magnons leads to non-quasiparticle excitations in the minority gap above the Fermi level [257]. Note that spin wave excitations lead to new states in the gap above and below the Fermi level, whereas at low temperatures the non-quasiparticle states introduce only additional states at and above E_F. Thirdly and most importantly several kind of defects are expected to lead to states in the gap as discussed above [258].

References

1. I. Žutić, J. Fabian, S. Das Sarma, Rev. Mod. Phys. **76**, 323 (2004)
2. A. Hirohata, K. Takanashi, J. Phys. D Appl. Phys. **47**, 193001 (2014)
3. S.A. Wolf, D.D. Awschalom, R.A. Buhrman, J.M. Daughton, S. von Molnár, M.L. Roukes, A.Y. Chtchelkanova, D.M. Treger, Science **294**, 1488 (2001)
4. G.A. Prinz, Science **282**, 1660 (1998)
5. G.A. Prinz, J. Magn. Magn. Mater. **200**, 57 (1999)
6. J. de Boeck, W. van Roy, J. Das, V. Motsnyi, Z. Liu, L. Lagae, H. Boeve, K. Dessein, G. Borghs, Semicond. Sci. Technol. **17**, 342 (2002)
7. J. de Boeck, W. van Roy, V. Motsnyi, Z. Liu, K. Dessein, G. Borghs, Thin Solid Films **412**, 3 (2002)
8. M. Bowen, A. Barthélèmy, M. Bibes, E. Jacquet, J.P. Contour, A. Fert, D. Wortmann, S. Blügel, J. Phys. Condens. Matter **17**, L407 (2005)
9. F. Heusler, Verh. Dtsch. Phys. Ges. **12**, 219 (1903)
10. P.J. Webster, K.R.A. Ziebeck, in *Alloys and Compounds of d-Elements with Main Group Elements. Part 2*. Landolt-Börnstein, New Series, Group III, vol 19c, ed. by H.R.J. Wijn (Springer, Berlin 1988) pp. 75–184
11. K.R.A. Ziebeck, K.U. Neumann, in *Magnetic Properties of Metals*. Landolt-Börnstein, New Series, Group III, vol 32/c, ed. by H.R.J. Wijn (Springer, Berlin 2001) pp. 64–414
12. J. Pierre, R.V. Skolozdra, J. Tobola, S. Kaprzyk, C. Hordequin, M.A. Kouacou, I. Karla, R. Currat, E. Lelièvre-Berna, J. Alloys Compd. **262–263**, 101 (1997)
13. J. Tobola, J. Pierre, S. Kaprzyk, R.V. Skolozdra, M.A. Kouacou, J. Phys. Condens. Matter **10**, 1013 (1998)
14. J. Tobola, J. Pierre, J. Alloys Compd. **296**, 243 (2000)
15. J. Tobola, S. Kaprzyk, P. Pecheur, Phys. St. Sol. (b) **236**, 531 (2003)
16. M. Gillessen, R. Dronskowski, J. Comput. Chem. **30**, 1290 (2009)
17. M. Gillessen, R. Dronskowski, J. Comput. Chem. **31**, 612 (2010)
18. K. Watanabe, Trans. Jpn. Inst. Met. **17**, 220 (1976)
19. R.A. de Groot, F.M. Mueller, P.G. van Engen, K.H.J. Buschow. Phys. Rev. Lett. **50**, 2024 (1983)
20. I. Galanakis, P.H. Dederichs, N. Papanikolaou, Phys. Rev. B **66**, 134428 (2002)
21. I. Galanakis, P.H. Dederichs, N. Papanikolaou, Phys. Rev. B **66**, 174429 (2002)
22. I. Galanakis, J. Phys. Condens. Matter **16**, 3089 (2004)
23. M. Zhang, X. Dai, H. Hu, G. Liu, Y. Cui, Z. Liu, J. Chen, J. Wang, G. Wu, J. Phys. Condens. Matter **15**, 7891 (2003)
24. M. Zhang, Z. Liu, H. Hu, G. Liu, Y. Cui, G. Wu, E. Brück, F.R. de Boer, Y. Li, J. Appl. Phys. **95**, 7219 (2004)
25. R.J. Soulen Jr, J.M. Byers, M.S. Osofsky, B. Nadgorny, T. Ambrose, S.F. Cheng, P.R. Broussard, C.T. Tanaka, J. Nowak, J.S. Moodera, A. Barry, J.M.D. Coey, Science **282**, 85 (1998)
26. H. Kato, T. Okuda, Y. Okimoto, Y. Tomioka, K. Oikawa, T. Kamiyama, Y. Tokura, Phys. Rev. B **69**, 184412 (2004)
27. T. Shishidou, A.J. Freeman, R. Asahi, Phys. Rev. B **64**, 180401 (2001)
28. I. Galanakis, Phys. Rev. B **66**, 012406 (2002)
29. I. Galanakis, Ph Mavropoulos, Phys. Rev. B **67**, 104417 (2003)
30. Ph. Mavropoulos, I. Galanakis, J. Phys. Condens. Matter **16**, 4261 (2004)
31. S. Sanvito, N.A. Hill, Phys. Rev. B **62**, 15553 (2000)
32. A. Continenza, S. Picozzi, W.T. Geng, A.J. Freeman, Phys. Rev. B **64**, 085204 (2001)
33. B.G. Liu, Phys. Rev. B **67**, 172411 (2003)
34. B. Sanyal, L. Bergqvist, O. Eriksson, Phys. Rev. B **68**, 054417 (2003)
35. W.-H. Xie, B.-G. Liu, D.G. Pettifor, Phys. Rev. B **68**, 134407 (2003)
36. W.-H. Xie, B.-G. Liu, D.G. Pettifor, Phys. Rev. Lett. **91**, 037204 (2003)
37. Y.Q. Xu, B.-G. Liu, D.G. Pettifor, Phys. Rev. B **68**, 184435 (2003)

38. M. Zhang et al., J. Phys. Condens. Matter **15**, 5017 (2003)
39. C.Y. Fong, M.C. Qian, J.E. Pask, L.H. Yang, S. Dag, Appl. Phys. Lett. **84**, 239 (2004)
40. J.E. Pask, L.H. Yang, C.Y. Fong, W.E. Pickett, S. Dag, Phys. Rev. B **67**, 224420 (2003)
41. J.-C. Zheng, J.W. Davenport, Phys. Rev. B **69**, 144415 (2004)
42. H. Akinaga, T. Manago, M. Shirai, Jpn. J. Appl. Phys. **39**, L1118 (2000)
43. M. Mizuguchi, H. Akinaga, T. Manago, K. Ono, M. Oshima, M. Shirai, J. Magn. Magn. Mater. **239**, 269 (2002)
44. M. Mizuguchi, H. Akinaga, T. Manago, K. Ono, M. Oshima, M. Shirai, M. Yuri, H.J. Lin, H.H. Hsieh, C.T. Chen, J. Appl. Phys. **91**, 7917 (2002)
45. M. Mizuguchi, M.K. Ono, M. Oshima, J. Okabayashi, H. Akinaga, T. Manago, M. Shirai, Surf. Rev. Lett. **9**, 331 (2002)
46. M. Nagao, M. Shirai, Y. Miura, J. Appl. Phys. **95**, 6518 (2004)
47. K. Ono, J. Okabayashi, M. Mizuguchi, M. Oshima, A. Fujimori, H. Akinaga, J. Appl. Phys. **91**, 8088 (2002)
48. M. Shirai, Physica E **10**, 143 (2001)
49. M. Shirai, J. Appl. Phys. **93**, 6844 (2003)
50. J.H. Zhao, F. Matsukura, K. Takamura, E. Abe, D. Chiba, H. Ohno, Appl. Phys. Lett. **79**, 2776 (2001)
51. J.H. Zhao, F. Matsukura, K. Takamura, E. Abe, D. Chiba, Y. Ohno, K. Ohtani, H. Ohno, Mat. Sci. Semicond. Process. **6**, 507 (2003)
52. M. Horne, P. Strange, W.M. Temmerman, Z. Szotek, A. Svane, H. Winter, J. Phys. Condens. Matter **16**, 5061 (2004)
53. A. Stroppa, S. Picozzi, A. Continenza, A.J. Freeman, Phys. Rev. B **68**, 155203 (2003)
54. H. Akai, Phys. Rev. Lett. **81**, 3002 (1998)
55. K. Kusakabe, M. Geshi, H. Tsukamoto, N. Suzuki, J. Phys. Condens. Matter **16**, S5639 (2004)
56. J.-H. Park, E. Vescovo, H.-J. Kim, C. Kwon, R. Ramesh, T. Venkatesan, Nature **392**, 794 (1998)
57. S. Datta, B. Das, Appl. Phys. Lett. **56**, 665 (1990)
58. K.A. Kilian, R.H. Victora, J. Appl. Phys. **87**, 7064 (2000)
59. C.T. Tanaka, J. Nowak, J.S. Moodera, J. Appl. Phys. **86**, 6239 (1999)
60. J.A. Caballero, Y.D. Park, J.R. Childress, J. Bass, W.-C. Chiang, A.C. Reilly, W.P. Pratt Jr, F. Petroff, J. Vac. Sci. Technol. A **16**, 1801 (1998)
61. C. Hordequin, J.P. Nozières, J. Pierre, J. Magn. Magn. Mater. **183**, 225 (1998)
62. M.M. Kirillova, A.A. Makhnev, E.I. Shreder, V.P. Dyakina, N.B. Gorina, Phys. Stat. Sol. (b) **187**, 231 (1995)
63. K.E.H.M. Hanssen, P.E. Mijnarends, Phys. Rev. B **34**, 5009 (1986)
64. K.E.H.M. Hanssen, P.E. Mijnarends, L.P.L.M. Rabou, K.H.J. Buschow, Phys. Rev. B **42**, 1533 (1990)
65. W. van Roy, M. Wojcik, E. Jdryka, S. Nadolski, D. Jalabert, B. Brijs, G. Borghs, J. De Boeck, Appl. Phys. Lett. **83**, 4214 (2003)
66. W. van Roy, J. de Boeck, B. Brijs, G. Borghs, Appl. Phys. Lett. **77**, 4190 (2000)
67. J.-P. Schlomka, M. Tolan, W. Press, Appl. Phys. Lett. **76**, 2005 (2000)
68. P. Bach, C. Rüster, C. Gould, C.R. Becker, G. Schmidt, L.W. Molenkamp, J. Cryst. Growth **251**, 323 (2003)
69. P. Bach, A.S. Bader, C. Rüster, C. Gould, C.R. Becker, G. Schmidt, L.W. Molenkamp, W. Weigand, C. Kumpf, E. Umbach, R. Urban, G. Woltersdorf, B. Heinrich, Appl. Phys. Lett. **83**, 521 (2003)
70. J. Giapintzakis, C. Grigorescu, A. Klini, A. Manousaki, V. Zorba, J. Androulakis, Z. Viskadourakis, C. Fotakis, Appl. Surf. Sci. **197**, 421 (2002)
71. J. Giapintzakis, C. Grigorescu, A. Klini, A. Manousaki, V. Zorba, J. Androulakis, Z. Viskadourakis, C. Fotakis, Appl. Phys. Lett. **80**, 2716 (2002)
72. C.E.A. Grigorescu, S.A. Manea, M. Mitrea, O. Monnereau, R. Rotonier, L. Tortet, R. Keschawarz, J. Giapintzakis, A. Klini, V. Zorba, J. Androulakis, C. Fotakis, Appl. Surf. Sci. **212**, 78 (2003)

73. S. Gardelis, J. Androulaki, P. Migiakis, J. Giapintzakis, S.K. Clowes, Y. Bugoslavsky, W.R. Branford, Y. Miyoshi, L.F. Cohen, J. Appl. Phys. **95**, 8063 (2004)
74. F.B. Mancoff, B.M. Clemens, E.J. Singley, D.N. Basov, Phys. Rev. B **60**(*R*12) 565 (1999)
75. W. Zhu, B. Sinkovic, E. Vescovo, C. Tanaka, J.S. Moodera, Phys. Rev. B **64**, R060403 (2001)
76. G.L. Bona, F. Meier, M. Taborelli, E. Bucher, P.H. Schmidt, Solid State Commun. **56**, 391 (1985)
77. S.K. Clowes, Y. Mioyoshi, Y. Bugoslavsky, W.R. Branford, C. Grigorescu, S.A. Manea, O. Monnereau, L.F. Cohen, Phys. Rev. B **69**, 214425 (2004)
78. J.A. Caballero, A.C. Reilly, Y. Hao, J. Bass, W.P. Pratt, F. Petroff, J.R. Childress, J. Magn. Magn. Mater. **198–199**, 55 (1999)
79. R. Kabani, M. Terada, A. Roshko, J.S. Moodera, J. Appl. Phys. **67**, 4898 (1990)
80. C.T. Tanaka, J. Nowak, J.S. Moodera, J. Appl. Phys. **81**, 5515 (1997)
81. A.N. Caruso, C.N. Borca, D. Ristoiu, J.P. Nozieres, P.A. Dowben, Surf. Sci. **525**, L109 (2003)
82. D. Ristoiu, J.P. Nozières, C.N. Borca, T. Komesu, H.-K. Jeong, P.A. Dowben, Europhys. Lett. **49**, 624 (2000)
83. D. Ristoiu, J.P. Nozières, C.N. Borca, B. Borca, P.A. Dowben, Appl. Phys. Lett. **76**, 2349 (2000)
84. T. Komesu, C.N. Borca, H.-K. Jeong, P.A. Dowben, D. Ristoiu, J.P. Nozières, Sh Stadler, Y.U. Idzerda, Phys. Lett. A **273**, 245 (2000)
85. I. Galanakis, S. Ostanin, M. Alouani, H. Dreyssé, J.M. Wills, Phys. Rev. B **61**, 4093 (2000)
86. E. Kulatov, I.I. Mazin, J. Phys. Condens. Matter **2**, 343 (1990)
87. S.V. Halilov, E.T. Kulatov, J. Phys. Condens. Matter **3**, 6363 (1991)
88. X. Wang, V.P. Antropov, B.N. Harmon, IEEE Trans. Magn. **30**, 4458 (1994)
89. S.J. Youn, B.I. Min, Phys. Rev. B **51**, 10436 (1995)
90. V.N. Antonov, P.M. Oppeneer, A.N. Yaresko, A. Ya, Perlov, T. Kraft, Phys. Rev. B **56**, 13012 (1997)
91. P. Larson, S.D. Mahanti, M.G. Kanatzidis, Phys. Rev. B **62**, 12574 (2000)
92. D. Orgassa, H. Fujiwara, T.C. Schulthess, W.H. Butler, Phys. Rev. B **60**, 13237 (1999)
93. I. Galanakis, J. Phys. Condens. Matter **14**, 6329 (2002)
94. I. Galanakis, J. Magn. Magn. Mater. **288**, 411 (2005)
95. M. Ležaic, I. Galanakis, G. Bihlmayer, S. Blügel, J. Phys. Condens. Matter **17**, 3121 (2005)
96. I. Galanakis, M. Ležaic, G. Bihlmayer, S. Blügel, Phys. Rev. B **71**, 214431 (2005)
97. I. Galanakis, K. Özdoğan, E. Şaşıoğlu, J. Appl. Phys. **104**, 083916 (2008)
98. S.J. Jenkins, D.A. King, Surf. Sci. **494**, L793 (2001)
99. S.J. Jenkins, D.A. King, Surf. Sci. **501**, L185 (2002)
100. G.A. Wijs, R.A. de Groot, Phys. Rev. B **64**, R020402 (2001)
101. A. Debernardi, M. Peressi, A. Baldereschi, Mater. Sci. Eng. C-Bio. S **23**, 743 (2003)
102. J. Kübler, Phys. Rev. B **67**, 220403 (2003)
103. I. Galanakis, E. Şaşıoğlu, K. Özdoğan, Phys. Rev. B **77**, 214417 (2008)
104. K. Özdoğan, E. Şaşıoğlu, I. Galanakis, J. Phys. D Appl. Phys. **42**, 085003 (2009)
105. P.J. Webster, J. Phys. Chem. Solids **32**, 1221 (1971)
106. K.R.A. Ziebeck, P.J. Webster, J. Phys. Chem. Solids **35**, 1 (1974)
107. J.C. Suits, Phys. Rev. B **14**, 4131 (1976)
108. J. Kübler, A.R. Williams, C.B. Sommers, Phys. Rev. B **28**, 1745 (1983)
109. S. Ishida, S. Akazawa, Y. Kubo, J. Ishida, J. Phys. F Met. Phys. **12**, 1111 (1982)
110. S. Ishida, S. Fujii, S. Kashiwagi, S. Asano, J. Phys. Soc. Jpn. **64**, 2152 (1995)
111. S. Fujii, S. Sugimura, S. Ishida, S. Asano, J. Phys. Condens. Matter **2**, 8583 (1990)
112. S. Fujii, S. Asano, S. Ishida, J. Phys. Soc. Jpn. **64**, 185 (1995)
113. P.J. Brown, K.U. Neumann, P.J. Webster, K.R.A. Ziebeck, J. Phys. Condens. Matter **12**, 1827 (2000)
114. F.Y. Yang, C.H. Shang, C.L. Chien, T. Ambrose, J.J. Krebs, G.A. Prinz, V.I. Nikitenko, V.S. Gornakov, A.J. Shapiro, R.D. Shull, Phys. Rev. B **65**, 174410 (2002)
115. T. Ambrose, J.J. Krebs, G.A. Prinz, Appl. Phys. Lett. **76**, 3280 (2000)
116. T. Ambrose, J.J. Krebs, G.A. Prinz, J. Appl. Phys. **87**, 5463 (2000)

117. T. Ambrose, J.J. Krebs, G.A. Prinz, J. Appl. Phys. **89**, 7522 (2001)
118. M.P. Raphael, B. Ravel, M.A. Willard, S.F. Cheng, B.N. Das, R.M. Stroud, K.M. Bussmann, J.H. Claassen, V.G. Harris, Appl. Phys. Lett. **70**, 4396 (2001)
119. M.P. Raphael, B. Ravel, Q. Huang, M.A. Willard, S.F. Cheng, B.N. Das, R.M. Stroud, K.M. Bussmann, J.H. Claassen, V.G. Harris, Phys. Rev. B **66**, 104429 (2002)
120. B. Ravel, M.P. Raphael, V.G. Harris, Q. Huang, Phys. Rev. B **65**, 184431 (2002)
121. L. Ritchie, G. Xiao, Y. Ji, T.Y. Chen, C.L. Chien, M. Chang, C. Chen, Z. Liu, G. Wu, X.X. Zhang, Phys. Rev. B **68**, 104430 (2003)
122. Y.J. Chen, D. Basiaga, J.R. O'Brien, D. Heiman, Appl. Phys. Lett. **84**, 4301 (2004)
123. U. Geiersbach, A. Bergmann, K. Westerholt, J. Magn. Magn. Mater. **240**, 546 (2002)
124. U. Geiersbach, A. Bergmann, K. Westerholt, Thin Solid Films **425**, 225 (2003)
125. K. Westerholt, U. Geirsbach, A. Bergmann, J. Magn. Magn. Mater. **257**, 239 (2003)
126. S. Picozzi, A. Continenza, A.J. Freeman, J. Appl. Phys. **94**, 4723 (2003)
127. S. Picozzi, A. Continenza, A.J. Freeman, J. Phys. Chem. Solids **64**, 1697 (2003)
128. I. Galanakis, J. Phys. Condens. Matter **16**, 8007 (2004)
129. I. Galanakis, J. Comput. Theor. Nanosci. **7**, 474 (2010)
130. I. Galanakis, J. Magn. Magn. Mater. **377**, 291 (2015)
131. E. Şaşıoğlu, L.M. Sandratskii, P. Bruno, I. Galanakis, Phys. Rev. B **72**, 184415 (2005)
132. I. Galanakis, E. Şaşıoğlu, J. Appl. Phys. **109**, 113912 (2011)
133. K. Özdoğan, I. Galanakis, J. Appl. Phys. **110**, 076101 (2011)
134. I. Galanakis, K. Özdoğan, B. Aktaş, E. Şaşıoğlu, Appl. Phys. Lett. **89**, 042502 (2006)
135. K. Özdoğan, B. Aktaş, E. Şaşıoğlu, I. Galanakis, Phys. Rev. B **74**, 172412 (2007)
136. K. Özdoğan, E. Şaşıoğlu, I. Galanakis, Phys. St. Sol. (RRL) **1**, 184 (2007)
137. I. Galanakis, K. Özdoğan, E. Şaşıoğlu, S. Blügel, J. Appl. Phys. **116**, 033903 (2014)
138. I. Galanakis, E. Şaşıoğlu, K. Özdoğan, S. Blügel, Phys. Rev. B **90**, 064408 (2014)
139. K. Özdoğan, I. Galanakis, E. Şaşıoğlu, B. Aktaş, Phys. St. Sol. (RRL) **1**, 95 (2007)
140. K. Özdoğan, I. Galanakis, E. Şaşıoğlu, B. Aktaş, Solid Stae Commun. **142**, 492 (2007)
141. S. Kämmerer, A. Thomas, A. Hütten, G. Reiss, Appl. Phys. Lett. **85**, 79 (2004)
142. J. Schmalhorst, S. Kr̈mmerer, M. Sacher, G. Reiss, A. Hütten, A. Scholl, Phys. Rev. B **70**, 024426 (2004)
143. K. Inomata, S. Okamura, R. Goto, N. Tezuka, Jpn. J. Appl. Phys. **42**, L419 (2003)
144. K. Inomata, N. Tezuka, S. Okamura, H. Kurebayashi, H. Hirohata, J. Appl. Phys. **95**, 7234 (2004)
145. S.H. Vosko, L. Wilk, N. Nusair, Can. J. Phys. **58**, 1200 (1980)
146. P. Hohenberg, W. Kohn, Phys. Rev. **136**, B864 (1964)
147. W. Kohn, L.J. Sham, Phys. Rev. **140**, A1133 (1965)
148. R. Zeller, P.H. Dederichs, B. Újfalussy, L. Szunyogh, P. Weinberger, Phys. Rev. B **52**, 8807 (1995)
149. N. Papanikolaou, R. Zeller, P.H. Dederichs, J. Phys. Condens. Matter **14**, 2799 (2002)
150. M.V. Yablonskikh, V.I. Grebennikov, Y.M. Yarmoshenko, E.Z. Kurmaev, S.M. Butorin, L.-C. Duda, C. Såthe, T. Kåmbre, M. Magnuson, J. Nordgren, S. Plogmann, M. Neumann, Solid State Commun. **117**, 79 (2001)
151. M.V. Yablonskikh, YuM Yarmoshenko, V.I. Grebennikov, E.Z. Kurmaev, S.M. Butorin, L.-C. Duda, J. Nordgren, S. Plogmann, M. Neumann, Phys. Rev. B **63**, 235117 (2001)
152. R.A. de Groot, A.M. van der Kraan, K.H.J. Buschow, J. Magn. Magn. Mater. **61**, 330 (1986)
153. L. Hedin, S. Lundqvist, in *Solid State Physics*, vol 23, ed. by F. Seitz, D. Turnbull, H. Ehrenreich (Academic Press, New York, 1969) pp 1–181
154. B.R.K. Nanda, I. Dasgupta, J. Phys. Condens. Matter **15**, 7307 (2003)
155. D. Jung, H.-J. Koo, M.-H. Whangbo, J. Mol. Struct. (Theochem) **527**, 113 (2000)
156. J.C. Slater, Phys. Rev. **49**, 931 (1936)
157. L. Pauling, Phys. Rev. **54**, 899 (1938)
158. J. Kübler, Physica B **127**, 257 (1984)
159. I. Galanakis, Phys. Rev. B **71**, 012413 (2005)

160. D. Brown, M.D. Crapper, K.H. Bedwell, M.T. Butterfield, S.J. Guilfoyle, A.E.R. Malins, M. Petty, Phys. Rev. B **57**, 1563 (1998)
161. R.A. Dunlap, D.F. Jones, Phys. Rev. B **26**, 6013 (1982)
162. S. Plogmann, T. Schlathölter, J. Braun, M. Neumann, YuM Yarmoshenko, M.V. Yablonskikh, E.I. Shreder, E.Z. Kurmaev, A. Wrona, A. Ślebarski, Phys. Rev. B **60**, 6428 (1999)
163. P.G. van Engen, K.H.J. Buschow, M. Erman, J. Magn. Magn. Mater. **30**, 374 (1983)
164. W. Pendl Jr., R.N. Saxena, A.W. Carbonari, J. Mestnik Filho, J. Schaff, J. Phys. Condens. Matter **8**, 11317 (1996)
165. Y. Feng, J.Y. Rhee, T.A. Wiener, D.W. Lynch, B.E. Hubbard, A.J. Sievers, D.L. Schlagel, T.A. Lograsson, L.L. Miller. Phys. Rev. B **63**, 165109 (2001)
166. C.S. Lue, J.H. Ross Jr, K.D.D. Rathnayaka, D.G. Naugle, S.Y. Wu, W.-H. Li, J. Phys. Condens. Matter **13**, 1585 (2001)
167. Y. Nishino, H. Kato, M. Kato, U. Mizutani, Phys. Rev. B **63**, 233303 (2001)
168. A. Matsushita, T. Naka, Y. Takanao, T. Takeuchi, T. Shishido, Y. Yamada, Phys. Rev. B **65**, 075204 (2002)
169. K. Özdoğan, I. Galanakis, E. Şaşıoğlu, B. Aktaş: J. Phys. Condens. Matter **18**, 2905 (2006)
170. H.M. Kandpal, G.H. Fecher, C. Felser, G. Schönhense, Phys. Rev. B **73**, 094422 (2006)
171. E. Şaşıoğlu, I. Galanakis, C. Friedrich, S. Blügel, Phys. Rev. B **88**, 134402 (2013)
172. C. Tsirogiannis, I. Galanakis, J. Magn. Magn. Mater. **393**, 297 (2015) arXiv:1501.00732
173. K. Özdoğan, I. Galanakis, J. Magn. Magn. Mater. **321**, L34 (2009)
174. G.D. Liu, X.F. Dai, H.Y. Liu, J.L. Chen, Y.X. Li, G. Xiao, G.H. Wu, Phys. Rev. B **77**, 014424 (2008)
175. M. Meinert, J.-M. Schmalhorst, G. Reiss, J. Phys. Condens. Matter **23**, 116005 (2011)
176. H. Luo, Z. Zhu, L. Ma, S. Xu, X. Zhu, C. Jiang, H. Xu, G. Wu, J. Phys. D Appl. Phys. **41**, 055010 (2008)
177. J. Li, H. Chen, Y. Li, Y. Xiao, Z. Li, J. Appl. Phys. **105**, 083717 (2009)
178. B. Xu, M. Zhang, H. Yan, Phys. St. Sol. (b) **248**, 2870 (2011)
179. M. Pugaczowa-Michalska, Intermetallics **24**, 128 (2012)
180. N. Kervan, S. Kervan, J. Phys. Chem. Solids **72**, 1358 (2011)
181. N. Kervan, S. Kervan, Solid State Commun. **151**, 1162 (2011)
182. N. Kervan, S. Kervan, J. Magn. Magn. Mater. **324**, 645 (2012)
183. E. Bayar, N. Kervan, S. Kervan, J. Magn. Magn. Mater. **323**, 2945 (2011)
184. J. Winterlik, G.H. Fecher, B. Balke, T. Graf, V. Alijani, V. Ksenofontov, C.A. Jenkins, O. Meshcheriakova, C. Felser, G. Liu, S. Ueda, K. Kobayashi, T. Nakamura, M. Wójcik, Phys. Rev. B **83**, 174448 (2011)
185. M. Meinert, J.-M. Schmalhorst, C. Klewe, G. Reiss, E. Arenholz, T. B?nert, K. Nielsch, Phys. Rev. B **84**, 132405 (2011)
186. P. Klaer, C.A. Jenkins, V. Alijani, J. Winterlik, B. Balke, C. Felser, H.J. Elmers, Appl. Phys. Lett. **98**, 212510 (2011)
187. V. Alijani, J. Winterlik, G.H. Fecher, C. Felser, Appl. Phys. Lett. **99**, 222510 (2012)
188. H. Luo, W. Zhu, L. Ma, G. Liu, Y. Li, X. Zhu, C. Jiang, H. Xu, G. Wu, J. Phys. D Appl. Phys. **42**, 095001 (2009)
189. I. Galanakis, E. Şaşıoğlu, Appl. Phys. Lett. **99**, 052509 (2011)
190. S. Skaftouros, K. Özdoğan, E. Şaşıoğlu, I. Galanakis, Phys. Rev. B **87**, 024420 (2013)
191. D. Xu, G. Liu, G.H. Fecher, C. Felser, Y. Li, H. Liu, J. Appl. Phys. **105**, 07E901 (2009)
192. V. Alijani, J. Winterlik, G.H. Fecher, S.S. Naghavi, C. Felser, Phys. Rev. B **83**, 184428 (2011)
193. S. Izadi, Z. Nourbakhsh, J. Supercond. Nov. Magn. **24**, 825 (2011)
194. G. Gökoğlu, Solid State Sci. **14**, 1273 (2012)
195. G.Z. Xu, E.K. Liu, Y. Du, G.J. Li, G.D. Liu, W.H. Wang, G.H. Wu, Europhys. Lett. **102**, 17007 (2013)
196. K. Özdoğan, E. Şaşıoğlu, I. Galanakis, J. Appl. Phys. **113**, 193903 (2013)
197. K. Özdoğan, B. Aktaş, I. Galanakis, E. Şaşıoğlu, J. Appl. Phys. **101**, 73910 (2007)
198. Y. Miura, K. Nagao, M. Shirai, Phys. Rev. B **69**, 144113 (2004)
199. Y. Miura, M. Shirai, K. Nagao, J. Appl. Phys. **95**, 7225 (2004)

200. K. Özdoğan, E. Şaşıoğlu, I. Galanakis, J. Appl. Phys. **103**, 023503 (2008)
201. C. Felser, B. Heitkamp, F. Kronast, D. Schmitz, S. Cramm, H.A. Dürr, H.-J. Elmers, G.H. Fecher, S. Wurmehl, T. Block, D. Valdaitsev, S.A. Nepijko, A. Gloskovskii, G. Jakob, G. Schonhense, W. Eberhardt: J. Phys. Condens. Matter **15**, 7019 (2003)
202. N. Auth, G. Jakob, T. Block, C. Felser, Phys. Rev. B **68**, 024403 (2003)
203. H.J. Elmers, G.H. Fecher, D. Valdaitsev, S.A. Nepijko, A. Gloskovskii, G. Jakob, G. Schonhense, S. Wurmehl, T. Block, C. Felser, P.-C. Hsu, W.-L. Tsai, S. Cramm, Phys. Rev. B **67**, 104412 (2003)
204. T. Block, C. Felser, G. Jakob, J. Ensling, B. Muhling, P. Gutlich, R.J. Cava, J. Solid State Chem. **176**, 646 (2003)
205. R. Kelekar, B.M. Klemens, J. Appl. Phys. **96**, 540 (2004)
206. X. Hu, Adv. Mater. **24**, 294 (2012)
207. S. Wurmehl, H.C. Kandpal, G.H. Fecher, C. Felser, J. Phys. Condens. Matter **18**, 6171 (2006)
208. H. van Leuken, R.A. de Groot, Phys. Rev. Lett. **74**, 1171 (1995)
209. M. Shaughnessy, C.Y. Fong, L.H. Yang, C. Felser, Preprint arXiv:1108.3651 (2011)
210. E. Şaşıoğlu, Phys. Rev. B **79**, 100406(R) (2009)
211. B. Balke, G.H. Fecher, J. Winterlik, C. Felser, Appl. Phys. Lett. **90**, 152504 (2007)
212. J. Winterlik, B. Balke, G.H. Fecher, C. Felser, M.C.M. Alves, F. Bernardi, J. Morais, Phys. Rev. B **77**, 054406 (2008)
213. I. Galanakis, K. Özdoğan, E. Şaşıoğlu, B. Aktaş, Phys. Rev. B **75**, 172405 (2007)
214. I. Galanakis, K. Özdoğan, E. Şaşıoğlu, B. Aktaş, Phys. Status Sol. (a) **205**, 1036 (2008)
215. I. Galanakis, K. Özdoğan, E. Şaşıoğlu, B. Aktaş, Phys. Rev. B **75**, 92407 (2007)
216. H. Luo, L. Ma, Z. Zhu, G. Wu, H. Liu, J. Qu, Y. Li, Physica B **403**, 1797 (2008)
217. W.E. Pickett, Phys. Rev. B **57**, 10613 (1998)
218. J.H. Park, S.K. Kwon, B.I. Min, Phys. Rev. B **65**, 174401 (2002)
219. M. Uehara, M. Yamada, Y. Kimishima, Solid. State Commun. **129**, 385 (2004)
220. Y.K. Wang, G.Y. Guo, Phys. Rev. B **73**, 064424 (2006)
221. V. Pardo, W.E. Pickett, Phys. Rev. B **80**, 054415 (2009)
222. M. Nakao, Phys. Rev. B **74**, 172404 (2006)
223. M. Nakao, Phys. Rev. B **77**, 134414 (2008)
224. H. Akai, M. Ogura, Phys. Rev. Lett. **97**, 026401 (2006)
225. N.H. Long, M. Ogura, H. Akai, J. Phys. Condens. Matter **21**, 064241 (2009)
226. M. Nakao, Phys. Rev. B **83**, 214404 (2011)
227. S. Picozzi, A. Continenza, A.J. Freeman, Phys. Rev. B **69**, 094423 (2004)
228. H. Kurt, K. Rode, M. Venkatesan, P. Stamenov, J.M.D. Coey, Phys. Rev. B **83**, 020405(R) (2011)
229. H. Kurt, K. Rode, M. Venkatesan, P. Stamenov, J.M.D. Coey, Phys. St. Sol. (b) **248**, 2338 (2011)
230. Q.F. Li, C.H. Yang, J.L. Su, Physica B **406**, 3726 (2011)
231. I. Galanakis, K. Özdoğan, E. Şaşıoğlu, Phys. Rev. B **86**, 134427 (2012)
232. S. Tirpanci, E. Şaşıoğlu, I. Galanakis, J. Appl. Phys. **113**, 043912 (2013)
233. J.S. Moodera, G.-X. Miao, T.S. Santos, Phys. Today **63**, 46 (2010)
234. G.-X. Miao, M. Münzenberg, J.S. Moodera, Rep. Prog. Phys. **74**, 036501 (2011)
235. I. Galanakis, K. Özdoğan, E. Şaşıoğlu, Appl. Phys. Lett. **103**, 142404 (2013)
236. I. Galanakis, K. Özdoğan, E. Şaşıoğlu, J. Phys. Condens. Matter **26**, 086003 (2014)
237. I.M. Tsidilkovski, *Electron Spectrum of Gapless Semiconductors*. Springer Series in Solid-State Sciences, vol. 116, ed. by von K. Klitzing (Springer, New York, 1996)
238. X.L. Wang, Phys. Rev. Lett. **100**, 156404 (2008)
239. X. Wang, G. Peleckis, C. Zhang, H. Kimura, S. Dou, Adv. Mater. **21**, 2196 (2009)
240. D.H. Kim, J. Hwang, E. Lee, K.J. Lee, S.M. Choo, M.H. Jung, J. Baik, H.J. Shin, B. Kim, K. Kim, B.I. Min, J.-S. Kang, Appl. Phys. Lett. **104**, 022411 (2014)
241. S. Skaftouros, K. Özdoğan, E. Şaşıoğlu, I. Galanakis, Appl. Phys. Lett. **102**, 022402 (2013)
242. G.Y. Gao, K.-L. Yao, Appl. Phys. Lett. **103**, 232409 (2013)
243. M. Meinert, J. Schmalhorst, G. Reiss, J. Phys. Condens. Matter **23**, 036001 (2011)

244. S. Ouardi, G.H. Fecher, C. Felser, J. Kübler, Phys. Rev. Lett. **110**, 100401 (2013)
245. M.E. Jamer, B.A. Assaf, T. Devakul, D. Heiman, Appl. Phys. Lett. **103**, 142403 (2013)
246. M.E. Jamer, B.A. Assaf, G.E. Sterbinsky, D.A. Arena, D. Heiman, J. Appl. Phys. **116**, 213914 (2014)
247. G.Z. Xu, Y. Du, X.M. Zhang, H.G. Zhang, E.K. Liu, W.H. Wang, G.H. Wu, Appl. Phys. Lett. **104**, 242408 (2014)
248. H.Y. Jia, X.F. Dai, L.Y. Wang, R. Liu, X.T. Wang, P.P. Li, Y.T. Cui, G.D. Liu, AIP Adv. **4**, 047113 (2014)
249. L. Wollmann, S. Chadov, J. Kübler, C. Felser, Phys. Rev. B **90**, 214420 (2014)
250. T. Moriya, *Spin Fluctuations in Itinerant Electron Magnetism*. Springer Series in Solid-State Sciences, No. 56 (Springer, Berlin, 1985)
251. I. Galanakis, K. Özdoğan, E. Şaşıoğlu, S. Blügel, J. Appl. Phys. **115**, 093908 (2014)
252. Y.J. Zhang, G.J. Li, E.K. Liu, J.L. Chen, W.H. Wang, G.H. Wu, J. Appl. Phys. **113**, 123901 (2013)
253. J. Kudrnovský, V. Drchal, I. Turek, Phys. Rev. B **88**, 014422 (2013)
254. Ph Mavropoulos, I. Galanakis, V. Popescu, P.H. Dederichs, J. Phys. Condens. Matter **16**, S5759 (2004)
255. P.A. Dowben, R. Skomski, J. Appl. Phys. **93**, 7948 (2003)
256. P.A. Dowben, R. Skomski, J. Appl. Phys. **95**, 7453 (2004)
257. L. Chioncel, M.I. Katsnelson, R.A. de Groot, A.I. Lichtenstein, Phys. Rev. B **68**, 144425 (2003)
258. I. Galanakis, Ph Mavropoulos, J. Phys. Condens. Matter **19**, 315213 (2007)
259. R. Zeller, J. Phys. Condens. Matter **16**, 6453 (2004)
260. K. Koepernik, H. Eschrig, Phys. Rev. B **59**, 1743 (1999)
261. K. Kopernik, *Full Potential Local Orbital Minimum Basis Bandstructure Scheme User's Manual*. http://www.fplo.de/download/doc.pdf

Chapter 2
Basics and Prospectives of Magnetic Heusler Compounds

C. Felser, L. Wollmann, S. Chadov, G.H. Fecher and S.S.P. Parkin

Abstract Manganese-rich Heusler compounds attract much interest in the context of spin transfer torque and rare-earth free hard magnets. Here we give a comprehensive overview of the magnetic properties of non-centrosymmetric cubic Mn_2-based Heusler materials, which are characterized by an antiparallel coupling of magnetic moments on Mn atoms. Such a ferrimagnetic order leads to the emergence of new properties that are absent in ferromagnetic centrosymmetric Heusler structures. In terms of the band structure calculations, we explain the formation of this magnetic order and the Curie temperatures. This overview is intended to establish guidelines for a basic understanding of magnetism in Mn_2-based Heusler compounds.

2.1 Basics and Prospectives of Magnetic Heusler Compounds

Heusler compounds form a remarkable class of materials with more than 1000 members and a wide range of extraordinary properties including half-metals, high temperature ferri- and ferromagnets, multiferroics, shape-memory alloys, and tunable topological insulators with a high potential for spintronics, energy technologies and magneto-caloric applications. Exceptional tunability of Heusler materials allows to

C. Felser (✉) · L. Wollmann · S. Chadov · G.H. Fecher
Max-Planck-Institut für Chemische Physik fester Stoffe,
Nöthnitzer Str. 40, 01187 Dresden, Germany
e-mail: claudia.felser@cpfs.mpg.de

L. Wollmann
e-mail: lukas.wollmann@cps.mpg.de

S. Chadov
e-mail: chadov@cpfs.mpg.de

G.H. Fecher
e-mail: fecher@uni-mainz.de

S.S.P. Parkin
Max-Planck-Institut für Mikrostrukturphysik,
Weinberg 2, 06120 Halle/saale, Germany
e-mail: stuart.parkin@mpi-halle.mpg.de

C. Felser and A. Hirohata (eds.), *Heusler Alloys*,
Springer Series in Materials Science 222, DOI 10.1007/978-3-319-21449-8_2

design almost any functionality. Whereas the Co_2-based Heusler compounds show high spin-polarization in tunnel junction devices and spin-resolved photoemission, the Mn-rich Heusler compounds attract much interest in the context of spin transfer torque, spin-Hall effect and rare earth free hard magnets. Many of the Mn_2-Heusler compounds crystallize in the inverse structure and are characterized by antiparallel coupling of magnetic moments on Mn atoms; the ferrimagnetic order and the lack of inversion symmetry leads to the emergence of new properties that are absent in ferromagnetic centrosymmetric Heusler structures, such as non-collinear magnetism, topological Hall effect and Skyrmions. Tetragonal Heusler compounds with large magnetocrystalline anisotropy can be easily designed by positioning the Fermi energy at the van Hove singularity in one of the spin channels. Here we give a comprehensive overview and a prospective on the magnetic properties of Heusler materials.

Materials for state-of-the-art applications become increasingly complex. Whereas the elemental semiconductors, such as Si and Ge, are still widely used, the binary, ternary and even quaternary semiconductors, such as GaN, $GaP_{1-x}As_x$, or Cu(GaIn) Se_2, play an important role in the nowadays electronics and energy conversion. Polyatomic compositions allow for more degrees of freedom, thereby leading to various multifunctionalities. Recent advances in chemistry, physics, and materials science enabled the rational design of these new functionalities for a plethora of advanced technologies. The control of disorder and defects in such systems and the availability of the required elements in the earths crust, makes their synthesis rather challenging. In the following, we will summarize the present knowledge on magnetic half-Heusler and Heusler compounds, with a focus on potential applications in spintronics, data storage, magnetocalorics and as permanent magnets.

Half-Heusler compounds (with the composition 1:1:1, formed by three interpenetrating fcc-lattices) and full Heusler (or just "Heusler") compounds (with the composition 2:1:1, formed by four interpenetrating fcc-lattices) were discovered in 1903 by Friedrich Heusler [1, 2]. There is no clear definition of a Heusler compound (see, e.g. [3]): the first Heusler compounds, namely CuMnSb, Cu_2MnAl and Cu_2MnSn, drew attention due to their ferromagnetism, though being formed by non-magnetic elements [1, 2]. These first Heusler compounds, all contained Mn, which plays a unique role in this class of materials. The emergence of the unexpected ferromagnetism, which does not evolve straightforwardly from any of these elements, is an important distinction from other magnetic alloys. The fact that, Heusler compounds are related to semiconductors, together with the unique role of Mn, enables us to formulate rules for their rational design.

Semiconducting half-Heusler compounds can be understood as "stuffed" variants of binary zinc-blende semiconductors, using simple electron counting (see Fig. 2.1a, b). A good starting point are Nowotny-Juza phases, which are relatives of the Heusler compounds [3]. This closely related set of compounds, with the formula LiYZ, is constituted by 3 interpenetrating fcc lattices; the sum of its valence electrons is eight, just like in the binary semiconductors. The Z atom is always a nonmagnetic semiconducting or metallic (main-group) element, as in the Heusler compounds. For instance, GaAs is a III-V semiconductor and LiMgAs is a I-II-V

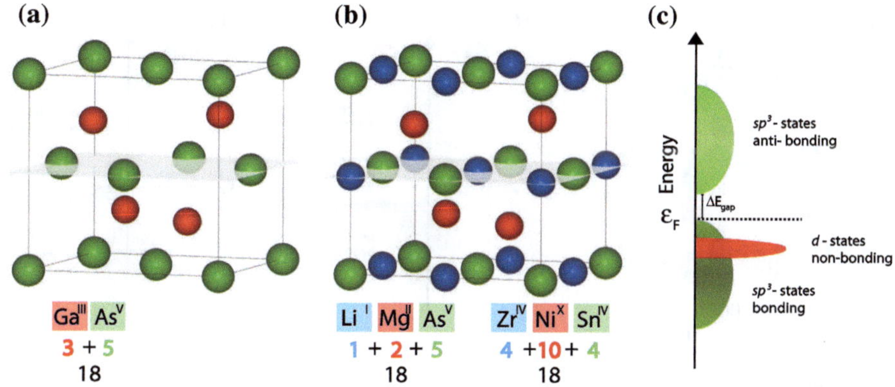

Fig. 2.1 **a** Zinc-blende crystal structure: *red spheres*—As, *green spheres*—Ga; **b** half-Heusler structure: *red spheres* – Mg (Ni), *green spheres*—As (Sn), *blue spheres*—Li (Zr); **c** schematic density of states (DOS) of ZrNiSn: *green*—bonding and antibonding sp^3-states, *red*—non-bonding d^{10}-states

semiconductor. The band gap in both compounds is formed similarly by bonding and antibonding *s*- and *p*-states and, thus, depends on the bonding strength between the atoms which comprise the zinc-blende lattice. These common features are visualized in Fig. 2.1: in half-Heusler (XYZ) compounds, X is the most electropositive element, which forms rock-salt (XZ) sublattice with the main-group atom Z, which indicates the ionic character of their bonding. In contrast, the Y and Z atoms form a zinc-blende type lattice, which corresponds to a more covalent bonding situation. Typical examples of nonmagnetic semiconductors are ZrNiSn and LaPtBi; both with 18 valence electrons. Stable transition metal complexes, such as $Ni(CO)_4$, also follow the 18 valence electrons rule, with the d^{10}, p^6 and s^2 closed shells. The scheme of the density of the states (DOS), including the occupied *d*-states, is shown in Fig. 2.1c. Although Zr, Ni and Sn are metals, the resulting compound is a semiconductor and, moreover, is a good thermoelectric material [3]. Its band gap is a result of strong covalent bonding. A rather recent example of the "binary semiconductor—half-Heusler compound" analogue are topological insulators. By analogy with the HgTe/CdTe quantum-well structure [4], topological insulators and the quantum spin-Hall effect have been predicted for the half-Heusler compounds as well [5, 6]. Many of these compounds are non-centrosymmetric superconductors and, therefore, candidates for Majorana quasiparticles [7]. Here, the challenge is the growth of quantum-well structures based on half-Heusler compounds.

The existing magnetic half-Heusler compounds contain either rare-earth elements or Mn, as e.g. GdNiSn or MnNiSb, YbPtBi, or PtMnBi (see Fig. 2.2) These rare earth atoms have a formal valence of RE^{3+} and contribute three electrons to the total amount of 18 valence electrons. Mn behaves similar to a rare earth element: Mn^{3+} in NiMnSb has a d^4-configuration, which leads to a strong localized magnetic moment of about 4 μ_B [3]. The magnetic moment of Ni is close to zero. Thus, magnetism in half-Heuslers is caused either by Mn or by a rare-earth element on the Y sites, and has

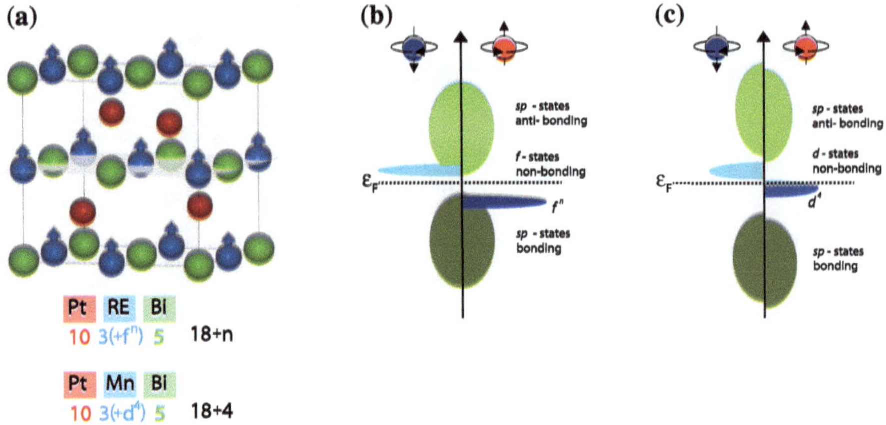

Fig. 2.2 **a** Crystal structure of the magnetic half-Heusler compounds. *Red spheres* indicate Pt, *green*—Bi, *blue*—Mn (or the rare-earth element). Schematic DOS for **b** GdNiSn, **c** PtMnBi. *Blue* color indicates *d*- and *f*-states

a localized type (see Fig. 2.2b, c). Their electronic structure and the electron counting rules are schematically shown in Fig. 2.2. In 1983, de Groot recognized that NiMnSb exhibits a peculiar band structure: seminconducting in the minority-spin, but metallic in the majority-spin channel [8]: This was named "half-metallic ferromagnetism". Kübler pointed out the similarity to the Co_2MnZ Heusler compounds [9].

As we have mentioned, Heusler compounds are ternary systems with the chemical formula X_2YZ which corresponds to the $L2_1$ structure, or Cu_2MnAl-type (defined in the Pearson Table). The X and Y atoms are transition metals, or lanthanides (rare-earth metals), and the Z position is occupied by a main group element. Often, Heusler compounds are named "full Heusler", in order to emphasize the filling of all voids of the fcc lattice (octahedral and tetrahedral), since only half of the tetrahedral sites are filled in half-Heusler compounds.

Semiconducting Heusler compounds exist with 8 (Li_2NaSb) or 18 valence electrons (Li_2CdGe) per formula unit. And just as for the case of half-Heusler compounds the number of valence electrons N_V is the sum of the *s*-, *p*- and *d*-electrons in the outer shell of the X, Y and Z elements. Even $N_V = 24$ semiconductors are possible in the case of three transition metals per formula unit [3]. For instance, Fe_2VAl is a non-magnetic semiconductor with non-magnetic iron [10] (see Fig. 2.3a. Co_2YZ Heusler compounds possess $N_V > 24$ (Co_2TiGa ($N_V = 25$)) and follow the so called Slater-Pauling rule [11–13]). If the number of valence electrons differs from 24, then these materials are magnetic, where the magnetic moment per formula unit is directly related to the number of valence electrons minus 24. An important distinction from the half Heusler compounds is the number of distinct magnetic sublattices, which is two or more for the full Heusler compounds. Kübler was the first to appreciate that the magnetic moment on the Y sites is of the localized type, whereas the local moment of cobalt is delocalized in nature. As an example, the magnetic structure and density

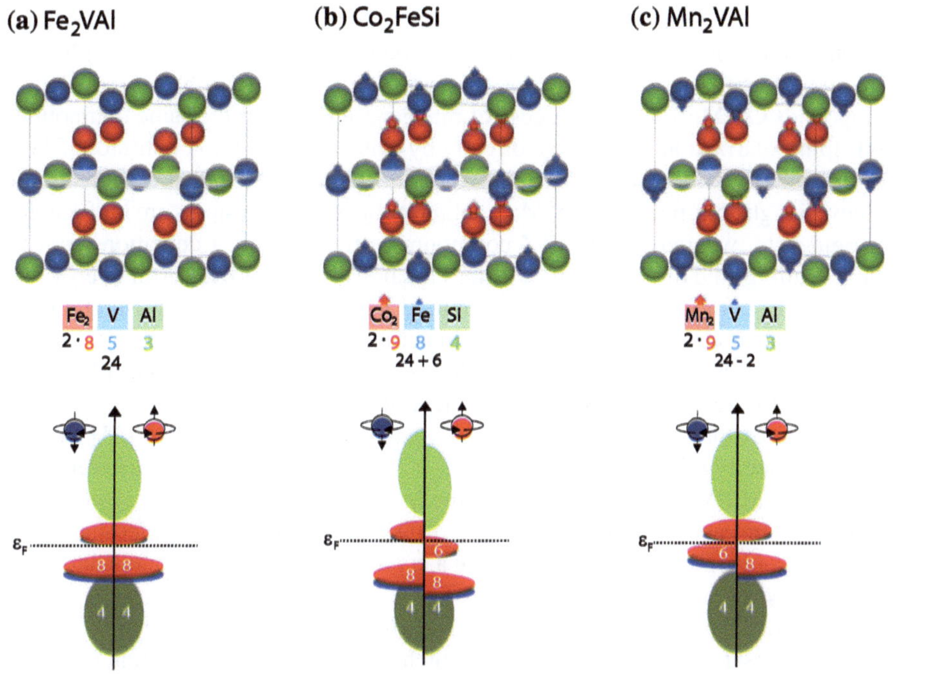

Fig. 2.3 Crystal structure of the magnetic full Heusler compounds **a** Fe_2VAl, **b** Co_2FeSi and **c** Mn_2VAl. *Red spheres* indicate the X_2, *blue*—Y, *green*—Z position of the X_2YZ stoichiometry. The schematic DOSs are shown below the corresponding crystal structures

of states of Co_2FeSi is illustrated in Fig. 2.3b: the exchange interaction between the cobalt and iron atoms leads to ferromagnetic coupling [11–13]. For these ferromagnetic, halfmetallic Heusler compounds the Curie temperature is proportional to the number of valence electrons per formula unit and increases by approximately 175 K per added electron [14, 15]. For example, Co_2TiSn (26 valence electrons with a saturation magnetization of 2 μ_B/formula unit) is ferromagnetic with a magnetic moment that mainly resides on the Co sites, that are identical in these compound. It exhibits a Curie temperature of 350 K [16]. Another example is Co_2FeSi. This systems is ferromagnetic as well, but the magnetic moments are found on both species, the Co and Fe atoms. It has the highest Curie temperature of any known Heusler compound, 1120 K [13]. Slater-Pauling behavior is a sign of bulk half-metallicity (100 % spin polarization) in that varying the number of valence electrons varies the filling of the metallic spin channel in a way that the moment is increased or decreased in direct proportion to the number of valence electrons [17–19]. This is true as long as the Fermi energy remains within the energy gap of the semiconducting spin channel. The compounds that follow the Slater-Pauling rule typically contain Co, Mn, or Fe atoms on the X and/or Y sites.

$Co_2(Cr,Fe)Al$ (CCFA) was the first compound for which a high magnetoresistance as a signature for high spin-polarization at room temperature was found [20, 21]. The highest tunnel magnetoresistance of more than 2000 % was observed in Heusler compounds Co_2MnSi with an excess of manganese [22]. For the same compound a high spin-polarization was proven directly by spin-polarized photoemission [23]. In the following it will become clear why Co and Mn are an excellent combination to achieve a high spin-polarization. Many of the cubic Co_2YZ systems are half metallic ferromagnets, whereas the Fe_2YZ compounds often exhibit a deviation from the Slater-Pauling rule due to disorder or less localized states at the Fermi energy. The Co_2-Heusler compounds are well understood and the interface with MgO and the local order is under control. To use the materials now in real devices and to inject spins into semiconductors, all-Heusler-devices are the future challenges [24].

One of the few examples of a half metallic Mn_2-Heusler compound with ferromagnetic coupling between the two manganese atoms is Mn_2VAl [25] that has $N_V < 24$ (see Fig. 2.3c). Most of the Mn_2YZ compounds are very different from Co_2YZ and Mn_2VAl, and crystallize in the inverse Heusler structure [3], where the manganese moments are located on two different sites and are coupled parallel to one another [26, 27] (each Mn resides on a crystallographically distinguishable site, as illustrated in Fig. 2.4).

Cubic Mn_3Ga, with 24 valence electrons is a "borderline compound" and is a fully compensated ferrimagnet [26–28] which is thus principally different from non-magnetic semiconductors like Fe_2VAl due to the special role of manganese in these Heusler compounds [9, 26]. The magnetic moments on the different sites show a different magnetization direction, which leads to a complete compensation of the magnetic moments, see Fig. 2.4a. Another example of a ferrimagnetic inverse Heusler compound is Mn_2CoAl, or written as (MnCo)MnAl, in which the local moment of manganese on the octahedrally coordinated Y site has a localized moment due to an approximate "d^4-configuration". The combined moment of Manganese and Cobalt (MnCo) on the tetrahedrally coordinated site is smaller with approximately 2 μ_B. The two manganese atoms on different sites are coupled antiparallel, and the Co atom on the tetrahedral site couples parallel to the Mn on the octahedral site which is similar to the situation in the Co_2MnZ compounds. However, the total moment still follows the Slater-Pauling rule with 26 valence electrons resulting in 2 μ_B for this ferrimagnetic compound. The explanation is that the octahedral site is always occupied by the most electropositive element (i.e. the earlier transition metal, so, for example, here Mn rather than Co), so that this arrangement leads to an occupancy of half of the Mn and the Co on the tetrahedral site. These special compounds in which Mn occupies both, the octahedral sites (Y) and half of the tetrahedral sites (X) in an ordered manner, have a lower space group symmetry (216 rather than 225) and are called inverse Heuslers (Fig. 2.4a). They have no inversion symmetry as compared to the Co_2-Heuslers compounds [3]. Mn_2CoAl is an exceptional compound, a spin gapless semiconductor with a Curie temperature of 720 K, that exhibits a comparatively low anomalous Hall effect which is explained by the symmetry properties of the Berry curvature [29].

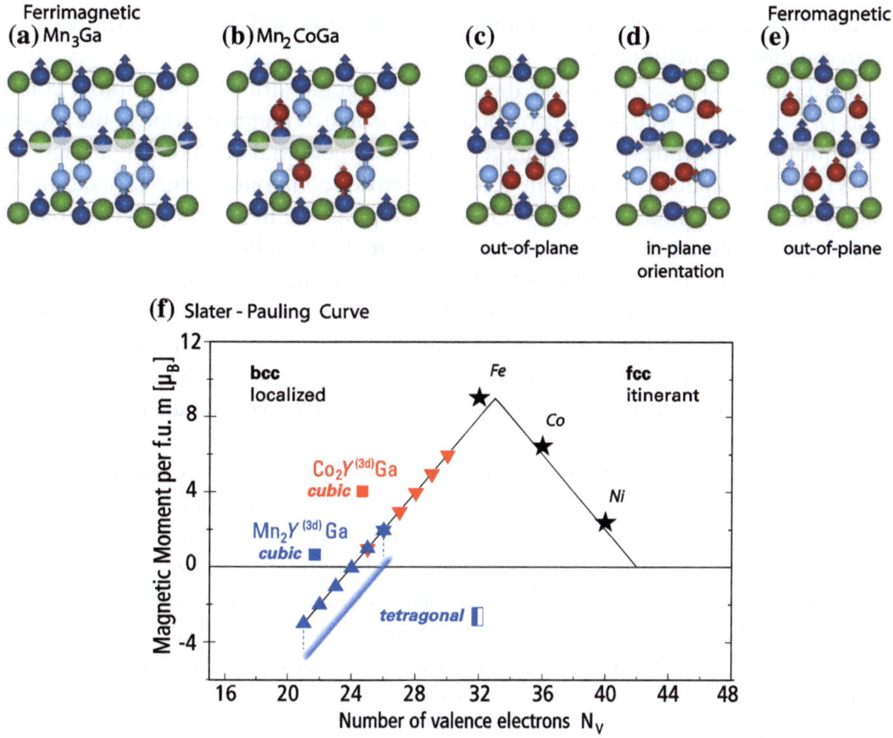

Fig. 2.4 Crystal structures of **a** cubic Mn_3Ga, **b** Mn_2CoGa. Different types of the magnetic ordering: **c** ferrimagnetic out-of-plane, **d** ferrimagnetic in-plane orientation; **e** ferromagnetic out-of-plane. *Red spheres* indicate a transition metal element, *green spheres*—main group element; *blue spheres*—Mn. **f** Slater-Pauling curve for cubic Co_2YZ (*red triangles*), cubic Mn_2YZ (*blue triangles*) and tetragonal Mn_2YZ (*blue line*)

Most of the cubic ferrimagnetic Mn_2YZ compounds still follow the Slater-Pauling rule [28, 30]. For the ferromagnetic, half-metallic Heusler compounds the Curie temperature is related to the number of valence electrons and scales approximately with the sum of the magnetic moments of the two antiparallel sublattices [14, 15]. All cubic Heusler compounds are naturally soft magnetic due to the vanishingly small magnetocrystalline anisotropy energy (MAE). At the same time, the magnetization values of the ferromagnetic Co_2- and Fe_2-Heusler compounds are high, whereas those of the ferrimagnetic Mn_2-Heusler compounds are comparatively low [30].

For the applications, such as spin transfer torque magnetic random access memory (STT-MRAM) [31, 32], ultrahigh density magnetic recording [33] and rare earth free permanent magnets, a high MAE is needed [34]. Anisotropic crystal structures such as tetragonally distorted Heusler compounds are a precondition for an uniaxial magnetocrystalline anisotropy. The martensitic to austenitic phase transition in shape memory alloys and magneto-caloric Heusler compounds are phase transitions from an anisotropic (tetragonal, orthorhombic or complex modulated structures) to an isotropic cubic structure [35].

Mn_3Ga and $Mn_{3-x}Ga$ possess the stable tetragonal structure at ambient conditions [27, 36, 37] as well as the perpendicularly oriented magnetization in the respective thin film electrodes, which suggests them as potential candidate materials for the STT RAM devices [38–40]. In general, Heusler compounds with low saturation magnetic moment, high MAE, low Gilbert damping, but high spin-polarization and high Curie temperature, are required to minimize the switching current and increase the switching speed in STT-MRAM devices according to the Slonczewski-Berger equation [31, 32]. Unfortunately, a suitable interface between the tunneling barrier and the electrode is still a challenge [41] and only low tunneling magneto-resistance effects have been observed so far [42, 43]. However, the material $M_{3-x}Ga$ exhibits a high MAE and THz oscillations, which makes it a potential candidate for spin torque oscillators [44].

Besides the Curie temperature, the quality of hard magnets is characterized by the maximum energy product defined through $B \times H_{\max}$. In order to maximize it, a strong coercive fields and a high magnetization is necessary. Large MAE in tetragonal Heusler compounds give the first important ingredient to realize such high coercive fields. The highest coercive fields are naturally provided by fully compensated tetragonally distorted ferrimagnets, as it is inversely proportional to the magnetization, which is vanishingly small, and directly proportional to the MAE, which is relatively large due to the tetragonal distortion [45].

From the Slater-Pauling curve (shown in Fig. 2.4), it follows that the rational design of compensated ferrimagnets in both cubic and tetragonal structures is possible, and the several examples habe been already synthesized [26, 46, 47]. There are at least three reasons why do the full Heusler compounds undergo a tetragonal distortion. First, is the presence of a Jahn-Teller ion in the Y site, such as, for example, Mn^{3+}. The Mn^{3+} has a d^4 configuration, which is a typical situation for the occurrence o a Jahn-Teller distortion [48, 49]. A Jahn-Teller electronic configuration is characterized by the degenerate orbitals which are partially filled such, that a structural distortion can lift the electronic degeneracy and lower the total energy of the system. The tetragonal distortion on the Y site affects the Z site by distorting the local tetrahedral environment. If Si occupies the Z site, tetragonal distortion of the unit cell is not observed because Si prefers a highly symmetric tetrahedral environment and a sp^3 hybridization. Typically, Heusler compounds with Al, Ga, Ge, Sn, Pb and Sb on the Z site allow for a tetragonal distortion in connection with Mn on the Y site. The octahedron can expand or shrink along one axis. Examples of Jahn-Teller driven tetragonally distorted Heusler systens are manganese-rich Heusler compounds [49]. So far, all ferromagnetic manganese-rich Heusler compounds and their relatives show a perpendicular magnetization in thin films, as predicted by theory (Fig. 2.4). The second reason for the occurrence of a tetragonal distortion in Heusler- or half Heusler systens is a van Hove singularity in the electronic band structure of the cubic magnetic compound: this is also sometimes called "band Jahn-Teller effect". The van Hove singularity is a saddle point in the electronic band structure at the Fermi energy which leads to a high density of states. Thus, a structural distortion can lower the energy of the system by opening a gap, or more often a pseudo-gap, in the band structure. By pseudo-gap describes the situation in that not all bands exhibit an energy gap after the distortion. Examples of known tetragonally distorted Heusler materials are the magnetic shape memory compounds: Ni_2MnGa and Mn_2NiGa [50, 51] and Rh_2FeSn, Rh_2CoSn and Rh_2FeSb [52]. However, films of the Rh_2-compounds show in plane magnetization [55], in accordance with theory [53]. At high temperature and in a high magnetic fields these materials are cubic (austenite phase) [54]. The Jahn-Teller effect and the van-Hove singularity, both can play a role in the same system since a degenerate state goes along with a high density of states as, for example, in Mn_3Ga. The third reason why there could be a tetragonal distortion is spin-orbit coupling. This is the case for 4*d* or 5*d* transition metals on the tetrahedral (X) sites because high Z atoms are needed for strong spin-orbit coupling. Known compounds are, for example, Mn_2PtSn and, Mn_2RhSn [55, 56] (see Fig. 2.4b). An example of a van-Hove singularity-driven, tetragonally distorted Heusler with spin-orbit coupling is Rh_2FeSn [52]. The known tetragonally distorted Heusler compounds crystallize, for example, in the following structural classes: 119 (I-4m2), 139 (I4/mmm) e.g. Ni_2MnGa and Rh_2CoSn, 129 (P4/mmm) and 131 (P42/mmc) [3].

The magnetic structure of the Mn_2-Heusler compounds is distinguished from the Co_2-Heusler compounds, although both families follow the Slater-Pauling curve, independent of the structure type, and the magnetization is continuous across the compensation point [30] (see Fig. 2.4e). The magnetism in Mn_2-Heuslers is different for compounds with $N_V \neq 24$. As mentioned above, Mn_2VAl with $N_V = 22$ is an itinerant ferromagnet, the manganese on the tetrahedral sites couple parallel to one another. Mn_3Ga with $N_V = 24$ is a compensated ferrimagnet. As discussed in detail in Wollmann et al. [30] the total magnetization is made up of contributions of different character (localized or itinerant) for the Heusler and the inverse Heusler structure types. Regarding the large constant localized moment, Mn is thus the only 3*d* transition metal element in inverse Heusler compounds that behaves like a rare-earth element. Wollmann et al. believe that the antiparallel coupling of the two types of Mn atoms in the inverse Heusler compounds is due to direct exchange being an atomic property of Mn as in the elementary metal (the half-filled *d* shell leads to antiparallel alignment). Due to the antiparallel coupling for most of the manganese-rich Heusler compounds, the total magnetic moment is rather low and independent of the structure. In the tetragonally distorted Mn_2-Heusler compounds the magnetic interaction does not change qualitatively, only the total magnetization is reduced, based on the over-compensation of the moments on the tetrahedral site versus the moments on the octahedral site. Therefore, the compounds do not fall on the Slater-Pauling curve. The curve of the total moments is shifted by approximately 1.5 μ_B, as can be seen in Fig. 2.4e.

The challenge for STT-MRAM and permanent magnets is the discovery of tetragonal Heusler compounds with high magnetic moments i.e. ferromagnetic coupling between the magnetic sub-lattices (see Fig. 2.4d). Based on our knowledge today, it should be possible to design new tetragonal compounds with larger magnetic moments, but it at first sight it might be difficult to design ferromagnetic Heusler compounds. In Mn_2-Heusler compounds ferromagnetism might be metastable: a hint is the large exchange bias effect that has recently been observed in the new tetragonal Heusler compound $Mn_{2-x}Pt_{1-x}Ga$, after zero-field cooling from the paramagnetic state. First-principles calculations and magnetic measurements revealed that Mn_2PtGa orders ferrimagnetically with some ferromagnetic inclusions [55]. The non-centrosymmetric Mn_2RhSn is a novel non-collinear tetragonal Heusler compound that exhibits an unusually strong canting of its magnetic sub-lattices [56]. In addition, Mn_2RhSn and related compounds are promising candidates for the realization ofthe Skyrmion state in the Heusler family [55, 56].

Of the large number of potential tetragonally distorted Heusler compounds, including quaternary alloys, only a few are known and the quest for new compounds and a deeper understanding of the relationship between their electronic structure, magnetism and crystal structure will likely have a high impact for all of mentioned technological applications discussed in this article.

Acknowledgments The authors gratefully acknowledge financial support from the Projects TP 2.3-A, TP 1.2-A of the research unit FOR 1464 "ASPIMATT") and the European Research Council (ERC) Grant (No. 291472) "Idea Heusler!".

References

1. F. Heusler, *Verh. DPG*, vol. 5, (Springer, Berlin, 1903), p. 219
2. F. Heusler, W. Starck, E. Haupt, *Verh. DPG*, vol. 5, (Springer, Berlin, 1903) p. 220
3. T. Graf, C. Felser, S.S.P. Parkin, Prog. Solid State Chem. **39**, 1 (2011)
4. M. König, S. Wiedmann, C. Brüne, A. Roth, H. Buhmann, L. Molenkamp, X.L. Qi, S.C. Zhang, Science **318**, 766 (2007)
5. S. Chadov, X.L. Qi, J. Kübler, G.H. Fecher, C. Felser, S.C. Zhang, Nat. Mater. **9**, 541 (2010)
6. H. Lin, L.A. Wray, Y. Xia, S. Xu, S. Jia, R.J. Cava, A. Bansil, M.Z. Hasan, Nat. Mater. **9**, 546 (2010)
7. L. Fu, C.L. Kane, Phys. Rev. Lett. **100**, 096407 (2008)
8. R.A. de Groot, F.M. Mueller, P.G. van Engen, K.H.J. Buschow, Phys. Rev. Lett. **50**, 2024 (1983)
9. J. Kübler, A.R. Williams, C.B. Sommers, Phys. Rev. B **28**, 1745 (1983)
10. Y. Nishino, M. Kato, S. Asano, K. Soda, M. Hayasaki, U. Mizutani, Phys. Rev. Lett. **79**, 1909 (1997)
11. H.C. Kandpal, G. Fecher, C. Felser, J. Phys. D: Appl. Phys. **40**(6), 1507 (2007)
12. S. Wurmehl, G.H. Fecher, H.C. Kandpal, V. Ksenofontov, C. Felser, H.J. Lin, Phys. Rev. B **72**, 184434 (2005)
13. C.G.F. Blum, C.A. Jenkins, J. Barth, C. Felser, S. Wurmehl, G. Friemel, C. Hess, G. Behr, B. Büchner, A. Reller, S. Riegg, S.G. Ebbinghaus, T. Ellis, P.J. Jacobs, J.T. Kohlhepp, H.J.M. Swagten, Appl. Phys. Lett. **95**(16), 161903 (2009)
14. J. Kübler, J. Phys.: Condens. Matter **18**(43), 9795 (2006)
15. J. Kübler, G.H. Fecher, C. Felser, Phys. Rev. B **76**, 024414 (2007)
16. H.C. Kandpal, V. Ksenofontov, M. Wojcik, R. Seshadri, C. Felser, J. Phys. D: Appl. Phys. **40**(6), 1587 (2007)
17. I. Galanakis, P.H. Dederichs, N. Papanikolaou, Phys. Rev. B **66**, 174429 (2002)
18. I. Galanakis, P. Mavropoulos, P.H. Dederichs, J. Phys. D: Appl. Phys. **39**(5), 765 (2006)
19. C. Felser, G.H. Fecher, B. Balke, Angew. Chem. Int. Ed. **46**, 668 (2007)
20. T. Block, C. Felser, G. Jakob, J. Ensling, B. Mühling, P. Gütlich, R.J. Cava, J. Solid State Chem. **176**(2), 646 (2003)

21. K. Inomata, S. Okamura, R. Goto, N. Tezuka, Jap. J. Appl. Phys. **42**(4B), L419 (2003)
22. H.X. Liu, Y. Honda, T. Taira, K.I. Matsuda, M. Arita, T. Uemura, M. Yamamoto, Appl. Phys. Lett. **101**(13), 132418 (2012)
23. M. Jourdan, J. Minár, J. Braun, A. Kronenberg, S. Chadov, B. Balke, A. Gloskovskii, M. Kolbe, H.J. Elmers, G. Schönhense, H. Ebert, C. Felser, M. Kläui, Nat. Comm. **5**, 3974 (2014)
24. S. Chadov, T. Graf, K. Chadova, X. Dai, F. Casper, G.H. Fecher, C. Felser, Phys. Rev. Lett. **107**, 047202 (2011)
25. C. Jiang, M. Venkatesan, J.M.D. Coey, Solid State Commun. **118**(10), 513 (2001)
26. S. Wurmehl, H.C. Kandpal, G.H. Fecher, C. Felser, J. Phys.: Condens. Matter **18**, 6171 (2006)
27. B. Balke, G.H. Fecher, J. Winterlik, C. Felser, Appl. Phys. Lett. **90**, 152504 (2007)
28. I. Galanakis, K. Özdogan, E. aioglu, B. Akta, Phys. Stat. Sol. (a) **205**(5), 1036 (2008)
29. S. Ouardi, G.H. Fecher, C. Felser, J. Kübler, Phys. Rev. Lett. **110**, 100401 (2013)
30. L. Wollmann, S. Chadov, J. Kübler, C. Felser, Phys. Rev. B **90**, 214420 (2014)
31. J. Slonczewski, Magn. Magn. Mater. **159**, L1 (1996)
32. L. Berger, Phys. Rev. B **54**, 9353 (1996)
33. D. Weller, A. Moser, IEEE Trans. Magn. **35**(6), 4423 (1999)
34. J.M.D. Coey, J. Phys.: Condens. Matter **26**(6), 064211 (2014)
35. P.J. Webster, K.R.A. Ziebeck, S.L. Town, M.S. Peak, Phil. Mag. B **49**(3), 295 (1984)
36. J. Winterlik, B. Balke, G.H. Fecher, C. Felser, M.C.M. Alves, F. Bernardi, J. Morais, Phys. Rev. B **77**, 054406 (2008)
37. E. Kren, G. Kadar, Solid State Commun. **8**(20), 1653 (1970)
38. F. Wu, S. Mizukami, D. Watanabe, H. Naganuma, M. Oogane, Y. Ando, T. Miyazaki, Appl. Phys. Lett. **94**(12), 122503 (2009)
39. M. Glas, D. Ebke, I.M. Imort, P. Thomas, G. Reiss, J. Magn. Magn. Mater. **333**, 134 (2013)
40. A. Köhler, I. Knez, D. Ebke, C. Felser, S.S.P. Parkin, Appl. Phys. Lett. **103**(16), 162406-1–162406-4 (2013)
41. C.E. Viol Barbosa, S. Ouardi, T. Kubota, S. Mizukami, G.H. Fecher, T. Miyazaki, X. Kozina, E. Ikenaga, C. Felser, J. Appl. Physics **116**(3), 034508 (2014)
42. T. Kubota, Y. Miura, D. Watanabe, S. Mizukami, F. Wu, H. Naganuma, X. Zhang, M. Oogane, M. Shirai, Y. Ando, T. Miyazaki, Appl. Phys. Express **4**(4), 043002 (2011)
43. T. Kubota, Q. Ma, S. Mizukami, X. Zhang, H. Naganuma, M. Oogane, Y. Ando, T. Miyazaki, Appl. Phys. Lett. **5**(4), 043003 (2012)
44. S. Mizukami, F. Wu, A. Sakuma, J. Walowski, D. Watanabe, T. Kubota, X. Zhang, H. Naganuma, M. Oogane, Y. Ando, T. Miyazaki, Phys. Rev. Lett. **106**, 117201 (2011)
45. J.M.D. Coey, *Magnetism and Magnetic Materials* (Cambridge University Press, 2010)
46. H. Kurt, K. Rode, P. Stamenov, M. Venkatesan, Y.C. Lau, E. Fonda, J.M.D. Coey, Phys. Rev. Lett. **112**, 027201 (2014)
47. A.K. Nayak, M. Nicklas, S. Chadov, P. Khuntia, C. Shekhar, M. Baenitz, Y. Skourski, V.K. Guduru, A. Puri, U. Zeitler, C. Felser, Design of compensated ferrimagnetic Heusler alloys for giant tunable exchange bias. Nat. Mater. **14**, 679 (2015). doi:10.1038/NMAT4248
48. P. Entel, V.D. Buchelnikov, V.V. Khovailo, A.T. Zayak, W.A. Adeagbo, M.E. Gruner, H.C. Herper, E.F. Wassermann, J. Phys. D: Appl. Phys. **39**(5), 865 (2006)
49. P.J. Brown, A.Y. Bargawi, J. Crangle, K.U. Neumann, K.R.A. Ziebeck, J. Phys.: Condens. Matter **11**(24), 4715 (1999)
50. G.D. Liu, J.L. Chen, Z.H. Liu, X.F. Dai, G.H. Wu, B. Zhang, X.X. Zhang, Appl. Phys. Lett. **87**(26), 262504 (2005)
51. G.D. Liu, X.F. Dai, S.Y. Yu, Z.Y. Zhu, J.L. Chen, G.H. Wu, H. Zhu, J.Q. Xiao, Phys. Rev. B **74**, 054435 (2006)
52. J. Suits, Solid State Commun. **18**(3), 423 (1976)
53. J. Winterlik, S. Chadov, A. Gupta, V. Alijani, T. Gasi, K. Filsinger, B. Balke, G.H. Fecher, C.A. Jenkins, F. Casper, J. Kübler, D. Liu, L. Gao, S.S.P. Parkin, C. Felser, Adv. Mater. **24**, 6283 (2012)
54. A.K. Nayak, C.S. Mejia, S.W. D'Souza, S. Chadov, Y. Skourski, C. Felser, M. Nicklas, Phys. Rev. B **90**, 220408(R) (2014)

55. A.K. Nayak, M. Nicklas, S. Chadov, C. Shekhar, Y. Skourski, J. Winterlik, C. Felser, Phys. Rev. Lett. **110**, 127204 (2013)
56. O. Meshcheriakova, S. Chadov, A.K. Nayak, U.K. Rössler, J. Kübler, G. André, A.A. Tsirlin, J. Kiss, S. Hausdorf, A. Kalache, W. Schnelle, M. Nicklas, C. Felser, Phys. Rev. Lett. **113**, 087203 (2014)

Part II
Properties

Chapter 3
Spin-Resolved Photoemission Spectroscopy of the Heusler Compound Co_2MnSi

Roman Fetzer, Martin Aeschlimann and Mirko Cinchetti

Abstract The Heusler compound Co_2MnSi (CMS) is probably the most extensively studied half-metallic material. Its peculiar spin-dependent electronic properties have been addressed both theoretically and experimentally; they also already found wide application in prototypical spintronics devices. In this chapter we review our spin-resolved photoemission studies conducted on the free surface of both stoichiometric and off-stoichiometric $Co_2Mn_\alpha Si$ samples. Our findings shed light on the peculiar dependence observed in the performance of CMS-based magnetic tunnel junctions on the Mn composition α. We also review our studies on the buried CMS/MgO interface, that could be directly investigated by spin-resolved low-energy photoemission spectroscopy. With this method we can then access the spin and symmetry properties of the interface electronic wave functions, which are ultimately responsible for the performance of CMS-based spintronic devices. In particular, the collected experimental data led us conclude that non-collinear spin moments at the CMS/MgO interface might be the main mechanism behind the controversially discussed temperature-dependent performance loss of state-of-the-art CMS-based magnetic tunnel junctions.

3.1 A Short Introduction to Spin-Resolved Photoemission Spectroscopy

Photoemission spectroscopy (PES) is one of the most powerful methods to study the electronic structure of solid state systems. In a PES experiment, a monochromatic light beam irradiates a solid leading to photoemission of former bound electrons. The

R. Fetzer (✉) · M. Aeschlimann · M. Cinchetti
Department of Physics and Research Center OPTIMAS, University of Kaiserslautern, Erwin-Schrödingerstr. 46, 67663 Kaiserslautern, Germany
e-mail: rfetzer@rhrk.uni-kl.de

C. Felser and A. Hirohata (eds.), *Heusler Alloys*,
Springer Series in Materials Science 222, DOI 10.1007/978-3-319-21449-8_3

photoemitted electrons can be subsequently analyzed with respect to their kinetic energy and/or momentum distribution. Comparison with theoretical photoemission or band structure calculations provides direct access to the electronic band structure of the surface of crystalline materials [1]. The surface sensitivity of conventional PES is given by the inelastic mean free path of electrons inside metals, which is lower than 1 nm for typical photon energies used in PES experiments (20–500 eV [2]); the detected photoelectrons stem thus mainly from the outermost sample surface.

In case of magnetic materials, direct investigation of a further electronic property, namely the spin, is desirable. In ferromagnetically ordered materials, for example, the exchange interaction favors the parallel alignment of the electron spins, leading to a spin-split density of states and thus also to a net magnetic moment even in absence of an external magnetic field. If we denote the total density of states of (occupied) majority and minority spin electrons respectively with $N^{\uparrow}$ and $N^{\downarrow}$, the magnetic moment will be proportional to $N^{\uparrow} - N^{\downarrow}$. Equivalently, to quantify the unbalance between majority and minority electrons, we can define the (total) spin polarization $SP_t = \frac{N^{\uparrow}-N^{\downarrow}}{N^{\uparrow}+N^{\downarrow}}$ [3, 4].

If magnetic materials are used in spintronics devices, like in the case of Heusler compounds discussed in this section, than their spin-dependent transport properties become of extreme importance. In this case, a more relevant physical observable is the imbalance between majority and minority electrons at the transport level. Depending on the specific device structure, the transport level will be at some particular energy close to the Fermi energy (E_F). This means that, in general, the SP should be measured in the energetic region close to E_F as a function of binding energy $SP(E) = \frac{N^{\uparrow}(E)-N^{\downarrow}(E)}{N^{\uparrow}(E)+N^{\downarrow}(E)}$, where now $N^{\uparrow,\downarrow}(E)$ are the number of majority and minority electrons at the specific binding energy (E). To determine the energy-resolved spin polarization, SP(E), spin-resolved photoemission spectroscopy (SR-PES) serves as the method of choice. Experimental access to the spin polarization of an ensemble of (photo)electrons is commonly realized by firstly accelerating the photoelectrons onto thin foils or single crystals made of high Z elements.

Afterwards the intensity of elastically scattered electrons is detected at two suitably chosen, different points in space. Due to the high spin-orbit coupling of the scattering targets the measured intensity will depend on its spin direction, i.e. a spin-dependent intensity asymmetry will occur in the detection channels [3, 4]. This asymmetry is proportional to the actual spin polarization, with the proportionality constant known as the Sherman factor which usually has to be determined separately. To overcome additional apparatus asymmetries which would falsify the determined spin polarization, measurements are conducted twice with respectively opposite remanent magnetization of the sample [3]. The combination of such a spin detector with an energy-resolving unit eventually allows to determine the desired majority and minority electron spectra $N^{\uparrow,\downarrow}(E)$ and the corresponding spin polarization.

3.2 SR-PES at Co-based Full Heusler Surfaces

In the last decade the material class of ferromagnetic Co-based full Heusler compounds (Co_2XY with Y: transition metal, Z: main group element) received large attention mainly because of the existence of a band gap at the Fermi energy for only one spin channel (this property being called half-metallicity [5]), that was predicted for several possible compositions [6]. The most prominent examples are Co_2MnSi, Co_2FeSi and $Co_2Cr_{0.6}Fe_{0.4}Al$ (the latter one having Cr and Fe atoms in the given distribution sitting on the Y lattice position) [7, 8]. The presence of the band gap in one spin channel results in a spin polarization $|SP(E_F)| = 100\,\%$ at the Fermi energy. Utilization of half-metallic materials is of great interest for the emerging research field of spintronics, as they are ideally suited as sources for spin-polarized currents, leading to a drastic performance increase for magnetoresistance and spin injection devices compared to the usage of conventional $3d$ ferromagnetic materials [9, 10].

In order to confirm these peculiar electronic properties at the surface of Co-based Heusler compounds, SR-PES experiments were performed very soon. Wang et al. grew Co_2MnSi (CMS) films on GaAs(001) substrates, which revealed not more than $+10\,\%$ SP at E_F using a photon energy of 70 eV [11]. The huge discrepancy between the experimental finding and the theoretically predicted very high SP was attributed to chemical disorder (i.e. the ideal $L2_1$ crystal structure [7] was not realized), as the finding of optimum growth conditions was hindered by the occurrence of chemical reactions at the GaAs interface when increasing the substrate temperatures. Later spin-resolved PES experiments focused on the compound $Co_2Cr_{0.6}Fe_{0.4}Al$, as its magnetic and electronic properties were predicted to be more stable against disorder [12]. While in fact surface spin polarization values of up to 45 % could be obtained [13], a more detailed study where SR-PES measurements were combined with LEED and Auger electron spectrosccopy revealed the stunning influence of substrate choice and post-growth annealing procedure on the spin properties of $Co_2Cr_{0.6}Fe_{0.4}Al$ [14].

Further notable SR-PES experiments were conducted in the group of M. Jourdan at Co_2MnGa [15, 16] using a Helium gas discharge lamp ($h\nu = 21.2\,\text{eV}$). Although this compound has not been predicted to exhibit complete spin polarization at E_F, it is known to show only slight sensitivity to oxidation in contrast to most other Co-based Heusler phases. In combination with *in-situ* sample fabrication, measurements at very clean and well-ordered surfaces could be granted resulting in detected spin polarization values of up to 55 %. Discussion of the obtained spectra was based on comparison with state-of-the-art ground state band structure and photoemission calculations, revealing an overall good agreement.

3.2.1 *Spin- and Symmetry-Resolved PES at the Off-Stoichiometric $Co_2Mn_\alpha Si$ Surface*

As already pointed out, PES is a surface-sensitive method, meaning that it provides information about the surface electronic structure. PES spectra will thus contain different contributions: from bulk-like bands that are also present at the surface region [17]; from surface specific contributions like surface states or surface resonances [18]; from crystallographic defects, that can be enhanced with respect to the bulk or even only appear at the surface [19]; and finally from effects related to the photoemission process itself, like final states effects [1]. In order to evaluate PES data, comparison with theoretical calculations is thus more than helpful and even necessary. In addition, complementary experiments revealing the crystalline structure and chemical composition of the surface help to interpret the photoemission results. In the following we review a comprehensive SR-PES study conducted on off-stoichiometric $Co_2Mn_\alpha Si$ compounds, that exemplarily shows how the features found in the SR-PES spectra could be assigned either to defect-induced contributions, bulk- or surface-like bands. The experiments were supported by state-of-the-art theoretical calculations done by the Ebert group (LMU Munich) [19]. Motivation for these experiments was given by the experiments of Ishikawa et al., who showed that for CMS-based magnetic tunnel junctions (MTJs) the performance does not inevitably decrease in case of non-stoichiometric composition [20]. While the tunnel magnetroresistance ratio (TMR) as the figure of merit drops for Mn-deficient compounds (i.e. $\alpha < 1$ for $Co_2Mn_\alpha Si$, with $\alpha = 1$ being the stoichiometric composition), it improves distinctively for $\alpha > 1$ up to a maximum around $\alpha = 1.3$, cf. left part of Fig. 3.1.

These findings have been explained by respective increase or supression of Co_{Mn} antisite defects (Co atoms occupying Mn lattice positions) by varying the chemical composition. This kind of defect has been found to be most detrimental to the half-

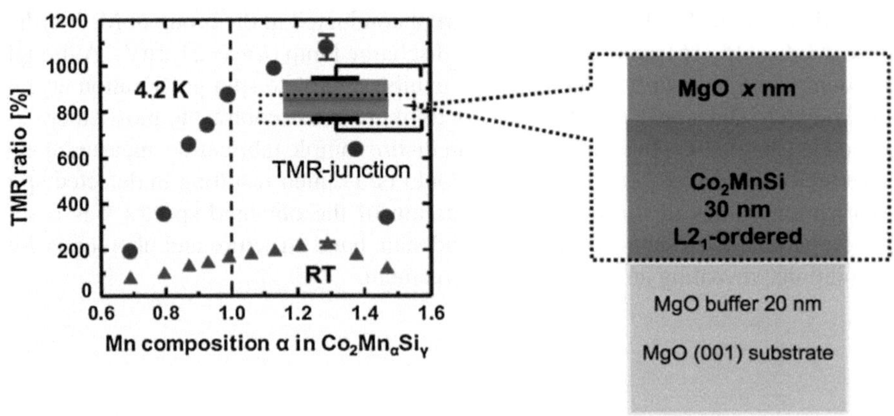

Fig. 3.1 TMR ratio in dependence of the Mn amount α in $Co_2Mn_\alpha Si$ (*left*, taken from [20]) and structure of the studied samples (*right*)

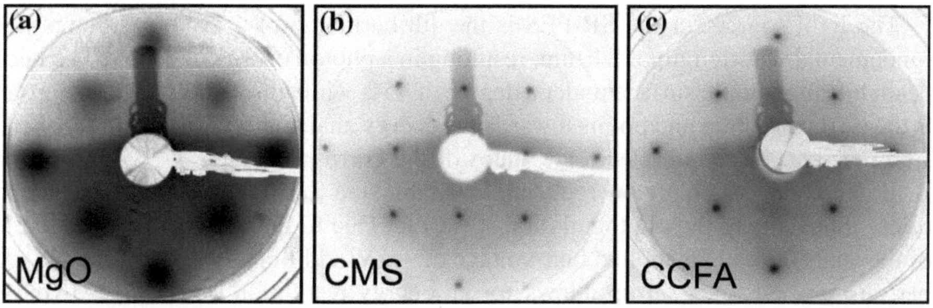

Fig. 3.2 LEED patterns of **a** the MgO top layer ($E = 82$ eV), **b** sample CMS069 ($E = 57$ eV) and **c** $Co_2Cr_{0.6}Fe_{0.4}Al$ ($E = 74$ eV). These data are already published in [19]

metallic property of CMS, as it leads to a finite density of states inside the minority band gap directly at E_F [21]. However, the TMR-experiments lack of a direct proof that changing the Mn amount in fact leads to the discussed variation of Co_{Mn} antisites.

In order to get a deeper insight, we performed SR-PES experiments at $Co_2Mn_\alpha Si$ thin films with $\alpha = 0.69$ (Mn-deficient case, henceforth called CMS069) and $\alpha = 1.19$ (Mn-enriched case, CMS119). The samples were fabricated by the Yamamoto group from Hokkaido University, Sapporo. They can be assumed as "half" TMR junctions (see right part of Fig. 3.1), as they are fabricated in the same manner as the MTJs described in [20], but with only one CMS layer (30 nm thick). The MgO tunneling barrier (2 nm) here serves as a protective capping layer for the underlying CMS film. Both epitaxial CMS and MgO films were grown as (001) layers. After introducing the samples into the experimental chamber, the MgO capping layer has been removed by gentle Ar^+ sputtering with $E_{kin} = 500$ eV. Subsequently the samples were annealed up to 500 °C to restore crystallinity of the outermost Co_2MnSi layers. Complete removal of MgO was controlled by Auger electron spectroscopy (AES), i.e. signals of elemental Mg and O vanished, while low energy electron diffraction (LEED) experiments revealed a sharp spot pattern which is in accordance with an assumed Mn-Si termination,[1] the thermodynamically most probable termination [22]. However, the pattern is also compatible to a vacancy-Si termination. Figure 3.2 shows such a spot pattern compared with the ones obtained prior to sputtering (i.e. with crystalline MgO on top) and for another Co-based Heusler compound, namely $Co_2Cr_{0.6}Fe_{0.4}Al$ (CCFA). The difference between the CMS and the CCFA patterns can be explained by the predominant B2 ordering of CCFA as compared to the predominant $L2_1$ ordering in CMS [19].

[1] Termination means the chemical and crystalline composition of the outermost atomic layer at the surface. In case of CMS, two different positions can be occupied by either Co, Mn or Si atoms. The bulk stack structure of CMS (100) films consists of alternating Co-Co and Mn-Si layers, hence these terminations might probably occur at the very surface. Nevertheless, further terminations such as vacancy-Si can occur due to induced defects and/or the respective surface properties which in general differ from the bulk.

The light source for the SR-PES is the 4th harmonic of a Ti:Sa oscillator with fundamental wavelength of 840 nm, resulting in a photon energy of 5.9 eV. The laser beam hits the sample surface under a degree of 45 °, while the sample surface normal is the (001) direction and points towards the energy analyzator with on-top SPLEED spin-detector. Hence all electronic states of the corresponding $\Gamma - X^2$ part of the Brillouin zone, which are energetically available and fulfill the symmetry selection rules, contribute to the photoemission spectrum. Due to the finite acceptance angle of the analyzer and an applied bias voltage of −4 V, additionally 40 % of the $\Gamma - K$ part are probed. This set-up allows for an energy resolution of $\Delta E = 210\,\mathrm{meV}$. Using a phase retarding plate the light polarization could be switched from *s-polarization* (electrical field vector perpendicular to the plane of incidence) to *p-polarization* (electrical field vector within the plane of incidence). Variation of the light polarization in the photoemission experiments allows to additionally determine the symmetry of the investigated electronic wave functions, as will be shown in the following.

3.2.1.1 SR-PES Results

Figures 3.3 and 3.4 show the obtained spin-resolved photoemission spectra (i.e. majority and minority spin, or $N^{\uparrow}$ and $N^{\downarrow}$) and the corresponding spin polarization (SP) for CMS069 and CMS119. The measurements were performed with p- and s-polarized light (left and right panels of Figs. 3.3 and 3.4, respectively). All found SP features are marked with capital letters, i.e. (A)...(E). Peaks occurring in the majority (minority) spectra are labeled M1...M3 (m1...m3). While the spectra and SP in general depend on the chemical composition, the most obvious difference is found directly at the Fermi energy (feature A), the most interesting energy region for transport. The Mn-rich compound exhibits positive SP values around 20 % at E_F with only slight dependence on the light polarization used. However, the obtained values are far away from the theoretically expected +100 %. The Mn-deficient sample even shows negative SP values of −10 to −20 % near the Fermi energy, depending on the light polarization. The energetic positions of all SP features and found peaks in the photoemission spectra are listed in Table 3.1. For the peaks M2 and M3 of the majority electron spectrum obtained at CMS119, the uncertainty regarding the energetic position is ±0.1 eV, since they are not well separated. In all other cases, the uncertainty is given by ±0.05 eV. The most probable origins of the spectral features will be discussed in the following sections.

3.2.1.2 Electronic Wave Function Symmetries

Usage of the two different light polarizations allows us to determine the wave function symmetry of the probed initial states. Regarding our experimental set-up, the crystal

[2]This Brillouin zone part is of great interest, since it exhibits the electronic states mainly involved into the tunneling process in case of a state-of-the-art MgO single-crystalline tunnel barrier.

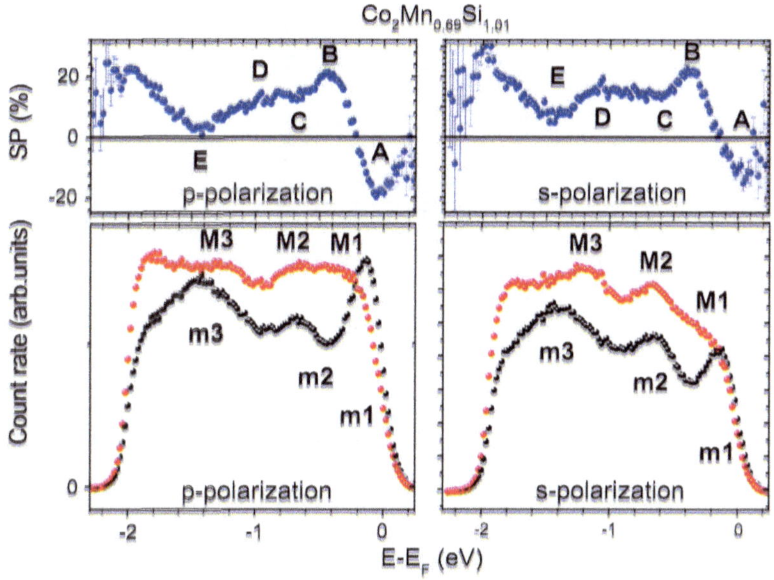

Fig. 3.3 Spin resolved photoemission spectra of the free CMS069 (100) surface for p-polarized excitation (*left*) and s-polarized excitation (*right*). *Upper panel* spin polarization (SP), *lower panel* majority and minority count rates. SP features are marked by A…E, Majority (minority) features are labeled M1…M3 (m1…m3). These data are already published in [19]

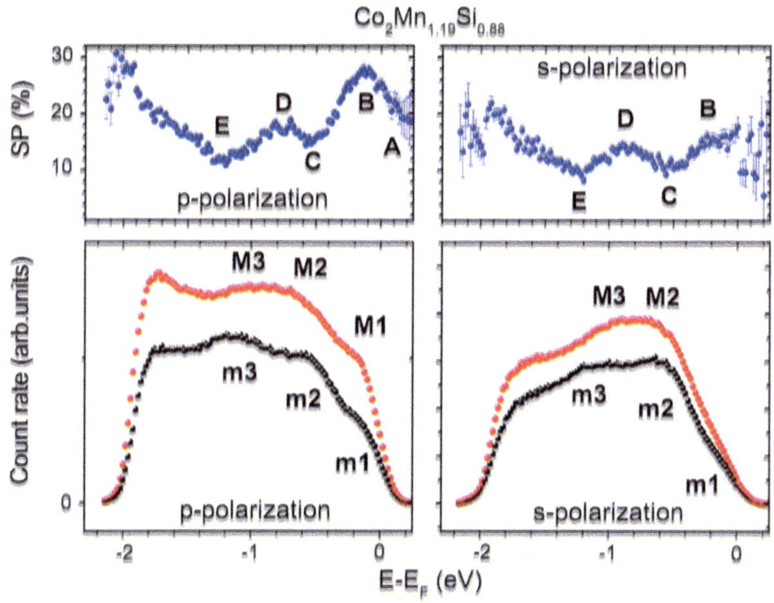

Fig. 3.4 Spin resolved photoemission spectra of the free CMS119 (100) surface for p-polarized excitation (*left*) and s-polarized excitation (*right*). *Upper panel* spin polarization (SP), *lower panel* majority and minority count rates. SP features are marked by A…E, Majority (minority) features are labeled M1…M3 (m1…m3). These data are already published in [19]

Table 3.1 Energetic positions of the described spin polarization features (A...E) and peaks within the minority (m) and majority(M) spectra, given in eV with respect to E_F

	A	B	C	D	E	m1	m2	m3	M1	M2	M3
CMS069-p	−0.20	−0.45	−0.70	−0.95	−1.45	−0.15	−0.70	−1.45	−0.20	−0.65	−1.30
CMS069-s	0.00	−0.40	−0.65	−1.10	−1.45	−0.15	−0.65	−1.40	−0.20	−0.70	−1.20
CMS119-p	0.00	−0.15	−0.55	−0.75	−1.25	−0.20	−0.60	−1.15	−0.20	−0.70	−0.90
CMS119-s	–	0.00	−0.50	−0.90	−1.25	−0.20	−0.65	−1.15	–	−0.60	−0.90

The respective peak positions correspond to their maximum relative intensity. The related statistical error is ±0.1 eV for the CMS119 peaks M2 and M3 and ±0.05 eV for all other peaks and features. These data are already published in [19] structure of CMS and the investigated part of the Brillouin zone, electronic states with the following symmetries are accessible by PES [19, 23, 24]:

- in case of s-polarized light, only Δ_5 states can be excited
- for p-polarized light, Δ_5 *and* Δ_1 states are probed. Additionally surface-related states can also only be excited by p-polarized light, since here an E-field vector component perpendicular to the surface is needed to excite these states, very similar to Δ_1 states.

While in general states with other symmetry can exist for this part of the Brillouin zone (i.e. Δ_2 and $\Delta_{2'}$ symmetry [25]), in case of CMS only electronic states with Δ_5 and Δ_1 symmetry are found near E_F [7, 26]. States with such symmetry are highly important for MTJ applications, since the coherent tunneling process in state-of-the-art MgO-based tunnel junctions is related to symmetry filtering especially of Δ_1 and Δ_5 states [25, 27].

If we now evaluate the p-s asymmetry $A^{\uparrow,\downarrow} = \frac{I_p^{\uparrow,\downarrow} - I_s^{\uparrow,\downarrow}}{I_p^{\uparrow,\downarrow} + I_s^{\uparrow,\downarrow}}$, we can determine the predominant wavefunction symmetry of the probed electronic states. $I_{p(s)}^{\uparrow,\downarrow}$ denotes the measured photoelectron intensity for majority ($\uparrow$) and minority ($\downarrow$) electrons, respectively, when p(s)-polarized laser light is used. For convenience all spectra are normalized to the respective m2 minority peak. Figure 3.5 shows the obtained majority and minority p-s asymmetry for CMS069 and CMS119. While the asymmetry background is formed by Δ_5 states (since both p- and s-polarized light probe states with this symmetry), an increase of the asymmetry means that additional Δ_1 states or surface-related states are excited (which can only be done with p-polarized light). In all cases the asymmetry is increased near the Fermi energy, which might be related to the respective spectral features M1 and m1. The Mn-deficient compound shows even a clear substructure near E_F, reaching an asymmetry minimum at the peaks M2 and m2. In case of the Mn-rich sample, the asymmetry is even more enhanced at the Fermi level, which indicates an increased amount of electronic states with Δ_1 symmetry near E_F.

3.2.1.3 Direct Bulk Transitions

Although the detected photoelectrons stem exclusively from the very surface region, bulk-related electronic features will be also detectable, as long as they are still present at the surface. In case of energy- and momentum-conserving interband transitions (i.e. $\Delta E = \hbar\omega (= 5.9\,\text{eV})$; $\Delta \mathbf{k} = 0$[3]) bulk features can even dominate a photoemission spectrum [17]. The possible existence of such direct bulk transitions in our spectra presented in Sect. 3.2.1.1 can be discussed with the help of band structure calculatons based on the LSDA+DMFT scheme. Figure 3.6 shows the occupied majority and minority electron band structure (black lines) along the $\Gamma - X$ crystal direction,

[3]The photon momentum is negligible compared to the electron momentum.

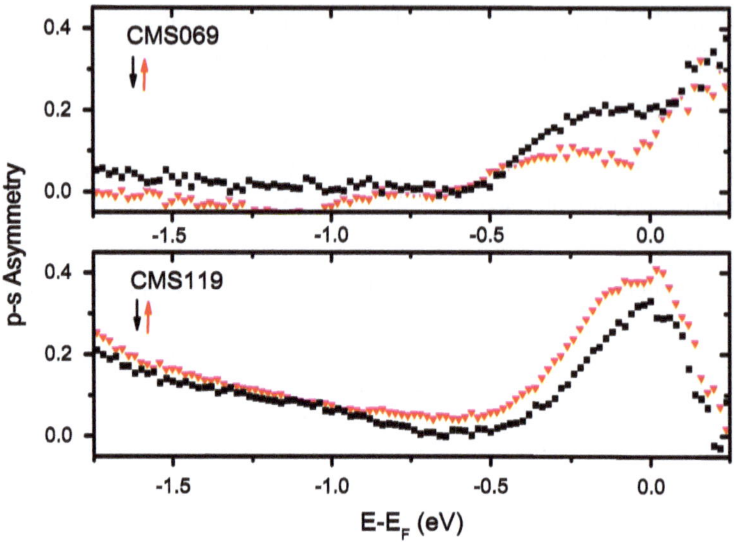

Fig. 3.5 Determined experimental asymmetry spectra for p- and s-polarized excitation in CMS069 (*upper panel*) and CMS119 (*lower panel*). Majority (minority) spin spectra are marked with red triangles (*black squares*). The spin resolved spectra of each sample and polarization were normalized on the respective minority m2 peak. These data are already published in [19]

assuming perfect $L2_1$ ordering and stoichiometric composition.[4] Additionally, unoccupied bands of this Brillouin zone part are shifted downwards by 5.9 eV (i.e. the used photon energy) and colored orange. This means that any crossing of a black and orange band in Fig. 3.6 can be the origin of a direct bulk transition, if the transition is actually allowed regarding matrix element effects [1].

Regarding the majority channel, in the very vicinity of E_F two occupied bands with Δ_1 and Δ_5 symmetry [7, 26] can be found. Since no crossing point exists there allowing for dominant bulk transitions, the majority peaks M3 and M2 in both the Mn-rich and Mn-deficient spectra might at least partly consist of surface resonances related to these bands. Going deeper in binding energy, at $E - E_F = -0.9$ eV a first direct bulk transition is allowed. As the initial states are given by a band with small dispersion, this transition will be quite robust and most probably is the origin for the broad M3 peak.

In the minority channel, a first possible bulk transition is found at -0.5 eV binding energy, which we assign to be the main contribution to the minority m2 peak. For energies nearer to E_F, either no fitting final state or no occupied initial state due to the minority band gap exists. Further interband transitions are located between -1.0 and -1.4 eV with highly localized initial states. These transitions most probably form the peak m3. We can conclude that in all cases the photoelectrons with initial states

[4] Although both investigated sample systems have non-stoichiometic composition, the induced randomly distributed crystalline defects do not affect significantly the bulk band structure.

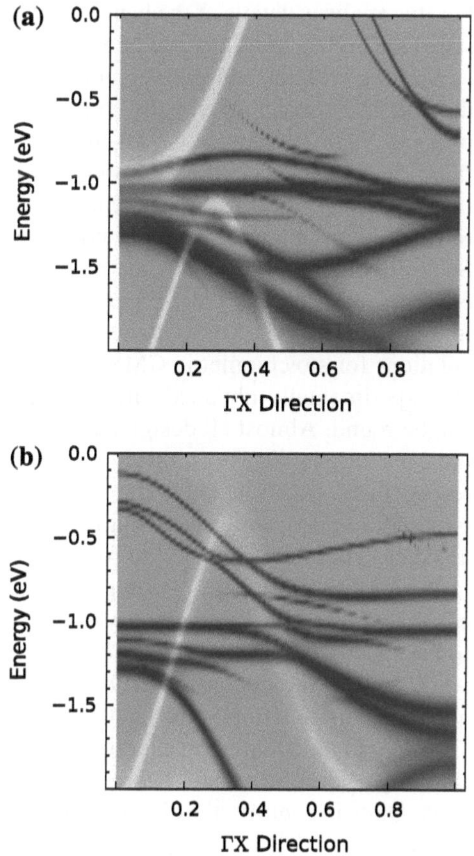

Fig. 3.6 Possible bulk transitions for majority and minority electrons along the ΓX momentum direction of CMS as calculated using LSDA+DMFT. Occupied bulk states (unoccupied final states) are shown as *black* (*orange*) lines. Crossing points correspond to energy and momentum allowed transitions at the photon energy of $\hbar\omega = 5.9\,\text{eV}$. These data are already published in [19]. **a** Majority. **b** Minority (Color figure online)

within the most interesting energy regime directly at E_F do not stem from direct bulk transitions. In general, such bulk transitions would dominate the photoemission spectra. Due to their absence in vicinity of E_F, the recorded photoemission spectra in this energy range will contain mainly information about surface electronic features, which is extremely valuable for the understanding of the performance of CMS-based spintronics devices.

3.2.1.4 Disorder Effects

In this section we discuss the influence of deliberate variation of the chemical composition in the surface electronic structure of $Co_2Mn_\alpha Si$. First of all, we observe that the photoemission spectra for the Mn-poor sample CMS069 in Fig. 3.3 and for the Mn-rich sample CMS119 in Fig. 3.4 show different features. In order to determine their origin, the spin-resolved ground state density of states (DOS) was calculated in [19] for both off-stoichiometric compounds, based on the LSDA+CPA scheme.

For this analysis details of the band structure are not necessary, as the induced defect states are highly localized and hence do not exhibit a noteworthy dispersion.

Starting with the Mn-poor compound ($\alpha = 0.69$), two extreme cases of induced disorder have been theoretically considered, while the crystal lattice itself keeps the perfect $L2_1$ order: Firstly, we assume that the lattice sites designated to Mn but not filled due to the Mn deficiency instead are occupied by Co and Si atoms in a 2:1 ratio (antisite model, cf. [28]). This leads to the chemical formula $Co_2(Co_{0.168}Mn_{0.748}Si_{0.084})Si$,[5] giving the respective composition of the X, Y and Z lattice positions. Such a behavior has been already proposed by the Yamamoto group to explain their experimental findings [28]. Figure 3.7a shows the the spin-resolved DOS and the inferred SP in case of the antisite model (dashed lines represent the respective calculations for stoichiometric CMS taken from [29]). Comparing the DOS data with the experimentally obtained ones, especially for the SP a quite good accordance can be found. Almost all designated features (A...E, cf. Fig. 3.3) are reproduced by the calculations. However, the energetic position of features (D) and (E) vary with respect to the experimental findings listed in Tab. 3.1. Furthermore the minority band gap is closed in accordance to our experimental results, most probably due to Co_{Mn} antisites. These are predicted to form in-gap states [21, 28] and strongly contribute to the CMS069 m1 peak, leading to even negative SP values.[6] Astonishingly, the calculated absolute value directly at E_F is underestimated especially with respect to the spectra obtained with p-polarized light ($-10\,\%$ instead of measured $-20\,\%$), while usually the opposite behavior is the case. We will explain the origin of this apparent contradiction in the next section.

Going to higher binding energies, we find excellent agreement between the broad minority peak centered at -1.44 eV binding energy and the experimentally found m3 peak. Quite in contrast, the distinct majority DOS peak at -0.92 eV is not reproduced by the measured data. This might be explained by final state effects (i.e. the excitation probability of this state is quite low due to very small transition matrix element values).

The second investigated kind of defect distribution assumes that the empty Mn sites are left unoccupied, i.e. we describe the composition as $Co_2(Mn_{0.69}Vacancy_{0.31})$ Si. Please note that formation of vacancies is less probable than the formation of Co_{Mn} and Si_{Mn} antisites, since the needed defect formation energy is much higher [28]. The respective DOS and SP can be found in Fig. 3.7b. Obviously, the introduction of vacancies at the Mn lattice positions leads to significant DOS changes in both spin channels compared to the perfectly ordered stoichiometric case. In general, agreement with our experimental data can be hardly found. Since the antisite model reproduces our experimental data far better than the vacancy model and additionally

[5]The total Mn amount value seems to be increased regarding this formula. Please note that if you sum up all relative Co and Si amounts independently of the crystal lattice position and normalize in order to achieve the $Co_2Mn_YSi_Z$ formula the originate Mn amount of 0.69 is found again.

[6]The impact of Si_{Mn} antisites on the electronic structure can be neglected, especially in the energy regime accessible by our experimental set-up [21].

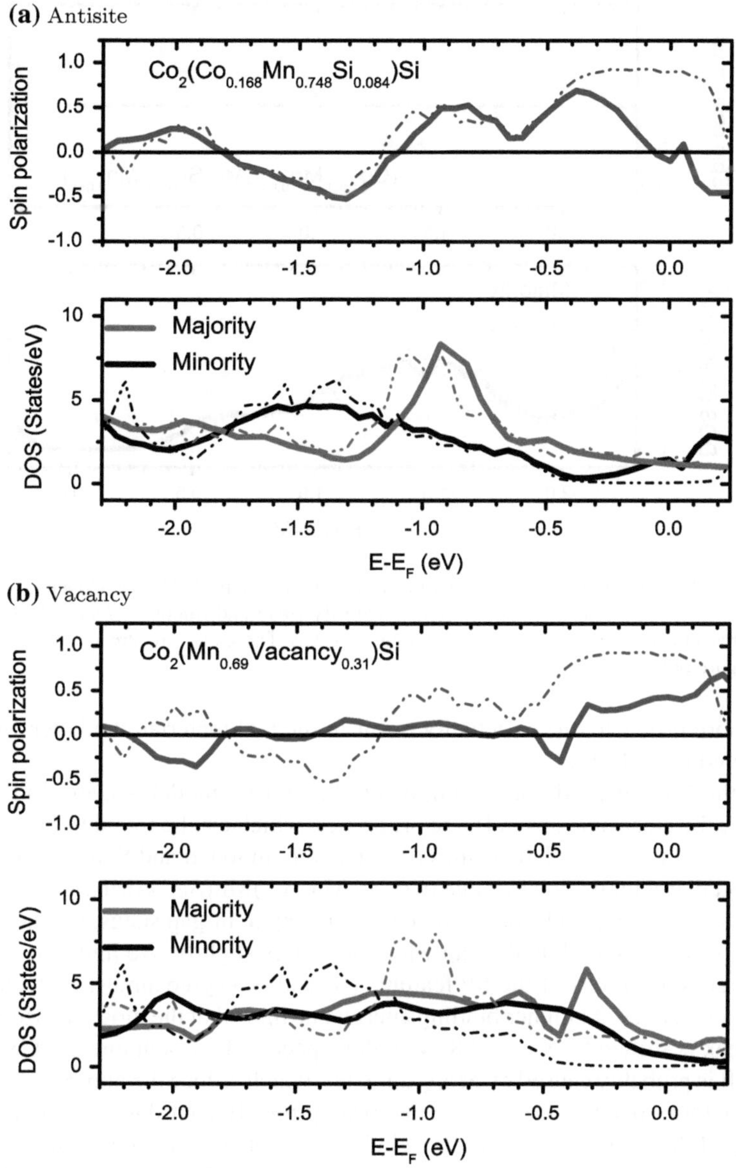

Fig. 3.7 CPA calculations for $\alpha = 0.69$. Missing Mn atoms are replaced by **a** Co and Si atoms and **b** vacancies. *Upper panel* spin polarization. *Lower panel* majority (*red*) and minority (*black*) DOS. *Dashed lines* represent pure LSDA calculations of stoichiometric CMS. These data are already published in [19] (Color figure online)

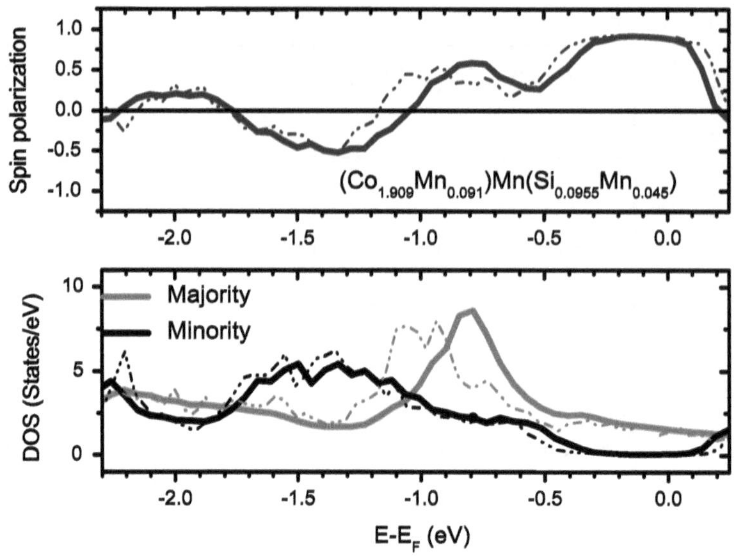

Fig. 3.8 CPA calculations for $\alpha = 1.19$. Surplus Mn atoms are hosted by the Co and Si sublattice. *Upper panel* spin polarization. *Lower panel* majority (*red*) and minority (*black*) DOS. *Dashed lines* represent LSDA calculations of stoichiometric CMS. These data are already published in [19] (Color figure online)

is the more probable one to actually occur in the real material, the latter one is not considered henceforth.

For the Mn-enriched compound, again the antisite model was used in order to calculate the electronic ground state properties, which can be found in Fig. 3.8. Here the excess Mn occupies lattice sites designated both to Co and Si atoms, leading to the respective composition formula $(Co_{1.909}Mn_{0.091})Mn(Si_{0.955}Mn_{0.045})$. Obviously the introduced Mn_{Co} and Mn_{Si} antisites do not create in-gap states, i.e. the minority band gap stays open while the gap width decreases slightly. We find that especially the energetic positions of the SP features (B) and (C) agree almost perfectly to the experimental ones. At higher binding energy one finds that the most intense majority DOS peak is shifted to $E = -0.8$ eV with respect to the stoichiometric CMS DOS. The shift is probably caused by Mn_{Co} antisites, which reduce the number of electrons in the majority bands. Furthermore this peak probably contributes to the broad M3 feature, while the already found direct bulk transition at this binding energy leads to the discrepancy regarding the energetic position ($E(M3) = -0.92$ eV). The broad peak around $E = -1.35$ eV might contribute to the experimentally found m3 feature, although the energetic position in the photoemission spectrum differs by 0.2 eV (cf. Table 3.1). Similarly to the M3 peak, the origin will be again a respective dominating direct bulk transition discussed in the latter section. Altough in general the SP features are reproduced quite well in a qualitative manner in case of Mn-rich CMS, especially the SP values near the Fermi level differ significantly. This results from the fact that despite the theoretical predictions we find finite minority electron contributions in our experiment and hence a clearly reduced, but nevertheless positive spin polarization at E_F. While these minority contributions originate at least partly from the disorder-induced band gap reduction found in the theoretical data, surface-related effects as a further reason will be discussed in the next section.

3.2.1.5 Surface Effects

Up to now we found that neither bulk transitions nor defects can fully explain the experimental spin-resolved spectra and SP values especially close to the Fermi level. This section will treat the influence of surface-related phenomena to the obtained measurements. For this purpose we will again make use of respective calculations, which will not present ground state properties but actual spin-resolved photoemission spectra. These calculations are based on the one-step photoemission model and will consider the used photon energy and light polarization as well as the sample surface termination. For the four most probable surface terminations (Co-Co, Mn-Si, Mn-Mn, vacancy-Si) regarding stoichiometric CMS [22], the respective energy-resolved spin polarization and the corresponding minority and majority spectra are shown in Figs. 3.9 and 3.10. In all cases the half-metallicity of bulk CMS can not be observed, since surfae-induced states within the minority band gap occur. At least for Mn-Mn termination this in in quite contrast to previously published theoretical work, where the Mn-Mn terminated surface was predicted to be the only one conserving full spin

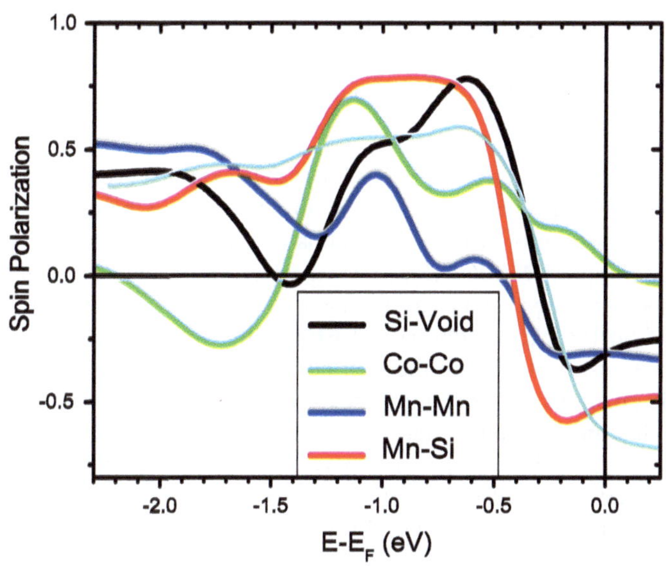

Fig. 3.9 Calculated spin polarization spectra ($h\nu = 5.9\,eV$; s-polarized light) of the CMS(100) surface, terminated by a final Co-Co, Mn-Mn, Si-void or Mn-Si layer respectively. These data are already published in [19]

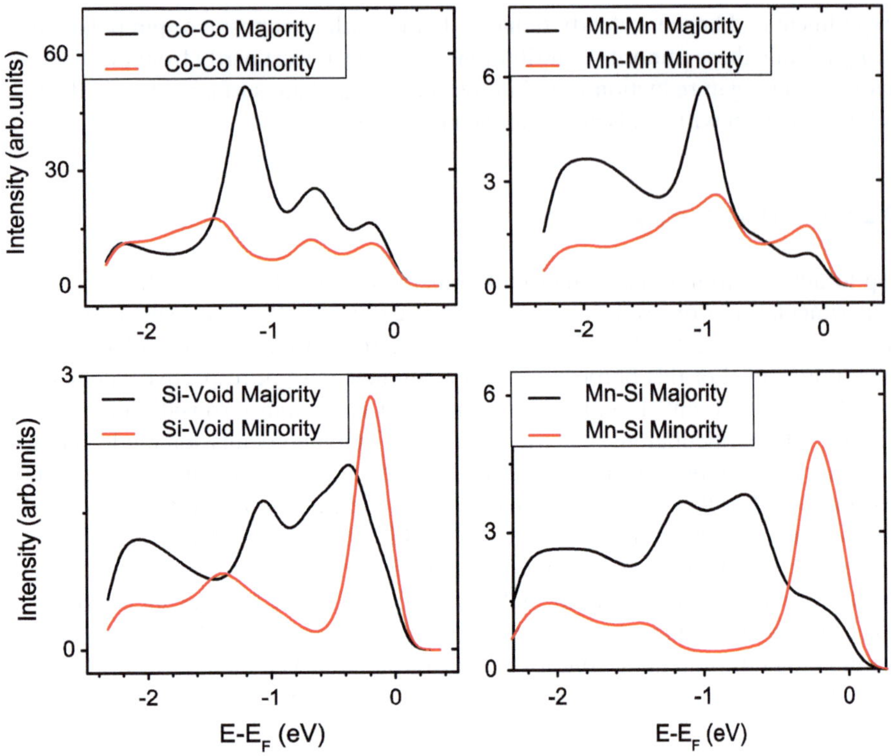

Fig. 3.10 Calculated majority and minority photoemission spectra ($h\nu = 5.9\,\text{eV}$; s-polarized light) of the CMS(100) surface, terminated by a final Co-Co, Mn-Mn, Si-void or Mn-Si layer respectively. These data are already published in [19]

polarization [22]. As depicted in Fig. 3.11, our data exhibit a prominent surface state for a Mn-Mn termination along the $\Gamma - X$ direction within the minority band gap.

Comparing the results of the calculations to the measured spin polarization of the Mn-poor sample CMS069 (Fig. 3.3), the vacancy-Si and the Mn-Si terminations correspond well especially to the features (B) to (E). As already noted in the beginning of Sect. 3.2.1, the observed LEED patterns are also in agreement with these terminations. Furthermore the spin polarization at the Fermi level as well as the energetic positions of the spin polarization features (D) and (E) are even better reproduced than in the ground state calculations. Hence the experimentally found negative spin polarization at E_F is not only due to the Co_{Mn} antisites as discussed in the latter section, but also caused by surface states. in case of the two further considered terminations (Co-Co and Mn-Mn), the calculated data are quite different to the measured spin polarization spectra, which emphasizes the strong dependence on the surface termination with respect to photoemission spectra.

Regarding the experimental results obtained at the Mn-rich compound CMS119 (cf. Fig. 3.4), the influence of surface states seems to be negligible since the measured

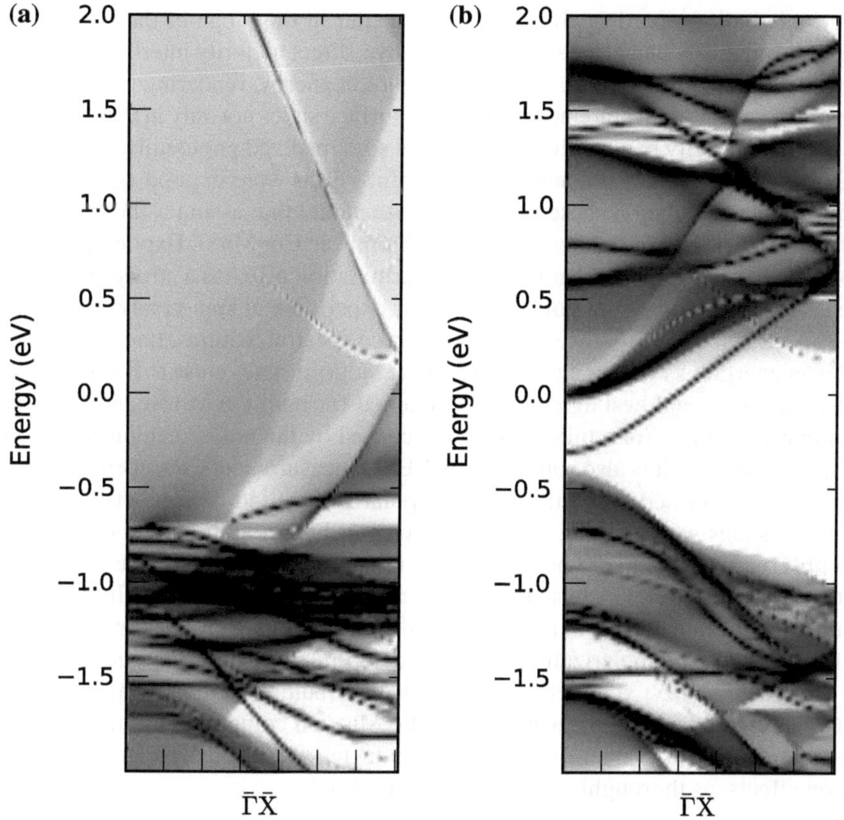

Fig. 3.11 Surface electronic structure of the Mn-Mn terminated CMS surface along the $\bar{\Gamma}\bar{X}$ direction, calculated in the LSDA scheme. These data are already published in [19]. **a** Majority. **b** Minority

spectra and SP curves do not fit to any of the theoretical ones. This is surprising because CMS119 should be electronically more similar to bulk CMS than CMS069. However, the experimentally found peaks M1 and m1 show strong dependence on the light polarization and almost vanish for s-polarized light. Therefore we can conclude that this behavior reflects the presence of surface resonances originating from bulk states with predominant Δ_1 symmetry at this energy (cf. Sect. 3.2.1.2).

3.2.1.6 Summary

In this section we reviewed the investigations reported in [19] of the free surface of off-stoichiometric $Co_2Mn_\alpha Si(100)$ films by spin-resolved photoelectron spectroscopy and electronic structure calculations. Due to the little dispersion of the bulk *d* bands of CMS, many features of the experimental spectra can directly be related to electronic bulk properties. However, in particular in the range of the Fermi energy where the majority bands are strongly dispersive, direct majority interband transitions are momentum-forbidden for the selected photon energy, rendering our experiment sensitive to even small amounts of defects or surface states not only in the minority but also in the majority bands. The experimental spin resolved photoemission spectra of the free surfaces of nonstoichiometric Co_2MnSi films were in good agreement with ground state calculations based on the antisite model that assumes the formation of antisites rather than vacancies for nonstoichiometric Co_2MnSi. Explicit photoemission calculations indicate that the surface termination also has a strong influence on the surface electronic spin polarization. Our experimental spin-resolved photoelectron spectra enable us to trace back the observed spectral features, finding bulk states at lower energies as well as surface and defect induced states close to E_F. For the Mn-deficient compound best agreement is obtained for both the Mn-Si and vacancy-Si terminated surfaces (or a mixture of both, caused by the non-stoichiometric sample composition) which is also supported by LEED measurements. Photoemission measurements performed with different light polarizations indicate majority as well as minority states with a strong Δ_1 symmetry contribution at the Fermi energy. Here, the observed negative spin polarization is related to minority surface states and to antisite defect states, which is also consistent with the antisite model. In the Mn-rich compound we find no sign of bulk defect induced minority gap states or surface states at E_F. Here the spin polarization is mainly determined by the contribution of majority electrons with Δ_1 symmetry. However we attribute the still existing discrepancy between theoretical and measured SP of the Mn-rich compound to the influence of finite temperature (all measurements are conducted at room temperature) and correlation effects, as thoroughly discussed in Sect. 3.3.2.

3.3 SR-PES at the Co_2MnSi/MgO interface

One of the main applications of half-metallic materials like CMS is the usage as highly spin-polarized electrodes in magnetoresistive devices, e.g. magnetic tunnel junctions (MTJs) [10, 30–32]. In such a multilayer structure, two ferromagnetic (FM) layers are separated by a thin (1–2 nm) tunneling barrier. When applying a finite voltage, the tunneling current will depend strongly on the relative magnetization directions of the ferromagnetic layers, hence serving as a magnetic sensor. The overall device performance depends crucially on the structural, chemical and electronic (spin) properties at the FM/insulator *interface*. The commonly used insulating material nowadays is epitaxially grown MgO due to its symmetry filtering property: As Butler et al. first predicted for Fe/MgO/Fe MTJs, the tunneling probability will depend additionally on the electron wave function symmetry in case of crystalline barriers due to $k_{||}$-conservation [27]. Using ferromagnetic materials with appropriate band structure, the preferential tunneling leads to significantly increased TMR ratios. This also holds for CMS/MgO/CMS MTJs [26].

To investigate directly the electronic properties of the CMS/MgO interface, SR-PES here again serves as the method of choice. Although we are not aware of such experiments performed at the CMS/MgO bilayer by other groups up to now, especially the interfaces between MgO and the 3*d* ferromagnetic materials Fe and Co have been investigated thoroughly [33–37]. Due to the large band gap of MgO, Fe bulk states and possible interface states are observable near E_F. However, the MgO coverage did not exceed more than 2 ML in these studies, since for thicker layers the photoemission yield stems mainly from MgO due to the high surface sensitivity of conventional photoemission, which furthermore results in a distinct drop of the detected spin polarization [38, 39], cf. Fig. 3.12. Hence the results can hardly be applied to MTJ properties, where the tunneling barrier usually has a thickness of at least 6 ML. Also pinholes can not be excluded for MgO layer thicknesses ≤ 2 ML, leading to a certain amount of metal surface contribution to the photoemission signal. In the following sections we will show that in case of an used photon energy similar to the MgO band gap width, the SR-PES technique exhibits true interface sensitivity, hence being a well-suited technique to investigate buried interfaces. Furthermore symmetry-resolved experiments allow us to discuss the most probable origin of the large temperature dependence of Heusler-based TMR devices.

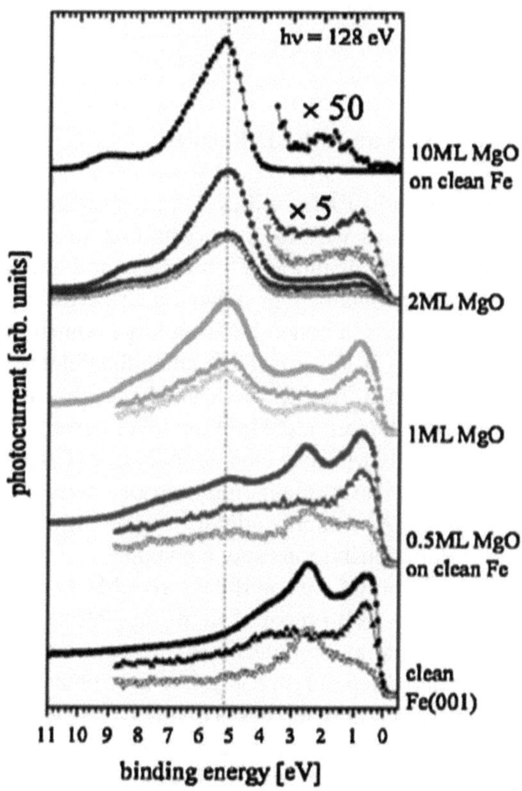

Fig. 3.12 Spin-resolved and spin-integrated spectra ($h\nu = 128$ eV) of Fe(001)/MgO in dependence of the MgO top layer thickness. The photoemission yield close to E_F drops distinctively already for 2 ML MgO. For 10 ML MgO, only MgO features are visible within the spectra. Taken from [39]

3.3.1 Low Energy SR-PES at the Off-Stoichiometric $Co_2Mn_\alpha Si/MgO$ Interface

Here we review our studies of the electronic and structural properties of CMS/MgO bilayers in dependence of the MgO thickness and of the annealing temperature [40]. These studies were motivated by the fact that he performance of MTJs depends critically on the tunnel barrier width [41] and heat treatment [14]. For these studies we have used LEED and Auger electron spectroscopy (AES) combined with SR-PES.

Sample fabrication was identical to the CMS119 one described in Sect. 3.2.1 (i.e. the ferromagnetic layer consisted of Mn-rich CMS), while the MgO layers on top varied in thickness (t_{MgO} = 4–50 ML; 1 ML = 0.21 nm) [40]. After the outermost MgO layer has been evaporated at RT, no further annealing took place in the preparation chamber. Hence a study of the annealing temperature dependence on the MgO top layer crystal structure and the PES experiments could be performed inside the experimental chamber. This is of high interest because typically an optimum annealing temperature (T_A) for MTJs has to be found, which results in a maximum TMR value [42]. After finishing all measurements for a certain annealing temperature step, the samples were subsequently heated for at least 30 min up to the next temperature step. The investigated T_A regime reaches from 400 to 600 °C. For devices consisting of CMS and MgO, the optimum annealing temperature is in the range between 450 and 600 °C [20, 43, 44].

3.3.1.1 AES and LEED Results

The AES and LEED studies were conducted in order to characterize the chemical composition as well as the crystalline structure of the outermost MgO top layer in dependence of the MgO thickness and annealing temperature. The Auger electron spectra obtained in dependence of the annealing temperature and MgO top layer thickness in general exhibit large contributions of the elements Mg and O (not shown). This is not surprising since the detected Auger electrons have kinetic energies between 100 and 2000 eV and hence this method is also highly surface-sensitive, while even the thinnest MgO top layer has a thickness of almost 1 nm. Nevertheless, for $t_{\mathrm{MgO}} \leq 10$ ML also clear signals from Co and Mn are visible. Regarding the T_A-dependent measurements, no sign of chemical interdiffusion due to the annealing process is found independently of the MgO top layer thickness, since the relative elemental contributions stay constant.

We will now focus on the relative Mg:O ratio which is normalized to the signals obtained at a well-prepared single crystalline bulk MgO sample as well as the carbon amount found at the outermost surface, both displayed in Fig. 3.13. The two values significantly depend on t_{MgO}, however in a opposite way.

The Mg:O ratio decreases distinctively when the MgO layer thickness is reduced. One simple origin is that the relative elemental contributions are evaluated from the height of element-specific peaks in the Auger spectra [45]. These peaks usually are

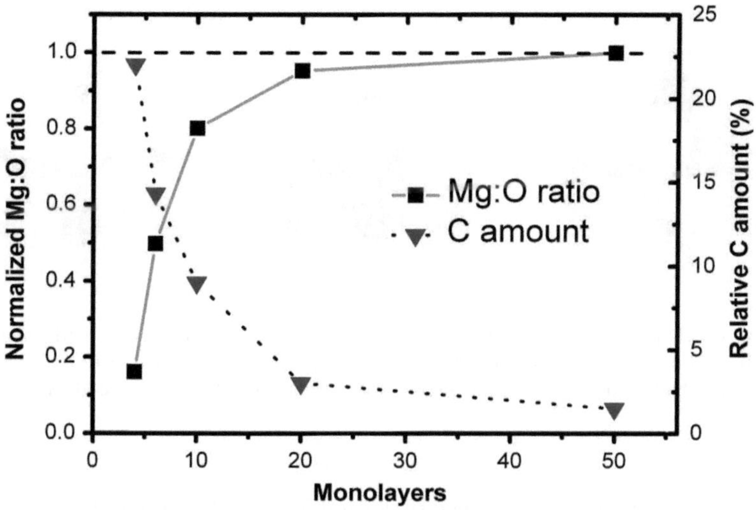

Fig. 3.13 Mg:O ratio and relative carbon amount for varying MgO top layer thickness for CMS/MgO samples annealed at $T_A = 600\,°C$. The connecting lines serve as a guide to the eye. These data are already published in [40]

found at different energies and hence the respective electrons will have different inelastic mean free paths (IMFPs)[46]. In case of MgO, the O KLL peak at 503 eV ($\lambda_{IMFP} \approx 13$ Å) and the Mg KLL peak at 1174 eV ($\lambda_{IMFP} \approx 25$ Å, almost twice as high as the IMFP of oxygen-related electrons) are used [2]. The measured relative amount of Mg will decrease when the MgO layer thickness is in the range of the IMFP, hence even for an actually stoichiometric MgO layer the evaluated composition changes for $t_{\text{MgO}} \leq 10\,\text{ML} \approx 20$ Å. However, this effect can not solely explain our results since we detect a ratio decrease down to less than 20 % for the 4 ML thick MgO overlayer, while for an only 1 ML thick but stoichiometric MgO film the decrease will be in the order of 50 % regarding the effect described above [45].

A further explanation is the enrichment of oxygen-containing adsorbates at the very surface in case of lower MgO coverages, leading to higher O signal with respect to Mg. Adsorption will take place when the samples are transferred *ex situ* into the experimental chamber. This effect would furthermore explain the increase of carbon signal in case of reduced MgO layer thickness (see Fig. 3.13), i.e. adsorption of carbon oxides takes place. We take these findings as a strong indication that crystalline defects are present at the very surface for $t_{\text{MgO}} \leq 10\,\text{ML}$, but not in case of thicker coverages. This we conclude from the fact that MgO defect surfaces tend to adsorb especially carbon oxides and molecular oxygen, in contrast to a defect free surface [47–54]. Finally we have to discuss why the thinner MgO top layers have a higher surface defect density than the thicker ones. The most probable origin is the inevitable occurrence of misfit dislocations at the CMS/MgO interface due to the finite lattice mismatch of 5 % [55]. These dislocations will also extend throughout

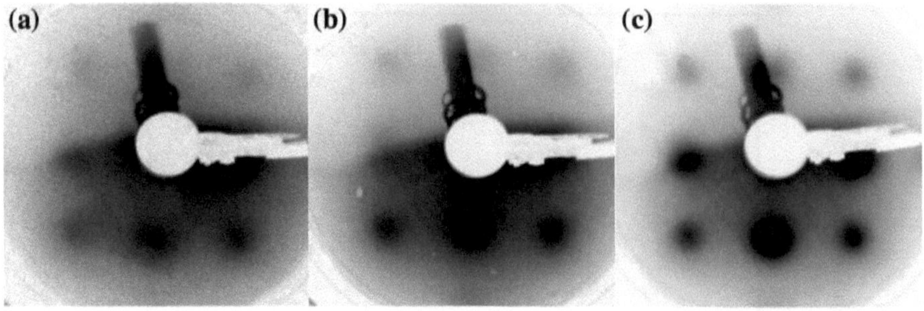

Fig. 3.14 LEED patterns at a primary electron energy of 90 eV of **a** $t_{MgO} = 10$ ML annealed to 400 °C, **b** $t_{MgO} = 10$ ML annealed to 600 °C and **c** $t_{MgO} = 20$ ML annealed to 400 °C. Please note that the upper half of the luminescent screen is less sensitive. These data are already published in [40]

the MgO layer for a certain distance of several ML and hence will lead to different amounts of surface defects depending on the MgO layer thickness.

Complementary to the AES studies, we performed LEED experiments in dependence of the annealing temperature within the most interesting energy regime (i.e. $T_A = 400$ °C, 500 °C, 600 °C).

Starting with the sample having the thinnest MgO top layer (4 ML), no LEED patterns were found for any annealing step. This might be due to the large amount of species adsorbed and arbitrarily arranged at the surface as discussed in the previous section. A further origin will be the very high surface defect density, being also the actual cause of the significant adsorption. However, for all other samples and annealing steps distinct LEED patterns could be observed, all in accordance with the MgO B1 crystal structure [19]. Three examples are shown in Fig. 3.14. In general the occurrence of annealing-induced LEED pattern changes depends on the MgO layer thickness. For $t_{MgO} \leq 10$ ML, the spot pattern quality clearly improves with increasing T_A, i.e. a diminished background signal as well as sharper and more distinct spots. In case of $t_{MgO} = 10$ ML this can be seen by comparing Fig. 3.14a ($T_A = 400$ °C) and b ($T_A = 600$ °C). The two samples with thickest MgO top layer $t_{MgO} = 20$ and 50 ML otherwise do not exhibit any improvement of the LEED pattern when the annealing temperature is increased, while the pattern quality itself is superior compared to the samples with thinner MgO top layer (cf. Fig. 3.14c). Our results can be easily explained again by assuming a thickness-dependent surface defect density as in the previous section. The surfaces of the samples with $t_{MgO} \leq 10$ ML are influenced by the misfit dislocations originating at the CMS/MgO interface. Hence the spot patterns change by increasing T_A due to partial defect healing. Nevertheless the pattern quality is lower than that of the samples with $t_{MgO} = 20$ and 50 ML, since here no defects induced by the misfit dislocations are present anymore. This is the actual reason why we do not find any dependence upon the annealing temperature for thicker MgO top layers.

3.3.1.2 SR-PES Results

The results presented in this section are obtained solely with p-polarized laser light ($h\nu = 5.9\,\text{eV}$). We will focus the discussion onto the two samples with $t_{\text{MgO}} = 4$ and 20 ML, i.e. the extreme cases regarding the MgO top layer thickness (the sample with $t_{\text{MgO}} = 50\,\text{ML}$ could not be investigated due to charging effects and is not considered henceforth). Figure 3.15 exhibits the spin-integrated photoemission spectra as well as the determined SP for both samples with $T_A = 600\,°\text{C}$, i.e. the highest used annealing temperature. Astonishingly, we are able to detect a finite positive SP especially near E_F. At first sight, this is quite surprising, since photoemission is known to be sensitive only to the outermost surface (cf. Sect. 3.1), which in our case consists of non-magnetic MgO. We explain this finding as follows: In our PES experiments, firstly, CMS valence band electrons near the interface will absorb the laser light, which in general can penetrate solid materials for several 10 nm and in our case furthermore will not be absorbed by the MgO top layer, since the photon energy of 5.9 eV is lower than the MgO band gap of around 7.8 eV [56, 57]. The excited electrons will now travel through the MgO layer to the sample surface within unoccupied conduction bands before they are photoemitted. Usually these electrons will scatter

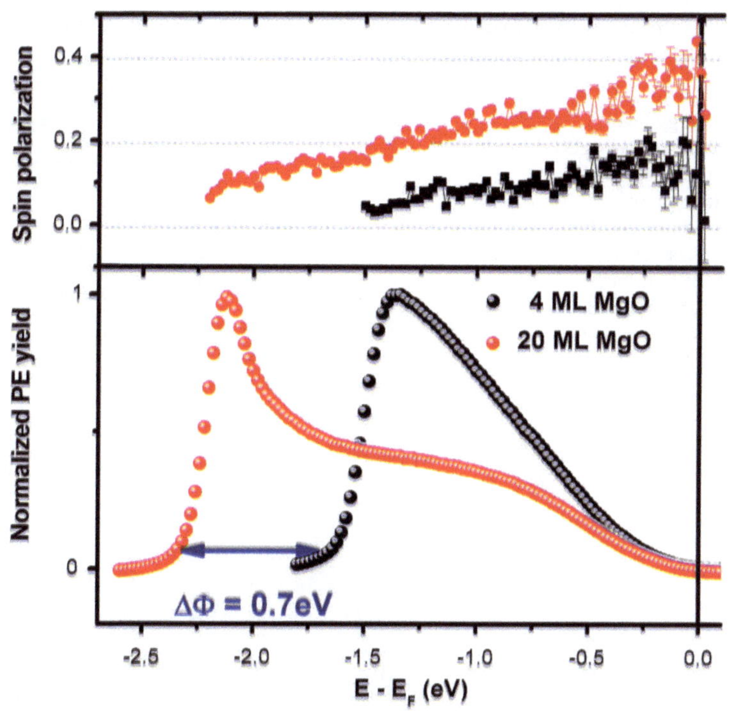

Fig. 3.15 Spin-integrated and normalized spectra (*lower part*) and spin polarization (*upper part*) for samples with $t_{\text{MgO}} = 4$ and 20 ML with $T_A = 600\,°\text{C}$. These data are already published in [40]

inelastically with the MgO valence band electrons, i.e. the former ones will lose energy and the latter ones will gain energy. Hence the excited electrons will not be detected anymore at least within the respective energy range. This is the actual origin why photoemission is surface-sensitive, as only electrons from the outermost layers will leave the sample without having scattered with other electrons. In our case, the traversing electrons have very little kinetic energy and therefore occupy only the lowest MgO conduction bands according to the low photon energy used. This furthermore means that inelastic scattering with an MgO valence band electron is not possible anymore, since there are no final scattering states due to the large MgO band gap. Fig. 3.16 represents a schematic of the above described process.

Returning to the spectra themselves shown in the lower part of Fig. 3.15, they do not exhibit significant features compared to the ones obtained at the free surface of Mn-rich CMS as shown in Fig. 3.4. We assume that electron scattering directly at the interface and induced by the lattice mismatch leads to a signifcant amount of secondary electrons, washing out the spectra. However, the electrons detected at the Fermi energy did not lose kinetic energy by definition and hence the SP obtained at E_F is not falsified due to the traversing of the MgO top layer. Therefore, low-energy photoemission turns out to be truly interface-sensitive in this case. However, the SP value at Fermi level is more than twice as high for $t_{MgO} = 20\,\text{ML}$ (35 %; averaged over an energy range of 300 meV below E_F) than for $t_{MgO} = 4\,\text{ML}$ (16 %), although the electrons cross an overlayer five times thicker. This apparent contradiction can

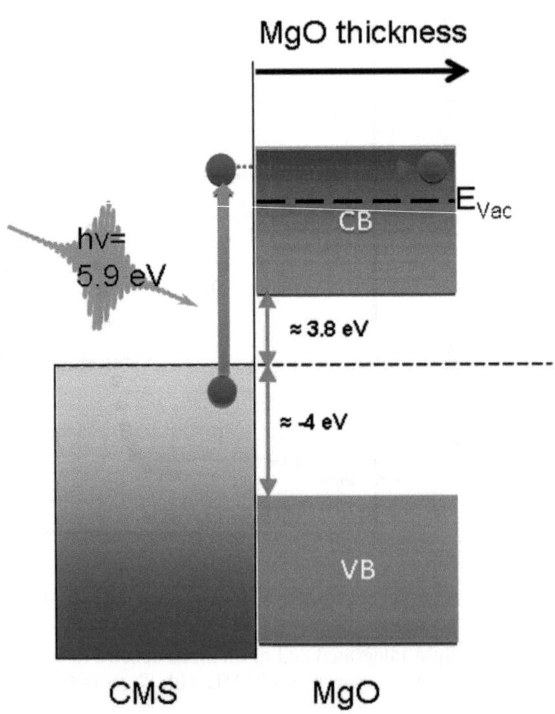

Fig. 3.16 Energy level alignment at the CMS/MgO interface and electrons excited by low-energy SR-PES traversing the MgO layer without scattering, resulting in the interface sensitivity of this method. The MgO valence band onset is about 4 eV below E_F, cf. inset of Fig. 3.20

be explained by additional strong non-spin-conserving scattering with the adsorbed species at the MgO surface in case of $t_{MgO} = 4\,ML$, as found in the previous section. This also explains the work function difference of 0.7 eV between the two samples, since the work function is a unique property of the sample surface. Otherwise, the SP value obtained at the interface is higher than the one detected at the free Mn-rich CMS surface. We attribute this to detrimental surface-related effects rather than to a kind of spin-filter effect. Furthermore the positive and relatively high obtained SP let us conclude that no detrimental interface states are present at least at the Mn-rich CMS/MgO interface. In case of the stoichiometric CMS/MgO interface, several theoretical works predict interface states with mainly minority contributions directly at E_F [58, 59].

We will now turn to the annealing-temperature-dependent photoemission studies. Except for the sample with thinnest MgO top layer (4 ML), all other ones show no distinct changes regarding the spectra as well as the SP. This means the results for $t_{MgO} = 6\,ML$ and 10 ML are quite similar to the ones of $t_{MgO} = 20\,ML$ presented in Fig. 3.15, independently of T_A. Especially a finite positive SP is always found already for $T_A = 400\,°C$. In case of $t_{MgO} = 4\,ML$, however, the changes of the photoemission spectra are remarkable and depicted in Fig. 3.17. Starting with the spectrum obtained at $T_A = 400\,°C$, no photoemssion signal exists down to $-1.5\,eV$ binding energy and therefor no SP can be defined near the Fermi level. As we found in Sect. 3.3.1.1, the surface of this sample is dominated by defects and adsorbed species. This leads to significantly increased inelastic electron scattering. If we now increase the annealing temperature to 500 °C and subsequently to 600 °C, we get finite photoemission yield up to the Fermi level and, similarly to all other samples, a positive SP value as already shown in Fig. 3.15. The origin of this behavior must be the defect healing inside the MgO top layer induced by higher annealing

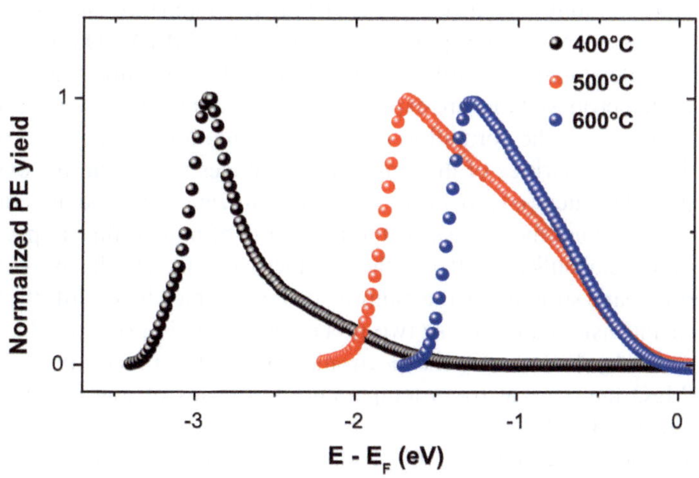

Fig. 3.17 Spin-integrated and normalized spectra for $t_{MgO} = 4\,ML$ in dependence of T_A. These data are already published in [40]

temperature. However, the LEED and AES measurements do not exhibit an annealing-temperature-dependent change of the crystalline and chemical surface structure in case of $t_{\mathrm{MgO}} = 4\,\mathrm{ML}$.

3.3.1.3 Summary

We presented a combined spin-resolved PES, AES and LEED study of the advanced MTJ interface consisting of half-metallic (non-stoichiometric) Co_2MnSi and MgO for varying annealing temperature and MgO top layer thickness. Distinct influence of interface defects onto the whole MgO layer is revealed by AES and LEED for thicknesses up to 10 ML, which only partly can be overcome by annealing. Spin-resolved low energy photoemission spectroscopy allowed to determine the interface spin polarization as the key property of MTJs, finding a distinct positive SP value in all cases. Remarkably, the CMS/MgO interface is accessible even for MgO thicknesses up to 4 nm, making low energy PES a unique method for the non-destructive characterization of buried spintronics interfaces.

3.3.2 Low Energy Spin- and Symmetry-Resolved PES at the Stoichiometric Co_2MnSi/MgO Interface

At low temperatures Heusler compounds used as MTJ electrodes have already been found to be superior to conventional $3d$-ferromagnetic materials regarding the TMR perfomance [42, 60, 61]. However, already far below room temperature, e.g. CMS-based MTJs show a puzzling drastic performance loss, as can be seen in Fig. 3.18 [62]. Although it is commonly assumed that such behavior should be connected to a strong temperature-dependence of the spin polarization of the CMS/MgO interface, a clear microscopic understanding is still lacking. In particular, two different main mechanisms have been proposed that are expected to play a significant role in half-metallic materials. The first is the formation of additional minority band gap states, arising either from peculiarities of the magnetic sublattices [63], from the existence of non-quasiparticle states [64], or from strong hybridization dependence on temperature [65]. The second mechanism is the formation of non-collinear spin moments inside the half-metal bulk as well as at the interface [65, 66]. This leads to a mirroring of the majority band structure in the minority channel, which hence fills the band gap. Due to their intrinsic nature, those two processes can be in principle distinguished from each other by determining the electronic spin- and symmetry properties of the Co_2MnSi/MgO interface at room temperature by means of low energy spin- and symmetry-resolved PES.

Here we review a recent work ([67]), where we addressed this important question by applying the spin- and symmetry-resolved low energy PES technique presented in the previous sections to the *stoichiometric* CMS/MgO interface. Samples have

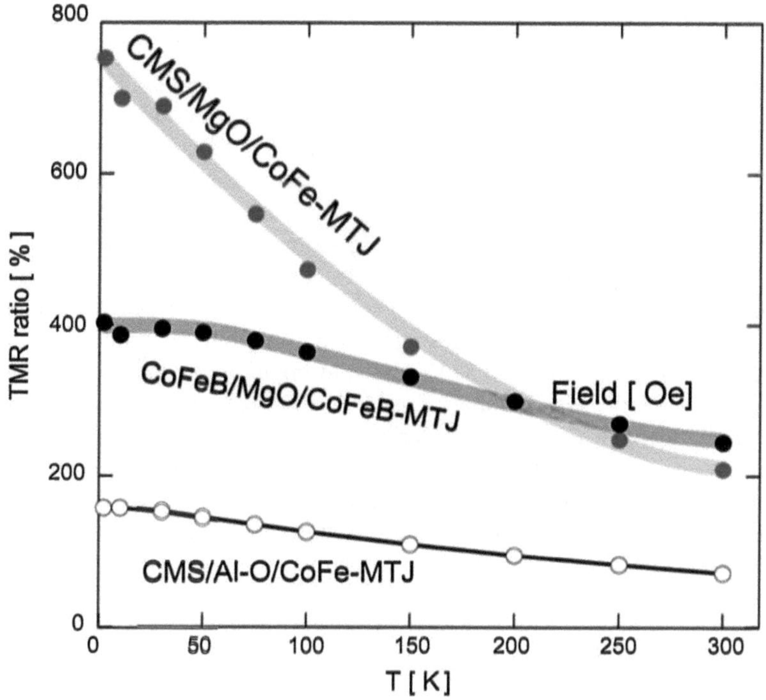

Fig. 3.18 Temperature dependence of the TMR ratio for CMS/MgO/CoFe MTJs, CMS/Al-O/CoFe MTJs and conventional CoFeB/MgO/CoFeB MTJs. Taken from [62]

been fabricated by Ando group (Tohoku University, Sendai) with 2 nm ($\approx$10 ML) MgO on top [68]. Prior to the photoemission experiment the samples have been gently sputtered with a removal of less than 1 ML MgO in order to remove residual adsorbates and afterwards annealed to 450 °C (cf. Sects. 3.3.1.1 and 3.2.1). Please note that this procedure was not applied to the samples used in Sect. 3.3.1.

Additionally, spin-resolved ultraviolet photoemission spectroscopy (UPS) spectra were recorded using a commercial He VUV gas discharge lamp providing unpolarized light with $h\nu = 21.2$ eV (He I line). Comparison to the low energy photoemission experiments again proves the true interface sensitivity of the latter technique.

Evaluating the light polarization asymmetry of the low energy photoemission experiments obtained at the CMS/MgO interface allows us to extract the electronic states present at the interface and exhibiting Δ_1 symmetry. Please note again that these states tunnel through epitaxial MgO(100) barriers with highest probability [26, 27]. Furthermore the asymmetry behavior in the two spin channels let us discuss the actual physical origin of the significant temperature dependence of Heusler-based TMR devices.

3.3.2.1 SR-PES Results

We start discussing the low energy PES results obtained from the bare stoichiometric CMS(100) surface, after removing the MgO top layer completely and annealing the sample afterwards to 450 °C. As shown in Fig. 3.19, the obtained spin- and symmetry-resolved photoemission spectra (lower panel) and the deduced spin polarization (upper panel) are different to both the Mn-deficient and the Mn-rich CMS surface (cf. Sect. 3.2.1.1). For s-polarized light, the $\Gamma - X$ bulk band structure (as given e.g. in [26]) is reproduced quite well near E_F, with a low photoemission yield directly at the Fermi energy (E_F) originating from dispersive bands and a distinct peak originating from the respective band bottoms dominating the spectra for binding energies around −0.5 eV. The spin polarization shows a maximum at $E - E_F = -0.2$ eV with a value of +45 %. Both spectra and spin polarization change dramatically at E_F when p-polarized light is used for excitation. In this case the minority channel exhibits a prominent peak which is cut by the Fermi-Dirac distribution at E_F. The corresponding state is centered at E_F, as indicated by the energy dependence of the spin polarization, which assumes its minimum of −20 % directly at the Fermi energy. This peak can be ascribed to a minority surface state with Δ_1 symmetry for four reasons: (i) it can only be excited by p-polarized light; (ii) several theoretical works [19, 22] predict minority surface states for CMS(001) and have

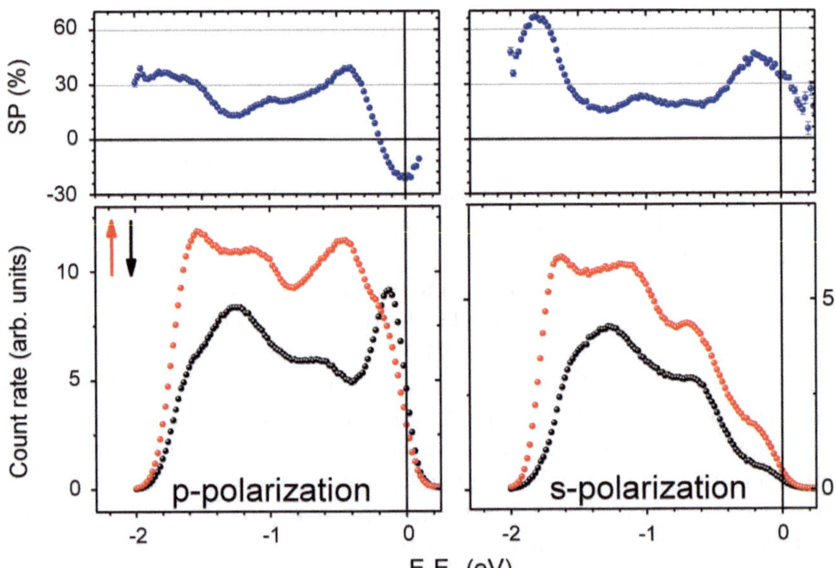

Fig. 3.19 Majority (*red*) and minority (*black*) electron photoemission spectra of the bare CMS(100) surface (*lower panel*) and deduced spin polarization (*upper panel*), measured by spin-resolved low energy PES with s-polarized (*left*) and p-polarized laser light (*right*). These data are already published in [67]

Table 3.2 Summary of the surface electronic features of stoichiometric and off-stoichiometric CMS close to the Fermi level as found by low energy SR-PES

Mn-deficient CMS ($\alpha = 0.69$)	Stoichiometric CMS ($\alpha = 1$)	Mn-rich CMS ($\alpha = 1.19$)
Strong influence of defects (Co_{Mn} antisites)	No defect states	No defect states
Minority surface state	Minority surface state	No surface state

been already confirmed experimentally for off-stoichiometric CMS (cf. Sect. 3.2.1.5 and [19]); (iii) the state is also present at the same binding energy in spin-resolved UPS spectra (not shown); (iv) the state vanishes when the surface is covered by MgO, as described in detail in the following. Furthermore the majority electron spectra show also an increased photoemission yield near E_F, which we attribute to a mixture of additionally excited Δ_1 bulk states and surface resonances. Table 3.2 summarizes the findings on the surface electronic structure of the stoichiometric compound as well as the off-stoichiometric ones discussed in Sect. 3.2.1.1.

Note that very recently a further spin-resolved PES study revealed almost full spin polarization at the free stoichiometric CMS surface around Fermi level [69]. This result was explained by enhanced majority signal due to a surface resonance occurring at the used photon energy. The authors did not find any sign of a minority surface state, in contradiction to our experimental findings and the theoretical results discussed above. Despite this striking difference, the results of [69] can be reconciled with our findings. In fact, the photoemission calculations presented in [69] showed that due to intrinsic correlation effects and to a finite energy resolution, the absence of a majority surface resonance would lead to the detection of a spin polarization of around 50 %. This prediction agrees with the values of the spin polarization we measure with s-polarized light, where the effect of the minority surface state is suppressed (cf. Fig. 3.19).

Let us now turn to the results obtained for the stoichiometric CMS/MgO(10 ML) system. The left part of Fig. 3.20 shows the UPS spectra (bottom panel) and the corresponding spin polarization (upper panel). In the UPS spectra almost no photoemission signal is detected close to E_F. For lower binding energies a clear feature at $E - E_F = -5\,\mathrm{eV}$ arising from the oxygen $2p$ peak in MgO is observed (inset in the left part of Fig. 3.20). This clearly shows that only the MgO top layer is spectroscopied by UPS, revealing the MgO band gap in the vicinity of E_F. The spin-resolved low energy PES results shown in the right part of Fig. 3.20 show a completely different behavior: The spectra themselves resemble the ones obtained at the CMS surface using s-polarized laser light (cf. Fig. 3.19), although the features are washed out. The inferred interface spin polarization also shows high similarity with a maximum value of +35–40 % in vicinity of the Fermi energy, dropping down to roughly 20 % at higher binding energies. The spectra recorded with s-polarized light (not shown) are almost identical to the spectra in the right panel of Fig. 3.20, even though for p-polarization the photoemission yield is increased in general due to excitation of additional states, as will be discussed in detail in the next paragraph.

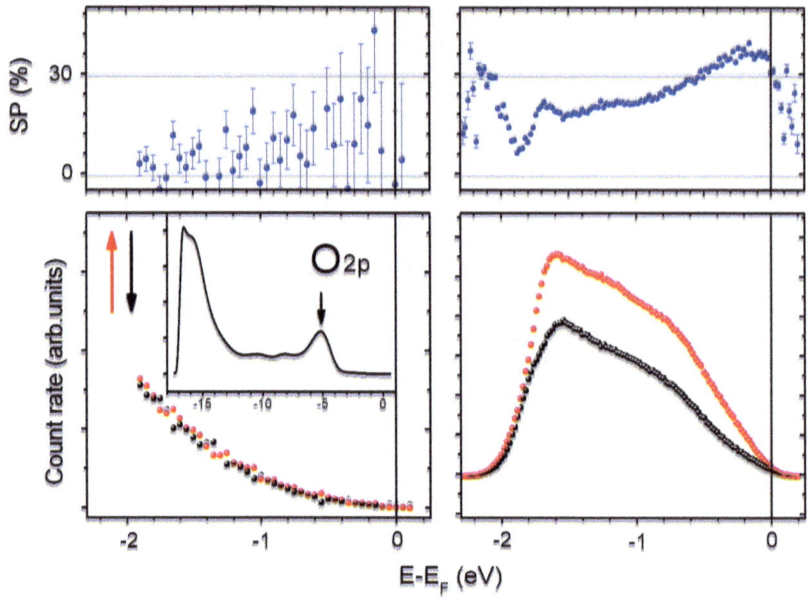

Fig. 3.20 Majority (*red*) and minority (*black*) electron photoemission spectra of the CMS/MgO(10 ML) sample (*lower panel*) and deduced spin polarization (*upper panel*), measured with spin-resolved UPS (*left*) and spin-resolved low energy PES with p-polarized light (*right*). *Inset* in the *left* part: full UPS spectrum of the CMS/MgO(10 ML) sample with spectral feature from the O $2p$ peak at $E - E_F = -5$ eV. These data are already published in [67]

The non-vanishing photoemission yield detected at E_F together with the striking similarity of the low energy PES spectra from the bare CMS surface to those from the CMS/MgO(10 ML) system clearly indicate that the low energy PES is extremely sensitive to the buried CMS/MgO interface, in strong contrast to UPS and as already pointed out in Sect. 3.3.1.2. Close inspection of the spectra in Figs. 3.19 and 3.20b shows that the Δ_1-like minority surface state does not convert into an interface state when the CMS/MgO interface is formed, as it is the case e.g. at the CoFe/MgO interface [37]. This would significantly influence the magnetoresistance properties since Δ_1 states contribute most to the tunneling current [27]. Furthermore, since no additional features appear in the interface spectra compared to the free surface, we can infer that no interface states with either Δ_1 or Δ_5 symmetry are formed at the CMS/MgO interface in the investigated energy range. To conclude, evaporation of MgO leads to a complete suppression of CMS surface contributions, and the CMS/MgO interface spectra resemble the spectra recorded from the CMS surface with s-polarized light, i.e. they also mimic the $\Gamma - X$ bulk band structure. We argue that the small spectroscopic discrepancies between the free CMS surface and the interface originate solely from interface spectra smearing due to both quasi-elastic and inelastic scattering at defects directly at the interface.

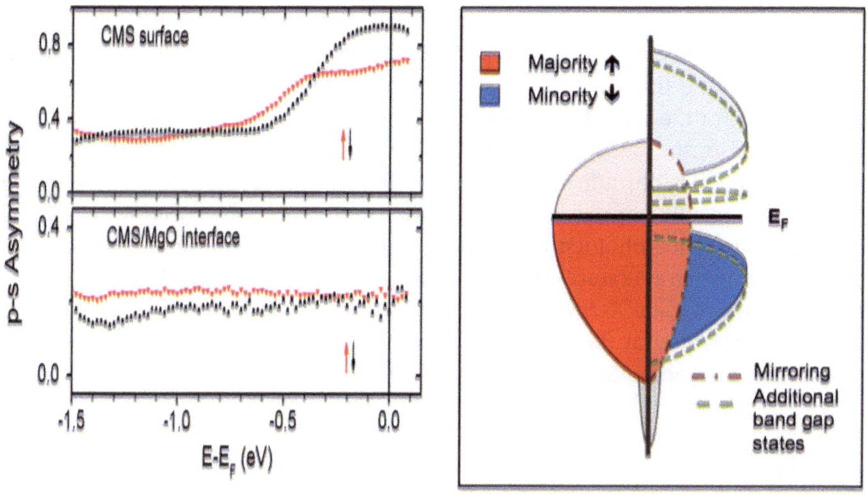

Fig. 3.21 *Left panel* Energy-resolved p-s asymmetry of the CMS free surface (*upper left panel*) and the CMS/MgO(10 ML) interface (*lower left panel*). The majority (minority) channel is marked with *red triangles* (*black squares*). These data are already published in [67]. *Right panel* Schematics of the two main mechanisms leading to thermally induced loss of spin polarization close to E_F. (i) Mirroring of majority states, and (ii) energetic shifts of existing states or formation of new minority states, like non-quasiparticle states

As final step we evaluate the p-s asymmetry $A = \frac{I_p - I_s}{I_p + I_s}$ as already described in Sect. 3.2.1.2. This allows us to discuss the relative Δ_1/Δ_5 contributions at the Fermi energy for minority and majority channel of both the stoichiometric CMS surface (upper part of Fig. 3.21) as well as the CMS/MgO(10 ML) interface (lower part). At the free surface both spin channels show enhanced asymmetry values near E_F compared to higher binding energies, respectively resulting from the minority surface state and majority surface resonances. On contrary, the interface asymmetry varies only marginally. Here, we recall that the strong temperature dependence of the spin polarization in CMS bulk and/or interface has been ascribed theoretically to two competing mechanisms, schematically depicted in the right part of Fig. 3.21. The first mechanism includes thermally activated shifts of bulk states into the minority band gap as well as the creation of new states inside the gap [64, 65, 70]. At room temperature, this mechanism leads to finite minority spectral contributions near E_F that differ in energy dispersion and symmetry compared to the majority states. The second process is given by thermal spin fluctuations, inducing non-collinear magnetic spin alignments which will cause the mirroring of the majority density of states into the minority channel [63]. While this second effect is assumed to be only marginal in the bulk of half-metals [65], latest theoretical considerations regarding the CMS/MgO interface found a distinctively weakened exchange coupling of Co interface atoms, leading to a projection of majority states onto the minority band gap at elevated room temperature and hence to a significant TMR ratio decrease compared to Fe/MgO-based MTJs [66, 71]. Crucially, our experiments reveal that both the minority and majority channel possess almost an identical spectral shape and same Δ_1/Δ_5 asymmetry near E_F and thus strongly indicate a dominant role played by non-collinear spin moments for the reduction of the spin polarization of the CMS/MgO interface. We note that a further reason for the diminished spin polarization values might be intrinsic photoemission-induced correlation effects, as discussed very recently [69]. However, our results are consistent with the recent work of Miyamoto et al. [72], where spin-integrated photoemission measurements showed no temperature-related changes in the spectral features of polycrystalline CMS samples, in line with the mirroring mechanism. The X-ray magnetic circular dichroism experiments by Tsunegi et al. [73] performed at the CMS/MgO interface are also in accordance with the presence of weakened interface moments.

3.3.2.2 Summary

We applied interface-sensitive low energy spin- and symmetry-resolved PES to access the electronic spin- and symmetry properties of the stoichiometric CMS/MgO interface buried below epitaxial MgO films with the same thickness implemented in state-of-the art Heusler-based magnetic tunnel junctions ($\approx$ 2 nm). We found a complete suppression of the minority surface state present at the CMS surface, and no indication for the formation of additional interface states. A detailed analysis of the spectral dependence, spin polarization and wave function symmetry of the interface-related photoemission spectra allowed us to identify thermally activated non-collinear spin moments at the interface as one main mechanism responsible for the strong temperature-related performance loss in CMS-based magnetic tunnel junctions.

Acknowledgments Financial support through the DFG/JST Research Unit 1464 ASPIMATT and FOR 1346, by the German ministry BMBF under contract 05KS10WMA is gratefully acknowledged. The work at Hokkaido University was partly supported by a Grant-in-Aid for Scientific Research (A) (Grant No. 23246055) from MEXT, Japan. We thank M. Jourdan and C. Herbort for providing us a CCFA sample for the comparison of the LEED patterns.

References

1. S. Hüfner, *Photoelectron Spectroscopy*. (Springer, Berlin, 1994)
2. A. Jablonski, C.J. Powell, NIST Electron Inelastic-Mean-Free-Path Database. Technical report, National Institute of Standards and Technology, 2000
3. J. Kessler, *Polarized Electrons*. Springer Series on Atoms and Plasmas, (Springer, Berlin, 1985)
4. H.C. Siegmann J. Stöhr. Magnetism: From Fundamentals to Nanoscale Dynamics. Springer Series in Solid-State Sciences, 2006
5. J.M.D. Coey, M. Venkatesan, M.A. Bari, *Half-Metallic Ferromagnets* (Springer, Berlin, 2002)
6. C. Felser, B. Hillebrands, New materials with high spin polarization: half-metallic Heusler compounds. J. Phys. D Appl. Phys. **40**(6), 1507 (2007)

7. B. Balke, G.H. Fecher, H.C. Kandpal, C. Felser, K. Kobayashi, E. Ikenaga, J.-J. Kim, S. Ueda, Properties of the quaternary half-metal-type Heusler alloy $Co_2Mn_{1-x}Fe_xSi$. Phys. Rev. B: (Condens. Matter Mater. Phys.) **74**(10), 104405 (2006)
8. V.N. Antonov, H.A. Dürr, Y. Kucherenko, L.V. Bekenov, A.N. Yaresko, Theoretical study of the electronic and magnetic structures of the Heusler alloys $Co_2Cr_{1-x}Fe_xAl$. Phys. Rev. B **72**(5), 054441 (2005)
9. S.A. Wolf, D.D. Awschalom, R.A. Buhrman, J.M. Daughton, S. von Molnar, M.L. Roukes, A.Y. Chtchelkanova, D.M. Treger, Spintronics: a spin-based electronics vision for the future. Science **294**(5546), 1488–1495 (2001)
10. I. Zutic, J. Fabian, S.D. Sarma. Spintronics: fundamentals and applications. Rev. Mod. Phys. **76**, 323 (2004)
11. W.H. Wang, M. Przybylski, W. Kuch, L.I. Chelaru, J. Wang, Y.F. Lu, J. Barthel, J. Kirschner, Spin polarization of single-crystalline Co_2MnSi films grown by PLD on GaAs(001). J. Magn. Magn. Mat. In: Proceedings of the 5th International Symposium on Metallic Multilayers **286**, 336 – 339, (2005)
12. Y. Miura, K. Nagao, M. Shirai, Atomic disorder effects on half-metallicity of the full-Heusler alloys $Co_2(Cr_{1-x}Fe_x)Al$: A first-principles study. Phys. Rev. B **69**, 144413 (2004)
13. M. Cinchetti, J.-P. Wüstenberg, M.S. Albaneda, F. Steeb, A. Conca, M. Jourdan, M. Aeschlimann, Towards a full Heusler alloy showing room temperature half-metallicity at the surface. J. Phys. D: Appl. Phys. **40**, 1544–1547 (2007)
14. J.-P. Wüstenberg, J. Fischer, C. Herbort, M. Jourdan, M. Aeschlimann, M. Cinchetti, Effects of post-growth annealing on structural and compositional properties of the $Co_2Cr_{0.6}Fe_{0.4}Al$ surface and its relevance for the surface electron spin polarization. J. Phys. D: Appl. Phys. **42**, 084016 (2009)
15. M. Hahn, G. Schönhense, E.A. Jorge, M. Jourdan. Significant spin polarization of Co_2MnGa Heusler thin films on MgO(100) measured by ultraviolet photoemission spectroscopy. Appl. Phys. Lett. **98**(23), 2352503 (2011)
16. M. Kolbe, S. Chadov, E. Arbelo, E.A. Jorge, G. Schönhense, C. Felser, H.-J. Elmers, M. Kläui, M. Jourdan, Test of band structure calculations for Heusler compounds by spin-resolved photoemission spectroscopy. Phys. Rev. B **86**, 024422 (2012)
17. A.M. Turner, A.W. Donoho, J.L. Erskine, Experimental bulk electronic properties of ferromagnetic iron. Phys. Rev. B **29**, 2986–3000 (1984)
18. A.M. Turner, J.L. Erskine, Surface electronic properties of Fe(100). Phys. Rev. B **30**, 6675–6688 (1984)
19. J.-P. Wüstenberg, R. Fetzer, M. Aeschlimann, M. Cinchetti, J. Minár, J. Braun, H. Ebert, T. Ishikawa, T. Uemura, M. Yamamoto, Surface spin polarization of the nonstoichiometric Heusler alloy Co_2MnSi. Phys. Rev. B **85**, 064407 (2012)
20. T. Ishikawa, H. Liu, T. Taira, K. Matsuda, T. Uemura, M. Yamamoto, Influence of film composition in Co_2MnSi electrodes on tunnel magnetoresistance characteristics of Co_2MnSi/MgO/Co_2MnSi magnetic tunnel junctions. Appl. Phys. Lett. **95**, 232512 (2009)
21. S. Picozzi, A. Continenza, A.J. Freeman, Role of structural defects on the half-metallic character of Co_2MnGe and Co_2MnSi Heusler alloys. Phys. Rev. B **69**, 094423 (2004)
22. S.J. Hashemifar, P. Kratzer, M. Scheffler, Preserving the half-metallicity at the Heusler alloy $Co_2MnSi(001)$ surface: a density functional theory study. Phys. Rev. Lett. **94**, 096402 (2005)
23. J. Hermanson, Final-state symmetry and polarization effects in angle-resolved photoemission spectroscopy. Solid State Commun. **22**(1), 9–11 (1977)
24. W. Eberhardt, F.J. Himpsel, Dipole selection rules for optical transitions in the fcc and bcc lattices. Phys. Rev. B **21**, 5572–5576 (1980)
25. W.H. Butler. Tunneling magnetoresistance from a symmetry filtering effect. Sci. Technol. Adv. Mater. 9(1), 014106 (17pp) (2008)
26. Y. Miura, H. Uchida, Y. Oba, K. Nagao, M. Shirai, Coherent tunnelling conductance in magnetic tunnel junctions of half-metallic full Heusler alloys with MgO barriers. J. Phys.: Condens. Matter **19**(36), 365228 (2007)

27. W.H. Butler, X.-G. Zhang, T.C. Schulthess, J.M. MacLaren, Spin-dependent tunneling conductance of $Fe|MgO|Fe$ sandwiches. Phys. Rev. B **63**(5), 054416 (2001)
28. M. Yamamoto, T. Ishikawa, T. Taira, G.F. Li, K.I. Matsuda, T. Uemura, Effect of defects in Heusler alloy thin films on spin-dependent tunnelling characteristics of $Co_2MnSi/MgO/Co_2MnSi$ and $Co_2MnGe/MgO/Co_2MnGe$ magnetic tunnel junctions. J. Phys.: Condens. Matter **22**(16), 164212 (2010)
29. S. Chadov, G.H. Fecher, C. Felser, J. Minar, J. Braun, H. Ebert, Electron correlations in Co_2Si Heusler compounds. J. Phys. D: Appl. Phys. **42**(8), 084002 (2009)
30. J.S. Moodera, L.R. Kinder, T.M. Wong, R. Meservey, Large magnetoresistance at room temperature in ferromagnetic thin film tunnel junctions. Phys. Rev. Lett. **74**, 3273–3276 (1995)
31. E.Y. Tsymbal, O.N. Mryasov, P.R. LeClair, Spin-dependent tunnelling in magnetic tunnel junctions. J. Phys.: Condens. Matter **15**(4), R109 (2003)
32. J.P. Velev, P.A. Dowben, E.Y. Tsymbal, S.J. Jenkins, A.N. Caruso, Interface effects in spin-polarized metal/insulator layered structures. Surf. Sci. Rep. **63**(9), 400–425 (2008)
33. M. Sicot, S. Andrieu, P. Turban, Y. Fagot-Revurat, H. Cercellier, A. Tagliaferri, C. De Nadai, N.B. Brookes, F. Bertran, F. Fortuna, Polarization of Fe(001) covered by MgO analyzed by spin-resolved x-ray photoemission spectroscopy. Phys. Rev. B **68**, 184406 (2003)
34. F. Matthes, L.-N. Tong, C.M. Schneider, Spin-polarized photoemission spectroscopy of the MgO/Fe interface on GaAs(100). J. Appl. Phys. **95**(11), 7240–7242 (2004)
35. M. Müller, F. Matthes, C.M. Schneider, Photoemission study of the Fe(001)/MgO interface for varying oxidation conditions of magnesium oxide. J. Appl. Phys. **101**(9), 09G519 (2007)
36. L.-N. Tong, C.-L. Deng, F. Matthes, M. Müller, C.M. Schneider, C.-G. Lee, Influence of MgO overlayers on the electronic states of bct Co(001) thin films grown on bcc Fe(001)/GaAs(001). Phys. Rev. B **73**, 214401 (2006)
37. F. Bonell, T. Hauet, S. Andrieu, F. Bertran, P. Le Fèvre, L. Calmels, A. Tejeda, F. Montaigne, B. Warot-Fonrose, B. Belhadji, A. Nicolaou, A. Taleb-Ibrahimi, Spin-polarized electron tunneling in bcc FeCo/MgO/FeCo(001) magnetic tunnel junctions. Phys. Rev. Lett. **108**, 176602 (2012)
38. Y.S. Dedkov, M. Fonin, U. Rüdiger, G. Güntherodt, Spin-resolved photoelectron spectroscopy of the MgO/Fe(110) system. Appl. Phys. A: Mater. Sci. Process. **82**, 489–493 (2006). doi:10.1007/s00339-005-3447-2
39. L. Plucinski, Y. Zhao, B. Sinkovic, E. Vescovo, MgO/Fe(100) interface: a study of the electronic structure. Phys. Rev. B **75**, 214411 (2007)
40. R. Fetzer, J.-P. Wüstenberg, T. Taira, T. Uemura, M. Yamamoto, M. Aeschlimann, M. Cinchetti, Structural, chemical, and electronic properties of the $Co_2MnSi(001)/MgO$ interface. Phys. Rev. B **87**, 184418 (2013)
41. J. Mathon, A. Umerski, Theory of tunneling magnetoresistance of an epitaxial Fe/MgO/Fe(001) junction. Phys. Rev. B **63**, 220403 (2001)
42. T. Ishikawa, S. Hakamata, K. Matsuda, T. Uemura, M. Yamamoto, Fabrication of fully epitaxial $Co_2MnSi/MgO/Co_2MnSi$ magnetic tunnel junctions. J. Appl. Phys. **103**, 07A919 (2008)
43. Y. Sakuraba, M. Hattori, M. Oogane, Y. Ando, H. Kato, A. Sakuma, T. Miyazaki, H. Kubota, Giant tunneling magnetoresistance in $Co_2MnSi/Al\text{-}O/Co_2MnSi$ magnetic tunnel junctions. Appl. Phys. Lett. **88**(19), 192508 (2006)
44. Y. Sakuraba, K. Takanashi, Y. Kota, T. Kubota, M. Oogane, A. Sakuma, Y. Ando, Evidence of Fermi level control in a half-metallic Heusler compound Co_2MnSi by Al-doping: comparison of measurements with first-principles calculations. Phys. Rev. B **81**(14), 144422 (2010)
45. S. Valeri, S. Altieri, A. di Bona, C. Giovanardi, T.S. Moia, Structural study of thin MgO layers on Ag(001) prepared by either MBE or sputter deposition. Thin Solid Films **400**, 16–21 (2001)
46. L. Davis, N. MacDonald, P. Palmberg, G. Riach, *Handbook of Auger Electron Spectroscopy: A Reference Book of Standard Data for Identification and Interpretation of Auger Electron Spectroscopy Data.* (Physical Electronics, Chanhassen, 1987)
47. V. Henrich, The surfaces of metal oxides. Rep. Prog. Phys. **48**, 1481–1541 (1985)
48. G. Geneste, J. Morillo, F. Finocchi, Adsorption and diffusion of Mg, O, and O_2 on the MgO(001) flat surface. J. Chem. Phys. **122**(17), 174707 (2005)

49. D. Ochs, M. Brause, B. Braun, W. Maus-Friedrichs, V. Kempter, CO_2 chemisorption at Mg and MgO surfaces: a study with MIES and UPS (He I). Surf. Sci. **397**, 101–107 (1998)
50. G. Pacchioni, J.M. Ricart, F. Illas, Ab initio cluster model calculations on the chemisorption of CO_2 and SO_2 probe molecules on mgo and cao (100) surfaces. a theoretical measure of oxide basicity. J. Am. Chem. Soc. **116**(22), 10152–10158 (1994)
51. G. Pacchioni, Physisorbed and chemisorbed CO_2 at surface and step sites of the MgO(100) surface. Surf. Sci. **281**, 207–219 (1993)
52. G. Pacchioni, T. Minerva, P.S. Bagus, Chemisorption of CO on defect sites of MgO. Surf. Sci. **275**(3), 450–458 (1992)
53. L.N. Kantorovich, M.J. Gillan, J.A. White, Adsorption of atomic oxygen on the MgO (100) surface. J. Chem. Soc. Faraday Trans. **92**, 2075–2080 (1996)
54. L.N. Kantorovich, M.J. Gillan, Adsorption of atomic and molecular oxygen on the MgO (001) surface. Surf. Sci. **374**, 373–386 (1997)
55. T. Miyajima, M. Oogane, Y. Kotaka, T. Yamazaki, M. Tsukada, Y. Kataoka, H. Naganuma, Y. Ando, Direct observation of atomic ordering and interface structure in Co_2MnSi magnetic tunnel junctions by high-angle annular dark-field scanning transmission electron microscopy. Appl. Phys. Express 2(9):093001 (2009)
56. D.M. Roessler, W.C. Walker, Electronic spectrum and ultraviolet optical properties of crystalline MgO. Phys. Rev. **159**, 733–738 (1967)
57. R. Pandey, J.E. Jaffe, A.B. Kunz, *Ab initio* band-structure calculations for alkaline-earth oxides and sulfides. Phys. Rev. B **43**, 9228–9237 (1991)
58. Y. Miura, H. Uchida, Y. Oba, K. Abe, M. Shirai, Half-metallic interface and coherent tunneling in Co_2YZ/MgO/Co_2YZ (YZ = MnSi, CrAl) magnetic tunnel junctions: A first-principles study. Phys. Rev. B **78**, 064416 (2008)
59. B. Hülsen, M. Scheffler, P. Kratzer, Thermodynamics of the Heusler alloy $Co_{2-x}Mn_{1+x}Si$: a combined density functional theory and cluster expansion study. Phys. Rev. B **79**, 094407 (2009)
60. S. Yuasa, A. Fukushima, H. Kubota, Y. Suzuki, K. Ando, Giant tunneling magnetoresistance up to 410% at room temperature in fully epitaxial Co/MgO/Co magnetic tunnel junctions with bcc Co(001) electrodes. Appl. Phys. Lett. **89**(4), 042505 (2006)
61. S.S.P. Parkin, C. Kaiser, A. Panchula, P.M. Rice, B. Hughes, M. Samant, S.-H. Yang, Giant tunnelling magnetoresistance at room temperature with MgO (100) tunnel barriers. Nat. Mat. **3**(12), 862–867 (2004)
62. S. Tsunegi, Y. Sakuraba, M. Oogane, K. Takanashi, Y. Ando, Large tunnel magnetoresistance in magnetic tunnel junctions using a Co_2MnSi Heusler alloy electrode and a MgO barrier. Appl. Phys. Lett. **93**(11), 112506 (2008)
63. P.A. Dowben, R. Skomski, Finite-temperature spin polarization in half-metallic ferromagnets. J. Appl. Phys. **93**(10), 7948–7950 (2003)
64. L. Chioncel, Y. Sakuraba, E. Arrigoni, M.I. Katsnelson, M. Oogane, Y. Ando, T. Miyazaki, E. Burzo, A.I. Lichtenstein, Nonquasiparticle states in Co_2MnSi evidenced through magnetic tunnel junction spectroscopy measurements. Phys. Rev. Lett. **100**, 086402 (2008)
65. M. Ležaić, P. Mavropoulos, J. Enkovaara, G. Bihlmayer, S. Blügel, Thermal collapse of spin polarization in half-metallic ferromagnets. Phys. Rev. Lett. **97**, 026404 (2006)
66. Y. Miura, K. Abe, M. Shirai, Effects of interfacial noncollinear magnetic structures on spin-dependent conductance in Co_2MnSi/MgO/Co_2MnSi magnetic tunnel junctions: A first-principles study. Phys. Rev. B **83**, 214411 (2011)
67. R. Fetzer, B. Stadtmüller, Y. Ohdaira, H. Naganuma, M. Oogane, Y. Ando, T. Taira, T. Uemura, M. Yamamoto, M. Aeschlimann, M. Cinchetti, Probing the electronic and spintronic properties of buried interfaces by extremely low energy photoemission spectroscopy. Sci. Rep. **5**, 8537 (2015)
68. S. Tsunegi, Y. Sakuraba, M. Oogane, H. Naganuma, K. Takanashi, Y. Ando, Enhancement in tunnel magnetoresistance effect by inserting CoFeB to the tunneling barrier interface in Co_2MnSi/MgO/CoFe magnetic tunnel junctions. Appl. Phys. Lett. **94**(25), 252503 (2009)

69. M. Jourdan, J. Minar, J. Braun, A. Kronenberg, S. Chadov, B. Balke, A. Gloskovskii, M. Kolbe, H.J. Elmers, G. Schönhense, H. Ebert, C. Felser, M. Kläui, Direct observation of half-metallicity in the Heusler compound Co_2MnSi. Nat. Comm. **5**, 3974 (2014)
70. N.H. Long, M. Ogura, H. Akai, Effects of spin-wave excitations in half-metallic materials. Phys. Rev. B **85**, 224437 (2012)
71. A. Sakuma, Y. Toga, H. Tsuchiura, Theoretical study on the stability of magnetic structures in the surface and interfaces of Heusler alloys, Co_2MnAl and Co_2MnSi. J. Appl. Phys. **105**(7), 07C910 (2009)
72. K. Miyamoto, A. Kimura, Y. Miura, M. Shirai, M. Ye, Y. Cui, K. Shimada, H. Namatame, M. Taniguchi, Y. Takeda, Y. Saitoh, E. Ikenaga, S. Ueda, K. Kobayashi, T. Kanomata, Absence of temperature dependence of the valence-band spectrum of Co_2MnSi. Phys. Rev. B, 79(10), 2009
73. S. Tsunegi, Y. Sakuraba, K. Amemiya, M. Sakamaki, E. Ozawa, A. Sakuma, K. Takanashi, Y. Ando, Observation of magnetic moments at the interface region in magnetic tunnel junctions using depth-resolved x-ray magnetic circular dichroism. Phys. Rev. B **85**, 180408 (2012)

Chapter 4
Structural Order in Heusler Compounds

S. Wurmehl and M. Wójcik

Abstract Heusler compounds exhibit different electronic ground states and functionalities, making them attractive for studies of their fundamental properties and for their technological exploitation. A key tool in the rational design of this class of materials is the precise control of the relationships between structure and physical properties since Heusler compounds are known to order in various structure types. This review describes the different types of order in Heusler compounds, identifies methods to analyze the local crystallographic and magnetic order as well as illustrates the effect of order on the properties of Heusler compounds by discussing recent examples of studies addressing order in Heusler compounds.

4.1 Heusler Compounds—A Versatile Class of Functional Materials

Heusler compounds are a subject of research for more than a century [1]. Remarkably, the Heusler compound that was discovered as the first one, Cu_2MnAl, [1], is a soft ferromagnet made from non ferromagnetic elements. Heusler compounds are, per definition, ternary intermetallic X_2YZ-compounds or XYZ-compounds, with X and Y typically being transition metals and Z being a main group element. So far, more than 1000 Heusler compounds have been reported [2–4]. Heusler compounds show a plethora of physical phenomena, e.g., spin density wave ordering (Mn_3Si [5]), heavy fermion behaviour (Cu_2CeIn [6]), superconductivity (Pd_2RSn

S. Wurmehl (✉)
Leibniz Institute for Solid State and Materials Research Dresden IFW, 01069 Dresden, Germany
e-mail: s.wurmehl@ifw-dresden.de

S. Wurmehl
Institut für Festkörperphysik, Technische Universität Dresden,
01062 Dresden, Germany

M. Wójcik
Institute of Physics, Polish Academy of Sciences, Aleja Lotników 32/46,
02-668 Warszawa, Poland
e-mail: wojci@ifpan.edu.pl

C. Felser and A. Hirohata (eds.), *Heusler Alloys*,
Springer Series in Materials Science 222, DOI 10.1007/978-3-319-21449-8_4

[7]), thermoelectricity (Fe_2VAl [8]), topological insulating properties (e.g., GdPtBi [9]), first-order phase transitions (e.g., Ni_2MnGa [10]), metamagnetic transitions (Ni-Mn-Co-In [11]), high magnetic anisotropy (Mn-Ga [12, 13]), high magnetic moments and Curie temperature (Co_2FeSi [14, 15]), as well as high spin polarization (e.g., NiMnSb [16], Co_2MnSi [17]). This diversity renders the class of Heusler compounds attractive materials to explore fundamental aspects of matter but also rises interest for the technological exploitation of their functionalities, e.g., in spintronic or ferroic cooling devices, in sensors and actuators or as permanent magnets.

Both theoretical (see, e.g., [18–24]) and experimental works on Heusler compounds agree, that their structural order has tremendous impact on their properties. This strong relation between structural order and functional properties calls for a thorough structural characterization whenever their physical properties are discussed. In the following, we will illustrate the different types of structural order in Heusler compounds.

4.2 Types of Structural Order in Heusler Compounds

The ordered X_2YZ Heusler compounds crystallize in the $L2_1$ structure (Strukturbericht designation, see Fig. 4.1) with spacegroup $Fm\overline{3}m$. The inorganic structure database (ICSD) notation is Cu_2MnAl type structure. The X element, typically the most electronegative transition metal, occupies the Wykhoff position 8c ($\frac{1}{4}\frac{1}{4}\frac{1}{4}$), Y, the less electronegative transition metal, is on 4b ($\frac{1}{2}\frac{1}{2}\frac{1}{2}$). In some cases, the Y ele-

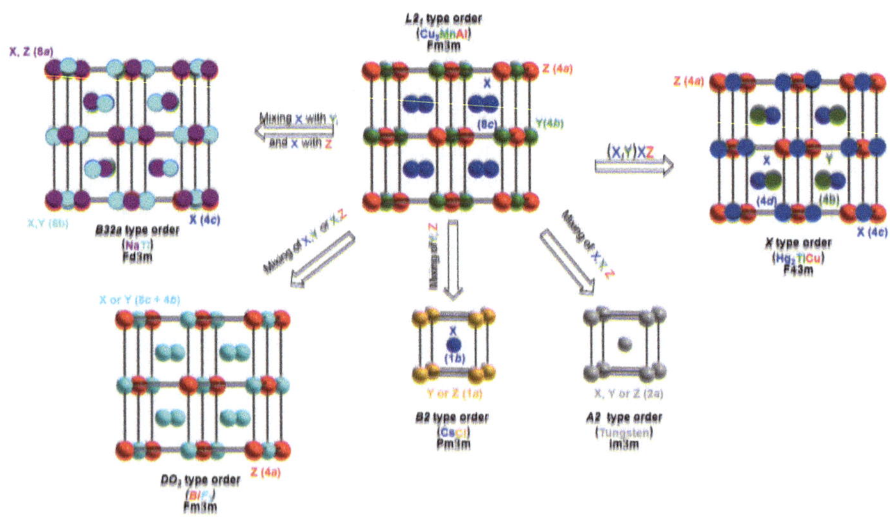

Fig. 4.1 Frequently occurring structure types for X_2YZ Heusler compounds. For details, see text

ment may also be a rare earth element or an alkaline earth metal. The Z atom, a main group element, is on 4*a* (0 0 0).

There are two different types of nomenclature, either based on the electronegativity of the constituents, in this case, the most electropositive element is quoted first or, as in most cases, the X element is named first. The Heusler structure may be visualized as a superstructure of the CsCl lattice, in which the lattice parameter of the CsCl lattice is doubled in all three dimensions. This leads to a cell consisting of eight cubes with one atom in each center. The 16 atoms in this supercell may be divided in four groups consisting of four atoms. The atoms in the center of each cube are the X atoms, while the Y and Z atoms occupy the corners of the eight cubes. Each atom is face centered cubic ordered, leading to four interpenetrating face centered cubic lattices. In case of only half filled centers of the eight cubes, the C_{1b}-Heusler structure with general formula XYZ is obtained ($C1_b$ Strukturbericht designation, space group $F\overline{4}3m$). This structure is sometimes referred to as half-Heusler structure.

Various variants of the $L2_1$ structure can be formulated, if the X, Y, and/or Z atoms are intermixed on the respective crystallographic positions, leading to different (local) symmetries and structure (also see Fig. 4.4) types [25]. In the following, we will describe the most common types of structures:

- DO_3-type structure
  Space group $Fm\overline{3}m$ is conserved but if X and Y or X and Z atoms are mixed on their crystallographic positions, the DO_3-type structure is obtained; the corresponding ICSD notation is BiF_3 structure type.
- $B2$-type structure
  If the Y and Z atoms are randomly intermixed on their crystallographic positions, the $B2$-type structure is obtained in which the Y and Z sites become equivalent. This structure may also be described based on a CsCl lattice, and as a result of this intermixing, a CsCl lattice with X on the center of the cube randomly surrounded by Y and Z atoms is obtained (Fig. 4.1). The symmetry is reduced and the resulting space group is ($Pm\overline{3}m$), ICDS notation is CsCl-type structure. All X atoms are on the 1*b* Wykhoff position and Z and Y atoms are randomly distributed on the 1*a* position.
- $A2$-type structure
  Complete disorder, viz., complete random intermixing on the Wykhoff 2*a* position, in the X_2YZ Heusler compounds between all sites results in the $A2$ type structure with reduced symmetry $Im\overline{3}m$. The X, Y and Z sites become equivalent leading to a body-centered cubic lattice, also known as the tungsten (W) structure-type according to ICSD notation.
- $B32a$-type structure
  This type of structure is formed if the X atoms on the two fcc lattices mix, but with different atoms, viz. if the X atoms of one fcc sublattice are randomly distributed with the Y atoms while the X atoms in the second fcc lattice are intermixed with the Z atoms (space group $Fd\overline{3}m$, ICSD notation NaTl). However, this type of order is very uncommon to be experimentally realized.

- X-type structure
  The X-type structure is also sometimes referred to as inverse Heusler structure (space group $F\overline{4}3m$, ICSD notation is $CuHg_2Ti$ structure). This type of structure is formed if the atomic number of the Y atom is higher than the number of the X atom. Half of X and all Z atoms form rock-salt lattice with the remaining X and Y atoms occupying the tetrahedral interstices, thereby forming a rock-salt lattice themselves. The difference to the $L2_1$ structure is highlighted by referring to the X structure as (X,Y)XZ compound.
- Y-type structure in quaternary Heusler compounds
  The Y-type structure is formed if each of the four interpenetrating fcc lattices of the former $L2_1$ type structure is occupied by a different atom. This type of structure is known as LiMgPdSn with space group $F\overline{4}3m$. Y-type quaternary compounds are formed e.g., in CoFeMnZ (Z = Al, Ga, Si, Ge) [26, 27].
- Tetragonally distorted Heusler compounds
  A tetragonal distortion is typically established for Mn_2YZ Heusler compounds by stretching the c axis in which the two Mn atoms are prone to two different local symmetries, cubic and tetragonal [4]. The resulting space group is $I4/mmm$ and the ICSD notation is Mn_2NiSn; the tetragonal unit cell is visualized by rotation of the cell axis by 45°. Here, the X atoms occupy the $4d$ Wyckoff position $(0\frac{1}{2}\frac{1}{4})$, Y is on $2b$ $(00\frac{1}{2})$ and Z is on $2a$ (000). A tetragonal distortion may also be observed for the inverse structure. Here one Mn is on $2b$, while the second Mn and the Y atom share the $4d$ position. The Z atom is on the $2a$ position, similar to the regular structures with tetragonal distortion. The most prominent member of this class of Heusler compounds is Mn_3Ga [12, 28]. If the Y atom in Mn_2YZ Heusler compounds is Fe, the inverse tetragonally distorted structure is formed with space group $I\overline{4}2\,m$.

Diffraction techniques such as x-ray and neutron diffraction are the basis for any structural characterization although not in all cases sufficient to clarify all structural issues as outlined in the following. The most pronounced differences in the powder pattern for Heusler compounds ordered in different structure types manifest in the intensity of the (111) and (200) superstructure reflections of the $L2_1$ type structure since the relative intensity of these superstructure reflections is increased, decreased or even fully suppressed compared to the $L2_1$ type ordering depending on the type of order (for details see [4]). Anomalous XRD using synchrotron radiation has some advantages and allows a more reliable structural analysis, however, is obviously not suitable for a fast characterization on a daily basis.

For the analysis of the type of order several aspects need to be considered:

- the intensity of the superstructure reflections may be small or even absent even in $L2_1$ ordered materials if the scattering coefficients of the constituents are very similar, as, e.g., in Co_2FeGa or Co_2FeGe [45]
- if different types of order are present concomitantly
- if the order of thin films is investigated where the quasi 2-dimensional character of the film samples leads to systematic enhancement or suppression of certain lattice reflections with certain Miller indices due to texture.

Similar hurdles apply to the structural characterization using neutron diffraction. As an example, it is not possible to distinguish between $L2_1$ and $B2$-type structures in Co-Cr-Al Heusler compounds due to the nearly equal scattering length of Cr and Al for neutrons.

Ordering in each of the different structure types as described above will lead to the formation of different local environments for each atom. In the next section, we will shortly introduce two methods which probe local order in Heusler compounds.

4.3 Methods for the Investigation of (local) Order in Heusler Compounds

4.3.1 Nuclear Magnetic Resonance Spectroscopy (NMR)

NMR is based on the interaction of the spin of a nucleus with external magnetic fields, neighbouring nuclei, the surrounding electrons of the nucleus itself and also the electrons of neighbouring atoms [30–32]. In contrast to paramagnetic and diamagnetic materials, where the externally applied magnetic field is the determining factor for the effective field, the hyperfine field makes the dominant contribution in ferromagnetic (spin polarized) materials. The hyperfine field refers to the (hyperfine) interaction of the nuclear magnetic moment with magnetic fields originating in the spin and orbital currents of the surrounding electrons. The hyperfine field is a sensitive probe of the local environments of the active nuclei due to the local contributions to the hyperfine field, specifically the transferred fields add to the hyperfine field which depends on the nearest and next nearest neighbour atoms and their magnetic moments. This renders NMR an ideal method to study structural properties of bulk samples as well as thin films of Heusler compounds on a very local scale. The NMR setup to measure ferromagnets is different from the common and conventional setup [31, 32] and requires different data analysis approaches and data correction, e.g., for the enhancement factor [31, 32].

4.3.2 Mößbauer Spectroscopy

Mößbauer spectroscopy is based on the resonant and recoil-free emission and absorption of γ-quants by atomic nuclei bound in solid matter: γ-quants which are emitted from a nucleus while relaxing back to a lower energy state can excite another nucleus of the same isotope if both nuclei are part of a crystal lattice which absorb the recoil energy. The mutual combination of emission and absorption in a given sample allows to probe very subtle energy differences in the sample compared to the well-known energy-levels of the source of γ-quants. These subtle changes in the energy levels typically originate from differences in the chemical environment of the probed material and, hence, Mößbauer spectroscopy is a valuable tool for studies of the local crystallographic, electronic and magnetic structure. In an experiment, the energy shifts arsing from the different chemical environments of the source of γ-radiation and the sample under study is compensated using the Doppler effect. The most common type of Mößbauer spectroscopy involves ^{57}Fe nuclei [33]. Three types of nuclear interaction manifest in Mößbauer spectroscopy, namely the isomer shift (or chemical shift), which allows to monitor changes in the electron density at the nucleus, quadrupole splitting, arising from changes in the electric field gradient in non-cubic environments and hyperfine splitting (or Zeeman splitting) originating from the presence of a magnetic field [33].

In the following, we will discuss experimental studies carried out mostly by the above sketched methods and addressing order-disorder phenomena in Heusler compounds sorted by their composition and structure.

4.4 Examples of Structural Order in Heusler Compounds

4.4.1 X_2YZ Heusler Compounds

4.4.1.1 Co_2MnSi

Co_2MnSi bulk single crystals were reported to exhibit $L2_1$-type order [34], while polycrystalline bulk samples were by the same group reported to have extensive chemical disordering, namely mixing of Co and Mn atoms. The order in those samples was studied combining XRD, neutron diffraction and extended x-ray-absorption fine-structure measurements and was, surprisingly, found to be rather independent of annealing and quenching conditions [35].

The technological exploitation of high spin polarization in Heusler compounds in spintronics typically requires thin films. The order in thin films strongly depends on the deposition temperature, but, as already addressed above, the proper differentiation of Co and Mn site disorder by XRD in thin films is challenging. A more indirect investigation of the degree of order by analysis of the residual resistivity ratios were performed by Ravel et al. [34] underlining the importance of appropriate annealing conditions to establish high structural order for the performance in spintronic devices. Along that line, there is a clear evolution of current-perpendicular-to-plane-giant-magnetoresistance (MR) upon annealing in Co_2MnSi Heusler films [36]. A first attempt to explain this evolution of MR as function of annealing was based on atomic structure of the films employing XRD [36]. A more detailed insight was gained by studying the impact of annealing by ^{59}Co NMR using the sensitivity of ^{59}Co and ^{55}Mn NMR to both magnetic and structural order of such films: In Co_2MnSi films which were annealed below 550 °C, no long-range $L2_1$-order is observed, while annealing above 550 °C leads to the formation of the ideal $L2_1$ Heusler structure, however, with a distinct degree of off-stoichiometry. Specifically, ^{59}Co NMR allows to detect about

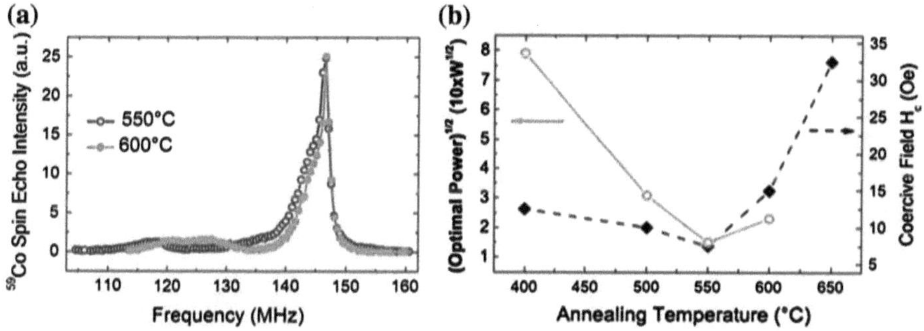

Fig. 4.2 ^{59}Co NMR spectra for Co_2MnSi films annealed at 550 and 600 °C, respectively. The low frequency tail originates in the off-stoichiometry. The line width is smaller for the film annealed at higher temperature indicating a higher structural order (**a**) Square root of the optimal power (proportional to restoring field) for the main line of the ^{59}Co spectrum in different films in comparison with the corresponding coercive fields (Hc) measured by magnetometry (**b**)

10 % of Mn sites being randomly occupied by excess Co atoms. The sudden decrease in MR for films annealed above 550 °C is not related the structural evolution since the degree of (local) order improves upon annealing, as evident from the decreasing NMR line width comparing the film annealed at 550 °C with the one annealed at 600 °C (Fig. 4.2a). Therefore, annealing these films at high temperature (up to 600 °C at least), seems to increasingly improve the overall structural order. Further evidence from NMR restoring field measurements hints that interdiffusion may account for the drop in MR observed for samples annealed at and above 600 °C (Fig. 4.2b). These results show that optimizing films for spintronics involves the identification of the best annealing temperature, high enough for long-range $L2_1$-type order to emerge, but simultaneously low enough to avoid interfacial interdiffusion or the emergence of unfavorable secondary phases and, hence, maintaining smooth interfaces [37].

4.4.1.2 Co_2FeAl

The Heusler compound Co_2FeAl is reported to occur in various structures, ranging from the completely ordered $L2_1$ to the completely disordered $A2$ structure type. The structure has been in detail investigated both in thin film and in bulk samples:

Inomata et al. [38] report thin films of Co_2FeAl either in the $B2$ type structure or in the $A2$ type structure with the local structure being probed by ^{59}Co NMR. Figure 4.3 illustrates the splitting of the ^{59}Co NMR line due to $B2$ type order (compare [38–40]). It turned out, that the type of order in Co_2FeAl films also depends on the substrate temperature. Films annealed below 473 K exhibit $A2$ type structure, while annealing above 473 K establishes $B2$ type order concomitantly with a decrease of the tunneling-magnetoresistance. This decrease is probably caused by an increase of the surface roughness at the interface related to temperature-activated interdiffusion [38].

A dependence of the structure type on an appropriate annealing procedure is also observed for bulk samples: As-cast Co_2FeAl bulk samples consist of a $B2$ type

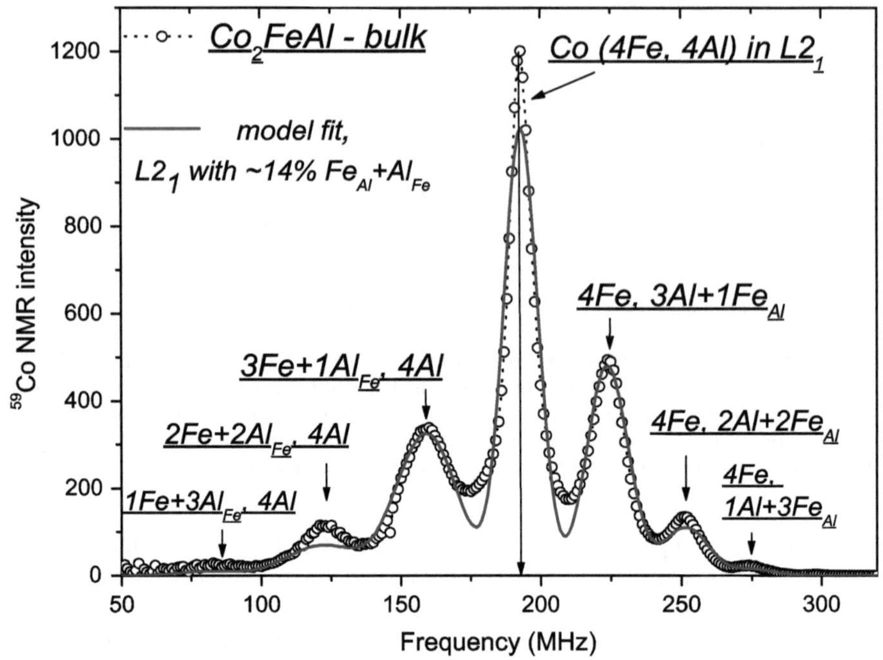

Fig. 4.3 ^{59}Co NMR of bulk Co_2FeAl. The different NMR lines correspond to the different local environments with different Fe and Al next neighbors in the first shell of Co due to $B2$ type order

structure [39] with $L2_1$ type structure contributions of about 10 %. Remarkably, these $L2_1$ contributions are not apparent using conventional x-ray, or neutron diffraction, or EXAFS [39], indicating that the $L2_1$ structure contributions of a few percent are eventually too small to be observed with these methods. The strength of the NMR technique becomes also obvious by having a closer look to the NMR lines: The ^{59}Co NMR lines for both bulk and thin films show the NMR line splitting originating from the mixing of Al and Fe in the first shell. In addition, the NMR main lines of the bulk samples show an additional modulation of the main lines, refection a higher degree of order in the bulk samples than in the thin films [39].

Bulk samples annealed at 1300 K order in a mixture of the $B2$ and $A2$ structures, in agreement with previous structural studies [39]. The pronounced difference of the structure of samples prepared under different conditions underlines the importance of finding optimal preparation conditions relating both structure and properties.

Sometimes, one finds materials ordered in a Heusler-type structure, where one is not expecting them. A systematic NMR study of the changes in the local atomic environments of Co in Co-FeAl alloys as a function of Al content revealed that a Co_2FeAl Heusler type structure is formed on a local scale [41]. The observed formation of a highly spin-polarized Heusler compound may explain the improved magnetotransport properties in Co-FeAl based current-perpendicular-to-the-plane spin-valves.

4.4.1.3 Co_2FeGa

Co_2FeGa is reported to order in the $L2_1$ type structure [42]. This high degree of order is established by appropriate annealing [42]: Mößbauer spectroscopy resolves two magnetic sextets with one hyperfine field but two different isomer shifts which merge after annealing, revealing that Fe has only one local environment after annealing. However, the ^{59}Co NMR line of the very same samples reveals a small contribution of about 7 % from $B2$ type order.

The phase equilibria and stability in the thermodynamic Co-Fe-Ga phase diagram were also subject of a study combining micro-probe analysis, differential scanning calorimetry and magnetometry [43]. According to this study, there is a transition from $B2$ to $L2_1$ order at 825 °C [43, 44]. Structure determination using extended x-ray absorption fine structure spectroscopy agrees on $L2_1$ ordering in Co_2FeGa [45]. Co_2FeGa has also been realized as nanoparticles or nanocrystals with high degree of structural order [46, 47].

4.4.1.4 Co_2FeSi

Co_2FeSi is predicted to be a half-metallic ferromagnet with an extraordinary high magnetic moment and Curie temperature [14, 15]. The half-metallic character of Co_2FeSi was recently confirmed by transport measurements as the resistivity data reveal a textbook-like exponential suppression of the electron-magnon scattering rate with decreasing temperature which provides strong evidence that Co_2FeSi indeed possesses perfect spin polarization at low temperature [48]. Annealed Co_2FeSi poly- and single crystalline bulk samples are among the Heusler materials which show the highest degree of order, manifesting, e.g., as a narrow ^{59}Co NMR line at 139 MHz [49, 50]. The observation of one line only is consistent with the pure $L2_1$ type ordering since in this structure type all Co nuclei have only one local environment and hence only one NMR line is expected. The exceptional high degree of order seen by NMR is in agreement with (i) magnetostructrual investigations using Mößbauer spectroscopy where one sextett only, no paramagnetic line and no quadrupole splitting was observed (ii) investigations of the short range order by extended x-ray absorption fine structure spectroscopy (EXAFS) (iii) as well as site-specific (x-ray magnetic circular dichroism) and macroscopic measurements of the magnetization [14, 15, 45]. The evolution of order in a $Fe_{3-x}Co_xFeSi$ substitution series was studied by NMR by Niculescu et al. [51, 52].

Co_2FeSi films typically have an off-stoichiometric composition with Fe excess caused by selective sputtering [53]: Using a local probe, as NMR, one observes a rather broad main resonance line corresponding to ^{59}Co nuclei in the $L2_1$ environment but also additional resonance lines at the high frequency site of the main line. The additional resonance lines correspond to ^{59}Co with more Fe next neighbors than expected for the $L2_1$ type ordering due to Fe excess. This off-stoichiometry may be difficult to be detected by standard bulk characterization methods such as XRD and scanning electron microscopy with compositional analysis.

The deviation from the 2:1:1 Heusler stoichiometry may be one explanation for the previously observed low TMR values corresponding to a lower spin polarization [54], in spite of the theoretical prediction of half-metallic properties for ideally ordered Co_2FeSi. Moreover, the off-stoichiometry may also be the origin of the smaller element-specific and macroscopic magnetic moments in thin film samples compared to the stoichiometric and highly ordered bulk compound [38, 55–57] since the Fe excess atoms occupy Si positions and hence, may affect the magnetic coupling and band structure significantly.

4.4.1.5 Co_2NbSn

Co_2NbSn undergoes a structural transition from the cubic Heusler phase ($Fm\overline{3}m$) into an orthorhombic low-temperature lattice ($Pmma$) at 235 K, similar to the Heusler compound Ni_2MnGa, but with the Curie temperature of 116 K below the structural transition [58]. The results stemming from combination of methods including neutron diffraction, XRD, Mößbauer and NMR spectroscopy were interpreted to confirm that the structure of Co_2NbSn is highly disordered [58].

4.4.1.6 Co_2TiSn

The Heusler compound Co_2TiSn is a very good example for the need of thorough structural characterization [59]. A mutual combination of XRD, Mößbauer and NMR spectroscopy was necessary to shed light on the origin of ambiguous results dealing with polycrystalline Co_2TiSn. Based on the data analysis by Kandpal et al., Co_2TiSn is only partially ordered with about 10 % of Co and Ti atoms mixing. While this type of order was not clearly resolved in XRD owing to the similar scattering coefficients of Co and Ti for x-rays, the different local hyperfine fields related to the mixing of Co and Ti were clearly revealed by Mößbauer and NMR spectroscopy [59].

4.4.1.7 Ni_2MnGa

The development of local structure and order in shape memory alloys upon the structural, martensitic transition in ferromagnetic shape memory alloys is still not fully understood. The paradigm ferromagnetic shape memory Heusler material is Ni_2MnGa [10], which was already subject of an NMR study [60] and the change in local crystallographic and magnetic structure of polycrystalline Ni_2MnGa was revisited [61] combining zero-field static and dynamic ^{55}Mn NMR experiments, X-ray powder diffraction and magnetization experiments. In particular the NMR spectra allow to monitor the sequence of structural phase transitions in this compound, from the high-T cubic austenitic structural phase to the pre-martensitic and finally down to the low-T martensitic phase. The NMR experiments also give access to the local magnetic anisotropy for each separate magnetic environment via a detailed investigation of the rf-enhancement factor. This study allowed to track down signals coming from austenitic and martensitic components and to follow their evolution with temperature. Powderization of bulk samples maintains a significant portion of martensitic traces inside the high-T austenitic region due to stress and strain. These traces can be subsequently removed by annealing.

4.4.2 Heusler Compopunds with Inverse Structure

4.4.2.1 $Co_{2-x}Fe_{1+x}Si$

While Co_2FeSi with $x = 0$ orders in $L2_1$ type structure, Fe_2CoSi is supposed to order in the inverse (X) Heusler structure. Both structure types are difficult to distinguish by XRD. A study by Jung et al. used Mößbauer spectroscopy to shed light on the type of order in a $Co_{2-x}Fe_{1+x}Si$ substitution series [62]. The main differences between the two structure types, namely $L2_1$ type and X type is the local environment of the Fe atoms. In the $L2_1$ type structure, only one Fe position with cubic symmetry exists, while there are two Fe positions in the X structure but with tetragonal symmetry. Fe-I has the same local environment as Fe in the $L2_1$ type structure (4Co+4Fe), Fe-II has 4Si+4Fe nearest neighbors and formally occupies a Co position of the $L2_1$ type structure. Hence, the observation of two magnetic sextets is expected for the series with $x \neq 0$. This evolution of the relative intensities of the two magnetic sextets as function of Fe content is shown in Fig. 4.4. The relative intensity of sextet-II with smaller magnetic splitting increases with increasing Fe content, since more and more Fe atoms occupy the former Co positions. At the same time, the relative intensity of sextet-I decreases with increasing Fe content, as the total number of Fe atoms increases. The changes in local environments affect the isomer shift and hyperfine fields in different ways [62] allowing to undergo a detailed structural analysis of the $Co_{2-x}Fe_{1+x}Si$ series [62].

4.4.2.2 Mn_2CoSn

The inverse Heusler structure ($CuHg_2Ti$, Xa type) was proposed for the Mn_2CoSn compound based on the theoretical band structure calculations and the refinement of an experimental powder XRD pattern [63]. However, as has been discussed above, powder XRD alone due to similar scattering factors of Mn and Co cannot distinguish between perfect Xa structure and its disordered variants involving mixing between Mn and Co atoms [64]. Two variants of Xa structure modification can be induced by disorder consisting of a mixing between Mn and Co atoms. In the first case, with DO_3 type structure, Mn and Co atoms share the $8c$ and $4b$ positions while the main group element (Sn) occupies $4a$ position. The second type of disorder leading to $L2_1b$ structure consists of a random occupation of $8c$ positions by equally populated Mn and Co atoms. The remaining $4b$ and $4a$ positions are exclusively occupied by

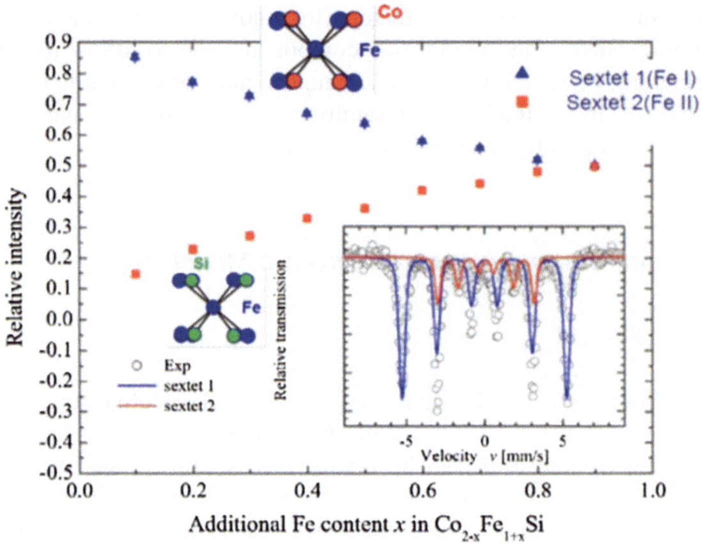

Fig. 4.4 Adapted with permission from [62]. ^{57}Fe Mößbauer spectroscopy of $Co_{2-x}Fe_{1+x}Si$. Shown are the relative intensities of the two hyperfine magnetic fields related to the sextets Fe-I and Fe-II along with the corresponding local environments. The *inset* displays the ^{57}Fe Mößbauer spectrum of $Co_{1.8}Fe_{1.2}Si$. Copyright 2008, American Physical Society

Mn and the main group element Sn. Differentiation between the perfect Xa order and two variants of the disordered structure has been performed by (^{117,119}Sn) Mößbauer effect and (^{55}Mn, ^{59}Co) NMR spectroscopic methods [64]. A single hyperfine field for ^{59}Co and ^{55}MnI nuclei along with the distributed hyperfine fields observed for ^{55}MnII and ^{117,119}Sn nuclei has been found from the spectra analysis. It has been concluded that these spectroscopic features can be only explained by a disorder on 8*c* positions leading to a distribution of the local environments around Mn in 4*b* and Sn in 4*a* positions and consequently $L2_{1b}$ structural order has been confirmed for Mn_2CoSn [64]. It is worth to mention that similar structural order has been concluded from a neutron diffraction experiment in related Mn_2NiGa compound classified as inverse Heusler [65]. In fact $L2_1b$ structure is a disorder type of regular Heusler $L2_1$ structure and as opposed to Xa structure it has a center of inversion characteristic for the regular Heusler compounds.

4.4.3 Pseudo-Ternary Heusler Compounds with 4 Elements

Half-metallic ferromagnets having a high stability of the minority band gap and, thus, a robust half-metallic behavior, are favored for spintronic applications [66, 67]. If the Fermi energy is close to one of the edges, the band gap may easily be affected by disorder or by quasi-particle excitations [66, 67]. Therefore, compounds with the Fermi energy located in the middle of the band gap are highly desirable [66, 67]. Such a stable gap is typically realized in pseudo-ternary Heusler compounds. From a structural point of view, such pseudo-ternary Heusler compounds are substitution series with two elements sharing one crystallographic position.

4.4.4 $Co_2Mn_{1-x}Fe_xSi$

The $L2_1$ structure of the complete substitutional series $Co_2Mn_{1-x}Fe_xSi$ is observed by XRD, EXAFS and Mössbauer spectroscopy [14, 45, 66] independent of the Fe concentration x. From a crystallographic point of view, the $L2_1$ structure type requires a random distribution of Mn and Fe at the $4b$ Wyckoff position. The investigation of such a local distribution of atoms requires a sophisticated local characterization method. ^{55}Mn spin-echo nuclear magnetic resonance (NMR) measurements revealed that the half-metallic Heusler compound $Co_2Mn_{0.5}Fe_{0.5}Si$ exhibits the $L2_1$ structure with random distribution of Mn and Fe on the $4b$ Wyckoff position [68]. The systematic analysis of the ^{55}Mn NMR data for the complete substitutional $Co_2Mn_{1-x}Fe_xSi$ series showed [69], that the spacing between adjacent NMR resonance lines is approximately constant within a compound, and decreases with increasing overall Fe concentration x. This observation supports the model that neighboring Fe in the 3rd nn shell contribute to a transferred hyperfine field. The decrease of the sub-line spacing is related to the gradual filling of different bands with different s character and different hybridization as well as to the changes in the Co, Mn and Fe magnetic moments with increasing Fe content x. The change in the density of states is also contributing to the decrease of the resonance frequencies for the Fe-rich environments, and to the shift of the whole spectrum, revealing a "bending" of the hyperfine field dependence on x, present in this compound. The observed bending may also reflect the transition from a more localized (Mn-rich compositions) to a more itinerant (Fe-rich compositions) electronic structure. Since the evolution of hyperfine fields is, in contrast to the $Co_2FeAl_{1-x}Si_x$ series [40], not linear as function of composition, a rigid band shift seems to be incompatible with the present NMR data suggesting a more complex change in the electronic structure when replacing Mn by Fe [69]. All modifications in the hyperfine field upon substitution of Mn by Fe can be uniquely assigned to an entirely random distribution of Mn and Fe limited to only one ($4b$) out of four fcc sublattices constituting the $L2_1$ Heusler structure, demonstrating that the complete substitutional series $Co_2Mn_{1-x}Fe_xSi$ is crystallographically very well defined. This very high degree of order sets stage for the observation of stable half-metallic ferromagnetism which is typically very sensitive to the structural order.

4.4.5 $Co_2Mn_{1-x}Fe_xAl$

$Co_2Mn_{1-x}Fe_xAl$ is found to order in a cubic structure type by XRD. However, a mixture of the cubic $B2$ and $L2_1$ structure types cannot be unambiguously distinguished by XRD as well as the expected random distribution of Mn and Fe on one crystallographic position of the Heusler lattice. Mößbauer spectroscopy was used to resolve the local structural contributions revealing a mixture of the $L2_1$ structure with addition of a large amount of $B2$ type structure [70, 71]. Differential scanning calorimetry measurements reveal two peaks upon cooling from the melt [71] consistent with an incongruent melting behavior and hence, a potential chemical phase segregation yielding two-phase samples by arc-melting. This chemical segregation may be consistent with the observed broad sextetts in Mößbauer spectroscopy besides contributions from the random distribution of Fe and Mn on the crystal lattice. This distribution of Fe and Mn on the lattice concomitantly with an $B2$ type order was implemented in a detailed fit model of the Mößbauer spectra taking into account the resulting different local environments [71].

4.4.6 $Co_2Cr_{1-x}Fe_xAl$

While the structural order of $Co_2Cr_{1-x}Fe_xAl$ bulk samples was controversially discussed and subject of various studies employing complementary methods [72–76], it turned out that samples prepared by arc-melting undergo a chemical segregation [74] into cubic phases making a detection of this segregation by XRD only very unlikely. However, this segregation has tremendous impact on the physical properties. Lower magnetic moments than expected from theory as well as disorder are reported for thin $Co_2Cr_{1-x}Fe_xAl$ films as well [38, 77–79]. Conversion electron Mößbauer spectroscopy measurements on $Co_2Cr_{1-x}Fe_xAl$ thin films show that annealing shifts spectral weight from the magnetic sextett to a central paramagnetic peak [80] finally recovering the bulk spectrum. This was partially interpreted to originate from an increase of Co-Fe disorder in the films induced by annealing along with interdiffusion of Fe from the buffer layer into the film [80].

Due to the phase segregation upon solidification from the melt, as-cast bulk samples do not allow one to study the intrinsic material properties of the $Co_2Cr_{1-x}Fe_xAl$ series. Samples grown using the FZ technique, as this method allows to grow phase-pure samples of incongruently melting compounds, however, revealed the presence of a phase transformation via a spinodal decomposition in the Cr-rich part of the pseudo-ternary Co-Cr-Fe-Al series [81]. This solid state miscibility gap is likely also present in $Co_2Cr_{1-x}Fe_xAl$ thin films, particularly after annealing, and, hence those recent results reconcile the various structure and magnetic anomalies reported in literature for $Co_2Cr_{1-x}Fe_xAl$ thin films as outlined above [81].

4.4.7 $Co_2Cr_{1-x}Fe_xGa$

The single phase region is much wider in the Co-Cr-Ga thermodynamic phase diagram than in the corresponding Co-Cr-Al case [82], suggesting a much more stable Co_2CrGa phase than the Al sister phase. The partial substitution of Cr by Fe is expected to raise the Curie temperature, demanding a thorough investigation of the $Co_2Cr_{1-x}Fe_xGa$ substitution series [82].

The $L2_1$ type structure is formed for all $Co_2Cr_{1-x}Fe_xGa$ samples (see Fig. 4.5). Annealing leads to the precipitation of additional phases while samples which were not annealed are single-phase. The formation of precipitates also lowers the saturation magnetization [82], underlining the strong link between structure, microstructure and physical properties in Heusler compounds.

4.4.8 $Co_2FeAl_{1-x}Si_x$

The local structure of $Co_2FeAl_{1-x}Si_x$ bulk samples were subject of two NMR studies [40, 83], revealing that the contribution of $L2_1$ type order evolves within the substitution series towards pure $L2_1$ order in Co_2FeSi [40, 49, 50]. The stronger tendency to the higher degree of ordering in Si-rich samples was assigned to the

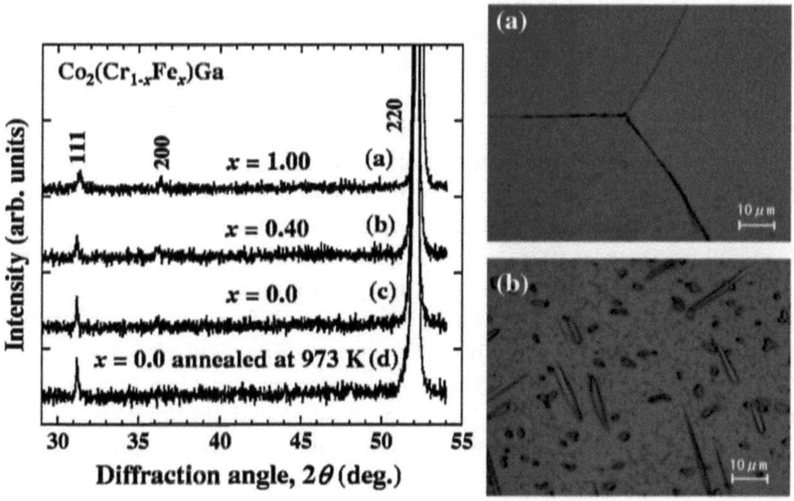

Fig. 4.5 Adapted with permission from [82]. *Left side* X-ray diffraction patterns for $Co_2Cr_{1-x}Fe_xGa$ compounds with increasing Fe content x, namely for $x = 0.0$, 0.40, and 1.00 quenched from 1373 K, respectively. **d** shows the diffractogram for the x = 0.0 sample annealed at 973 K for 1 week after quenching from 1373 K. *Rigth side* Optical micrographs of single phase Co_2CrGa obtained by quenching from 1373 K, and micrographs of Co_2CrGa annealed at 973 K for 1 week after quenching clearly exhibiting precipitates. Copyright 2005, American Physical Society

stronger hybridization of the Si 3(s, p) orbitals with the 3d transition-metal orbitals in $Co_2FeAl_{1-x}Si_x$ [40, 67]. Epitaxially grown $Co_2FeAl_{0.5}Si_{0.5}$ films on lattice-matched $MgAl_2O_4$ (001) substrates by a novel off-axis sputtering technique yielded films with an exceptionally high quality [84], as confirmed by NMR spectroscopy [84]. Nuclear magnetic resonance study of thin $Co_2FeAl_{0.5}Si_{0.5}$ Heusler films with varying thickness [85].

4.4.9 YXZ Heusler Compounds

Half-Heusler compounds ($C1_b$ structure) are known for the richness of their physical properties which include: half-metallic ferromagnetism, shape memory properties, semiconducting and thermoelectric properties as well as the most fascinating state of topological insulator [9, 86]. In the following, the role of structural order for this class of material is exemplified mainly in the context of their magnetic and half-metallic properties.

4.4.9.1 NiMnSb

NiMnSb is the first proposed half-metallic material [16] it has $C1_b$ structure. This material has been recognized from the very beginning as an important material for spintronic applications. However, while bulk measurements support the half-metallic behavior [87, 88], surface sensitive techniques did not show full spin polarization at the Fermi level [89]. In addition a low temperature anomaly was reported for NiMnSb: a crossover from half-metallic to normal ferromagnetic behaviour around 80 °C [90], which questioned the reported room temperature half-metallic state. Only very recently it has been shown that this is not an intrinsic property of NiMnSb but results from an excess of Mn on interstitial positions [91]. This highlights an extreme sensitivity of the half metallic state to crystal structure quality. Orgasa et al. have been the first to point out that structural disorder in NiMnSb above 5 % results in the loss of half-metallic properties [92]. An NMR study on bulk NiMnSb material carried out in external magnetic field by Hihara et al. [93], from times before spintronics era, has identified a single hyperfine field on $^{121,123}Sb$ and ^{55}Mn nuclei in accordance with one crystallographic position for Sb and Mn atoms, respectively, in the $C1_b$ structure. NMR study carried out recently did not reveal any traces of the fact structural disorder on bulk arc-melted stoichiometric NiMnSb, showing that despite that one of the fcc sublattice is empty in $C1_b$ structure, NiMnSb grown at high temperature is not prone to disorder [94].

The situation is more complicated in case of thin films. It has been found that NiMnSb films epitaxially grown on MnO show fewer defects the higher the temperature of the growth [95]. However, MBE growth on $III-V$ semiconducting substrates (e.g., GaAs) requires a low temperature (200–300 °C) to avoid an extensive mixing at the interface. In such growth conditions, the Sb content is not self-limiting and controlling the Sb flux during NiMnSb film growth turned out to be critical from point of

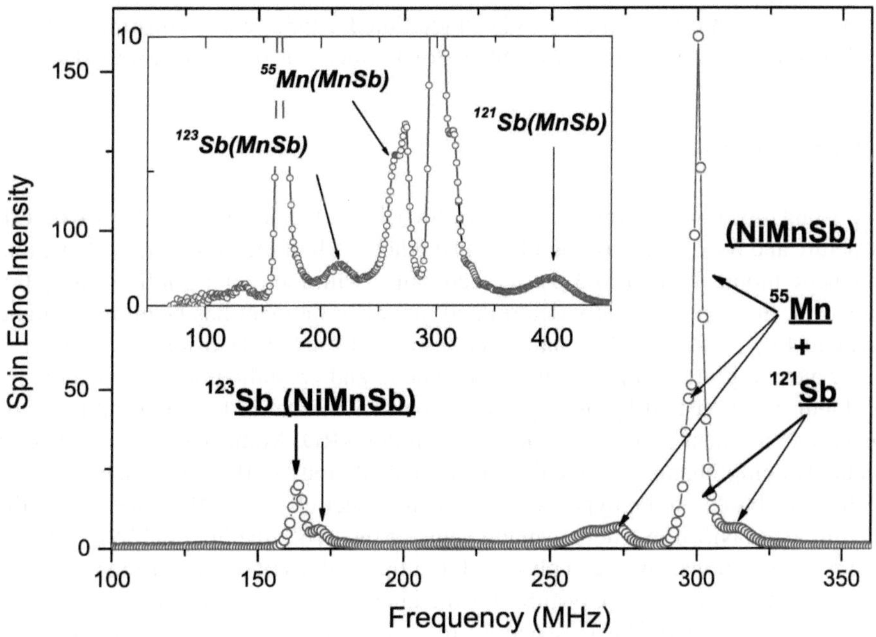

Fig. 4.6 NMR spectrum recorded in the NiMnSb/GaAs(001) thin film. The arrows indicate ^{123}Sb resonance at 164 MHz and overlapping ^{121}Sb, ^{55}Mn resonances at 300 MHz along with their satellites originating from main NiMnSb phase. A low intensity structure, shown in the inset, consisting of ^{123}Sb(217 MHz), ^{55}Mn(263 MHz) and ^{121}Sb (399 MHz) resonances reveals inclusions of MnSb

view of stoichiometry and amount of disorder the films exhibit [96]. NMR carried out on epitaxial NiMnSb films on GaAs(001) grown at temperature of 300 °C and for an optimal Sb flux corresponding to 1:1:1 stoichiometry, determined by a post growth Rutherford Backscattering Spectrometry (RBS) experiment, revealed two types of defects [97]. The satellite lines to ^{121,123}Sb and ^{55}Mn main lines have been interpreted as due to a small amount (below 4 %) of Mn and Sb antisites changing the second or the third nearest neighbor shell around Mn in $4b$ and Sb in $4a$ positions (Fig 4.6). Moreover, presence of a second phase was confirmed: hexagonal MnSb, amounting to 14 % and more, has been commonly found in all films characterized from the RBS experiment as stoichiometric and Sb rich compositions. This MnSb phase was absent in the films with a poor Sb stoichiometry [96]. For NiMnSb(001) oriented films, XRD was blind to MnSb phase due to formation of MnSb as a sequence of hcp stacking faults extending along [111] direction, however MnSb together with non-ferromagnetic NiSb was detected in XRD for NiMnSb(111) oriented films [98]. Optimization of the growth conditions for GaAs(111)B substrate led to NiMnSb(111) films with very low chemical disorder as revealed by NMR experiment [99]. It has been found that only around 1 % of Mn atoms and 0.2 % of Sb atoms was displaced from the original positions in $C1_b$ structure. In the interface to GaAs some diffusion of As (0.5 %) into Sb position has been identified. MnSb level of disorder detected by NMR was consistent with a 1:0.99:1 stoichiometry as obtained from RBS.

4.4.9.2 CoMnSb

For a long time it was assumed that CoMnSb belongs to half-Heusler alloys with $C1_b$ structure and band structure calculations predicted a half metallic state with the total magnetic moment equal to 3 μ_B in accordance with the Slater Pauling rule [100]. However, the experimentally determined magnetic moment has been much higher (around 4 μ_B) [101] and this challenged the postulated half-Heusler ($C1_b$) structural order and suggested a more complex magnetic and crystallographic structure. The problem was revisited by an experiment on stoichiometric bulk CoMnSb sample (magnetic moment equal to 3.8 μ_B) combining XRD, Mößbauer effect and NMR methods complimented by band structure calculations [94]. Unsurprisingly, these techniques based on the hyperfine fields (Mößbauer effect, NMR) have shown that in contrast to NiMnSb where a single value of the hyperfine field in ^{61}Ni, ^{121,123}Sb and ^{55}Mn positions have been observed, multiple values of the hyperfine fields have been identified in case of CoMnSb for the nuclei of each of the chemical elements (^{59}Co, ^{121,123}Sb and ^{55}Mn) revealing more complex than the $C1_b$ crystal structure. In fact, Mößbauer effect and NMR unambiguously identified two hyperfine fields at ^{121,123}Sb nuclei corresponding to two (Sb1, Sb2) nonequivalent crystallographic positions of Sb atoms populated as 1:3. ^{55}Mn NMR allowed to identify three nonequivalent crystallographic positions of Mn atoms (Mn1, Mn2, Mn3) populated according to the ratio 1:1:6. ^{59}Co NMR has found one hyperfine field and correspondingly a single crystallographic position has been assigned for Co. Cumulated NMR spectrum intensities corresponding to three groups of the hyperfine fields were found equal to 1:1:1 ratio reflecting the stoichiometry of Co, Mn and Sb [94]. These results fully conformed to a refinement of the superstructure in the XRD pattern based on the $Fm\overline{3}m$ space group. Accordingly, a new superstructure has been assigned to CoMnSb with a supercell incorporating eight elementary $C1_b$ cells with displaced Sb and Mn atoms. This supercell can be seen as build of alternating Co_2MnSb and MnSb structural units in all directions. It contains two antimony, three manganese nonequivalent positions and one cobalt position as observed experimentally. Band structure calculations based on this new structure predicated for CoMnSb a ferromagnetic ground state with the total magnetic moment of 4 μ_B in agreement with the experimental observations [94].

4.4.9.3 CoVSb

Band structure calculations predicted different ground states for CoVSb depending on the atomic order. For Co atoms occupying 4*b* sites in the $C1_b$ structure, a ferromagnetic half-metallic state with a magnetic moment equal to 1 μ_B per formula unit associated with vanadium was obtained while a ferrimagnetic state without gap and magnetization equal to 0.17 μ_B prevailed in calculations when the position of Co and V was inverted in $C1_b$ structure [100, 102]. On the other hand, though neutron powder diffraction experiments confirmed the $C1_b$ crystal structure and $4b$ position of Co in CoVSb, the magnetization data on the same sample annealed at 850 °C indicated a week itinerant ferromagnetic state with magnetization around 0.18 μ_B instead of a strong half-metallic ferromagnetic ground state [103]. NMR study has shown complex zero field and field swept spectra at low temperature revealing coexistence of nonmagnetic and magnetic Co and V atoms in CoVSb [104]. It has been, thus, concluded that weak ferromagnetism indicated from the macroscopic magnetization is complex on the microscopic scale and the magnetic moment of 0.18 μ_B obtained from the static magnetization measurements has to be treated as an average moment of magnetic and nonmagnetic sites of cobalt and vanadium atoms [104].

4.4.9.4 ScTSb

Ternary antimonides ScTSb (with T = Ni, Pd, Pt) crystallize in half-Heusler MgAgAs type structure. These compounds have been studied due to their interesting thermoelectric properties and it has been found that defect formation on the T sites as well as substitutions on the T and Sb sites strongly modify the Seebeck coefficients [105, 106]. Comprehensive solid state ^{45}Sc and ^{121}Sb NMR and ^{121}Sb Mößbauer effect study of the local environments in the antimonides ScTSb with T = Ni, Pd, Pt have been carried out on annealed and quenched samples in order to assess the effect of incomplete site occupancy and site distortions [107]. It has been concluded that majority of as quenched samples are characterized by a disorder consisting of defects on the transition metal sites, e.g., a shift to a vacancy position. The fact that these defects can be removed by annealing has an important impact on their thermoelectric properties.

4.5 Summary

The Heusler compounds are a fascinating class of materials, however, crystallographic structure, phase dynamics and order need to be taken into account if properties and functionalities are being discussed.

Acknowledgments We thank J. T. Kohlhepp, E. Jedryka, B. Koopmans, H. J. M. Swagten, A. Alfonsov, M. Belesi, S. Rodan, T. Ellis, P. J. Jacobs, V. Ksenofontov, G.H. Fecher, C. Felser, C.G.F. Blum, A. Omar, G. Jakob, H. Schneider, D. Ebke, G. Reiss, B. Balke, S. Maat, M. J. Carey, and J. R. Childress, for fruitful discussions. We also like to thank all authors who gave permission to show their figures in this review and as well as the corresponding journals. This work has been supported in part by a grant from the Ministry of Sciences and High Education of Poland under the research Project No. 4531/B/T02/2010/38 for the years 2010-2013 and by Deutsche Forschungsgemeinschaft DFG in project WU595/3-1 and WU595/3-2.

References

1. Fr Heusler, Verh. Dtsch. Phys. Ges. **12**, 219 (1903)
2. P.J. Webster, K.R.A. Ziebeck, Heusler alloys, in *Alloys and Compounds of d-Elements with Main Group Elements*, Landolt-Börnstein—Group III Condensed Matter, Part 2, vol. 19C, ed. by H.P.J. Wijn (Springer, Heidelberg, 1988), pp. 104–185
3. K.R.A. Ziebeck, K.U. Neumann, Heusler alloys, in *Alloys and Compounds of d-Elements with Main Group Elements*, Landolt-Börnstein—Group III Condensed Matter, Part 2, vol. 32C, ed. by H.P.J. Wijn (Springer, Heidelberg, 2001), pp. 64–314
4. T. Graf, C. Felser, S.S.P. Parkin, Prog. Solid State Chem. **39**, 1–50 (2011)
5. C. Pfleiderer, J. Boeuf, Phys. Rev. B **65** (2002)
6. H. Nakamura, Y. Kitaoka, K. Asayama, Y. Nuki, T. Komatsubara, J. Phys. Soc. Jpn. **57**, 2276 (1988)
7. M. Ishikawa, J.L. Jorda, A. Junod, Heusler alloys, in *Superconductivity in d- and f-Band Metals*, Landolt-Börnstein - Group III Condensed Matter, ed. by W. Buckel, W. Weber (Springer, Kernforschungszentrum, Karlsruhe, 1982)
8. Y. Nishino, M. Kato, S. Asano, K. Soda, M. Hayasaki, U. Mizutani, Phys. Rev. Lett. **79**, 1909 (1997)
9. S. Chadov, X. Qi, J. Kuebler, G.H. Fecher, C. Felser, S.C. Zhang, Nat. Mater. **9**, 541 (2010)
10. P.J. Webster, K.R.A. Ziebeck, S.L. Town, M.S. Peak, Philos. Mag. B **49**, 295 (1984)
11. R. Kainuma, Y. Imano, W. Ito, H. Morito, Y. Sutou, K. Oikawa, A. Fujita, K. Ishida, S. Okamoto, O. Kitakami, T. Kanomata, Appl. Phys. Lett. **88**, 192513 (2006)
12. B. Balke, G.H. Fecher, J. Winterlik, C. Felser, Appl. Phys. Lett. **90**, 152504 (2007)
13. J. Winterlik, B. Balke, G.H. Fecher, C. Felser, Phys. Rev. B **77**, 054406 (2008)
14. S. Wurmehl, G.H. Fecher, H.C. Kandpal, V. Ksenofontov, C. Felser, H.-J. Lin, J. Morais, Phys. Rev. B **72**, 184434 (2005)
15. S. Wurmehl, G.H. Fecher, H.C. Kandpal, V. Ksenofontov, C. Felser, H.-J. Lin, Appl. Phys. Lett. **88**, 032503 (2006)
16. R.A. de Groot, F.M. Müller, P.G. van Engen, K.H.J. Buschow, Phys. Rev. Lett. **50**, 2024 (1983)
17. B. Ravel, M.P. Raphael, V.G. Harris, Q. Huang, Phys. Rev. B **65**, 184431 (2003)
18. J. Pierre, R.V. Skolozdra, J. Tobola, S. Kaprzyk, C. Hordequin, M.A. Kouacou, I. Karla, R. Currat, E. Lelievre-Berna, J. Alloy. Compd. **262–263**, 101 (1997)
19. Y. Miura, K. Nagao, M. Shirai, Phys. Rev. B **69**, 144413 (2004)
20. Y. Miura, K. Nagao, M. Shirai, J. Appl. Phys. **95**, 7225 (2004)
21. J. Magn, Magn. Mater. **272276**, 315 (2004)
22. S. Picozzi, A. Continenza, A.J. Freeman, Phys. Rev. B **69**, 094423 (2004)
23. S. Wurmehl, G.H. Fecher, K. Kroth, F. Kronast, H.A. Dürr, Y. Takeda, Y. Saitoh, K. Kobayashi, H.-J. Lin, G. Schönhense, J. Phys. D: Appl. Phys. **39**, 803 (2006)
24. G.H. Fecher, H.C. Kandpal, S. Wurmehl, G. Schönhense, C. Felser, J. Appl. Phys. **17**, 7237 (2005)
25. G.E. Bacon, J.S. Plant, J. Phys. F: Metal Phys. **1**, 524 (1971)
26. Vajiheh Alijani, Juergen Winterlik, Gerhard H. Fecher, S.S. Naghavi, C. Felser, Phys. Rev. B **83**, 184428 (2011)
27. V. Alijani, S. Ouardi, G.H. Fecher, J. Winterlik, S.S. Naghavi, X. Kozina, G. Stryganyuk, C. Felser, E. Ikenaga, Y. Yamashita, S. Ueda, K. Kobayashi, Phys. Rev. B **84**, 224416 (2011)
28. J. Winterlik, B. Balke, G.H. Fecher, C. Felser, M.C.M. Alves, F. Bernardi, J. Morais, Phys. Rev. B **77**, 054406 (2008)
29. B. Balke, G.H. Fecher, C. Felser, Appl. Phys. Lett. **90**, 242503 (2007)
30. A. Abragam, in *Principles of Nuclear Magnetism* (Oxford University Press, Oxford, 1996)
31. P. Panissod, in *Structural and Magnetic Investigations by NMR. Application to Magnetic Multilayers*, vol. 48, NATO ASI Series High Tech, ed. by V.G. Baryakhtar, P.E. Wigen, N.A. Lesnik (Kluwer Academic, Dordrecht, 1997), p. 225
32. S. Wurmehl, J.T. Kohlhepp, J. Phys. D: Appl. Phys. **41**, 173002 (2008)

33. Philipp Gütlich, Chemie in unserer Zeit **4**, 133 (1970)
34. M.P. Raphael, B. Ravel, M.A. Willard, S.F. Cheng, B.N. Das, R.M. Stroud, K.M. Bussmann, J.H. Claassen, V.G. Harris, Appl. Phys. Lett. **79**, 4396 (2001)
35. M.P. Raphael, B. Ravel, Q. Huang, M.A. Willard, S.F. Cheng, B.N. Das, R.M. Stroud, K.M. Bussmann, J.H. Claassen, V.G. Harris, Phys. Rev. B **66**, 104429 (2002)
36. Y. Sakuraba, K. Izumi, T. Iwase, S. Bosu, K. Saito, K. Takanashi, K. Abe, Y. Miura, K. Futatsukawa, M. Shirai, Phys. Rev. B **82**, 094444 (2010)
37. S. Rodan, A. Alfonsov, M. Belesi, F. Ferraro, J.T. Kohlhepp, H.J.M. Swagten, B. Koopmans, Y. Sakuraba, S. Bosu, K. Takanshi, B. Büchner, S. Wurmehl, Appl. Phys. Lett. **102**, 242404 (2013)
38. K. Inomata, S. Okamura, A. Miyazaki, N. Tezuka, M. Wójcik, E. Jedryka, J. Phys. D: Appl. Phys. **39**, 816 (2006)
39. S. Wurmehl, J.T. Kohlhepp, H.J.M. Swagten, B. Koopmans, J. Phys. D: Appl. Phys. **41**, 115007 (2008)
40. M. Wójcik, E. Jedryka, H. Sukegawa, T. Nakatani, K. Inomata, Phys. Rev. B **85**, 100401 (2012)
41. S. Wurmehl, P.J. Jacobs, J.T. Kohlhepp, H.J.M. Swagten, B. Koopmans, M.J. Carey, S. Maat, J.R. Childress, Appl. Phys. Lett. **98**, 12506 (2011)
42. N.K. Jaggi, K.R.P.M. Rao, A.K. Grover, L.C. Gupta, R. Vijayaraghavan, L.D. Khoi, Hyperfine Interact. **4**, 402 (1978)
43. R. Ducher, R. Kainum, I. Ohnumaa, K. Ishida, J. Alloy. Compd. **437**, 93 (2007)
44. K. Kobayashi, R. Kainuma, K. Fukamichi, K. Ishida, J. Alloy. Compd. **403**, 161 (2005)
45. B. Balke, S. Wurmehl, G.H. Fecher, C. Felser, M.C.M. Alves, F. Bernardi, Appl. Phys. Lett. **90**, 172501 (2007)
46. L. Basit, C. Wang, C.A. Jenkins, B. Balke, V. Ksenofontov, G.H. Fecher, C. Felser, E. Mugnaioli, U. Kolb, S.A. Nepijko, G. Schönhense, M. Klimenkov, J. Phys. D: Appl. Phys. **42**(8), 084018 (2009)
47. M. Gellesch, M. Dimitrakopoulou, M. Scholz, C.G.F. Blum, M. Schulze, J. van den Brink, S. Hampel, S. Wurmehl, B. Büchner, Cryst. Growth Des. **13**(7), 2707 (2013)
48. D. Bombor, C.G.F. Blum, O. Volkonskiy, S. Rodan, S. Wurmehl, C. Hess, B. Büchner, Phys. Rev. Lett. **110**, 066601 (2013)
49. K. Inomata, M. Wojcik, E. Jedryka, N. Ikeda, N. Tezuka, Phys. Rev. B **77**, 214425 (2008)
50. C.G.F. Blum, C. Jenkins, J. Barth, C. Felser, S. Wurmehl, G. Friemel, C. Hess, G. Behr, B. Buchner, A. Reller, S. Riegg, S.G. Ebbinghaus, T. Ellis, P.J. Jacobs, J.T. Kohlhepp, H.J.M. Swagten, Appl. Phys. Lett. **95**, 161903 (2009)
51. V.A. Niculescu, J.I. Budnick, W.A. Hines, K. Raj, S. Pickart, S. Skalski, Phys. Rev. B **19**, 452 (1979)
52. V.A. Niculescu, T.J. Burch, J.I. Budnick, J. Magn. Magn. Mater. **39**, 223 (1983)
53. S. Wurmehl, J.T. Kohlhepp, H.J.M. Swagten, B. Koopmans, C.G.F. Blum, V. Ksenofontov, H. Schneider, G. Jakob, D. Ebke, G. Reiss, J. Phys. D: Appl. Phys. **42**, 084017 (2009)
54. H. Schneider, G. Jakob, M. Kallmayer, H.-J. Elmers, M. Cinchetti, B. Balke, S. Wurmehl, C. Felser, M. Aeschlimann, H. Adrian, Phys. Rev. B **74**, 174426 (2006)
55. N. Tezuka, S. Okamura, A. Miyazaki, M. Kikuchi, K. Inomata, J. Appl. Phys. **99**, 08T314 (2006)
56. Z. Gercsi, A. Rajanikanth, Y.K. Takahashi, K. Hono, M. Kikuchi, N. Tezuka, K. Inomata, Appl. Phys. Lett. **89**, 082512 (2006)
57. M. Kallmayer, H. Schneider, G. Jakob, H.-J. Elmers, B. Balke, S. Cramm, J. Phys. D: Appl. Phys. **40**, 1552 (2007)
58. A.U.B. Wolter, A. Bosse, D. Baabe, I. Maksimov, D. Mienert, H.H. Klauß, F.J. Litterst, D. Niemeier, R. Michalak, C. Geibel, R. Feyerherm, R. Hendrikx, J.A. Mydosh, S. Süllow, Phys. Rev. B **66**, 174428 (2002)
59. H.C. Kandpal, V. Ksenofontov, M. Wojcik, R. Seshadri, C. Felser, J. Phys. D.: Appl. Phys. **40**, 1587 (2006)
60. K. Ooiwa, K. Endo, A. Shinogi, J. Magn. Magn. Mater. **2011**, 104–107 (1992)

61. M. Belesi, L. Giebeler, C.G.F. Blum, U.K. Rößler, B. Büchner, S. Wurmehl, Phys. Rev. B **91**, 134415 (2014)
62. V. Jung, B. Balke, G.H. Fecher, Appl. Phys. Lett. **93**, 042507 (2008)
63. G.D. Liu, X.F. Dai, H.Y. Liu, J.L. Chen, Y.X. Li, G. Xiao, Phys. Rev. B **77**, 014423 (2008)
64. J. Winterlik, G.H. Fecher, B. Balke, T. Graf, V. Alijani, V. Ksaenofontov, C.A. Jenkins, O. Meshcheriakova, C. Felser, G. Liu, S. Ueda, K. Kobayashi, T. Nakamura, M. Wojcik, Phys. Rev. B **83**, 174448 (2011)
65. P.J. Brown, T. Kanomata, K. Neumann, K.U. Neumann, B. Ouladdiaf, A. Sheikh, K.R.A. Ziebeck, J. Phys.: Condens. Matter **22**, 506001 (2010)
66. B. Balke, G.H. Fecher, H.C. Kandpal, C. Felser, K. Kobayashi, E. Ikenaga, J.-J. Kim, S. Ueda, Phys. Rev. B **74**, 104405 (2006)
67. G.H. Fecher, C. Felser, J. Phys. D. Appl. Phys. **40**, 1582 (2007)
68. S. Wurmehl, J.T. Kohlhepp, H.J.M. Swagten, B. Koopmans, M. Wojcik, B. Balke, C.G.H. Blum, V. Ksenofontov, G.H. Fecher, C. Felser, Appl. Phys. Lett. **91**, 052506 (2007)
69. S. Wurmehl, A. Alfonsov, J.T. Kohlhepp, H.J.M. Swagten, B. Koopmans, M. Wojcik, B. Balke, V. Ksenofontov, C.G.F. Blum, B. Büchner, Phys. Rev. B **88**, 134424 (2013)
70. V. Jung, B. Balke, G.H. Fecher, V. Ksenofontov, C. Felser, Hyperfine Interact. **184**, 15 (2008)
71. V. Jung, G.H. Fecher, B. Balke, V. Ksenofontov, C. Felser, J. Phys. D: Appl. Phys. **42**, 084007 (2009)
72. C. Felser, B. Heitkamp, F. Kronast, D. Schmitz, S. Cramm, H.A. Dürr, H.J. Elmers, G.H. Fecher, S. Wurmehl, T. Block, D. Valdaitsev, S.A. Nepijko, A. Gloskovskii, G. Jakob, G. Schönhense, W. Eberhardt, J. Phys.: Condens. Matter **15**(41), 7019 (2003)
73. H.J. Elmers, G.H. Fecher, D. Valdaitsev, S.A. Nepijko, A. Gloskovskii, G. Jakob, G. Schönhense, S. Wurmehl, T. Block, C. Felser et al., Phys. Rev. B **67**, 104412 (2003)
74. J.M. De Teresa, D. Serrate, R. Córdoba, S.M. Yusuf, J. Alloy. Compd. **450**, 31 (2008)
75. S. Wurmehl, G.H. Fecher, K. Kroth, F. Kronast, H.A. Dürr, Y. Takeda, Y. Saitoh, K. Kobayashi, H.J. Lin, G. Schönhense, C. Felser, J. Phys. D: Appl. Phys. **39**(5), 803 (2006)
76. T. Block, C. Felser, G. Jakob, J. Ensling, B. Mühling, P. Gütlich, R.J. Cava, J. Solid State Chem. **176**, 646 (2003)
77. S. Okamura, A. Miyazaki, N. Tezuka, S. Sugimoto, K. Inomata, Mater. Trans. **47**, 15 (2006)
78. G. Jakob, F. Casper, V. Beaumont, S. Falk, N. Auth, H.J. Elmers, C. Felser, **290**, 1104 (2005)
79. K. Matsuda, T. Kasahara, T. Marukame, T. Uemura, M. Yamamoto, J. Cryst. Growth **286**, 389 (2006)
80. V. Ksenofontov, C. Herbort, M. Jourdan, C. Felser, Appl. Phys. Lett. **92**, 262501 (2008)
81. A. Omar, M. Dimitrakopoulou, C.G.F. Blum, H. Wendrock, S. Rodan, S. Hampel, W. Löser, B. Büchner, S. Wurmehl, Cryst. Growth Des. **13**, 3925 (2013)
82. R.Y. Umetsu, K. Kobayashi, A. Fujita, K. Oikawa, R. Kainuma, K. Ishida, N. Endo, K. Fukamichi, A. Sakuma, Phys. Rev. B **72**, 214412 (2005)
83. S. Wurmehl, J.T. Kohlhepp, H.J.M. Swagten, B. Koopmans, J. Appl. Phys. **111**, 043903 (2012)
84. B. Peters, A. Alfonsov, C.G.F. Blum, P.M. Woodward, S. Wurmehl, B. Büchner, F.Y. Yang, Appl. Phys. Lett. **103**, 162404 (2013)
85. A. Alfonsov, B. Peters, F. Y. Yang, B. Büchner, and S. Wurmehl Phys. Rev. B **91**, 064421 (2015)
86. F. Casper, T. Graf, S. Chadov, B. Balke, C. Felser, Semicond. Sci. Technol. **27**, 063001 (2012)
87. P.A.M. van der Heide, W. Baelde, R.A. De Groot, J. Phys. F: Met. Phys. **15**, L75 (1985)
88. K.E.H.M. Hanssen, P.E. Mijnarends, L.P.P. Rabou, Phys. Rev. B **42**, 1533 (1990)
89. W. Zhu, B. Sinkovic, E. Vescovo, C. Tanaka, J.S. Moodera, Phys. Rev. B **64**, 060403 (2001)
90. C. Hordequin, D. Ristoiu, L. Ranno, J. Pierre, Eur. Phys. J. B **16**, 287 (2000)
91. B. Zhang, J.A. Heuver, F. Wang, J. Baas, G.A. de Wijs, T. Fukuhara, T.T.M. Palstra, R.A. de Groot, Phys. Rev. B **88**, 014418 (2013)
92. D. Orgassa, H. Fujiwara, T.C. Schulthess, W.H. Butler, Phys. Rev. B **60**, 13237 (1999)
93. T. Hihara, M. Kawakami, M. Kasaya, H. Enokiya, J. Phys. Soc. Jpn **26**, 1061 (1969)
94. V. Ksenofontov, G. Melnyk, M. Wójcik, S. Wurmehl, K. Kroth, S. Reimann, P. Blaha, C. Felser, Phys. Rev. B **74**, 134426 (2006)

95. P. Turban, S. Andrieu, E. Snoeck, C. Teodorescu, A. Traverse, Phys. Rev. B **64**, 134417 (2002)
96. W. Van Roy, J. De Boeck, B. Brijs, G. Borghs, Appl. Phys. Lett. **77**, 4190 (2000)
97. M. Wojcik, W. van Roy, E. Jedryka, S. Nadolski, G. Borghs, J. De Boeck, J. Magn. Magn. Mater. **240**, 414 (2002)
98. W. Van Roy, G. Borghs, J. De Boeck, J. Magn. Magn. Mater. **242–245**, 489 (2002)
99. W. van Roy, M. Wójcik, E. Jedryka, S. Nadolski, D. Jalabert, B. Brijs, G. Borghs, J. De Boeck, Appl. Phys. Lett. **83**, 4214 (2003)
100. J. Tobola, J. Pierre, J. Alloy. Compd. **296**, 243 (2000)
101. K. Kaczmarska, J. Pierre, J. Tobola, R.V. Skolozdra, Phys. Rev. B **373**, 60 (1999)
102. K. Kaczmarska, J. Pierre, J. Beille, J. Tobola, R.V. Skolozdra, G.A. Melnik, J. Magn. Magn. Mater. **210**, 187 (1998)
103. L. Heyne, T. Igarashi, T. Kanomata, K.U. Neumann, B. Ouladdiaf, K.R.A. Ziebeck, J. Phys.: Condens. Matter **17**, 4991 (2005)
104. H. Nishihara, T. Kanomata, Y. Furutani, T. Igarashi, K. Koyama, T. Goto, Phys. Status Solidi **3**, 2779 (2006)
105. J. Oestreich, U. Probst. in *Proceedings ICT'02. 21st International Conference on Thermoelectrics*, p. 135 (2002)
106. J. Oestreich, U. Probst, F. Richardt, E. Bucher, J. Phys.: Condens. Matter **15**, 635 (2003)
107. T. Harmening, H. Eckert, R. Pöttgen, Solid State Sci. **11**, 900 (2009)

Chapter 5
Heusler Compounds Go Nano

Judith Meyer, Niclas Teichert, Alexander Auge, Changhai Wang, Andreas Hütten and Claudia Felser

Abstract This chapter is addressing the physical impact of ferromagnetic Heusler entities when approaching the nanoscale, e.g. as nanoparticles or as very small grains in magnetic shape Heusler alloys, on resulting magnetic as well as microstructural properties. Based on the soft magnetic behavior of Co_2FeGa and Co_2FeSi as two representatives of the full Heusler family their superparamagnetic potential is projected to applications in biotechnology. These applications can now be pictured due to the progress which has been made in synthesizing Heusler nanoparticles. Taken Co_2FeGa as a candidate the chemical preparation avenue to achieve nanoparticles with reliable physical properties is demonstrated leading to a nanoparticular GMR-effect. It is shown that magnetic nanoparticles can be embedded in agarose as a biogel when employing external magnetic fields so as to configure the nanoparticle arrangements for optimizing the GMR-effect. Possible consequences in case of a nanoparticular TMR-effect are pictured. The very small grain size in magnetic shape Heusler alloys is determining the austenite-martensite transformation in ultra-thin films which might play a major role for spintronic applications also bridging two research field in addition. The principle microstructural influences on the austenite-martensite transformation in thin films are discussed in terms of epitaxial growth, phase compatibility, crystal quality and size scale effects. Thereafter, details concerning the martensitic transformation in a film thickness range from 10 to 100 nm are discussed for two off-stoichiometric NiMnSn Heusler compositions.

The triumphant advance of Heusler compounds can mainly be attributed to their unique band structure enabling the realization of different physical properties such as ferromagnetism, semi- and superconductivity in one material class [1]. Moreover,

J. Meyer · N. Teichert · A. Auge · A. Hütten (✉)
Department of Physics, Thin Films and Physics of Nanostructures,
Bielefeld University, 100131, 33501 Bielefeld, Germany
e-mail: huetten@physik.uni-bielefeld.de

C. Wang · C. Felser
Max Planck Institute for Chemical Physics of Solids, NoethnitzerStraße 40,
01187 Dresden, Germany

C. Felser and A. Hirohata (eds.), *Heusler Alloys*,
Springer Series in Materials Science 222, DOI 10.1007/978-3-319-21449-8_5

the theoretical prediction of 100 % spin polarization in an entire class of materials, the half-Heusler XYZ [2–5] as well as the full-Heusler X_2YZ [6–9] alloys (X being a transition metal element such as Fe, Ru, Co, Rh, Ni, Pd, Pt, or Cu, Y being another transition metal element such as Ti, Zr, V, Nb, Cr, or Mn, and Z being a group III, IV, or V element such as Al, Ga, In, Si, Ge, Sn, As, or Sb) are currently the driving force for spintronic applications in form of thin film devices such as magnetic tunneling junctions (MTJ). A Heusler-MTJ generally is a layered thin film structure with a tunneling barrier, e.g. Al_2O_3 [10], MgO [11, 12] or BaO [13], sandwiched in between two ferromagnetic electrodes whereby at least one electrode is made of a Heusler compound. The interface quality between the ferromagnetic electrodes and the tunneling barrier as well as the crystallinity of the barrier critically determine the resulting tunneling magneto resistance (TMR) effect amplitude. The thickness of the ferromagnetic electrodes has not a marked impact on the TMR effect amplitude.

However, the central question of this chapter is aiming at small ferromagnetic Heusler entities approaching the nanoscale such as nanoparticles or grains where size effects as well as constrains by the surroundings are drastically influencing the physical properties.

5.1 Identifying the Future Role of Heusler Nanoparticles for Applications

Our discussion is focused on two special Heusler compounds Co_2FeGa and Co_2FeSi which are characterized among the Heusler alloys by a high saturation magnetization as well as a high Curie temperature. When comparing their intrinsic magnetic properties with those of conventional magnetic phases at room temperature, see Fig. 5.1, it becomes obvious that they are magnetically unique since they have a fairly high saturation magnetization combined with very low magneto crystalline anisotropies and fairly high Curie temperatures. Their magneto crystalline anisotropies are comparable to Fe_3O_4 and Fe_2O_3 which are very important for biotechnological applications because of their biocompatibility but about 12 times larger than that of the soft magnetic Permalloy ($Ni_{81}Fe_{19}$).

The superparamagnetic limit at room temperature $T_{300\,K}$ of all magnetic phases summarized in Fig. 5.1 can be determined by assuming spherical nanoparticles and applying the equilibrium condition given by 5.1 where the product of magneto crystalline anisotropy K times the nanoparticle volume V is balancing thermal fluctuation at room temperature. Considering nanoparticles which are magnetically stable against thermal switching for one year, 5.1 yields:

$$KV = 38k_BT, \tag{5.1}$$

k_B being the Boltzmann constant. Figure 5.2 compares the resulting superparamagnetic limits D_{SPM} at room temperature when applying 5.1. Both Heusler compounds

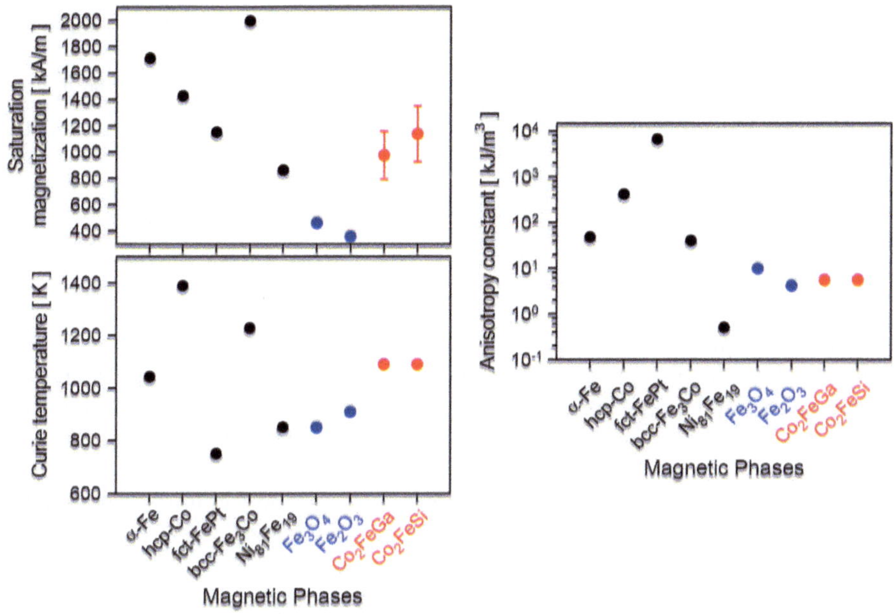

Fig. 5.1 Comparison of intrinsic magnetic properties of different magnetic phases. Co_2FeGa and Co_2FeSi show medium large saturation magnetization combined with very low magneto crystalline anisotropies but large Curie temperatures. The data are collected from [14–17]

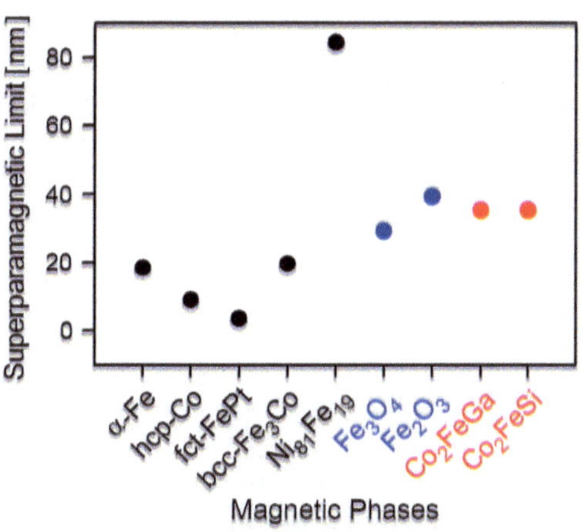

Fig. 5.2 Calculated superparamagnetic limits of all magnetic phases summarized in Fig. 5.1. The critical nanoparticle diameter for superparamagnetic behavior of Co_2FeGa and Co_2FeSi is comparable to those of Fe_3O_4 and Fe_2O_3

have superparamagnetic limits comparable to the value for magnetite (Fe_3O_4) and maghemite (Fe_2O_3) but a magnetization, compare with Fig. 5.1, which is about 2 to 3 time higher. This combination will have an impact on the corresponding magnetization reversal as will now be shown.

Since D_{SPM} is separating ferromagnetic from superparamagnetic behavior, related nanoparticle size distributions $<D> \pm \sigma$ above and below D_{SPM}, with $<D>$ as the

Table 5.1 Particle size distributions for calculating the magnetization reversal in the superparamagnetic and ferromagnetic regime applying 5.2 and 5.3

Magnetic phase:	Color code	D_{SPM} [nm]	$<D>_{SPM} \pm \sigma$	$<D>_{ferro} \pm \sigma$
hcp-Co	black	9.0	7.5 ± 0.75	50 ± 5
Fe_3Co	blue	19.6	15 ± 1.5	50 ± 5
Fe_3O_4	pink	30.1	23 ± 2.3	50 ± 5
Co_2FeSi	red	36.3	28 ± 2.8	50 ± 5

mean particle diameter and σ being the standard deviation, can be defined so as to explore both magnetic regimes. The magnetization reversal of the superparamagnetic regime can calculated by using a Langevin function and taking into account the nanoparticle size distributions over a $<D> \pm 5\sigma$ range. For our discussion four magnetic phases will be considered and are given in table 5.1.

The magnetization reversal within the superparamagnetic regime can be evaluated using 5.2 [18]:

$$M(H_{ext}) = M_S \int_{<D>_{SPM}-5\sigma}^{<D>_{SPM}+5\sigma} \frac{1}{\sqrt{2\pi}\sigma} \exp\left[-\frac{(D-<D>_{SPM})^2}{2\sigma^2}\right] \times \left[\coth\left[\frac{\frac{4}{3}\pi\left(\frac{D}{2}\right)^3 M_S H_{ext}}{k_B T}\right] - \frac{k_B T}{\frac{4}{3}\pi\left(\frac{D}{2}\right)^3 M_S H_{ext}}\right] dD \tag{5.2}$$

M_S denotes the saturation magnetization which is given in Fig. 5.1. The magnetization reversal within the ferromagnetic regime can be determined by a model [19] which is given by 5.3:

$$M(H_{ext}) = \frac{2M_S}{\pi} \text{atan}\left[\frac{H_{ext} \pm H_C}{H_C} \tan\left(\frac{\pi S}{2}\right)\right] \text{ with}$$
$$H_C = 0.556 H_A \left\{1 - 0.977\left[\frac{k_B T}{KV} \ln\left(\frac{f_0 \tau}{\ln(2)}\right)\right]^{\frac{2}{3}}\right\} \tag{5.3}$$

The coercivity H_C is estimated assuming that the magnetic nanoparticles are randomly oriented in a self-assembled monolayer like grains in a thin film plane [20]. The squareness S of the magnetization loop is assumed to be 0.75. S is defines as the ration of the remanence divided by the saturation magnetization. The so called frequency factor f_0 has a value of 10^9 Hz and τ is the measurement time, here considered for one year. The anisotropy field H_A can be estimated by a Stoner-Wohlfahrt-Ansatz [14]. The resulting superparamagnetic magnetization curves are shown in Fig. 5.3. The magnetization reversal of Co_2FeSi ascends steeply leading to a rectangular profile when compared to the other three magnetic phases. In the ferromagnetic regime the magnetization reversal of Co_2FeSi given in Fig. 5.4b is extremely narrow and is

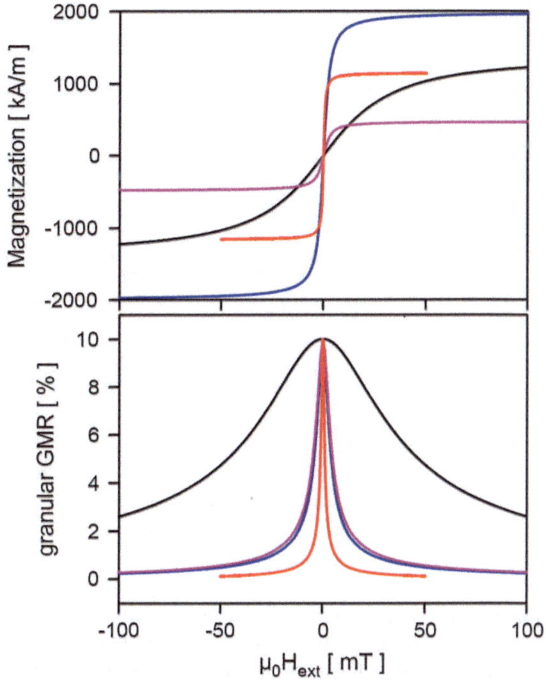

Fig. 5.3 The *upper* plot shows the calculated superparamagnetic magnetization reversal at room temperature applying 5.2. The underlying color coded magnetic phases and particle size distributions are given in table 5.1. The *lower* plot compares the resulting granular GMR behavior at room temperature for non-interaction magnetic nanoparticle when embedded in a well conducting matrix. The GMR characteristic for Co_2FeSi is very narrow due to a large D_{SPM} and a high M_S. This relates to a very high sensitivity when Co_2FeSi would be applied as granular GMR-sensors

characterized by a coercivity of about 3mT only. This is because of the low magneto crystalline anisotropy. A similar behavior can be expected for Co_2FeGa with about the same value for K.

The potential of these Heusler nanoparticles can directly be identified when assuming that they are forming a self-assembled monolayer and are electrical connected by being embedded in a well conducting matrix at a concentration where their dipolar strayfields do not interact. This nanoparticular monolayer can then be used as a nanogranular giant magnetoresistive (GMR) sensor. The related sensor characteristic can be calculated using the magnetization reversal taken from Fig. 5.3 together with 5.4:

$$\frac{\Delta R}{R}(H_{ext}) = A_{GMR}\left[1 - \left(\frac{M(H_{ext})}{M_S}\right)^2\right] \tag{5.4}$$

A_{GMR} is the GMR effect amplitude which is assumed to be 10 % for our ongoing discussion. The resulting GMR sensor characteristics are summarizes in Fig. 5.3. The GMR characteristic for Co_2FeSi is very narrow due to a large D_{SPM} and a high M_S. This relates to the highest GMR sensor sensitivity among the group of regarded magnetic phases. A comparable behavior can be expected for Co_2FeGa as well due to similar values for D_{SPM} and M_S. Consequently, the high potential of superparamagnetic Co_2FeSi and Co_2FeGa Heusler nanoparticles is identified when

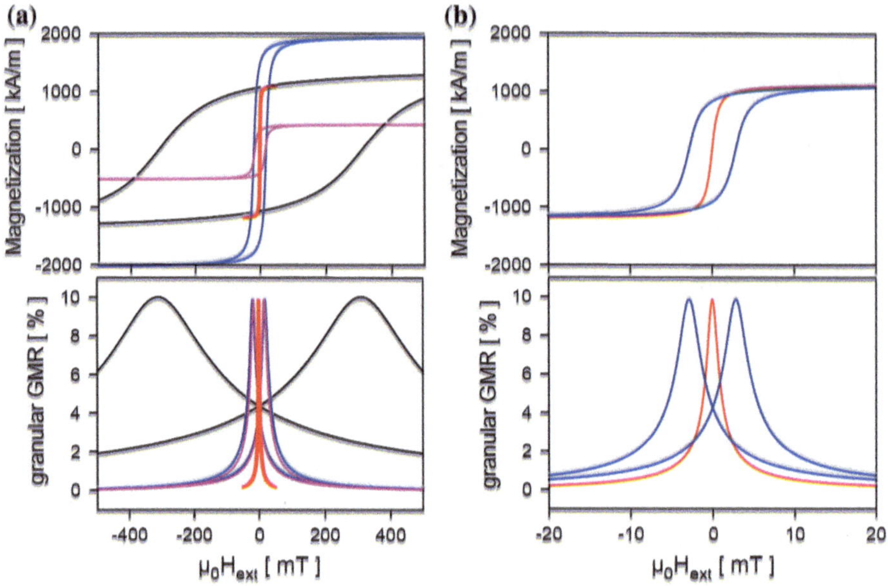

Fig. 5.4 **a** Comparing the magnetization reversal and the corresponding GMR characteristic of all four magnetic phases. **b** Magnetization and GMR curves of Co_2FeSi in both magnetic regimes, superparamagnetic behavior in *red* and ferromagnetic behavior in *blue*

applying them as nanogranular GMR-sensors. Imagining the conducting matrix gel-like then the sensor fabrication could be done purely by printing on various substrates. Due to the very low coercivity of these Heusler nanoparticles their GMR behavior is slightly hysteretic in the ferromagnetic regime and hence also interesting for sensor applications as is shown in Fig. 5.4b.

With a reliable synthesis route of superparamagnetic or ferromagnetic Heusler nanoparticles with superior magnetic moments the question arises whether these nanomagnets could replace the iron oxide particles as magnetic carriers in in vitro separation technology. The potential of Heusler nanoparticles for biotechnology can directly be evaluated, comparing their magnetic moment of single nanoparticles and the resulting magnetic force which can be exerted on them when applying an external magnetic field gradient. These considerations are shown in Fig. 5.5. Indeed, Co_2FeSi is superior to magnetite due to a 4 times higher moment and a magnetic force which is three orders of magnitude larger. Again, similar data can be expected for Co_2FeGa as well. Thus, biotechnological application can also be envisaged in the future.

However, in order to benefit in technological applications from Heusler nanoparticles some major challenges have to be faced. Above all, stoichiometry has to be controlled. The smaller the size of the particles, the more difficult it becomes to obtain the desired atomically well-ordered Heusler phase.

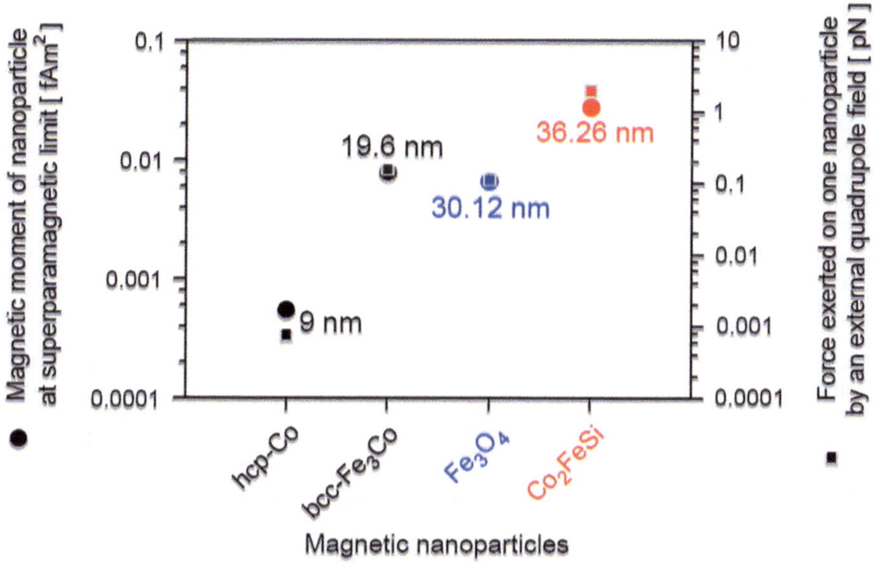

Fig. 5.5 Resulting magnetic moments of one nanoparticle of all regarded magnetic phases given in table 5.1 at the superparamagnetic limit together with the magnetic force exerted on one nanoparticle by an external quadruple field. Co_2FeSi is superior to magnetite due to a four times higher moment and a magnetic force which is three orders of magnitude larger

5.1.1 Progress in Synthesizing Heusler Nanoparticles

For the synthesis of Heusler nanoparticles, different methods have been reported so far [21–24]. A common method is the ball-milling technique, which has successfully been used by Wang et al. for the preparation of 10 nm Ni_2MnGa Heusler nanoparticles [22]. Applying this method, nanoparticles are mechanically prepared by placing a bulk material as starting material in a mill. Macroscopic ceramic balls colliding with the initial bulk material grind it to powder. Jing et al. used a modified gas condensation sputtering technique for the preparation of Fe_3Si nanoparticles [23]. The formation of the DO_3 ordered Fe_3Si phase was indicated by X-ray diffraction (XRD) analysis. With a newly developed colloidal approach Dahal et al. succeeded in the synthesis of 4 to 7 nm Fe_3Si nanoparticles by reacting the preformed iron nanoparticles with silicon tetrachloride at 220 to 250 °C [21]. Recently, the successful synthesis of ternary Co_2FeGa Heusler nanoparticles from precursors was reported and their size-related structures and magnetic properties were analyzed [24–26]. The particles were synthesized in the presence of silica supports by a modified impregnation-calcination method [25]. In this technique, silica spheres are added to the precursor solution of the precursor salts dispersed in methanol. After the removal of the methanol by a rotary evaporator, the obtained solid is dried, grounded to a powder and annealed under H_2–atmosphere. While this approach yielded good results for the Co_2FeGa particles, this approach could not be used to realize Co_2FeAl,

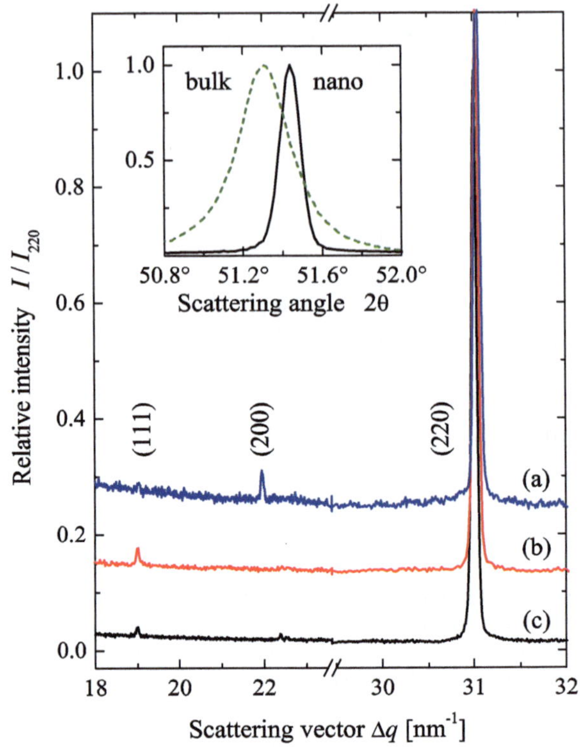

Fig. 5.6 The high resolution diffraction patterns of the Co_2FeGa nanoparticles were taken by anomalous XRD at excitation energies close to the Co (**a**) or Fe (**b**) absorption K-edges using photon energies of 7.7077 keV or 7.112 keV, respectively [24]. The off resonant diffraction data (**c**) were taken at 7.05 keV. The inset shows the (2 2 0) reflections taken from bulk material compared to the nanoparticles (7.05 keV). The broadening of the bulk reflection is a result of the sample preparation

Co_2FeSi or Co_2MnGa particles. Figure 5.6 displays anomalous XRD data measured after synthesis of the Co_2FeGa nanoparticles for different excitation energies that were chosen close to the K absorption edges of cobalt and iron [24]. The (100) and (200) superstructure reflections are evident and proof the $L2_1$ structure of the particles. The lattice parameter at room temperature was determined by Rietveld refinement to a = 0.573 nm, which is slightly smaller than the value for polycrystalline bulk material [27]. This difference may be explained by increased intrinsic pressure in the nanoparticles caused by the comparably high ratio of surface to bulk atoms [24]. The anomalous XRD data additionally confirm their chemical purity and a high crystalline quality as can be concluded from the sharpness of the (200) reflection.

As a general comment we stress here that for a high degree of spin polarization, a well-ordered $L2_1$ or B_2 structure is required. X-ray diffraction, however, is not very suitable to unambiguously determine the crystal structure of Heusler nanoparticles. The nearly equal scattering factors of the constituents may result in particle-size induced broadening or even disappearance of the typical reflections of the face centered cubic lattice such as the characteristic (111) and (200) peak. To overcome these difficulties it has become feasible to use techniques such as extended x-ray absorption fine structure (EXAFS) spectroscopy and anomalous x-ray diffraction (AXD) to clearly identify the crystalline structure of Heusler compound nanoparticles.

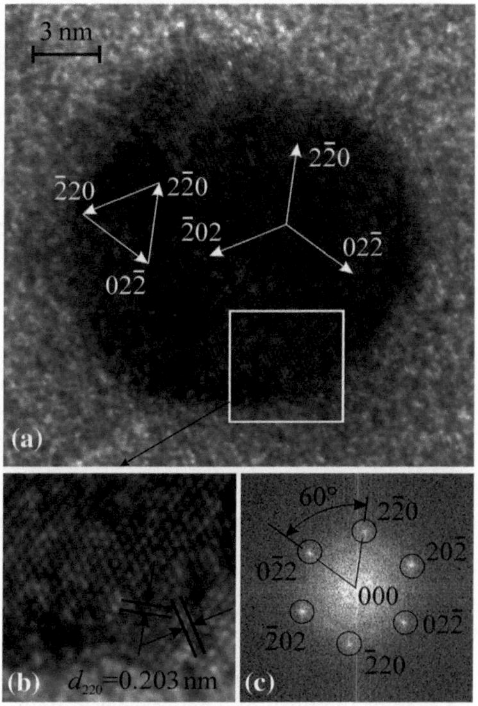

Fig. 5.7 A high resolution TEM image of the Co_2FeGa nanoparticles with a diameter of 18 nm is shown in subfigure (**a**). A part of the nanoparticle can be seen on enlarged scale in subfigure (**b**) and the Fourier transform of the image in subfigure (**c**) [24]

A high resolution transmission electron microscopy (TEM) image of a single 18 nm Co_2FeGa nanoparticle and the Fourier transformation can be seen in Fig. 5.7 [24]. At the given resolution, this data agree with the results of the anomalous XRD measurements.

The size distribution of the particles was obtained by TEM and features its maximum at a diameter of 10–15 nm by comparing Fig. 5.8. A mean diameter of (26.6 ± 47.3) nm was revealed by statistical analysis. The low temperature magnetization at T = 5 K is displayed in Fig. 5.8, where the soft magnetic behavior of the Co_2FeGa nanoparticles is indicated by the small hysteresis, which can also be found on an enlarged scale in the inset (a) [24]. The remanence of the measurement amounts to B_r = 77 mT and the coercivity to H_c = 7.7 kAm^{-1}. The temperature dependence of the magnetization is given in inset (b). At low temperature the value is close to 5 μ_B, which is in good agreement with the value found for polycrystalline bulk material. With increasing temperature it slightly decreases and reaches 4.77 μ_B at room temperature. The fact that the magnetization value corresponds to the expected one for Co_2FeGa in the half-metallic ferromagnetic state suggests that the half-metallic ferromagnetic properties of Heusler compounds are conserved also when dealing with nanoparticles. Moreover, a Curie temperature far above room temperature was found which suggest that these nanoparticles are magnetically stable above room temperature.

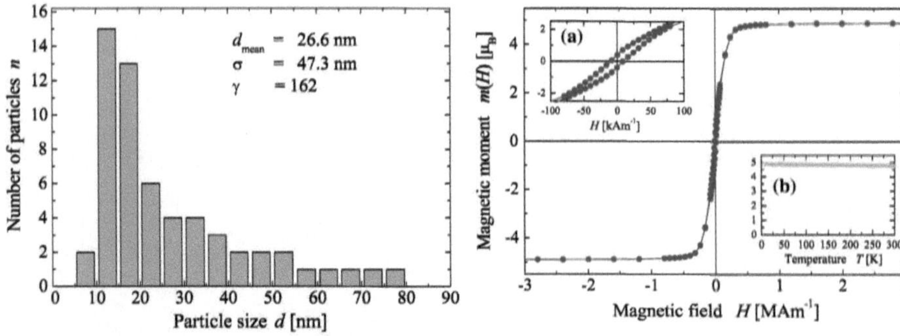

Fig. 5.8 *Left* plot, size distribution of Co_2FeGa nanoparticles determined by transmission electron microscopy [24]. *Right* plot, low temperature magnetization of the Co_2FeGa nanoparticles at T = 5 K is displayed. Inset (**a**) gives the hysteresis close to the origin on an enlarged scale and inset (**b**) the temperature dependence of the magnetization [24]

5.2 Nanoparticular GMR-Effect Based on Co_2FeGa Nanoparticles

The discovery of the giant magneto resistance (GMR) effect in 1988 [28, 29] by Grünberg and Fert, for which they were awarded with the Nobel Prize in 2007, can be seen as the birth of a new research field, namely spintronics. The large change of resistivity in response to an applied magnetic field was originally found in multilayer structures where ferromagnetic layers (Fe) are separated by a nonmagnetic conducting spacer (Cr) with thickness reduced to nanoscale. The electric resistance of the device depends on the relative orientation of the magnetizations in the two layers. It is low for a parallel and high for an antiparallel configuration due to an increased scattering probability of the conducting electrons in the latter case. In 1992, the GMR effect was also found in granular systems with magnetic particles embedded in a metallic matrix [30, 31]. Here, the scattering probability of the conducting electrons is high if the magnetizations of the single granules are statistically distributed and a low degree of magnetic order is found along the assembly, whereas the scattering probability is low when an external magnetic field is aligning all magnetic moments of the nanoparticles.

With the progressive miniaturization of functional devices, the superparamagnetic limit is approached. On these scales thermal fluctuations become significant, which is why larger anisotropies are necessary to obtain stable magnetic equilibrium states. This means that larger magnetic fields are required for switching as well as for writing the magnetizations. For magnetic field sensors based on magnetoresistive effects, this corresponds to decreasing sensor sensitivity. A possibility to overcome this problem is the change to granular material, where magnetic domains are separated by nonmagnetic but well-conducting metallic matrices [30, 31].Instead of conventional metallic matrices granular GMR devices can also be realized by embedding magnetic nanoparticles in conductive gels [32].

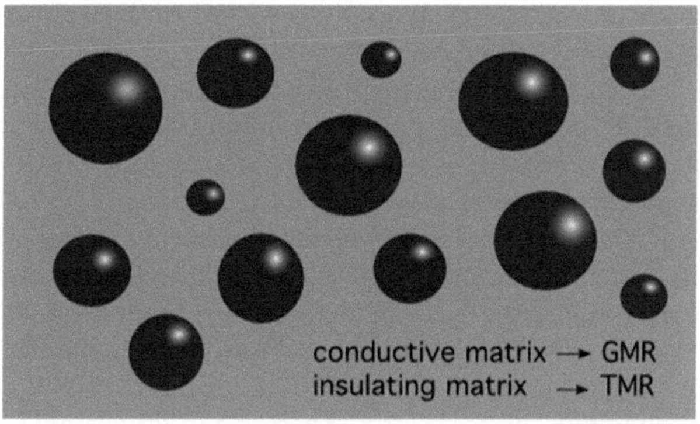

Fig. 5.9 Incorporation of nanoparticles in a conductive matrix yields a granular GMR system. Exchanging the matrix by an insulating one leads to a granular tunneling TMR system

Incorporation of Heusler compound nanoparticles in those granular systems could bring further progress as one would profit from the qualities of Heusler compounds already known from layer systems. The soft magnetic behavior of Heusler compounds, for example, makes them ideal candidates for the employment in GMR devices, see above. In transport measurements a soft switching leads to narrow peaks in the GMR curve. In regard of magnetoresistive sensors, this is of high technological relevance since it results in high detector sensitivity [33]. As a first step in this direction we embedded Co_2FeGa nanoparticles which were already described above in a conductive hydrogel commercially available in pharmacies. The particles arrangement is sketches in Fig. 5.9. After this preparation step particle-gel samples on a silicon dioxide wafers can be investigated. Transport measurements using a 4-point-probe method in an in plane configuration were carried out at room temperature. The measurements revealed magneto resistance effects of up to 120 % [34] as is shown in Fig. 5.10. This is far above common values for granular systems measured at room or even at low temperatures [31, 35, 36]. Hence, this result is a proof of principle for

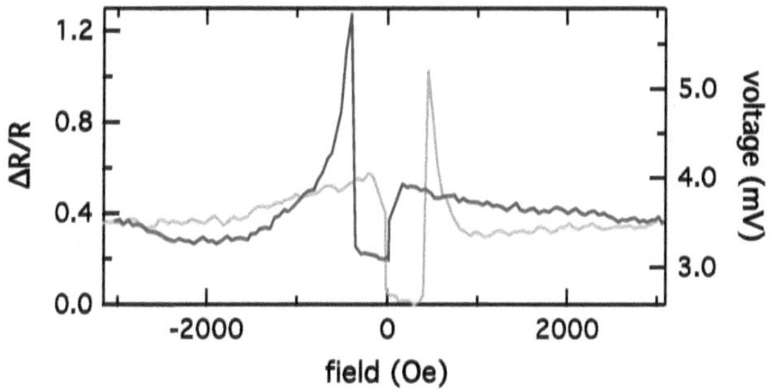

Fig. 5.10 Transport measurement of Co_2FeGa nanoparticles in a hydrogel matrix with an effect amplitude of 120 % at room temperature. In this context, *red* color indicates increasing and *blue* color decreasing magnetic field. The hysteresis is attributed to the ferromagnetic behavior of the particles (Color figure online)

nanoparticular GMR-Heusler-sensors which even puts our considerations in Sect. 5.1 in a proper perspective.

Using the D_{SPM} of Co_2FeGa as a reference, it becomes clear that the particles used for determining the nanoparticular GMR show a very brood size distribution as can be seen in Fig. 5.8. Thus, the resulting GMR characteristic is based on a superparamagnetic as well as a ferromagnetic contribution. In contrary to the calculation of Sect. 5.1 dipolare interactions between neighboring nanoparticles cannot be neglected anymore and will also contribute to the overall GMR characteristics. This clearly suggests that the particle size distribution is another crucial issue in preparing Heusler nanoparticles for the realization of nanoparticular GMR-sensors.

The gel matrix additionally offers the possibility to deliberately arrange the nanoparticles in the matrix. Agarose gel, for example, gets liquid when it is heated and nanoparticles can be introduced. Cooling down the particle-gel mixture initializes the gelling process. When the gelling process is finished the nanoparticles are fixed with respect to neighboring particles. Applying a homogenous magnetic field during preparation when the matrix is still in its liquid state, however, allows to arrange particles in chain fragments as is presented in Fig. 5.11. Chain fragments deliberately orientated in the direction of the current during transport measurements, for instance, lead to increased spin-dependent scattering along the chains and is associated with higher GMR-effect amplitudes. In Fig. 5.12 this is exemplary presented using Co nanoparticles in an agarose gel.

The implementation of Heusler compounds (TMR) stacks has led to a drastic increase of the magneto resistance effect amplitudes. Although the TMR effect, where electrons tunnel through an insulating barrier, has been established in artificial layered structures in ferromagnetic metal—insulator—metal junctions, it can also be utilized in granular films consisting of ferromagnetic clusters separated by an insulating matrix [37–39]. An advantage of granular TMR systems compared to conventional stack devices is their electrical stability. In particular, the breakdown of

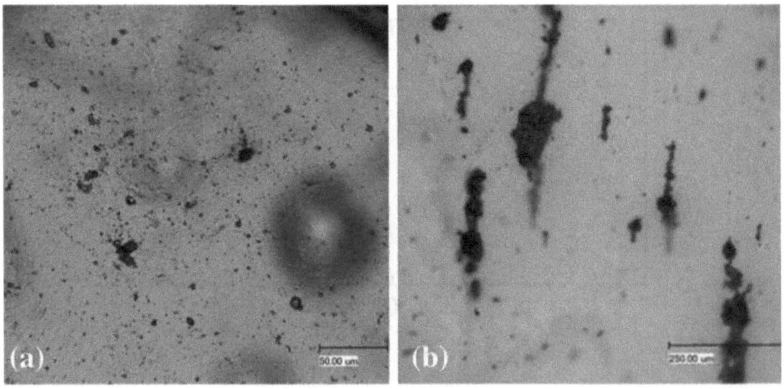

Fig. 5.11 Co nanoparticles in agarose gel (**a**) and Co nanoparticle chain fragments in agarose gel that have been created by applying a homogenous magnetic field during sample preparation (**b**) [23]

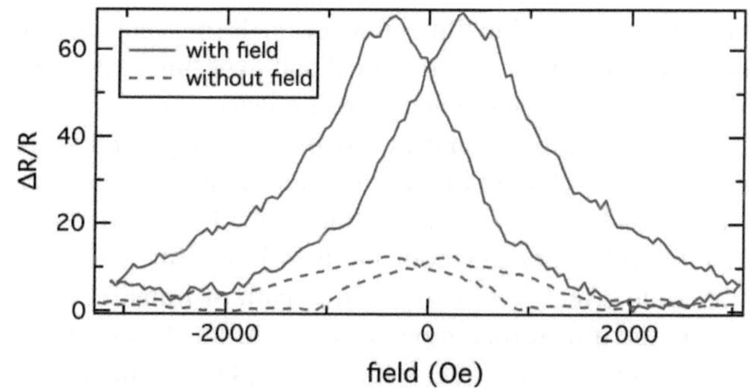

Fig. 5.12 GMR measurements of Co nanoparticles in agarose matrix. For the sample with a magnetic field applied during preparation, nanoparticle chain fragments could evolve and an effect amplitude of more than 60 % was achieved, whereas for the sample with no field applied during preparation, no chains could be found in the sample and an effect amplitude of only 12 % was measured [23]

the tunneling barrier, which poses problems in thin films, can be avoided by switching to granular materials. Granular TMR systems based on Heusler nanoparticles realized by embedding Heusler nanoparticles in an isolating environment, see Fig. 5.9, are expected to significantly raise the conventionally achieved TMR effect amplitudes. Regarding magnetoresistive sensors this would again result in higher sensitivities giving rise to substantial potential for technical applications. Therefore, there is still a lot of room for further experimental as well as theoretical studies of Heusler based nanoparticles in the fields of magnetoresistive effects.

5.3 The Limits for the Austenite-Martensite Transformation in Ultra-Thin Films

Thin films made of Heusler compounds are not only relevant for spintronic applications. There is also great interest in the class of shape-memory compounds like Ni-Mn-X (X = Ga, Sn, or In) in thin films. For applications such as actuators, magnetic cooling and hybrid systems, thin films are often advantageous or mandatory. It was shown that shape memory alloys in thin films can be used as microfluidic valves, micro pumps and micro grippers [40]. For magnetic cooling devices, thin films offer the advantage of a high surface to volume ratio. This allows a fast heat transfer.

Shape-memory Heusler alloyed thin films have also high potential in spintronic applications: in a multilayer structure, a shape-memory thin film can induce stress into the adjacent layers due to the martensitic transition. This stress changes the band-structure and thus could be used to switch half-metallicity on and off by magnetic field or temperature changes.

A general knowledge of the thin film influence on the martensitic transition (MT) is required for all these applications. Theoretical and experimental studies have shown that the transformation behavior as well as the microstructure changes significantly in thin films [41, 42]. Investigations on the transformation behavior in thin films reveal a broader transition and a change in the transition temperatures when compared to thicker films. The broadening of the transition as well as a reduced transition temperature can be caused by substrate constraints [43], confinement of the martensitic nucleus and size scale effects on the mean free path of transformation dislocations [42].An increase in the transition temperature can be attributed to stress induced by the lattice mismatch between substrate and thin film [41]. In thin films the substrate as well as induced texture plays a major role. The degree and kind of texture has a large influence on the allowed martensitic structure variants [44]. A rigid substrate leads to symmetry breaking effects in the sense that not all orientations of the austenite-martensite interfaces allow coarsening of nano twinned martensite to macroscopic non modulated structure variants [45]. Hence, the focus here will be on sorting out all influences on the martensitic transformation as a function of the thin film thickness

As for most Heusler compound thin films aiming for spintronic applications, the preferred fabrication method for shape memory alloyed thin films is magnetron sputtering. The lattice constants for most Ni-Mn based shape-memory compounds is approximately in the range between 5.8 and 6.0 Å which makes single-crystal MgO (001) the preferred substrate material to obtain epitaxial films. Also, depending on the material, Cr is often used as a seed layer. Epitaxy is also reported on Al_2O_3 [46] and GaAs with a $Sc_{0.3}Er_{0.7}As$ buffer [47].

5.3.1 The Martensitic Transformation in Real Crystals and Thin Films

In thin films, the most important peculiarities are the dimension of the film and the presence of a rigid substrate. The substrate prohibits any macroscopic length changes of the film and clamps the atoms close to the interface at the substrate. Thus, to transform the material close to the substrate, additional elastic energy is required for the displacement of atoms.

The transformation is suppressed close to the substrate until the necessary driving force in form of chemical energy is provided which requires undercooling. That means for lower parts of the film that the martensitic transition starts at significantly lower temperatures. This leads to a delayed completion of the transformation and to a residual austenite layer in the vicinity of the substrate even at low temperatures. The thickness of this austenite layer is not dependent on the film thickness but on the lattice mismatch between the film and the substrate and the phase compatibility between martensite and austenite. Thus, both factors are more important in thin films than in bulk material due to the required presence of a substrate. Another, usually desired, side effect of the substrate is epitaxial growth if the interatomic spacing of the thin film and substrate material is comparable.

Besides the influence of the substrate there are also size scale effects which purely originate from the film thickness and resulting grain size.

5.3.1.1 The Influence of Epitaxy Films

In general, epitaxial thin film are favored because one major advantage of epitaxial thin films over bulk material is the fixed reference frame determined by the single crystalline substrate which allows to study crystallographic orientations in absolute coordinates. In addition, the geometrical constraints at the interface to the substrate were found to stabilize otherwise thermodynamically unstable phases [48].

These features were used by Fähler et al. to study the influence of the martensitic microstructure on the shape-memory effect in Ni-Mn-Ga [45, 48, 49].

It was found that the martensite forms an exact interface, the habit plane, to the residual austenite near the substrate by an adaptive, modulated microstructure. It was also ascertained that modulated monoclinic martensite and tetragonal martensite are formed by identical tetragonal building blocks. Within this concept, the modulated phase is a nano twinned tetragonal phase which forms to assure a coherent interface to the austenite. The occurrence of different modulation periodicities (6 M, 10 M, 14 M) is observed in bulk material depending on the valence electron density e/a. These can also be regarded as adaptive, tetragonal phases. Since the tetragonality of the martensite increases with the valence electron concentration, different modulation periods are required to form a coherent interphase to the austenite.

Nonmodulated martensite (NM) forms created the modulated one by a coarsening process where the variant width is subsequently doubled to reduce the twin-boundary energy. This process starts at the habit plane near the substrate and leads up to micrometer sized variants at the film surface. This coarsening process does not change the variant ratio which is fixed by the prohibition of length changes of the thin film.

Due to geometrical constraints there are variants of the modulated phase which form a coherent interface to the austenite but do not allow coarsening. Here, the metastable 14 M is stabilized over the thermodynamically preferred NM. This results in nano twinned microstructure visible at the film surface. The coexistence of 14 M and NM at the sample surface is shown in Fig. 5.13. Such behavior is attributed to the symmetry breaking effect of the rigid substrate and it is not expected for bulk single crystals. So in this case the experimental advantages of thin films over bulk single crystals allow a better understanding on the martensitic microstructure itself.

5.3.1.2 The Influence of Phase Compatibility

The phase compatibility, also called geometric compatibility between martensite and austenite is a measure for how well the martensite fits to the austenite. In other words,

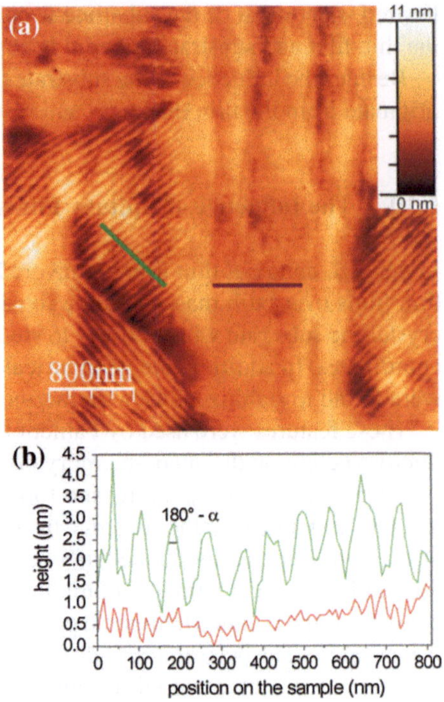

Fig. 5.13 **a** Atomic force micrograph of the surface topography of a 420 nm thick $Ni_{54.8}Mn_{22.0}Ga_{23.2}$ film. It reveals two types of features. Finely twinned regions of the 14 M phase (*green line*) and large variants of the nonmodulated martensite (*red*). In **b** the height profiles along both lines are plotted. The pictures edges are along the MgO<100> directions. Taken from [45]

if the phases are compatible, only one single variant of martensite is required to fulfill the boundary condition to the austenite.

The phase compatibility is given if $\lambda_2 = 1$, where λ_2 is the eigenvalue of the transformation stretch or Bain matrix U with $\lambda_1 \leq \lambda_2 \leq \lambda_3$, [44]. If martensite and austenite are not compatible, $\lambda_2 \neq 1$, a fine mixture of laminate twins is generated as shown in Fig. 5.14. This fine laminate mixture is approximately compatible with the austenite, but it also involves elastic energy and interfacial energy between the variants. This energy barrier has to be overcome during the forward and reverse transformation, thus giving rise to a transformation hysteresis. The phase compatibility

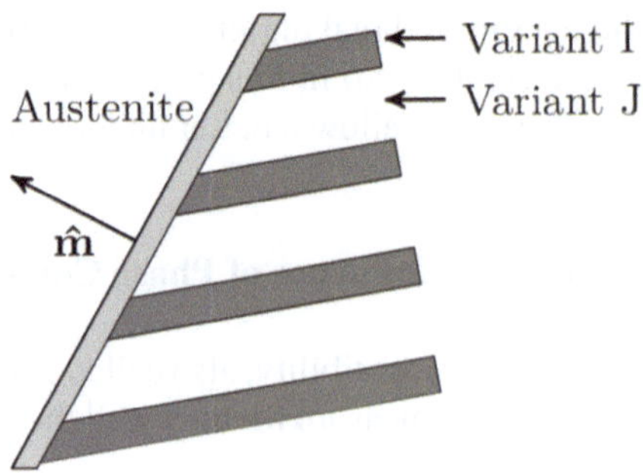

Fig. 5.14 A typical austenite-martensite interface. I and J describe two martensite Variants. $\hat{m}$ is the orientation of the austenite/martensite interface. A better phase compatibility leads to a coarser variant structure

is dependent on both lattice parameters of the martensite and austenite phase and hence has strong compositional dependence.

One of the six transformation stretch matrices for the cubic to orthorhombic transformation is given by [44]:

$$\mathbf{U} = \begin{pmatrix} \beta & 0 & 0 \\ 0 & \frac{\alpha+\gamma}{2} & \frac{\alpha-\gamma}{2} \\ 0 & \frac{\alpha-\gamma}{2} & \frac{\alpha+\gamma}{2} \end{pmatrix}$$

where $\alpha = a/a_0$, $\beta = b/a_0$ and $\gamma = c/a_0$. a_0 is the lattice constant of the cubic phase and a, b and c are the lattice constants of the orthorhombic phase.

The influence of the phase compatibility can be understood in the following way: A better phase compatibility leads to a coarser laminate structure. A lower number of interfaces reduce the probability of interaction with defects. Thus the transition is less hindered by defects. Furthermore, a coarser laminate structure requires less energy for its formation due to the lower number of interfaces. This effect becomes important at low film thicknesses: Here a high chemical energy is required to overcome the additional energy contributions of the substrate. Thus, the importance of a higher energy due to a finer laminate structure increases with decreasing film thickness. Both, the additional energy and the probability to interact with defects can have a large influence on the critical film thickness, residual austenite, undercooling and the hysteresis width.

5.3.1.3 The Influence of the Crystal Quality

The martensitic transformation in real crystals depends to a large degree on its quality. It is determined by the number of defects which can be of point (0D), line (1D), planar (2D) or precipitate (3D) type. Generally, such defects affect the martensitic nucleation and growth. The activation barrier for martensitic nucleation is reduced by defects due to their local strain fields. This leads to local variations of the transition temperature as e.g. determined by Malygin for dislocations [50] and precipitates [51]. The nucleation of martensite especially occurs at complex structural defects like interphase boundaries, free surfaces of aggregations or point defects [52]. Grain boundaries itself hinder the nucleation, however they are sources of dislocations which itself promote the martensitic nucleation [53].

While defects promote the nucleation, the growth process is hindered by it. In a perfect crystal a nucleation center could transform the entire volume of the crystal, if the bulk energy is larger than the surface energy. However, in a real crystal the growth is stopped at defects and additional energy, corresponding to a small temperature change, is required to overcome the obstacle.

In epitaxial thin films, the lattice mismatch between the substrate and the film has a high impact on the crystal quality. A lattice mismatch leads to dislocations near the interface to the substrate which occur because the film adapts the lattice constant

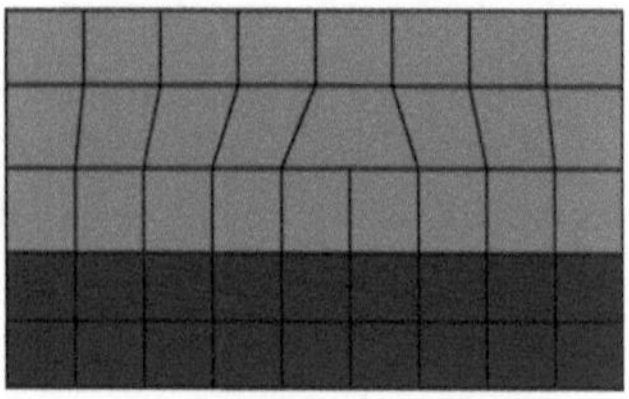

Fig. 5.15 In epitaxial films, the lattice constant relaxes under the incorporation of dislocations. The substrate is depicted in *blue* and the film in *red* (Color figure online)

from the substrate at the interface leading to strain. This strain is reduced within the film by the incorporation of misfit dislocations as is sketched in Fig. 5.15.

Above a critical thickness where the relaxation process starts, the density of misfit dislocations decreases with increasing film thickness due to the progressing relieve of the lattice misfit [54].

5.3.1.4 The Influence of Size Scale Effects

Size scale effects denote the general influence of the grain size or film thickness on the martensitic transition where other effects such as substrate influences are not taken into account. According to a theoretical consideration by Malygin [42] the change of the transformation behavior can be attributed to two different effects: On the one hand, the kinetics of the transformation is influenced by the mean free path of transition dislocations λ. On the other hand, the thermodynamics of the transition is affected by the confinement of the martensitic nucleus assumed disk shaped with diameter l. Both, film thickness and grain size become important when the size scales of these become similar to λ and/or to l leading to a hindered martensitic transformation.

Meng et al. determined the influence of the grain size on the nucleation barrier and critical nucleation size by consideration of grain boundary energy and interface boundary energy. Below 100 nm grain size a rise of the nucleation barrier and a larger critical nucleus was found [55]. Thus both approaches predict lower martensite start temperatures (M_s) as can be seen in Fig. 5.16 and an increasing transformation range for increasing confinement.

5.3.2 The Martensitic Transformation in Ultra-Thin Films

Systematic studies on the influence of the film thickness on the martensitic transition (MT) are sparse in literature. Epitaxial Ni-Mn-Ga films have been investigated in the thickness range of 150 to 500 nm with varying composition [45]. An increasing influence of the substrate with decreasing thickness has been found due to stress induced martensite at the interface. Polycrystalline Ni-Mn-Sn films grown on Si have been investigated in the thickness range from 120 nm up to 2.5 μm by Vishnoi

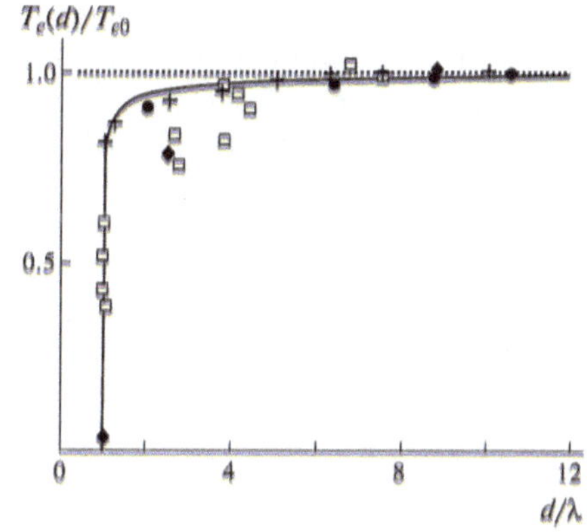

Fig. 5.16 Characteristic temperature of the transition T_C versus critical grain size d in Cu-Zn-Sn and Fe-Ni-C alloys according to Malygin. T_{C0} is the characteristic temperature in the absence of size effects and λ the mean free path between transformation dislocations. Taken from [42]

et al. [56]. They found a suppression of the MT below 410 nm and an increasing transition temperature with increasing film thickness. Recently, Ranzieri et al. [57] investigated Ni-Mn-Ga films with thicknesses down to 10 nm and observed a MT in films thicker than 40 nm. Most other studies on the MT in thin films investigated a single film thickness in the range of several 100 nm up to several μm [41, 58–61].

Here, some experimental results of Auge and Teichert et al. [62, 63] on the martensitic transition in Ni-Mn-Sn ultra-thin films down to 10 nm, are presented. The two compositions $Ni_{51.6}Mn_{32.9}Sn_{15.5}$ (SERIES A) and $Ni_{51.6}Mn_{34.9}Sn_{13.5}$ (SERIES B) have been investigated.

The result of electrical resistivity over temperature measurements are depicted in Fig. 5.17. It is clearly visible that the resistivity change and thus the relative amount of transforming material decreases with decreasing film thickness. This is attributed

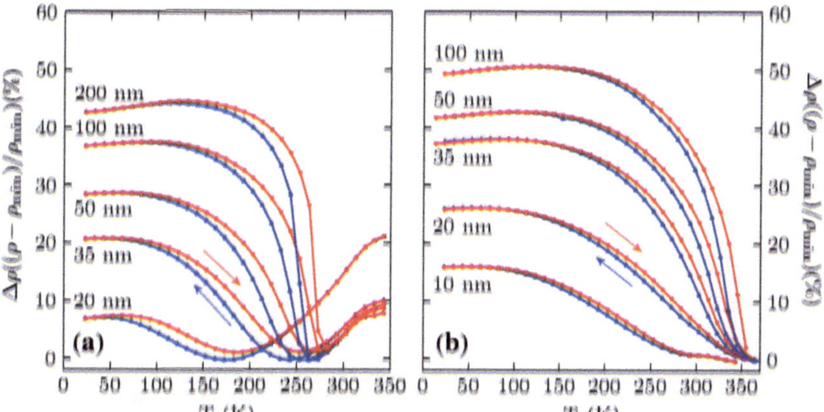

Fig. 5.17 The relative resistivity change for two Ni-Mn-Sn thin film thickness series is shown. The martensite phase has a higher resistivity than austenite. The *arrows* indicate the direction of temperature change. **a** SERIES A and **b** SERIES B. Taken from [63]

to a layer of residual austenite on the substrate which does not transform and shunts the electrical transport in the film. This layer is independent of the film thickness and thus the effect is more pronounced in thinner films. The thickness of this layer is determined to 8.4 nm for SERIES A and 2.5 nm for SERIES B.

Another indication of the presence of the substrate is a slow-down of the transition at the low temperature end of the hysteresis. This is due to the required undercooling to transform material close to the substrate.

The hysteresis width of SERIES B is smaller than in SERIES A. This, together with the lower amount of residual austenite indicates a better phase compatibility of SERIES B. This is supported by the eigenvalue λ_2 which is 1.005 and very close to 1 for SERIES B.

M_s decreases with decreasing film thickness and the transition range extends over a larger temperature range. Apart from substrate influences this is induced by size scale effects. The martensitic nucleus is confined by the small film thickness what delays and smears out the transition.

The constraints of the rigid substrate can be discarded by the fabrication of free-standing films. For this purpose, the film is usually deposited onto a sacrificial layer [47, 64, 65] or a dissolvable substrate [66]. Also focused ion beam (FIB) can be employed to cut out freestanding cantilevers [67].

Deposition of Ni-Mn-Ga on dissolvable (110) NaCl single crystal substrates turned out to be rather difficult because of the hygroscopic behavior of the material. FIB is a very time consuming technique and also it is only suitable to release small cantilevers.

A common technique to obtain freestanding Ni-Mn-Ga films is to use standard single crystal MgO substrates with a Cr buffer layer of up to 100 nm thickness. The Ni-Mn-Ga grows epitaxial on the Cr layer. The film is released by selective chemical wet etching with a suitable Cr etchant.

In contrast to Heusler compound thin films designed for spintronic which do not show a marked impact on the TMR effect amplitude we have identified layer thickness and other constrains for shape memory Heusler compounds down to a layer thickness of 10 nm.

Acknowledgments The authors affiliated with Bielefeld University would like to thank the FOR 945, the SPP 1599, and International Office of BMBF for financial support in the framework of the project 3, A6 and TUR09/I01, respectively.

References

1. T. Graf, C. Felser, S.S.P. Parkin, Prog. Solid State Chem. **39**, 1 (2011)
2. R. de Groot, F. Mueller, P. van Engen, K. Buschow, Phys. Rev. Lett. **50**, 2024 (1983)
3. J. Tobola, J. Pierre, S. Kaprzyk, R.V. Skolozdra, M.A. Kounacou, J. Phys.: Condens. Matter **10**, 1013 (1998)
4. J. Tobola, J. Pierre, J. Alloys Compd. **296**, 243 (2000)
5. I. Galanakis, P.H. Dederichs, N. Papanikolaou, Phys. Rev. B **66**, 134428 (2002)

6. S. Ishida, T. Masaki, S. Fujii, S. Asano, Physica B **245**, 1 (1998)
7. A. Ayuela, J. Enkovaara, K. Ullakko, R.M. Nieminen, J. Phys.: Condens. Matter **11**, 2017 (1999)
8. A. Deb, Y. Sakurai, J. Phys.: Condens. Matter **12**, 2997 (2000)
9. I. Galanakis, P.H. Dederichs, N. Papanikolaou, Phys. Rev. B **66**, 174429 (2002)
10. C.T. Tanaka, J. Nowak, J.S. Moodera, J. Appl. Phys. **86**, 6239 (1999)
11. S. Yuasa, T. Nagahama1 A. Fukushima, Y. Suzuki, K. Ando. Nat. Mater. **3**, 868 (2004)
12. S.S. Parkin, C. Kaiser, A. Panchula, P.M. Rice, B. Hughes, M. Samant, S.H. Yang, Nat. Mater. **3**, 862 (2004)
13. J. Rogge, P. Hedwig, C. Sterwerf, A. Hütten, IEEE Trans. Magn. **48**, 3825 (2012)
14. R.C. O'Handley, *Modern Magnetic Materials: Principles and Applications* (Wiley, New York, 2000)
15. S. Trudel, O. Gaier, J. Harmle, B. Hillebrands, J. Phys. D. Appl. Phys. **43**, 193001 (2010)
16. M. Hashimoto, J. Herget, H.-P. Schönherr, K.H. Ploog, Appl. Phys. Lett. **87**, 102506 (2005)
17. C. Wang, L. Basit, Y. Khalavka, Y. guo, F. Casper, T. Gasi, V. Ksenofontov, B. Balke, G.H. Fecher, C. Sönnichen, Y.-K. Hwu, J.-J. Lee, C. Felser. Chem. Mater. **22**, 6575 (2010)
18. A. Hütten, D. Sudfeld, K. Wojczykowski, P. Jutzi, G. Reiss, J. Magn. Magn. Mater. **262**, 23 (2003)
19. M.B. Stearns, Y. Cheng, J. Appl. Phys. **75**, 6894 (1994)
20. D.J. Sellmyer, C.P. Luo, Y.Qiang, *Handbook of Thin Film Devices Magnetic, 5 , Nanomaterials And Magnetic Thin Films*, vol 37 (Academic Press, New York, 2000)
21. N. Dahal, V. Chikan, Chem. Mater. **22**, 2892 (2010)
22. Y.D. Wang, Y. Ren, H.Z. Nie, D.M. Liu, L. Zuo, H. Choo, H. Li, P.K. Liaw, J.Q. Yan, R.J. Mcqueeney, J.W. Richardson, A. Huq, J. Appl. Phys. **101**, 063530 (2007)
23. Y. Jing, Y.H. Xu, J.P. Wang, J. Appl. Phys. **105**, 07B520 (2009)
24. L. Basit et al., J. Phys. D Appl. Phys. **42**, 084018 (2009)
25. C. Wang et al., Chem. Mater. **22**, 6575–6582 (2012)
26. C.H. Wang, Y.Z. Guo, F. Casper, B. Balke, G.H. Fecher, C. Felser, Y. Hwu, Appl. Phys. Lett. **97**, 103106 (2010)
27. K.H.J. Buschow, P.G. van Engen, J. Magn. Magn. Mater. **25**, 90–6 (1981)
28. M.N. Baibich, J.M. Broto, A. Fert, F. Nguyen Van Dau, F. Petroff, Phys. Rev. Lett. **61**(21), 2472–2475 (1988)
29. G. Binasch, P. Grünberg, F. Saurenbach, W. Zinn, Phys. Rev. B **39**(7), 4828–4830 (1989)
30. A.E. Berkowitz, J.R. Mitchell, M.J. Carey, A.P. Young, S. Zhang, F.E. Spada, F.T. Parker, A. Hütten, G. Thomas, Phys. Rev. Lett. **68**(25), 3745–3748 (1992)
31. J.Q. Xiao, J. Samuel Jiang, C.L. Chien, Phys. Rev. Lett. **68**(25), 3749–3752 (1992)
32. J. Meyer, T. Rempel, M. Schäfers, F. Wittbracht, C. Müller, A.V. Patel, A. Hütten, Smart Mater. Struct. **22**, 025032 (2013) (accepted for publication)
33. A. Weddemann et al., Beilstein J. Nanotechnol. **1**, 75–93 (2010)
34. C. Wang, J. Meyer, N. Teichert, A. Auge, E. Rausch, B. Balke, A. Hütten, G.H. Fecher, C. Felser, J. Vac. Sci. Technol. B **32**(2), 020802 (2014)
35. P. Allia, Phys. Rev. B **52**, 15398 (1995)
36. Y. Chen, J. Appl. Phys. A **73**, 103–106 (2001)
37. H. Fujimori, S. Mitani, S. Ohnuma, J. Magn. Magn. Mater. **156**, 311–314 (1996)
38. J. Inoue, S. Maekawa, Phys. Rev. B **53**(18), R11 927 (1996)
39. M. Holdenried, B. Hackenbroich, H. Micklitz, J. Magn. Magn. Mater. **231**, L13–L19 (2001)
40. B. Winzek, S. Schmitz, H. Rumpf., T. Sterzl, R. Hassdorf, S. Thienhaus, J. Feydt, M. Moskeand E. Quandt. Mater. Sci. Eng. A **378**, 40 (2004)
41. D.C. Dunand, P. Müllner, Adv. Mater. **23**, 216 (2011)
42. G. Malygin, Tech. Phys. **54**, 1782 (2009)
43. A. Roytburd, T. Kim, Q. Su, J. Slutsker, M. Wuttig, Acta Mater. **46**, 5095 (1998)
44. K. Bhattacharya, *Microstructure of Martensite*, (Oxford University Press, Oxford, 2003)
45. S. Kaufmann, R. Niemann, T. Thersleff, U.K. Rösler, O. Heczko, J. Buschbeck, B. Holzapfel, L. Schultz, S. Fähler, New J. Phys. **13**, 053029 (2011)

46. G. Jakob, H.J. Elmers, J. Magn. Magn. Mater **310**, 2779 (2007)
47. J.W. Dong, L.C. Chen, C.J. Palmstrøm, R.D. James, S. McKernan. Appl. Phys. Lett. **75**, 1443 (1999)
48. S. Kaufmann, U. Rößler, O. Heczko, M. Wuttig, J. Buschbeck, L. Schultz, S. Fähler, Phys. Rev. Lett. **104**, 145702 (2010)
49. M. Thomas, O. Heczko, J. Buschbeck, U.K. Rößler, J. McCord, N. Scheerbaum, L. Schultz, S. Fähler, New J. Phys. **10**, 023040 (2008)
50. G. Malygin, Phys. Solid State **45**, 345 (2003)
51. G. Malygin, Phys. Solid State **45**, 1566 (2003)
52. A. Potekaev, A. Klopotov, V. Kulagina, V. Gyunter, Steel Transl. **40**, 881 (2010)
53. B. Li, X.M. Zhang, P.C. Clapp, J.A. Rifkin, J. Appl. Phys. **95**, 1698 (2004)
54. M. Ohring, *Materials Science of Thin Films*, (Academic Press, New York, 2001)
55. O. Meng, Y. Rong, T.Y. Hsu, Phys. Rev. B **65**, 174118 (2002)
56. R. Vishnoi, R. Singhal, D. Kaur, J. Nanopart. Res. **13**, 3975 (2011)
57. P. Ranzieri, S. Fabbrici, L. Nasi, L. Righi, F. Casoli, V.A. Chernenko, E. Villa, F. Albertini, Acta Mater. **61**, 263 (2013)
58. V. Recarte, J.I. Pérez-Landazábal, V. Sánchez-Alárcos, V.A. Chernenko, M. Ohtsuka, Appl. Phys. Lett. **95**, 141908 (2009)
59. R. Vishnoi, D. Kaur, Surf. Coat. Technol. **204**, 3773 (2010)
60. R. Niemann, O. Heczko, L. Schultz, S. Fähler, Appl. Phys. Lett. **97**, 222507 (2010)
61. K. Załęski, J. Dubowik, I. Gościańska, B. Andrzejewskiand, T. Toliński, Cent. Eur. J. Phys. **9**, 558 (2011)
62. A. Auge, N. Teichert, M. Meinert, G. Reiss, A. Hütten, E. Yüzüak, İ. Dinçer, Y. Elerman, Phys. Rev. B. **85**, 214118 (2012)
63. N. Teichert, A. Auge, E. Yüzüak, I. Dincer, Y. Elerman, B. Krumme, H. Wende, O. Yildirim, K. Potzger, A. Hütten, Acta Mater. **86**, 279 (2015)
64. J. Tillier, D. Bourgault, B. Barbara, S. Pairis, L. Porcar, P. Chometon, D. Dufeu, N. Caillault, L.J. Carbone, Alloys Comp. **489**, 509 (2010)
65. A. Backen, S.R. Yeduru, M. Kohl, S. Baunack, A. Diestel, B. Holzapfel, L. Schultz, S. Fähler, Acta Mater. **58**, 3415 (2010)
66. M. Thomas, O. Heczko, J. Buschbeck, Y.W. Lai, J. McCord, S. Kaufmann, L. Schultz, S. Fähler, Adv. Mater. **21**, 3708 (2009)
67. C.A. Jenkins, R. Ramesh, M. Huth, T. Eichhorn, P. Pörsch, H.J. Elmers, G. Jakob, Appl. Phys. Lett. **93**, 234101 (2008)

Chapter 6
Chemical Bonding in MgAgAs-Type Compounds

D. Bende, Yu. Grin and F.R. Wagner

Abstract The chemical bonding in MgAgAs-type compounds of the main-group elements is analyzed with the aid of quantum chemical direct-space techniques. These phases show a variety of different bonding patterns from highly covalent interactions as in diamond to mostly ionic bonding as in the zinc-blende type compound BeS. The formal concept of the $8 - N$ rule (octet rule) is related to the direct-space techniques to rationalize the results and compare the MgAgAs-type compounds with classical valence compounds like elements of the diamond structure, zinc-blende and Zintl phases.

6.1 Crystal Structure and Structural Relations

Members of the MgAgAs-type family are ternary compounds with a 1:1:1 stoichiometric ratio of the components. The cubic face-centered crystal structure is characterized by the space group $F\bar{4}3m$ (Fig. 6.1). Lattice parameters are found to lie between five and seven Angstrom. Magnesium and silver constitute a rock-salt partial structure, arsenic forms zinc-blende partial structures with both other components. Thus, it has eight nearest neighbours, the coordination polyhedron is a cube. As the vertices of this cube are formed by alternating Mg and Ag atoms, this position is called the heterocubic site (HC), hereafter. Mg and Ag each have four As atoms in the first coordination sphere forming a tetrahedron. There is another set-up of the MgAgAs structure type that illustrates its relation to the fluorite crystal structure. Within the face-centered lattice of arsenic, Mg and Ag populate all its tetrahedral voids (Fig. 6.1). Thus, the MgAgAs-type is a superstructure (colouring variant) of the CaF_2 type.

D. Bende · Y. Grin (✉) · F.R. Wagner
Max-Planck-Institut für Chemische Physik fester Stoffe,
Nöthnitzer Straße 40, 01187 Dresden, Germany
e-mail: juri.grin@cpfs.mpg.de

D. Bende
e-mail: david.bende@cpfs.mpg.de

F.R. Wagner
e-mail: frank.wagner@cpfs.mpg.de

C. Felser and A. Hirohata (eds.), *Heusler Alloys*,
Springer Series in Materials Science 222, DOI 10.1007/978-3-319-21449-8_6

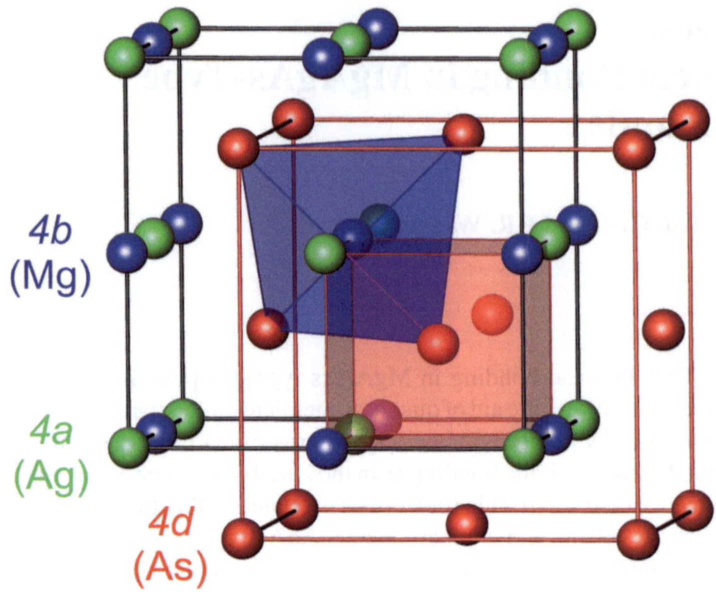

Fig. 6.1 Atomic arrangement in the structure type of MgAgAs with two interpenetrating face-centered partial structures. The coordination polyhedra of Ag and As are also shown

Concerning only the number of nearest neighbours, there are two qualitatively different positions in the MgAgAs structure type—one with cubic and two with tetrahedral coordination sphere. This results in three possible atomic arrangements of a compound XYZ according to the three possibilities to occupy the heterocubic site. Although they are configurationally isotypic [1] one may distinguish three chemical subtypes with either the most electronegative or one of the less electronegative elements at the heterocubic site. For practical reasons the following notation for the three arrangements is introduced—a compound XYZ with the MgAgAs-type crystal structure and the Z component at the heterocubic site is written as XYZ^{HC}.

MgAgAs-type compounds crystallize with a variety of different elements. Compounds containing three main-group elements $A'AE$ are known with A' and A being early main elements of group 1, 2 or 13. E is typically a late main-group element of group 14 or 15. $A'AE$ compounds have 8 valence electrons per formula unit. MgAgAs-type compounds with up to two transition metals (T' and T) and rare-earth elements (R) are known: ATE, $T'TE$ and RTE. The semiconducting representatives of the transition metal containing compounds have 18 valence electrons per formula unit.

The energies of 648 hypothetic 8-electron MgAgAs-type compounds were calculated in all three atomic arrangements. The majority of the compounds was found to be most stable when the late main-group element occupies the heterocubic site [2]. The MgAgAs structure type is among the most common ternary structure types with a stoichiometric ratio 1:1:1. A variant of the "Global Space Group Optimisation" algorithmhas been applied to assign 488 ternary compounds to 41 structure types.

Only main-group elements and elements of groups 11 and 12 have been investigated in stoichiometries with 8 or 18 valence electrons. The MgAgAs structure type was found to be among the top four most common structure types lying behind the LiGaGe structure type with space group $P6_3/mmc$, the PbClF structure type ($P4/nmm$) and the TiNiSi structure type ($Pnma$) [3].

6.2 Notions on Chemical Bonding in MgAgAs-Type Structures

The chemical bonding of compounds with MgAgAs structure type has been subject to many investigations since their discovery. The first attempt to classify the chemical interactions in terms of ionic or metallic features was made in the pioneer paper from 1941 [4]. The authors characterize the substances MgCuAs, MgCuSb, MgCuBi and MgAgAs as valence compounds with ionic bonding. Referring to [5] MgAgAs is described by the charge balance $Mg^{2+}Ag^{1+}As^{3-}$ indicating the formal charge distribution in the structure. The constant composition and the absence of defects indicated ionic interactions in MgAgAs-type compounds, whereas metallic bonding would manifest in a significant homogeneity range and a mixed occupation of lattice sites [4]. However, it had already been shown that disorder can also occur in salt-like compounds when the electrostatic energy is insensitive to a random distribution of some components like in $LiFeO_2$ [6]. The lithium and iron atoms statistically occupy the cation positions in the rock-salt lattice formed with the oxygen anions.

The ionicity of compounds with MgAgAs structure type was ranked between Mg_3As_2 which has a higher ionicity and Cu_2Sb with lower ionicity by comparing colour, melting points and interatomic distances. It was also anticipated that the ratio of the atomic radii may play a role in the formation of a certain atomic arrangement [4].

This discussion was continued in a publication about the MgAgAs-type nitrides LiMgN and LiZnN [7]. The authors point out that ionicity cannot play an important role in MgAgAs-type compounds when the transition metal occupies the heterocubic sites. Then, two cations face each other with the shortest interatomic distance of $d = \sqrt{3}/4\,a$ in the structure which is not favorable for electrostatic interactions. In LiMgN and LiZnN however, nitrogen occupies the heterocubic site and is surrounded by eight cations. This would indicate ionic interactions explaining the chemical properties i.e. high melting points, easy hydrolisability and the light colour of LiMgN [7].

However, none of the early authors considered covalent interactions to explain the chemical differences of MgAgAs-type phases to other intermetallic compound families. This was first recognized in the analysis of the volume contractions which are usually observed when metals form intermetallic phases. The latter are virtually absent in MgAgAs-type compounds. Thus, it was emphasized that the realisation of high coordination numbers and efficient (high) space filling can usually be interpreted as manifestations of metallic bonding in intermetallic phases whereas the absence of these characteristics in the MgAgAs-type compounds "can be taken as an indication for directional bonding" [8]. This assumption was previously confirmed in the analysis of the structural transformation in TiPtGe. The low-temperature semiconducting phase with the MgAgAs structure type transforms at elevated temperatures into the metallic one with the TiNiSi-type of crystal structure. Thereby, the volume reduces by 10 % after the phase transition [9].

The increasing demand on semiconductor materials for microelectronics and photovoltaic applications was an important development that greatly enhanced the interest in the family of MgAgAs-type compounds. Theoretical investigations were carried out how noble gas atoms affect the band structure of zinc-blende type binary semiconductors when they enter the vacancies. A shift from an indirect to a direct gap semiconductor was found [10]. The change in the electronic properties can be explained as the insertion of noble gas atoms into a vacant site of the zinc-blende crystal structure influences each band energy independently proportional to the charge density of the corresponding band at the insertion site [11]. Moreover, LiZnP was predicted to be a stable direct band gap semiconductor and the formation of a $[ZnP]^-$ polyanionic network with Li^+ as positive counter-ions was reasoned to create the analogy to the noble gas infused binary semiconductors. This concept was extended to an "interstitial insertion rule" which explains the indirect gap to direct gap transition according to the afore mentioned model [12].

The concept to consider MgAgAs-type compounds of the main-group elements as polyanionic networks with the most electropositive cations providing solely electrons for covalent bonding within the polyanion became very popular and is used until today. The concept of the polyanionic network is deduced from the generalized $8 - N$ rule for valence compounds with heteroatomic tetrahedral partial structure which predicts structure patterns for a given material composition [13]. For example, the calculated phonon spectrum of LiMgAs is in agreement with the presence of the polyanionic network. The calculated Born effective charges from the splitting of longitudinal and transverse optical phonons indicate a strong similarity of the [MgAs] substructure in LiMgAs to AlAs. Thus, LiMgAs is described as a polyanionic compound according to $Li^{1+}[MgAs]^{1-}$ [14].

The concept of the polyanionic network was also supported by the analysis of the electronic density of states (DOS) and diagrams of the crystal orbital hamiltonian population (COHP), the distribution of valence charge density and the valence part of the electron localisation function (valence ELF) [15]. For compounds solely comprised of main-group elements $A'AE$ the authors found strong covalent interactions between the two most electronegative elements of a compound. Describing these compounds as filled zinc-blende structures, $A'^{n+}[AE]^{n-}$, would reflect the bonding properties of the compounds. The comparison of the main-group MgAgAs-type compounds with the diamond structure of silicon showed strong similarities in the bonding indicators which additionaly underlined their chemical relation.

The compounds with 18 valence electrons per formula unit $T'TE$ showed the same trends in the analysis of valence charge density and valence-ELF maps. Additionally, COHP diagrams revealed significant bonding between the transition metal at the heterocubic site of the crystal structure and all its eight neighbours. This finding is confirmed in another study applying the same quantum chemical tools to a different set of MgAgAs-type compounds. The authors point out that compounds with two transition elements, "bonding-wise, belong to a separate branch of the half-Heusler family" than MgAgAs-type phases with three main-group or group 11, 12 elements [16]. Indeed, the analysis of the topology of the electron localizability indicator (ELI-D) in TiPtGe revealed ELI-D attractors that indicate covalent interactions between Pt at the heterocubic site and all eight neighbours. The corresponding ELI-D basins were interpreted as two-center Ti–Pt bonds and three-center Ti–Pt–Ge bonds, respectively [9].

The semiconducting 18-electron compounds were predicted to be more stable than phases with a different valence electron count because the COHP diagrams of the valence bands showed a transition from bonding to antibonding states at a filling level of 18 electrons. The phases with an electron number different from this ideal value should therefore be more likely to crystallize in another structure type and often show cooperative magnetism or metallic behavior [15]. Overall, the majority of MgAgAs-type compounds has 18 valence electrons per formula unit [16]. It was also found that the existence of unoccupied bonding or the occupation of antibonding states can be related to the homogeneity range of compounds with MgAgAs structure type [17]. The importance of the valence electron count in MgAgAs-type compounds was first recognized in [18]. Furthermore, ferromagnetism was predicted for phases with 17 and 19 valence electrons which show a sharp peak at the Fermi level of the non-spin-polarized DOS [19, 20].

Aiming to identify new compounds for optoelectronic applications, the chemical mechanism responsible for the formation of a band gap in MgAgAs-type compounds with 8 and 18 valence electrons was investigated extensively. The existence of a band gap is usually attributed to the presence of covalent interactions. In phases with three main-group or group 11, 12 elements the gap is found between bonding and antibonding sp^3 states within the polyanion of the polyanionic network. Additionally, the width of the gap can be related to the electronegativities of the three components. A higher electronegativity difference between the elements of the polyanion decreases the band dispersion and therefore increases the width of the band gap [15]. A decreasing electronegativity of the remaining cation supports the complete transfer of its valence electrons to the polyanion, which additionaly increases the width of a compound's band gap [21]. However, LiZnN displays a smaller band gap than the less polar LiZnP [22]. The analysis of the band structure and DOS showed that nitrogen p states lie quite close to the zinc d states in LiZnN. Thus, the smaller band gap arises from strong p-d coupling which overcompensates for the polarity effect. Furthermore, the smaller band gap in the compounds LiZnX (X = N, P, As) compared to their binary analoga is due to the volume increase of the ternary compounds [22].

Compounds with two transition metals exhibit a so called "d-d gap" between bonding and antibonding states of the transition metals [23, 24]. Also the symmetry reduction (with respect to the rock-salt partial structure) by the occupation of tetrahedrally coordinated lattice sites was found to induce a necessary band anticrossing for the gap to be opened [23]. The concept of electronegativity differences influencing the band gap was also formulated for compounds containing two transition metals. The gap grows with increasing electronegativity difference of the components [25]. The bonding concepts for MgAgAs-type semiconductors as well as the models explaining the formation of a band gap were summarized in a recent review article about Heusler compounds [26].

Extensive studies were also carried out to explain the energetic order of the three atomic arrangements in MgAgAs-type structures. The bond lengths for each arrangement in LiZnAs were compared to the ideal bond lengths defined in the semi-classical valence-force-field model [27]. The deviation of the calculated from the ideal bond lengths increased with decreasing stability of the arrangements, and the value of the Madelung constants was consistent with the calculated stability sequence [28]. Additionally, the comparison of valence charge density maps showed covalent interactions between zinc and arsenic. Therefore, LiZnAs was described as a "half covalent half ionic" compound with covalence between zinc and arsenic and ionic interactions induced by the presence of the positively charged lithium. The least stable arrangement $\mathrm{Li^{HC}ZnAs}$ (MgAgAs) was predicted to be metallic [28].

For MgAgAs-type compounds containing transition elements, the stability order of the three atomic arrangements was investigated. For this purpose they were decomposed into the corresponding binary partial structures. In particular, for the compounds T'NiSn ($T' =$ Ti, Zr, Hf) the rock-salt and zinc-blende like arrangements of the binary partial structures T'Ni, T'Sn and NiSn were compared. The ZrSn rock-salt partial structure contained in the most stable ternary arrangement turned out to be much more stable than the zinc-blende partial structure ZrSn contained in the less stable arrangements whereas the energy differences between the remaining pairs of binary partial structures were negligible. It was concluded that the relative stability of $\mathrm{ZrNi^{HC}Sn}$ was dominantly determined by the interactions in the ZrSn partial structure [23]. In the study on the MgAgAs-type compounds $T'\mathrm{Ni}E$ ($T' =$ Sc, Ti, Zr, Hf, $E =$ Sn, Sb) it was emphasized that the stability of the atomic arrangements $T'\mathrm{Ni^{HC}}E$ is due to interactions in all three binary partial structures [29].

Integrated COHP values for the interaction between the late transition metal and the main-group element were found to be largest for the most stable atomic arrangement in a representative series of MgAgAs-type compounds with two transition metals. Therefore, the increased covalence was made responsible for the relative stability of an arrangement [24].

Quantum chemical real-space techniques were also utilized to systematically investigate the energetic sequence of the atomic arrangements in MgAgAs-type compounds [30]. All known stable compounds with three main-group elements $A'AE$ as well as a representative set of 18-electron phases with one or two transition metals ATE and $T'TE$ were analyzed. Two characteristic quantities were identified - one reflecting the contributions of ionic interactions to the total energy, the other represents the total covalence in a compound. These quantities are the Madelung energy $\mathrm{E_M^{QTAIM}}$ calculated from QTAIM effective charges and the summed delocalization indices per formula unit between all nearest neighbours ς_{nn}. The most stable arrangement of the compounds containing three main-group elements $A'AE^{\mathrm{HC}}$ was characterized by the largest $E_{\mathrm{M}}^{\mathrm{QTAIM}}$ values. The covalent bonding was found to be confined to the zinc-blende partial structure of the two most electronegative components [AE] and was not necessarily largest for the most stable atomic arrangement. The $T'TE$ compounds showed a different bonding pattern. Here, ionic bonding was smaller and not necessarily largest for the most stable arrangement $T'T^{HC}E$. The covalence was largest for the most stable atomic arrangement. In contrast to the $A'AE$ compounds, $T'TE$ phases showed increased covalence within both zinc-blende partial structures [$T'T$] and [TE] and significant $T' - E$ bonding. The corresponding $A' - A$ bonding was virtually absent in the $A'AE$ compounds. ATE compounds showed bonding patterns in-between the $A'AE$ and $T'TE$ phases. Here, the interplay between ionic and covalent bonding determined the most stable atomic arrangement, either $AT^{HC}E$ or ATE^{HC}.

6.3 Real-Space Analysis of Chemical Bonding in MgAgAs-Type Compounds of the Main-Group Elements

This chapter addresses chemical bonding in MgAgAs-type compounds containing exclusively main-group elements [31]. The most stable atomic arrangements of ten compounds $A'AE^{HC}$ were analyzed in position space by means of the QTAIM and ELI-D approaches: $LiAlSi^{HC}$, $LiAlGe^{HC}$, $LiInGe^{HC}$ and $LiInSn^{HC}$ as well as $LiMgN^{HC}$, $LiMgP^{HC}$, $LiMgAs^{HC}$, $LiMgSb^{HC}$, $LiMgBi^{HC}$ and $AlBeB^{HC}$. The sequence of the elements was chosen according to Sanderson's electronegativity scale [32]. For the sake of brevity, the indicator of the atomic arrangement (HC) is omitted from now on.

The real-space analysis of chemical bonding links quantum chemical calculations with chemical concepts and can be applied within every theoretical framework that provides access to the electron density and to the pair density. Thus, quantum chemical real-space analysis is independent of the basis-set type.

6.3.1 QTAIM Analysis

The Quantum Theory of Atoms in Molecules (QTAIM) utilizes the electron density $\rho(\boldsymbol{r})$ in position space to relate chemical concepts like atomic size or charge to quantum mechanics [33]. The zero-flux surfaces S fulfilling equation (6.1) partition the space into space-filling and non-overlapping regions (QTAIM basins). In most cases such a basin includes a nucleus, and is called atomic basin (QTAIM atom). The case of non-atomic basins is very rare in general and was not observed in the MgAgAs-type compounds, yet. The zero-flux surface of $\rho(\boldsymbol{r})$ represents the boundary between two atoms (interatomic surface),

Table 6.1 The critical points of the electron density in MgAgAs-type phases

Critical point	Curvatures	Number	Wyckoff site	Coordinates
MAX	$- - -$	3×4	$4a, 4b, 4d$	$(0, 0, 0), (\frac{1}{2}, \frac{1}{2}, \frac{1}{2}), (\frac{3}{4}, \frac{3}{4}, \frac{3}{4})$
MIN	$+ + +$	4	$4c$	$(\frac{1}{4}, \frac{1}{4}, \frac{1}{4})$
BCP	$- - +$	2×16	$16e, 16e$	(x, x, x)
RCP	$- + +$	24	$24f$	$(x, 0, 0)$

Curvatures can be positive (+) or negative (−)

$$\nabla \rho(\boldsymbol{r}) \cdot \boldsymbol{n}(\boldsymbol{r}) = 0, \ \forall \boldsymbol{r} \in S. \tag{6.1}$$

Quantity $\boldsymbol{n}(\boldsymbol{r})$ denotes the normal vector of the zero-flux surface S at position $\boldsymbol{r}$. The zero-flux condition ensures the validity of the virial theorem for each atomic basin. This uniquely defines the energy of an atom within a chemical system (molecule, solid).

Within the QTAIM approach a compound can be characterized by the set of critical points (CPs) of the electron density in the unit cell. A critical point at position $\boldsymbol{r}_c$ is characterized by the zero gradient of the electron density. Depending on the sign of the three eigenvalues of the Hessian matrix at $\boldsymbol{r}_c$, four different (topologically stable) types of CPs are distinguished: maxima (MAX), minima (MIN) and two types of saddle points, the bond CP (BCP) and the ring CP (RCP).

The set of CPs (number and Wyckoff positions) in the unit cell is found to be the same for all analyzed compounds with MgAgAs-type structure (Table 6.1). The maxima are located at the atomic positions, the minima at the heterocubic vacancies. There are two types of bond critical points with a different free parameter value of the corresponding Wyckoff position (Table 6.1). They occur between the atom at the heterocubic site and its 2×4 neighbours. The ring critical points octahedrally surround each atom within the rock-salt partial structure of the crystal structure (Fig. 6.2). The numbers of critical points (N) fulfill the Poincaré-Hopf theorem for systems with periodic boundary conditions [34],

$$N(\text{MAX}) - N(\text{BCP}) + N(\text{RCP}) - N(\text{MIN}) = 0. \tag{6.2}$$

For atom X with atomic number $Z(X)$ the integration of the electron density within an atomic basin $\Omega_X^{\text{QTAIM}} \equiv \Omega_X$ yields the average electronic population $\bar{N}(\Omega_X)$ and the QTAIM effective charge $Q^{\text{eff}}(\Omega_X)$:

$$\bar{N}(\Omega_X) = \int\limits_{\Omega_X} \rho(\boldsymbol{r}) \mathrm{d}\boldsymbol{r} \tag{6.3}$$

$$Q^{\text{eff}}(\Omega_X) = Z(X) - \bar{N}(\Omega_X). \tag{6.4}$$

For the sake of brevity, QTAIM atomic basins Ω_X and QTAIM atoms X are no longer distinguished: $\Omega_X \equiv X$.

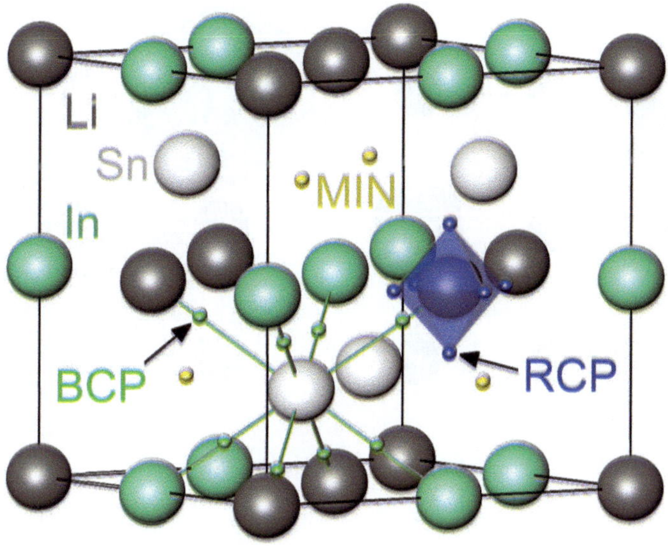

Fig. 6.2 The set of critical points for LiInSn. The maxima (MAX) coincide with the positions of the atoms. The set is the same for all main-group MgAgAs-type compounds

In all investigated MgAgAs-type compounds the most electronegative element carries a negative charge (Table 6.2). The anion charge ranges from −0.84 for tin in LiInSn to −3.50 for boron in AlBeB. In all compounds—except LiInSn—the remaining species serve as cations. Each anion is surrounded by 8 close neighbours to optimize ionic interactions [28, 30]. This arrangement and the calculated effective charges suggest that ionic interactions play an important role in main-group MgAgAs-type compounds.

Table 6.2 Experimental and optimized lattice parameters a_{exp} and a_{opt} (in Å), QTAIM atomic charges Q^{eff} and QTAIM atomic volumes V (in Å^3) of the MgAgAs-type compounds $A'AE$ with three main-group elements

$A'AE$	a_{exp}	a_{opt}	$Q^{eff}(A')$	$V(A')$	$Q^{eff}(A)$	$V(A)$	$Q^{eff}(E)$	$V(E)$
LiMgN	4.995 [35]	5.003	+0.81	3.2	+1.57	5.8	−2.38	22.3
LiMgP	6.003 [36]	6.014	+0.83	4.0	+1.49	7.7	−2.32	42.7
LiMgAs	6.181 [37]	6.212	+0.83	4.2	+1.46	8.4	−2.29	47.4
LiMgSb	6.60 [38]	6.672	+0.84	4.6	+1.39	9.7	−2.23	60.0
LiMgBi	6.73 [38]	6.865	+0.84	4.9	+1.31	11.3	−2.15	64.7
LiAlSi	5.922 [39]	5.937	+0.82	3.7	+1.40	12.3	−2.22	36.3
LiAlGe	5.977 [39]	6.020	+0.83	3.9	+1.34	13.2	−2.16	37.4
LiInGe	6.304 [40]	6.404	+0.84	4.4	+0.18	28.9	−1.02	32.4
LiInSn	6.676 [40]	6.820	+0.84	4.9	0.00	33.1	−0.84	41.3
AlBeB	4.93 [41]	4.962	+2.08	5.3	+1.42	2.8	−3.50	22.4

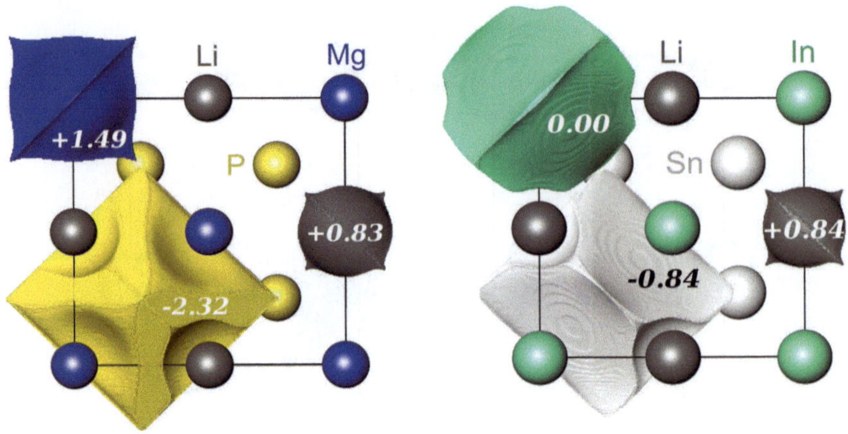

Fig. 6.3 The QTAIM basins and charges in LiMgP and LiInSn

In the MgAgAs-type phases the atomic basins of the anions have a larger volume than the atomic basins of the cations (Table 6.2 and Fig. 6.3).

6.3.2 Electron Localizability Analysis

The Electron Localizability Indicator (ELI) is a family of functionals used for the analysis of atomic interactions. It may be seen as a generalization of the traditional Electron Localization Function (ELF) [42, 43] at a correlated level of theory. In the framework of ELI the ω-restricted space partitioning is applied to combine two position-dependent quantities, the electron density and the pair density [44]. Depending on the spin coupling of the analyzed pairs and how the electron- and the pair density are utilized in the ω-restricted space partitioning, there exist a number of different ELI functionals [45–49].

The functional ELI-D $\Upsilon_D^{\sigma}(\boldsymbol{r})$ can be written as the product of the spin-selective electron density $\rho_{\sigma}(\boldsymbol{r})$ and the pair-volume function $\tilde{V}_D^{\sigma}(\boldsymbol{r})$ for same-spin (σ) electron pairs:

$$\Upsilon_D^{\sigma}(\boldsymbol{r}) = \rho_{\sigma}(\boldsymbol{r}) \cdot \tilde{V}_D^{\sigma}(\boldsymbol{r}). \tag{6.5}$$

The ELI-D can be interpreted as the charge necessary to locally form a fixed fraction of a same-spin electron pair. Loosely speaking, ELI-D is related to the spatial avoidance of same-spin electrons due to the Pauli principle. A highly localized electron repells other same-spin electrons from its region which is reflected by a large ELI-D value. The pair-volume function is proportional to the volume necessary to form a fixed fraction of a same-spin electron pair around position $\boldsymbol{r}$,

$$\tilde{V}_D^{\sigma}(\boldsymbol{r}) = \left[\frac{12}{g_{\sigma}(\boldsymbol{r})}\right]^{\frac{3}{8}} \tag{6.6}$$

with $g_{\sigma}(\boldsymbol{r})$ being the Laplacian of the $\sigma\sigma$-spin pair density at coalescence.

The local maxima of ELI-D (attractors) are located inside the ELI-D basins (analog to the QTAIM space partitioning). In an isolated atom there is a spherical ELI-D attractor for each atomic shell and the corresponding basins contain the number of electrons according to the Aufbau principle ($\pm 0.2e^-$) [50, 51]. In the following analysis the inner shells of an atom X are combined into one ELI-D core basin set $\Omega_X^{\text{core}} \equiv C^X$. ELI-D attractors in the valence regions of a compound are interpreted as signatures of chemical bonding i.e. bonds and lone pairs. The population of the corresponding basins $\Omega^{\text{valence}} \equiv B$ indicates how many electrons belong to a particular bond or lone-pair region.

The ELI-D displays two different topologies in main-group MgAgAs-type semiconductors $A'AE$ describing the same bonding scenario (Fig. 6.4). There are either ELI-D maxima between the anion E and the more electronegative cation A or between the anion E and all its eight neighbours $(4A' + 4A)$. These two topologies result from a continuous development of the Hessian eigenvalues at the ELI-D critical point on the $A' - E$ connection line of each compound. When it changes sign (RCP$[- + +] \longrightarrow$ MAX$[- - -]$) the change between the two ELI-D topologies is observed. In this case the occurence of the ELI-D maxima between A' and E does not indicate a change in the chemical bonding scenario [31]. The ELI-D valence basin populations $\bar{N}(B)$ in compounds with four ELI-D maxima around each anion are around two electrons (Fig. 6.4 left and Table 6.3). For the other ELI-D topology a pair

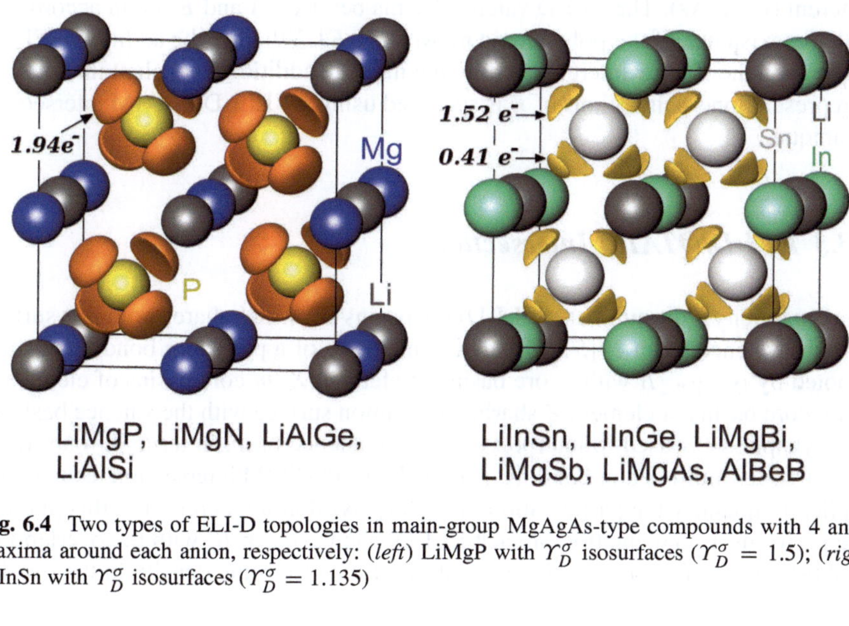

Fig. 6.4 Two types of ELI-D topologies in main-group MgAgAs-type compounds with 4 and 8 maxima around each anion, respectively: (*left*) LiMgP with Υ_D^{σ} isosurfaces ($\Upsilon_D^{\sigma} = 1.5$); (*right*) LiInSn with Υ_D^{σ} isosurfaces ($\Upsilon_D^{\sigma} = 1.135$)

Table 6.3 Synapticity and ELI-D/QTAIM intersections of the main-group MgAgAs-type compounds: $\bar{N}(B)$—electron population of the ELI-D valence basin; $p(B^X)$—bond fractions from the ELI-D/QTAIM intersection technique

$A'AE$	Υ_D^σ basin B	$\bar{N}(B)$	$p(B^{A'})$ (%)	$p(B^A)$ (%)	$p(B^E)$ (%)
LiMgN	4 × ${}_{3\text{Li,Mg,N}}B$	$1.94e^-$	1.8	4.5	93.7
LiMgP	4 × ${}_{3\text{Li,Mg,P}}B$	$1.94e^-$	1.6	5.5	92.9
LiMgAs	4 × ${}_{3\text{Li,Mg,As}}B$	$1.92e^-$	1.4	6.0	92.6
	4 × ${}_{\text{Li,As}}B$	$0.10e^-$	4.5	0.0	95.5
LiMgSb	4 × ${}_{3\text{Li,Mg,Sb}}B$	$1.64e^-$	0.7	8.0	91.3
	4 × ${}_{\text{Li,Sb}}B$	$0.39e^-$	5.4	0.0	94.6
LiMgBi	4 × ${}_{3\text{Li,Mg,Bi}}B$	$1.54e^-$	0.3	9.8	89.9
	4 × ${}_{\text{Li,Bi}}B$	$0.52e^-$	4.9	0.0	95.1
LiAlSi	4 × ${}_{3\text{Li,Al,Si}}B$	$1.95e^-$	1.7	19.5	78.8
LiAlGe	4 × ${}_{3\text{Li,Al,Ge}}B$	$2.02e^-$	1.5	19.6	78.9
LiInGe	4 × ${}_{3\text{Li,In,Ge}}B$	$1.85e^-$	1.2	34.3	64.5
	4 × ${}_{\text{Li,Ge}}B$	$0.14e^-$	5.4	0.0	94.6
LiInSn	4 × ${}_{3\text{Li,In,Sn}}B$	$1.52e^-$	0.6	43.0	56.4
	4 × ${}_{\text{Li,Sn}}B$	$0.41e^-$	5.2	0.0	94.8
AlBeB	4 × ${}_{3\text{Al,Be,B}}B$	$0.99e^-$	3.5	12.5	84.0
	4 × ${}_{\text{Al,B}}B$	$0.95e^-$	18.2	0.0	81.8

of the two different basins also contains two electrons (Fig. 6.4 right). The electron populations of the two different basins can either be very similar (AlBeB) or very different (LiMgAs). The ELI-D valence basins between A and E are in accordance with the existence of the polyanionic network $[AE]$. Whether the additional ELI-D maxima on the $A' - E$ interconnection line indicate additional covalent interactions or represent lone-pair regions of E is specified using the ELI-D/QTAIM intersection technique.

6.3.3 ELI-D/QTAIM Intersections

The synapticity is the number of ELI-D core basins C^X which share a common surface with the valence basin B [52, 53]. The synapticity of a particular bonding basin is denoted by ${}_{lX,mY,nZ}B$ with l core basins of element X, m core basins of element Y and n core basins of element Z sharing a common surface with the valence basin B. The synapticity is used to interpret an ELI-D valence basin in terms of one- (lone pair), two- or multi-center bonding, and the ELI-D/QTAIM intersection technique is applied to quantify the participation of each QTAIM atom therein. For this purpose one performs a segmentation of each ELI-D valence basin B_i with the overlapping QTAIM basins of atoms X. The integration of the electronic density in the overlap

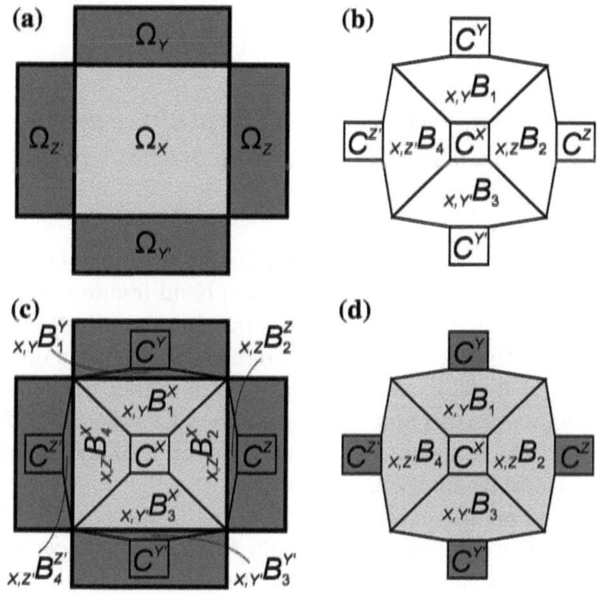

Fig. 6.5 The ELI-D/QTAIM intersection technique: **a** the QTAIM basins $\Omega_X, \Omega_Y, \Omega_Z, \Omega_{Y'}$ and $\Omega_{Z'}$ of the atoms X, Y, Z, Y' and Z'; **b** the ELI-D topology with atomic core basins $C^X \ldots C^{Z'}$, and with the disynaptic valence basins ${}_{X,Y}B_1 \ldots {}_{X,Z'}B_4$, the index X, Y indicates that the valence basin ${}_{X,Y}B_1$ touches the two core basins C^X and C^Y; **c** for the ELI-D/QTAIM intersection, each point of a valence ELI-D basin is assigned to an atom, resulting in the intersected valence regions ${}_{X,Y}B_1^X \ldots {}_{X,Z'}B_4^X$ and ${}_{X,Y}B_1^Y \ldots {}_{X,Z'}B_4^{Z'}$; **d** within the *ELIBON* concept, the dominating contribution of atom X in all intersected ELI-D valence basins yields the complete assignment of these basins to atom X (*yellow*)

volumina B_i^X yields the contribution of atom X to the ELI-D bonding basin (Fig. 6.5). In the scenario of a non-polar bond, all atoms contribute the same number of electrons to the bonding basin. A polar bond is characterized by a disbalance of the atomic contributions to the electron count in an ELI-D bonding basin.

The bond fraction $p(B_i^X)$ of the QTAIM atom X for the valence basin B_i ranges between zero and one:

$$p(B_i^X) = \frac{\bar{N}(B_i^X)}{\bar{N}(B_i)}. \tag{6.7}$$

A zero value means that the QTAIM atom X does not overlap with the ELI-D valence basin B_i, whereas a value of one implies that the ELI-D valence basin completely resides within the QTAIM atom X and represents a lone-pair of X. For a disynaptic basin $p(B_i^X) = 0.5$ represents the case of a non-polar bond.

The ELI-D/QTAIM intersection technique also plays a role in the recovering of the classical concept of oxidation numbers. To calculate the oxidation numbers in a particular compound classically, the valence electrons are assigned to the most electronegative element. The ELI-D based oxidation numbers (*ELIBON*) assign the valence electrons to the atoms of a compound depending on the bond fractions:

$$ELIBON(X) = Z(X) - \bar{N}(C^X) - \sum_i \bar{N}(B_i^X) \cdot \Theta(B_i^X), \tag{6.8}$$

$$\Theta(B_i^X) = \begin{cases} 0, \text{ if there exists an atom } Y \text{ with } p(B_i^Y) > p(B_i^X) \\ \frac{1}{m}, \text{ if there exist } (m-1)X' \text{ with } p(B_i^{X'}) = p(B_i^X) = max \\ 1, \text{ if all } p(B_i^Y) < p(B_i^X) \end{cases}$$

For a given valence basin B_i this means that the electron population is completely counted to the *ELIBON* of X if it has the highest bond fraction (Fig. 6.5d). If there are m atoms with the same—highest—bond fraction the electrons are distributed equally among them. The electron population of an ELI-D basin does not influence the *ELIBON* of X if there is another atom with a higher bond fraction. For the MgAgAs-type compounds this procedure yields the ELI-D based oxidation numbers $A'^{+m}A^{+n}E^{-(m+n)}$ which are in agreement with the classical oxidation numbers calculated from electronegativity differences.

There are di- and pentasynaptic valence basins in the analyzed MgAgAs-type compounds $A'AE$. The compounds with eight ELI-D maxima per anion show four disynaptic and four pentasynaptic basins whereas the remaining phases display four pentasynaptic basins (Fig. 6.6 and Table 6.3).

The disynaptic basins ${}_{A',E}B$ can be considered as effectively monoatomic ${}_{(A',)E}B$ since the bond fraction of the anion E is close to 100 % in all cases - except ${}_{\mathrm{Al,B}}B$ in AlBeB (Table 6.3). The pentasynaptic basins ${}_{3A',A,E}B$ are centered between the anion E and the most electronegative cation A, with the bond fraction of the less electronegative cation being always very small. Thus, depending on the bond fraction of cation A these basins can be considered as effectively monoatomic or diatomic. The bond fraction of A ranges between 4.5 and 43 %. In the cases where the bond fractions of both cations are low the pentasynaptic basins effectively have monoatomic character.

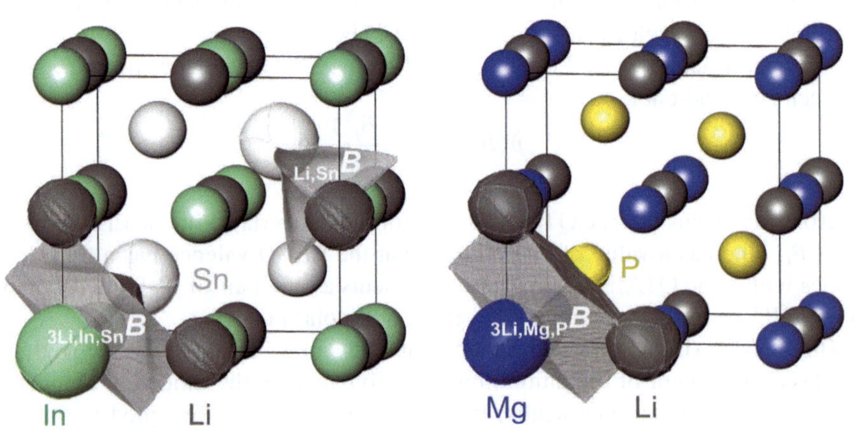

Fig. 6.6 Synapticity of the ELI-D valence basins in LiMgP and LiInSn. Each ELI-D core basin that touches one of the displayed valence basins B is shown

LiInSn and LiInGe with similar bond fractions of the pentasynaptic basin, very small bond fractions of the most electropositive element and a comparably small charge difference between the two more electronegative components realize a bonding situation close to the ideal polyanionic network. The pentel compounds with very small bond fractions of both cations and large negative charges on the most electronegative component get close to the ideal situation of pure ionic bonding. The remaining MgAgAs-type compounds lie in-between the two prototype bonding scenarios as polyanionic networks with highly polar bonding within the polyanion or as ionic compounds with highly polar bonding between the anion and its nearest neighbors. This has been quantified using a newly defined quantity, the connection index, which is calculated solely on the basis of bond fractions and electronic basin populations [31].

6.4 Bonding Analysis of Some Related Compounds

Real space bonding characteristics of the related elemental and binary phases carbon, silicon, germanium, GaAs, AlP, BN and BeS were calculated for comparison with the MgAgAs type. The zinc-blende structure type of the binary phases belongs to the same space group as the MgAgAs structure type, $F\bar{4}3m$. Both structure types represent superstructures (colouring variants) of the diamond structure type.

Figure 6.7 displays the diamond and the zinc-blende structures of silicon and BeS, respectively. The charge transfer patterns range from the neutral tetrel elements with $Q^{\mathrm{eff}} = 0$ to ionic compounds like BeS with an effective charge $Q^{\mathrm{eff}}(\mathrm{S}) = -1.57$ (Table 6.4). Again, the anions possess a larger volume than the cations.

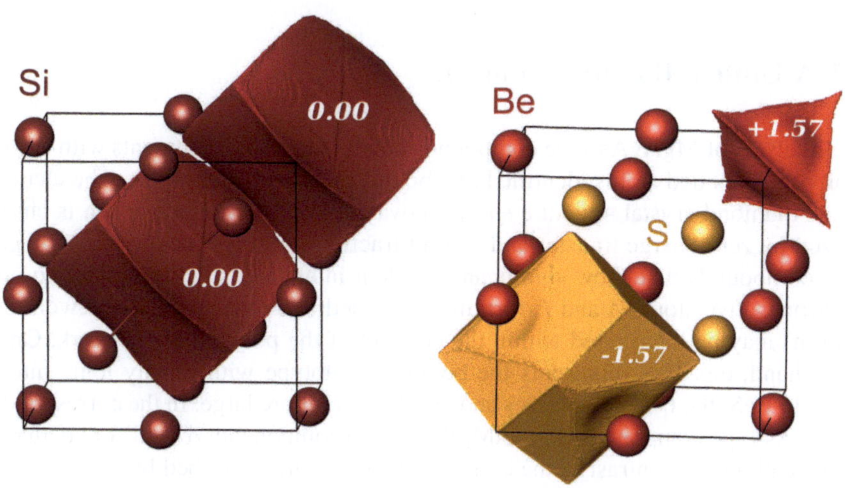

Fig. 6.7 Silicon with the diamond structure and the zinc-blende crystal structure of BeS. The QTAIM basins and effective charges of silicon, beryllium and sulfur are also shown

Table 6.4 The real space bonding analysis of diamond, silicon, germanium, GaAs, AlP, BN and BeS: Q^{eff}—QTAIM charge; $\bar{N}(B)$—average electron population of the ELI-D valence basin; $p(B^X)$—bond fraction via the ELI-D/QTAIM intersection technique

AE	$Q^{\text{eff}}(A)$	$Q^{\text{eff}}(E)$	Υ_D^{σ} basin B	$\bar{N}(B)$	$p(B^A)$ (%)	$p(B^E)$ (%)
Ge	–	0.00	$4 \times {}_{\text{Ge,Ge}}B$	$2.12e^-$	50.0	50.0
Si	–	0.00	$4 \times {}_{\text{Si,Si}}B$	$1.95e^-$	50.0	50.0
C	–	0.00	$4 \times {}_{\text{C,C}}B$	$1.94e^-$	50.0	50.0
GaAs	+0.62	−0.62	$4 \times {}_{\text{As,Ga}}B$	$2.12e^-$	31.1	68.9
AlP	+1.97	−1.97	$4 \times {}_{\text{P,Al}}B$	$1.96e^-$	12.1	87.9
BN	+2.06	−2.06	$4 \times {}_{\text{N,B}}B$	$1.94e^-$	10.9	89.1
BeS	+1.57	−1.57	$4 \times {}_{\text{S,Be}}B$	$1.96e^-$	4.6	95.4

The same ELI-D topology is found for the elements and zinc-blende-type compounds. Each atom is surrounded by four ELI-D maxima, each pointing to one of its four tetrahedrally arranged nearest neighbours. This is consistent with the expected covalent bonds within the tetrahedral networks of the diamond and zinc-blende crystal structures. However, a wide range of bond fractions is found from 50 % indicating non-polar bonds in the elemental compounds to highly polarized valence basins (95.4 %) in BeS (Table 6.4). Two different bonding patterns can be identified. Not surprisingly, the elements bind covalently and show zero charge separation while the QTAIM charges and bond fractions of the compounds AlP, BN and BeS indicate predominant ionic interactions. GaAs with a smaller charge transfer and an intermediate bond fraction is somewhere in-between the purely covalent and ionic bonding scenario.

6.5 A Unified Bonding Concept

The analysis of MgAgAs-type compounds and binary phases/elements with tetrahedrally four-bonded atoms identified two boundary bonding scenarios. The elements in the diamond crystal structure show a covalent bonding pattern which is characterized by zero charge transfer and a bond fraction of 50 %. The ternary MgAgAs-type compounds that show significant covalent interactions between the two more electronegative atoms A and E forming a tetrahedral diamond-like framework (e.g. LiInSn) may be interpreted within the concept of the polyanionic network. On the other hand, e.g. BeS represents the bonding prototype with mainly ionic interactions. In BeS, the QTAIM charges and bond fractions are large. In the corresponding MgAgAs-type compounds (e.g. LiMgP) the two components A' and A play the role of the cations. In contrast to the compounds which are described by the concept of the polyanionic network the two cations show here similarly small bond fractions (Table 6.3).

In order to describe the different bonding situations quantitatively, and to categorize the compounds between the two prototype bonding scenarios, three characteristic quantities are employed. The effective valence population $\bar{N}_{\mathrm{val}}^{\mathrm{ELI}}(X)$ indicates the effective electron filling of the valence shell of a QTAIM atom X:

$$\bar{N}_{\mathrm{val}}^{\mathrm{ELI}}(X) = \bar{N}(X) - \bar{N}(C^X) \approx \sum_{i}^{s_X} p(B_i^X) \cdot \bar{N}(B_i) \tag{6.9}$$

The access electron set s_X is the set of ELI-D valence basins that surround/share a common surface with the ELI-D core basin C^X. The approximate equality indicates that atom X may contribute electrons to ELI-D valence basins that do not have a common surface with C^X. Such contributions are negligible for the analyzed phases. The superscript "ELI" serves to remind, that $\bar{N}_{\mathrm{val}}^{\mathrm{ELI}}(X)$ is based on the shell structure electron counts given by ELI. The access electron number $\bar{N}_{\mathrm{acc}}^{\mathrm{ELI}}(C^X)$ is calculated as

$$\bar{N}_{\mathrm{acc}}^{\mathrm{ELI}}(C^X) = \sum_{i=1}^{s_X} \bar{N}(B_i). \tag{6.10}$$

The genuine charge claim $P_X(X)$ of a QTAIM atom X is defined as the average bond fraction of X in the valence basins B_i of the access electron set s_X [31]:

$$P_X^{\mathrm{ELI}}(X) = \frac{\sum_i^{s_X} p(B_i^X) \cdot \bar{N}(B_i)}{\sum_i^{s_X} \bar{N}(B_i)} = \frac{\sum_i^{s_X} p(B_i^X) \cdot \bar{N}(B_i)}{\bar{N}_{\mathrm{acc}}^{\mathrm{ELI}}(C^X)} \approx \frac{\bar{N}_{\mathrm{val}}^{\mathrm{ELI}}(X)}{\bar{N}_{\mathrm{acc}}^{\mathrm{ELI}}(C^X)} \tag{6.11}$$

A charge claim $P_X^{\mathrm{ELI}}(X) = 0$ indicates that the QTAIM atom X is not participating in the covalent bonding and plays the role of a cation, i.e. the QTAIM atom X is identical with the ELI-D core basin C^X (Fig. 6.8). An atom with $P_X^{\mathrm{ELI}}(X) = 0.5$ participates on average in non-polar bonds. A genuine charge claim of $P_X^{\mathrm{ELI}}(X) = 1$ indicates that the atom possesses all electrons in the bonding basins that share a common surface with its core basin.

In Table 6.5, the analyzed phases are categorized according to the genuine charge claims $P_E^{\mathrm{ELI}}(E)$. The elements with diamond structure (top) show a completely covalent bonding pattern, BeS (bottom) indicates large ionicity of the bonding. The compounds in-between can be considered less covalent than the elements and less ionic than BeS.

The ELI-D valence basins are decomposed into polar and non-polar contributions using the ELI-D/QTAIM intersection technique (Fig. 6.9). Closed-shell interactions are represented by monoatomic basins i.e. lone-pair like valence regions. Homo- and heteropolar interactions are represented by di- or multiatomic ELI-D valence basins. Homopolar bonding is characterized by an equal electronic contribution of each atom that participates in the bond. These contributions are considered as being non-polar in character (Fig. 6.9 left). The opposite scenario to homopolar bonding is the closed-shell interaction. Here, one QTAIM atom contributes the whole electron population

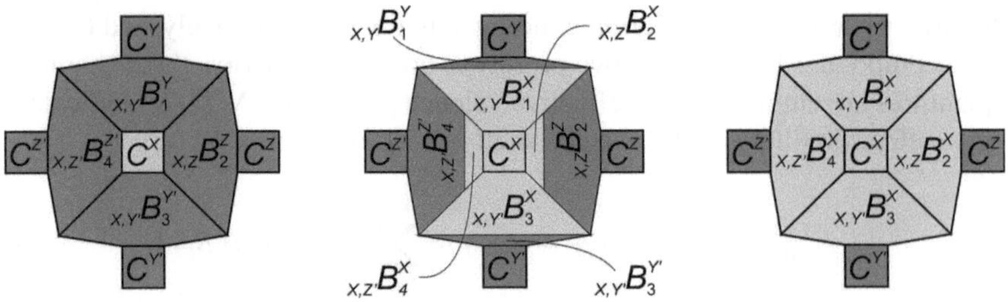

$p(B_i^X) = 0$ $\quad$ $p(_{X,Y}B_1^X) = p(_{X,Y'}B_3^X) = 0.75$ $\quad$ $p(B_i^X) = 1$

$p(_{X,Z}B_2^X) = p(_{X,Z'}B_4^X) = 0.25$

$\rightarrow P_X^{\text{ELI}}(X) = 0$ $\quad$ $\rightarrow P_X^{\text{ELI}}(X) = 0.5$ $\quad$ $\rightarrow P_X^{\text{ELI}}(X) = 1$

Fig. 6.8 Charge claim scenarios of a QTAIM atom *X*

Table 6.5 Anion genuine charge claims $P_E^{\text{ELI}}(E)$ in tetrahedral compounds. The bonding pattern becomes more polar with increasing charge claim. The anion's valence electron number $\bar{N}_{\text{val}}(E)$, access electron number $\bar{N}_{\text{acc}}^{\text{ELI}}(C^E)$ and the calculated number of covalent bonds/lone pairs $N_{\text{cb}}(E)/N_{\text{lp}}(E)$ (6.13) and (6.14) are shown

Compound	$\bar{N}_{\text{val}}^{\text{ELI}}(E)$	$\bar{N}_{\text{acc}}^{\text{ELI}}(C^E)$	$P_E^{\text{ELI}}(E)$ (%)	$N_{\text{cb}}(E)$	$N_{\text{lp}}(E)$
C	3.88	7.76	50.0	3.88	0.00
Si	3.90	7.80	50.0	3.90	0.00
Ge	4.24	8.48	50.0	4.24	0.00
LiInSn	4.99	7.72	64.6	2.73	1.13
LiInGe	5.28	7.96	66.6	2.68	1.32
GaAs	5.85	8.48	68.9	2.63	1.61
LiAlSi	6.14	7.80	78.8	1.66	2.25
LiAlGe	6.43	8.08	78.9	1.65	2.34
AlBeB	6.42	7.76	83.0	1.34	2.56
AlP	6.89	7.84	87.9	0.94	2.98
BN	6.93	7.80	89.1	0.86	3.04
LiMgBi	7.52	8.24	91.2	0.72	3.39
LiMgSb	7.47	8.12	92.0	0.65	3.40
LiMgAs	7.53	8.08	92.7	0.55	3.45
LiMgP	7.24	7.76	92.9	0.52	3.33
LiMgN	7.26	7.76	93.7	0.50	3.39
BeS	7.50	7.84	95.4	0.39	3.53

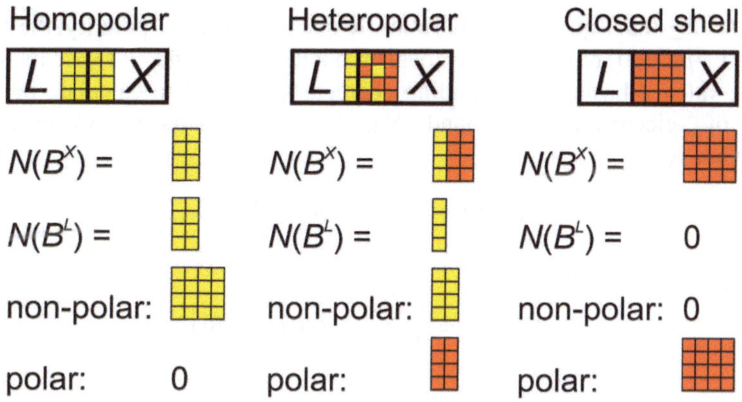

Fig. 6.9 Homopolar, heteropolar and closed-shell interactions from the ELI-D/QTAIM intersection perspective: the *fat black lines* represent the atomic boundaries of the QTAIM atoms L and X; the ELI-D core basins appear as *white squares*; the *yellow* and *orange colored regions* represent the ELI-D valence basins with the total electron population symbolized by 16 mini-squares. *Yellow mini-squares* symbolize the covalent part, *orange mini-squares* symbolize the lone-pair like part of the basin population

to the valence region (Fig. 6.9 right). The bond has ultimate polar character. This situation is characteristic for compounds with ionic or van-der-Waals interactions.

The case of heteropolar bonding is characterized by different electron contributions of the participating atoms (Fig. 6.9 middle). The minority contributing species A with $p(B^A) \leq 0.5$ determines the non-polar part of the heteropolar bond. The remaining electrons in the majority contributing region represent the polar part of the heteropolar bond. This view is related to L. Pauling's ideas about "partially ionic bonds" in zinc-blende semiconductors [54]. For Pauling, the "hypoelectronic" species (atom with less than four valence electrons) determines the covalence of the bond. In Fig. 6.9 the decomposition (yellow and orange mini-squares) of the majority contributing region is only made to illustrate the electron populations and does not mean that non-polar and polar parts are spatially separated from each other.

The non-polar part of the heteropolar bond may be termed the covalent part; the polar part of the heteropolar bond may be termed the "hidden lone-pair" part. The covalent character (cc) and the lone-pair character (lpc) of an ELI-D valence region B is calculated according to

$$\begin{aligned} cc(B) &= -2p(B^X) + 2 \\ lpc(B) &= 2p(B^X) - 1 \qquad (6.12) \\ lpc(B) + cc(B) &= 1 \end{aligned}$$

with $p(B^X) \geq 0.5$. Equations (6.12) allow to quantify the bond polarity: the situation of $cc = 1$ ($lpc = 0$) represents the covalent bond, $cc = 0$ ($lpc = 1$) occurs for lone-pair like valence basins. The bonding characters are used to calculate the overall number of 2-electron covalent bonds $N_{\rm cb}(X)$ and lone pairs $N_{\rm lp}(X)$ for an access electron set s_X.

$$N_{\rm cb}(X) = \frac{1}{2}\sum_{i=1}^{s_X} cc(B_i) \cdot \bar{N}(B_i) \tag{6.13}$$

$$N_{\rm lp}(X) = \frac{1}{2}\sum_{i=1}^{s_X} lpc(B_i) \cdot \bar{N}(B_i) \tag{6.14}$$

Equivalently, the charge claims can be used to obtain $N_{\rm cb}(X)$ and $N_{\rm lp}(X)$ (Fig. 6.10).

$$N_{\rm cb}(X) = \left[1 - P_X^{\rm ELI}(X)\right] \cdot \bar{N}_{\rm acc}^{\rm ELI}(C^X) \approx \bar{N}_{\rm acc}^{\rm ELI}(C^X) - \bar{N}_{\rm val}^{\rm ELI}(X) \tag{6.15}$$

$$N_{\rm lp}(X) = \left[P_X^{\rm ELI}(X) - \frac{1}{2}\right] \cdot \bar{N}_{\rm acc}^{\rm ELI}(C^X) \approx \bar{N}_{\rm val}^{\rm ELI}(X) - \frac{1}{2}\bar{N}_{\rm acc}^{\rm ELI}(C^X) \tag{6.16}$$

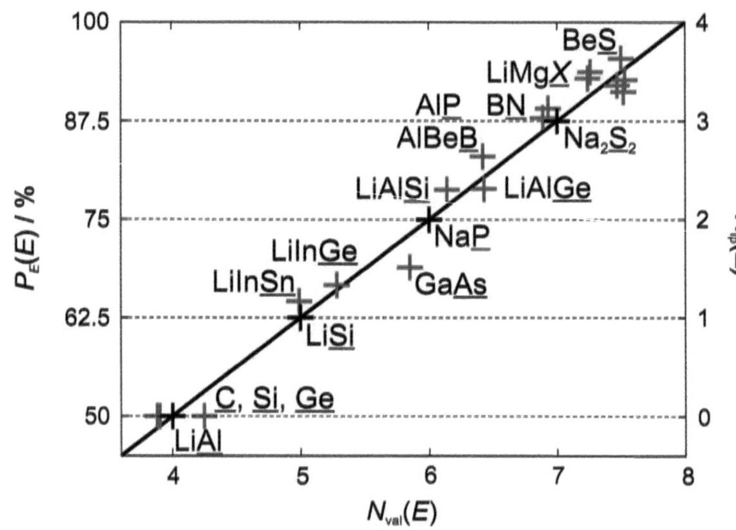

Fig. 6.10 The 8 − N rule from the ELI-D/QTAIM perspective: correlation between the valence electron number $\bar{N}_{\rm val}(E)$, the atomic charge claim $P_E(E)$ and $N_{\rm lp}(E)$. The *red crosses* represent the ELI-D charge claim and ELI-D valence electron number of the underlined element. For LiMgX, X = N, P, As, Sb, Bi. The *black crosses* mark the PTE fix points derived for the Zintl anions in LiAl, LiSi, NaP and Na_2S_2

$$2\left[N_{\mathrm{cb}}(X)+N_{\mathrm{lp}}(X)\right]=\bar{N}_{\mathrm{acc}}^{\mathrm{ELI}}(C^X) \tag{6.17}$$

Relation (6.17) is obtained by combination of (6.15) and (6.16). The calculated $N_{\mathrm{cb}}(E)$ and $N_{\mathrm{lp}}(E)$ are not necessarily integer. Table 6.5 shows that $N_{\mathrm{cb}}(E)$ ranges between 0.39 (BeS) and 4.25 (Ge). Equation (6.15) represents the $8-N$ rule within the ELI-D/QTAIM framework for $\bar{N}_{\mathrm{acc}}^{\mathrm{ELI}}(C^E)=8$ and $\bar{N}_{\mathrm{val}}^{\mathrm{ELI}}(E)\rightarrow N$. Accordingly, $\bar{N}_{\mathrm{acc}}^{\mathrm{ELI}}(C^E)=8$ represents the octet rule meaning that an electron octet is achieved when the valence basins (access electron set s_X) touching the corresponding ELI-D core basin contain 8 valence electrons in total.

The black line in Fig. 6.10 represents the $8-N$ rule from an idealized ($\bar{N}_{\mathrm{acc}}^{\mathrm{ELI}}(C^E)=8$) ELI-D/QTAIM view. The Zintl phases LiSi, NaP and Na_2S_2 with 1, 2, 3 lone pairs and 5, 6, 7 valence electrons per anion formally follow this trend (Fig. 6.10, black crosses). For example, in NaP phosphorous atoms form 1D spiral chains which are surrounded by Na atoms. Corresponding to the Zintl count, P takes the valence electron of Na and realizes a bonding pattern typical for its electronic configuration $(2b)P^{1-}$ i.e. two covalent bonds and two separate lone pairs. The covalent bonds are formed between P ions within the spirals and are non-polar. One can use the formal electron shell populations from the periodic table of the elements (PTE) and formal bond fractions—1 for lone pairs and 0.5 for non-polar bonds—to derive PTE valence electron populations and bond fractions.

$$N_{\mathrm{val}}^{\mathrm{PTE}}(X)=\left[N(X)-N(C^X)\right]_{\mathrm{formal}} \tag{6.18}$$

$$p^{\mathrm{PTE}}(X)=\left[\frac{\sum_i^{s_X} N(B_i)\cdot p(B_i^X)}{\sum_i^{s_X} N(B_i)}\right]_{\mathrm{formal}} \tag{6.19}$$

For P in NaP one obtains $N_{\mathrm{val}}^{\mathrm{PTE}}(\mathrm{P})^{1-})=6e^-$ and

$$p^{\mathrm{PTE}}(\mathrm{P}^{1-})=\underbrace{\frac{2e^-\cdot 1+2e^-\cdot 1}{8e^-}}_{\text{lone pairs}}+\underbrace{\frac{2e^-\cdot 0.5+2e^-\cdot 0.5}{8e^-}}_{\text{bonds}}=0.75\,\widehat{=}\,75\,\%.$$

Analog to this procedure, the $N_{\mathrm{val}}^{\mathrm{PTE}}$ and p^{PTE} values of the anions in LiSi and Na_2S_2 appear as fix points in Fig. 6.10.

The calculated values for the MgAgAs-type and binary compounds as well as the elements deviate only slightly from the ideal line. The deviations originate from the ELI-D shell populations which do not sharply match the integer values from the Aufbau principle [50, 51].

The occurrence of MgAgAs-type compounds with polar bonding and Zintl phases with non-polar covalent bonding within the same diagram illustrates an interesting analogy between the bonding patterns (Fig. 6.11). Zintl phases realize the respective bonding patterns (4b), (3b, 1lp), (2b, 2lp), (1b, 3lp), (4lp) where b symbolizes a 2-center-2-electron covalent bond and lp a 2-electron lone pair. The atoms E in

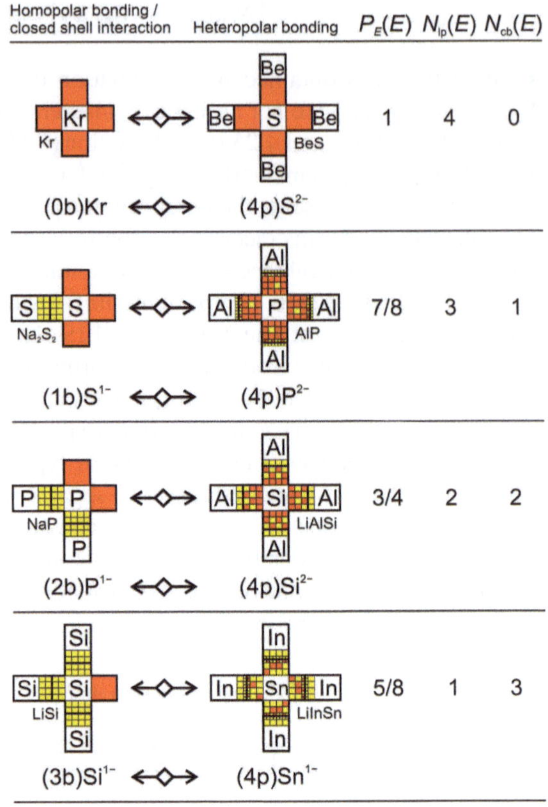

Fig. 6.11 The $8 - N$ rule from the ELI-D/QTAIM perspective: conceptual correspondence of compounds with covalent bonding/lone-pair (0b–3b, *left*) and polar bonding scenarios (4p, *right*). The correspondence is indicated by the special *arrow symbol*. Only the bonding regions of the central atom is shown completely. The displayed $P_X^{\text{ELI}}(X)$, $N_{\text{cb}}(X)$ and $N_{\text{lp}}(X)$ are rounded for illustration

MgAgAs-type compounds realize the bonding pattern (4p) with four equal heteropolar bonds. These are interpreted as being made up of a covalent part b′ and a lone-pair like part lp′:

$$(4p) = ([4x]\,b',\ 4[1 - x]\,lp') \tag{6.20}$$

The parameter x depends on the particular features of the contributing elements, e.g. electronegativity difference ($x = 1$ for the covalent bonding scenario 4b′ and $x = 0$ for the lone-pair like valence basins 4lp′). Four lone pairs per atom are found either for the noble gas atom (van-der-Waals interactions with the environment) or for sulfur in the ionic zinc-blende compound BeS (4p with very small x). Three lone pairs and one covalent bond occur for sulfur in the phase Na_2S_2 (1b, 3lp). An equivalent bonding scenario is found for AlP where three lone pairs and one covalent bond are mixed in four polar bonds (4p with larger x as in BeS). Accordingly, the remaining compounds in Fig. 6.11 illustrate the correspondence between covalent bonding/lone-pair and polar bonding scenarios for cases with two and one (effective) lone pairs per anion.

6.6 Conclusion

The bonding patterns of main-group MgAgAs-type compounds $A'AE$ are categorized between the polyanionic and the ionic bonding model by two real-space quantities, the valence electron population $\bar{N}_{\text{val}}^{\text{ELI}}(E)$ and the genuine charge claim $P_{E}^{\text{ELI}}(E)$ of the QTAIM atom E.

According to the classical views of valence compounds composed of main-group elements an electronic octet is achieved for all constituent atoms. Within formal electron counting, the least electronegative atoms achieve this situation by transferring their valence electrons to the more electronegative partners and become the cations. The most electronegative atoms receive the electrons and according to the resulting formal count of 4, 5, 6, 7 or 8 electrons achieve the octet by forming tetrahedral patterns of 2-center-2-electron anion-anion bonds b and 2-electron lone pairs lp, i.e., (4b), (3b, 1lp), (2b, 2lp), (1b, 3lp), and (4lp), respectively, where the number of bonds b can be inferred from the $8 - N$ rule. In main-group compounds adopting the zinc-blende type or MgAgAs type of crystal structure, no separate lone pairs and covalent bonds are formed, but a partial covalent (non-polar) contribution b′ and a partial lone pair (polar) contribution lp′ is defined within each polar bond p according to (p) = (x b′, $[1 - x]$ lp′). The balance between covalent and lone-pair contributions (the value of x) for a particular valence region is quantified with the aid of the ELI-D/QTAIM intersection technique at a quantum chemical level. Although the anions in MgAgAs-type compounds show the same coordination the effective number of covalent bonds changes with the polarity.

References

1. J. Lima-de Faria, E. Hellner, F. Liebau, E. Makovicky, E. Parthe, A. Crystallogr. A **46**, 1
2. T. Gruhn, Phys. Rev. B **82**, 125210 (2010)
3. X. Zhang, L. Yu, A. Zakutayev, A. Zunger, Adv. Funct. Mater. **22**, 1425 (2012)
4. H. Nowotny, W. Sibert, Z. Metallkd. **33**, 391 (1941)
5. E. Zintl, H. Kaiser, Z. Anorg. Allg. Chem. **211**, 113 (1933)
6. E. Posnjak, T. Barth, Phys. Rev. **38**, 2234 (1931)
7. R. Juza, F. Hund, Z. Anorg. Allg. Chem. **257**, 1 (1948)
8. W. Jeitschko, Metall. Trans. **1**, 3159 (1970)
9. S.-V. Ackerbauer, A. Senyshyn, H. Borrmann, U. Burkhardt, A. Ormeci, H. Rosner, W. Schnelle, M. Gamza, R. Gumeniuk, R. Ramlau, E. Bischoff, J.C. Schuster, F. Weitzer, A. Leithe-Jasper, L.H. Tjeng, Yu. Grin, Chem. Eur. J. **18**, 6272 (2012). 5, 6
10. H.W.A.M. Rompa, M.F.H. Schuurmans, F. Williams, Phys. Rev. Lett. **52**, 675 (1984)
11. D.M. Wood, A. Zunger, R. de Groot, Phys. Rev. B **31**, 2570 (1985)
12. A.E. Carlsson, A. Zunger, D.M. Wood, Phys. Rev. B **32**, 1386 (1985)
13. E. Parthé, *Elements of Inorganic Structural Chemistry* (K. Sutter Parthé, Petit-Lancy, 1996)
14. A. Mellouki, B. Bennecer, F. Kalarasse, J. Phys. C: Condens. Matter **21**, 305402 (2009)
15. H.C. Kandpal, C. Felser, R. Seshadri, J. Phys. D: Appl. Phys. **39**, 776 (2006)
16. L. Offernes, P. Ravindran, A. Kjekshus, J. Alloys Compd. **439**, 37 (2007)
17. L. Offernes, P. Ravindran, C. Seim, A. Kjekshus, J. Alloys Compd. **458**, 47 (2008)

18. J. Pierre, R. Skolozdra, J. Tobola, S. Kaprzyk, C. Hordequin, M. Kouacou, I. Karla, R. Currat, E. Lelievre-Berna, J. Alloys Compd. **101**, 262 (1997)
19. E.C. Stoner, Proc. R. Soc. Lond. A **165**, 372 (1938)
20. D. Jung, H.-J. Koo, M.-H. Whangbo, J. Mol. Struct. **527**, 113 (2000)
21. D. Kieven, R. Klenk, S. Naghavi, C. Felser, T. Gruhn, Phys. Rev. B **81**, 075208 (2010)
22. F. Kalarasse, B. Bennecer, J. Phys. Chem. Solids **67**, 1850 (2006)
23. S. Ögüt, K.M. Rabe, Phys. Rev. B **51**, 10443 (1995)
24. B.R.K. Nanda, I. Dasgupta, J. Phys. C: Condens. Matter **15**, 7307 (2003)
25. J. Tobola, J. Pierre, J. Alloys Compd. **296**, 243 (2000)
26. T. Graf, C. Felser, S.S.P. Parkin, Prog. Solid State Chem. **39**, 1 (2011)
27. P.N. Keating, Phys. Rev. **145**, 637 (1966)
28. S.-H. Wei, A. Zunger, Phys. Rev. Lett. **56**, 528 (1986)
29. P. Larson, S.D. Mahanti, M.G. Kanatzidis, Phys. Rev. B **62**, 12754 (2000)
30. D. Bende, Yu. Grin, F.R. Wagner, Chem. Eur. J. **20**, 9702 (2014)
31. D. Bende, F.R. Wagner, Yu. Grin, Inorg. Chem. **54**, 3970 (2015)
32. R.T. Sanderson, *Inorganic Chemistry* (Reinhold Publishing, 1967)
33. R.F.W. Bader, *Atoms in Molecules* (Oxford University Press, 2003)
34. V. Guillemin, A. Pollack, *Differential Topology* (AMS Chelsea Publishing, 2010)
35. K. Kuriyama, Y. Yamashita, T. Ishikawa, K. Kushida, Phys. Rev. B **75**, 233204 (2007)
36. K. Kuriyama, K. Kushida, Solid State Commun. **112**, 429 (1999)
37. K. Kuriyama, Y. Yamashita, Y. Suzuki, K. Matsumoto, K. Kushida, AIP Conf. Proc. **1199**, 67 (2010)
38. H. Nowotny, F. Holub, Monatsh. Chem. **91**, 877 (1960)
39. J. Barth, G.H. Fecher, M. Schwind, A. Baleanu, C. Felser, A. Shkabko, A. Weidenkaff, J. Hanss, A. Reller, M.J. Köhne, J. Electron. Mater. **2010**, 39 (1856)
40. W. Bockelmann, H.-U. Schuster, Z. Anorg. Allg. Chem. **410**, 241 (1974)
41. H.J. Becher, Z. Anorg. Allg. Chem. **317**, 346 (1962)
42. A.D. Becke, K.E. Edgecombe, J. Chem. Phys. **92**, 5397 (1990)
43. A. Savin, O. Jepsen, J. Flad, O.K. Andersen, H. Preuss, H.G. von Schnering, Angew. Chem. **104**, 186 (1992)
44. A. Martin Pendas, M. Kohout, M.A. Blanco, E. Francisco, *Modern Charge-Density Analysis* (Springer, 2012)
45. M. Kohout, K. Pernal, F.R. Wagner, Yu. Grin, Theor. Chem. Acc. **113**, 287 (2005)
46. M. Kohout, F.R. Wagner, Yu. Grin, Int. J. Quantum Chem. **106**, 1499 (2006)
47. M. Kohout, Faraday Disc. **135**, 43 (2007)
48. F.R. Wagner, V. Bezugly, M. Kohout, Yu. Grin, Chem. Eur. J. **13**, 5724 (2007)
49. M. Kohout, F.R. Wagner, Yu. Grin, Theor. Chem. Acc. **119**, 413 (2008)
50. M. Kohout, A. Savin, Int. J. Quantum Chem. **60**, 875 (1996)
51. A.I. Baranov, J. Comput. Chem. **35**, 565 (2014)
52. B. Silvi, A. Savin, F.R. Wagner, *Modelling of Minerals and Silicated Materials* (Springer, 1997), p. 179
53. B. Silvi, J. Mol. Struct. **614**, 3 (2002)
54. L. Pauling, *The Nature of the Chemical Bond* (Cornell University Press, 1960)

Chapter 7
Magnetic and Electronic Properties of Thin Films of Mn-Ga and Mn-Ge Compounds with Cubic, Tetragonal and Hexagonal Crystal Structures

Huseyin Kurt and J.M.D. Coey

High density magnetic memory driven by spin transfer torque requires novel materials combining high spin polarization and high uniaxial magnetic anisotropy. In thin films the easy axis should be perpendicular to the film plane. Perpendicularly-magnetized films may also find applications in magnetic memory. This chapter reviews recent developments in Mn-Ga and Mn-Ge based Heusler alloys which can meet these requirements. The zero moment half metallic behavior observed in cubic $Mn_2Ru_{0.5}Ga$ is of particular interest. Other crystalline forms of Mn-Ga and Mn-Ge alloys where triangular antiferromagnetism is used to demonstrate exchange bias are also reviewed.

7.1 Introduction

Manganese is one of the most distinctive magnetic elements in the periodic table. The *d* band is more than half full, yet manganese is not ferromagnetic. It has the largest and most complex unit cell of any element. The cubic cell of αMn, contains 48 atoms in four inequivalent sites with moments ranging from 0.5 μ_B to 2.8 μ_B that order in a complex triangular antiferromagnetic structure below 90 K [1, 2]. The direct exchange between Mn atoms becomes ferromagnetic when the Mn atoms are ordered on a lattice with sufficient interatomic separation. The first example of such a material was Cu_2MnAl, discovered by Friedrich Heusler in 1903 [3, 4], long before the development of quantum mechanics. His discovery that the first 'Heusler alloy'

H. Kurt (✉)
Department of Engineering Physics, Istanbul Medeniyet University,
Ünalan Mah Ünalan Cd, Üsküdar, 34700 Istanbul, Turkey
e-mail: huseyin.kurt@medeniyet.edu.tr; kurtphys@gmail.com

J.M.D. Coey
Department of Physics, Trinity College Dublin, Dublin 2, Ireland
e-mail: jcoey@tcd.ie

C. Felser and A. Hirohata (eds.), *Heusler Alloys*,
Springer Series in Materials Science 222, DOI 10.1007/978-3-319-21449-8_7

was ferromagnetic at room temperature came as a complete surprise since nobody at the time expected that magnetism could emerge in an alloy of nonmagnetic elements. The ever-growing Heusler family now includes compounds which exhibit a wide variety of magnetic and electronic properties, including antiferromagnets, ferrimagnets, ferromagnetic shape- memory alloys, superconductors [5, 6], half metals, semiconductors, semi-metals [7] and topological insulators [8, 9].

Much of the recent interest in Heusler alloys was stimulated by the electronic structure calculations of de Groot et al. [10], who predicted that the Mn-based half Heusler alloys NiMnSb and PtMnSb should be *half-metallic*. A half metal has an electronic density of states that is metallic for one spin sub-band, but insulating, with a spin gap, for the other. Since only one spin channel is conducting, one can imagine a device where the electronic conduction could be simply turned on and off by a magnetic field in a spin valve (SV) or magnetic tunnel junction (MTJ), analogous to the electric field-effect devices, where the conduction is turned on and off by a gate electrode. This could have significant implications for information technology, where nonvolatile magnetic memories can be integrated in the CMOS logic circuits [11], thereby saving much of the energy that is consumed by portable consumer electronics.

The magnetic moments of many ferromagnetic $3d$ transition elements and their alloys that are half-metals or strong ferromagnets follow a modified Slater-Pauling rule, which is a simple relation between the magnetic moment and the average number of valence electrons per atom. The difference between a half metal and a strong ferromagnet is that the half metal has a spin gap in the *total* density of states, whereas the spin gap in the strong ferromagnet involves just the $3d$ density of states, because $4s$ electrons of both spins are present at the Fermi level E_F. The spin moment m in Bohr magnetons per atom is $m = n_\uparrow - n_\downarrow$ and the total number n_v of valence electrons per atom is $n_v = n_\uparrow + n_\downarrow$, where $n_\uparrow$ and $n_\downarrow$ are the numbers of majority and minority spins in the conduction bands. We can therefore write m in one of two ways:

$$m = 2n_\uparrow - n_v \tag{7.1}$$

$$m = n_v - 2n_\downarrow \tag{7.2}$$

The first equation is appropriate for a nearly-full $3d$ band, where all the $3d_\uparrow$ states are full. Hence $n_\uparrow \approx 5.3$ (the number is slightly greater than 5 because some $4s$ states are also occupied) and the descending branch in Fig. 7.1 has slope -1. When the $3d$ band contains fewer than about eight electrons, the $4s$ band can be empty, and the material then becomes a half-metal, with $3d$ electrons of only one spin at the Fermi level. If the spin splitting of the $3d$ band were really large, half metals with $m = n_v = n_\uparrow$ could be envisaged for $n_v < 5$. In fact, three $3d_\downarrow$ bands lie below E_F and two above. Hence, from 7.2 we find $m = n_v - 6$; when $n_v > 6$, there is a rising branch in Fig. 7.1 with slope 1. When the number of atoms per formula ν is greater than one, the moment m per formula shows a similar variation as a function of the number of valence electrons per formula when $\nu n_v > 6\nu$. The data for Co-based cubic Heusler alloys X_2YZ with X = Co in Fig. 7.1 ($\nu = 4$) follow such a curve when

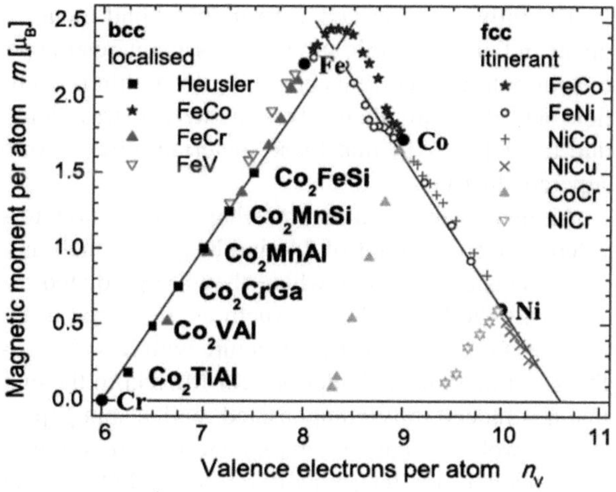

Fig. 7.1 The Slater-Pauling curves for 3*d* metals and alloys with bcc and fcc structures. Agreement is found between the measured magnetic moments of Co-based Heusler alloys and the half-metallic values predicted by Slater-Pauling rule $m = \nu n_v - 24$ [16]. Figure from [16]

$\nu n_v > 24$, just as expected if they are half metals. Moments for half-metallic half Heuslers XYZ such as NiMnSb ($\nu = 3$) will fall on a straight line when $\nu n_v > 18$. For a stoichiometric half metal, νn_v is an integer, as are $\nu n_\downarrow$ and $\nu n_\uparrow$, because of the spin gap. It follows that the spin moment per formula unit is an integer number of Bohr magnetons (which could be zero).

The predicted integral spin moment per formula unit when $n_v < 8$ is strictly true only at zero temperature in the absence of spin-orbit coupling, which tends to mix the spin states. Whereas the predicted magnetic moments agree rather well with experimental values in many cases [12], the half-metallic spin polarization is never perfect. The spin polarization of NiMnSb or Co_2MnSi that is observed experimentally by point contact Andreev reflection spectroscopy turns out to be about 0.5 [13, 14]. Higher values of about 0.9 are obtained for the tunneling spin polarization in magnetic tunnel junctions (MTJs) [15], but these values are sensitive to stoichiometry and atomic disorder in the structure.

While these half-metals with $n_v < 8$ (properly termed *ferrimagnetic* half-metals, because there are oppositely-directed magnetic moments on different chemical sublattices, and it is only the resultant spin moment that must be an integral number of Bohr magnetons per formula) are attractive for spin electronic devices, the electronic and magnetic structure of alloys with just the critical number of electrons, 24 for Heuslers, 18 for half-Heuslers, is a particularly interesting topic for investigation. They may be *zero-moment half-metals* with equal and opposite moments on crystallographically-inequivalent magnetic sublattices, a possibility first envisioned by van Leuken and de Groot in 1995 [17] who called the effect ‘half-metallic antiferromagnetism’. Here we prefer to reserve ‘antiferromagnetic’ for crystallographically-equivalent sublattices. Of course, the absence of a net magnetic moment is not sufficient evidence of a zero-moment half metal. The atoms themselves may simply be nonmagnetic, or else exhibit very weak itinerant ferromagnetism without a spin gap. Both these possibilities have been suggested for Fe_2VGa [18].

The fully-compensated, zero-moment half metal, where there are large moments on the individual atomic sublattices is an interesting possibility, because it could lead to spin-polarized currents emerging from thin films that are insensitive to magnetic field and the films themselves create no stray field. There is also the prediction due to Pickett [19] that these materials might exhibit an unusual type of triplet superconductivity.

An underlying focus in this chapter is on how to stabilize Mn_3Ga, which has a valence electron count of 24, and Mn_2Ge, which has a valence count of 18, in the cubic Heusler structures where they are predicted to be zero-moment half-metals. These are the $L2_1$ and $C1_b$ structures illustrated in Fig. 7.2a, c, with X = Y = Mn and Z = Ga or Ge. The $L2_1$ structure with X = Y is also known as the $D0_3$ structure. Density functional calculations for Mn_3Ga in the cubic $D0_3$-structure [20] suggest that it is indeed fully-compensated, with a spin gap.

The problem is that the bulk materials are tetragonal; bulk Mn_3Ga, for example, crystallizes in the tetragonally-deformed $D0_{22}$ variant of the $L2_1$ structure, which has a quite different electronic structure and an uncompensated moment of 0.4 μ_B per formula [21]. Both Mn-Ga compounds can be stabilized in a cubic structure in thin film form but neither behaves quite as expected. Nevertheless, they suggest a way to create a zero-moment ferromagnet with a spin gap [20].

It is possible to grow oriented tetragonal thin films of these Mn-based binary alloys with perpendicular magnetic anisotropy in either the $L1_0$ or $D0_{22}$ structures. The optimized films are of interest because of their potential application in perpendicular spin-electronic device structures. The preparation and properties of these anisotropic films is the main focus of the chapter.

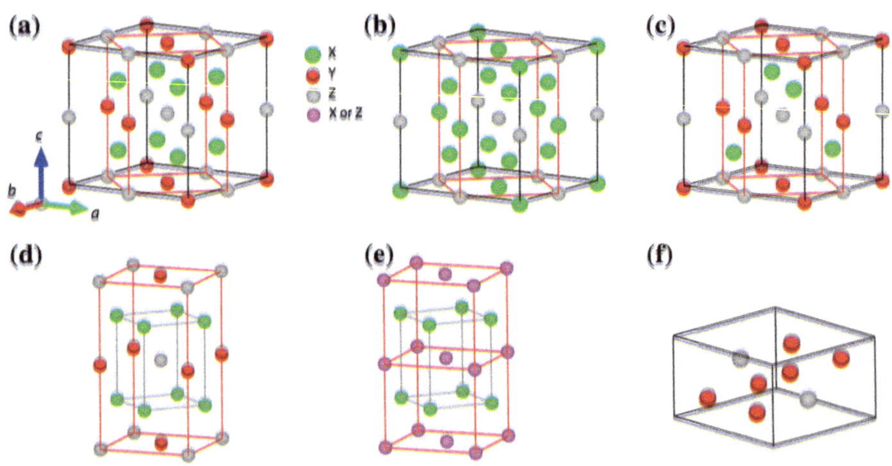

Fig. 7.2 Perfectly ordered crystal structures of **(a)** $L2_1$ full Heusler X_2YZ **(b)** $D0_3$ X_3Z and **(c)** $C1_b$ half-Heusler XYZ compounds. *Red lines* show the portion of the unit cell, which distorts to form the tetragonal unit cell of the $D0_{22}$ structure **(d)** and The $L1_0$ structure (two unit cells) **(e)**. **f** shows the hexagonal $D0_{19}$ structure

7.2 Crystal Structure and Magnetic Order

Both the Mn-Ga [22] and Mn-Ge [23] phase diagrams are complex, with as many as 15 or 16 distinct binary phases, respectively. Furthermore, the low-temperature region $T < 500\,^{\circ}C$, where thin films are usually prepared, is often uncertain because of the long times necessary for bulk material to attain equilibrium. In addition, substrates, seed/buffer layers and film thickness can all influence the phases obtained. Here we are concerned with two composition ranges where interesting magnetic properties are observed. One is slightly Mn-rich, but close to equiatomic MnZ, Z = Ga, Ge, where the tetragonal, ordered $L1_0$ structure appears. The other is more manganese-rich, between Mn_2Z and Mn_3Z, where the structure is derived from $D0_{22}$.

Relevant structures, and the relations between them, are illustrated in Fig. 7.2. The ordered cubic $L2_1 X_2YZ$ Heusler structure is shown in Fig. 7.2a, where the X atoms occupy 8*c* positions and the Y and Z atoms occupy 4*a* and 4*b* positions. The $L2_1$ structure can be viewed as four interpenetrating fcc lattices, forming a filled diamond-like structure with each X atom positioned at 1/4, 1/4, 1/4 from the eight corners of the unit cell. If the site occupancy is entirely random, the structure becomes body-centred cubic, and if X = Y, the cubic $D0_3$ structure of Fig. 7.2b is obtained. The half Heusler $C1_b$ structure (Fig. 7.2c) is obtained by eliminating one of the two X sublattices, removing one X atom from the formula, or four from the unit cell, which resembles a diamond structure except for the presence of the Y sublattice. Otherwise, the atom can be taken from the Y sublattice, or from some combination of 4*a* and 8*c* positions. Note that the X atoms are tetrahedrally coordinated by Z atoms, whereas the Y atoms are octahedrally coordinated by Z atoms. Ordering of the X, Y and Z atoms in the 4*a*, *4b* and 8*c* positions is often less than ideal.

The tetragonal Heuslers are formed by stretching a smaller cell (shown by red lines) of these $L2_1$ and $C1_b$ Heusler structures along c-axis. The binary $D0_3$ structure may be similarly deformed. The deformation is substantial; the c/a ratio of the red cell increases from $\sqrt{2}$ to about 1.8 in the $D0_{22}$ structure, shown in Fig. 7.2d. The binary $L1_0$ structure, two unit cells of which are shown in Fig. 7.2e, can be regarded as a variant of $D0_{22}$ with Y = Z. This uniaxial stretch of the big cubic cell leads to an overall change of coordination number from 8 to 12. The tetragonal variants are ordered face-centred structures, whereas the cubic parents are basically body-centred structures.

A property of the tetragonal Mn-Ga compounds is that they adopt the tetragonal $D0_{22}$ structure over a wide range of stoichiometry, from Mn_2Ga to Mn_3Ga [24]. The compounds are ferrimagnetic, with an uncompensated net moment of Mn in 2*a* and 4*c* sublattices. Compositions with a Mn:Ga ratio less than 1.9 crystallize in the $L1_0$ structure. Ideally, this as an equiatomic face-centred tetragonal phase with alternating atomic layers of Mn and Ga as indicated in Fig. 7.2e, but excess Mn is accommodated on the Ga planes. Materials in this structure are usually Mn rich, because the equiatomic structure is thermodynamically unstable [25, 26], as is also the case for $L1_0$-MnAl [27]. The $D0_{22}$ and $L1_0$ structures differ only in the degree of atomic order in the 2*a*/2*b* sites. This is something which is not easy to determine experimentally, especially in thin films where the structural attribution is often ambiguous.

The tetragonal $D0_{22}$ $Mn_{3-x}Ga$ compounds ($0 \leq x \leq 1$) undergo a first order phase transition at about 750 K, depending on x, where they transform into the hexagonal $D0_{19}$ phase [24] (Fig. 7.2f). The sub-stoichiometric $x > 0$ compositions crystallizing in a variant of the $D0_{22}$ or $D0_{19}$ structures have empty atomic positions, while maintaining a nearly constant unit cell volume, thereby reducing the density of the alloy [24].

Like Mn_3Ga, Mn_3Ge crystallizes in the $D0_{22}$ structure, and is also a ferrimagnet. In both cases, the Curie temperature would be above 865 K, where the tetragonal material transforms to the hexagonal $D0_{19}$ structure [28]. Unlike Mn_3Ga, $D0_{22}$-Mn_3Ge is stabilized easily for Mn rich portion of the binary phase diagram [29].

The magnetic structure of these manganese-rich compounds has been determined by neutron diffraction. An early study by Krén and Kádár [21] on a sample of composition $Mn_{2.85}Ga_{1.15}$ found moments of 1.6(2) μ_B on 4*d* sites (MnII) and −2.8(3) μ_B on 2*b* sites (MnI), giving a resultant of 0.4(2) μ_B per formula. A new neutron study of Mn_3Ga by Rode et al. [30] gives comparable values of the c-axis components, 2.08(2) μ_B on 4*d* sites and −3.07(4) μ_B on 2*b* sites, but the data reveal a very interesting magnetic twist; the ordered magnetic structure includes an in-plane component of the 2*b* moment of 1.19(24) μ_B so the 2*b* moments are of magnitude 3.29 μ_B and are tilted at 21° away from the c-axis [30], as suggested by the circles in Fig. 7.3a. This noncollinear ferrimagnetic structure was attributed to anisotropy of opposite sign at the two sites, easy-axis at the tetrahedrally-coordinated 4*d* sites, and easy-plane at the octahedrally-coordinated 2*b* sites, which when combined with an oscillating exchange coupling between nearest and next nearest neighbours allows

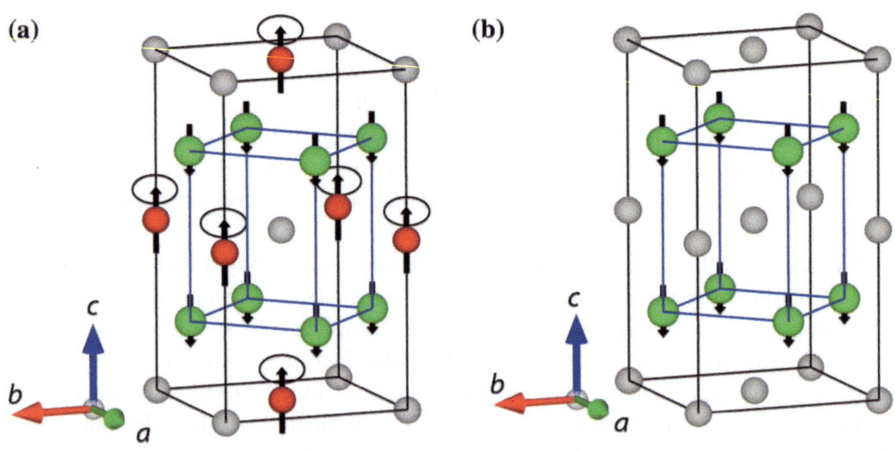

Fig. 7.3 **a** The ferrimagnetic structure of $D0_{22}$ Mn_3Ga or $D0_{22}$ Mn_3Ge. There is overall c-axis anisotropy but the structure is noncollinear and the 2*b* sublattice has a soft, in-plane component, which is indicated by *small circles*. **b** Ferromagnetic structure of $L1_0$ MnGa. The Mn atoms in the 4*d* positions all couple ferromagnetically

for a non collinear magnetic ground state. The smaller moments are found on the tetrahedral sites, which have the shorter Mn-Ga bond lengths. The extrapolated Curie temperature lies above the temperature (~800 K) where the compound transforms to the hexagonal $D0_{19}$ structure.

The ferrimagnetic structure of $D0_{22}$ Mn_3Ge is similar, with c-axis moments of about 1.9(2) μ_B on $4d$ (MnII) sites, and moments that are less than twice as large on $2b$(MnI) sites [28, 29, 31]. Electronic structure calculations [5, 30, 32, 33] confirm the observation of larger moments on $2b$ sites than on $4d$ sites, and reproduce the observed magnitudes of the moments moderately well. Furthermore, the calculations suggest that cubic $D0_3$ Mn_3Ga should be a full-compensated zero-moment half metal [20], and that $D0_3$ Mn_3Ge should be a half-metal with a moment of 1 μ_B, since it has one extra valence electron [33, 34]. However, unlike Fe_3Al or Fe_3Si, these compounds do not naturally crystallize in a cubic structure.

On passing from tetragonal Mn_3Ga to Mn_2Ga in the series $Mn_{3-x}Ga$, the magnetization increases uniformly with increasing x [24]. There is no minimum or compensation point, which indicates that the $4d$ (MnII) sublattice remains dominant. It follows that the Mn is preferentially removed from the $2b$ sites as the Mn content is reduced, thereby increasing the magnetization. With further removal of Mn (or addition of Ga), the vacant $2b$ sites fill up with Ga. There seems to be a continuum between the $D0_{22}$ and $L1_0$ structures, at least from a magnetic point of view. The $L1_0$ structure is essentially ferromagnetic, with the net moment per Mn reduced by antiparallel coupling of the $4d$ sublattice to any Mn remaining in $2b$ sites. The greatest value achievable is around 2 μ_B/Mn. These magnetic structures are illustrated in Fig. 7.3.

The ferrimagnetic structures in tetragonal Mn-Ga or Mn-Ge can be rationalized in terms of the Mn-Mn bond lengths. Direct exchange between manganese atoms mainly depends on the Mn-Mn distance, and it is usually antiferromagnetic below 2.9 Å [29] and ferromagnetic at greater distances. In $D0_{22}$ Mn_3Ga, where Mn occupies $4d$ and $2b$ positions, this leads to the ferrimagnetic structure. The interaction between the $4d$ Mn atoms on planes at $z = 1/4$ and $z = 3/4$ is ferromagnetic ($d = 3.55$ Å), as is the second-neighbour $4d$-$4d$coupling ($d = 3.91$ Å) but the nearest-neighbor $4d$ coupling ($d = 2.77$ Å), like the nearest-neighbor $4d$-$2b$ interaction ($d = 2.65$ Å) is antiferromagnetic. The latter strengthens the tendency towards ferromagnetic order within the $4d$ planes, which orient antiparallel to the $2b$ moments, creating the ferrimagnetic structure with alternating up and down moments along c-axis, as shown in Fig. 7.3a. In the case of the ideally ordered $L1_0$ phase, Mn only occupies the $4d$ positions and the net coupling is ferromagnetic (Fig. 7.3b). In the stable Mn-rich $L1_0$ alloys some of Ga atoms are replaced with Mn, which couple antiferromagnetically with the Mn atoms in $4d$ positions thereby reducing the net magnetization. The highest magnetization would be achieved in the ideal $L1_0$ structure, due to the absence of these antiferromagnetically coupled $2b$ Mn atoms, but it seems that it is not stable.

The magnetic structure of the hexagonal $D0_{19}$ Mn_3Ga is quite different [21, 35]. It has a triangular antiferromagnetic structure with a Néel temperature of 470 K, and a Mn moment of 2.4(2) μ_B, with a weak uncompensated component of 0.016 μ_B. Hexagonal Mn_3Ge shows a similar, in-plane triangular antiferromagnetic structure, with a weak uncompensated moment [28, 36]. Frustration of the antiferromagnetic interactions of Mn atoms gives rise to spin triangles with opposite chiralities [21] in the z = 1/4 and z = 3/4 planes. The triangular magnetic structure is illustrated in Fig. 7.4. A weak moment measured in powder materials prepared for the neutron diffraction experiments [21, 28] is thought to be associated with the canting of the spins in a triangle along c-axis. The magnetic moment in the hexagonal phase disappears when the antiferromagnetic Néel temperature is reached, which proves that this moment is associated with the Mn moments in the structure, and is not due to a secondary phase.

The crystal structures and lattice parameters of the bulk materials are listed in Table 7.1.

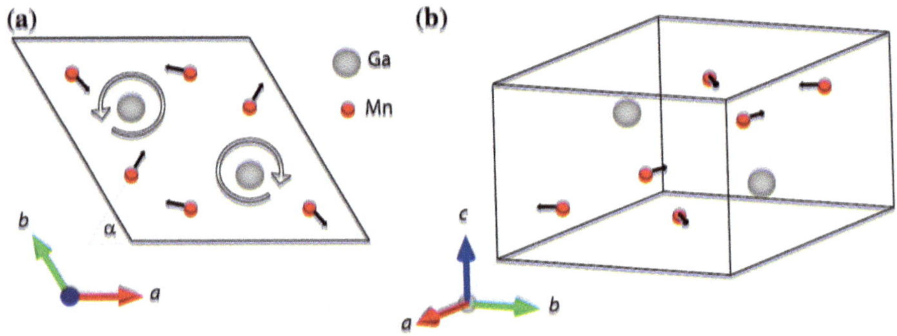

Fig. 7.4 **a** Basal plane of the hexagonal $D0_{19}$ structure of Mn_3Ga and Mn_3Ge. The magnetic moment of the first Mn points at an angle $\alpha = 45°$ from the crystalline a axis and the rest of the magnetic moments arrange in a *triangle*. The chiralities of the spin triangles in adjacent planes are opposite, as indicated by the *circular arrows*. **b** Three-dimensional view of the $D0_{19}$ unit cell. From [37]

Table 7.1 Representative lattice constants of bulk polycrystalline Mn-Ga and Mn-Ge compounds

Material	Phase	Space group	*a* (Å)	*c* (Å)
$Mn_{3.0}Ga$ [24]	$D0_{22}$	I4/mmm	3.909	7.098
$Mn_{2.5}Ga$ [24]	$D0_{22}$	I4/mmm	3.909	7.130
$Mn_{2.33}Ga$ [24]	$D0_{22}$	I4/mmm	3.909	7.130
$Mn_{2.0}Ga$ [24]	$D0_{22}$	I4/mmm	3.905	7.193
Mn_3Ga [20][a]	$L2_1$	Fm-3m	5.823	5.823
Mn_3Ga [18][b]			3.758	3.822
$Mn_{2.85}Ga_{1.15}$[21]	$D0_{19}$	P63/mmc	5.363	4.327
$Mn_{1.7}Ga$ [38]	$L1_0$	P4/mmm	3.897	3.625
$Mn_{3.1}Ge_{0.9}$[28]	$D0_{22}$	I4/mmm	3.816	7.261
$Mn_{60}Ga_{40}$ [39]	δ		2.750	3.642
$Mn_{3.1}Ge_{0.9}$[28]	$D0_{19}$	P63/mmc	5.363	4.327
$Mn_{3.4}Ge$ [40]	$D0_{22}$	I4/mmm	3.803	7.236

[a]Theoretical value

[b] Nanocrystalline, phase unknown

7.3 Thin Film Preparation

The materials whose bulk properties we have just discussed have been prepared in thin film form by a variety of methods. The structure and magnetic properties of the films may differ from those of the bulk, since they depend on the substrate used, the method of preparation, growth temperature, the seed/buffer layer and cap layer (if used) and the film thickness. One of the useful features of thin film growth is the ability to strain the structure, by epitaxy on a single crystal substrate. Here we are interested to see which compositions grow in a cubic or tetragonal form with c-axis orientation. A summary of the in-plane lattice parameters of some relevant substrates and seed layers is given in Table 7.2. The Mn-Ga and Mn-Ge films can be grown on metals, semiconductors and insulators. Their growth and characterization has been reviewed by Zhu et al. [41, 42].

The films are usually grown on lattice-matched single crystal substrates or polycrystalline thin films with a specific texture that has very small lattice mismatch with the basal plane of the material to be grown. For oriented c-axis growth of $D0_{22}$ or $L1_0$ films, a substrate or buffer with an in-plane lattice spacing close to 3.90 Å is required. For cubic $L2_1$ or $D0_3$ films, the spacing should be about 4.12 Å. The hexagonal $D0_{19}$ material can be grown on another hexagonal material such as Ru (0001) or on the (111) face of Pt or Pd, which crystallize in fcc structure. The lattice constant of an fcc material can match with that of a hexagonal material on the (111) face if $a_{fcc}/\sqrt{2} \cong a_{hex}$. Depending on the substrate or seed lattice parameter being smaller or larger than that of the required material, very thin films can be grown with compressive or tensile strain. As the film thickness increases beyond some tens of nanometers, the strain in the film is relaxed by defects. Strain can affect the magnetic properties, especially the magnetic anisotropy in thin films. Depositions on metallic seeds are usually of higher quality than oxide seeds and substrates.

Methods that have been used to produce Mn-Ga and Mn-Ge films are now reviewed briefly.

Evaporation (MBE): Thin film deposition is carried out in ultrahigh vacuum by resistive or electron-beam heating of the material placed in a boat or crucible with which it does not form an alloy. Evaporation is a low-energy deposition method

Table 7.2 In-plane lattice spacings in Ångstroms for various substrates and seed layers

InAs[b]	4.28	AlAs[b]	4.00
V[a]	4.28	Pt	3.92
MgO	4.21	$SrTiO_3$	3.90
Cr[a]	4.11	Pd	3.89
Au	4.08	Ru[a]	3.82
Al	4.05	Si[b]	3.84
GaAs[b]	4.00	Cu	3.61

[a] $a_0\sqrt{2}$
[b] $a_0/\sqrt{2}$

since the energy of the evaporated atoms/molecules is small (~100 meV) when they condense on the substrate. Growth by molecular beam epitaxy (MBE) is carried out slowly, normally with Knudsen cells, each emitting a constant flux of atoms, while the substrate is maintained at a pre-defined temperature. The rate of growth is monitored by reflection high energy electron diffraction (RHEED), which is used to count the number of atomic layers deposited; MBE is well established for making semiconductor stacks for electronic and optoelectronic devices. The first tetragonal $L1_0$ Mn-Ga films with perpendicular anisotropy were grown this way on (001) GaAs [25, 26, 43]. A tetragonal zinc blende structure with a high manganese moment (3.2 μ_B) has been grown on (111) GaAs [44], and there are reports of $L1_0$ films on GaSb [45, 46], GaN [47, 48], ScN [49], Si [45] and sapphire [45, 48]. The choice of substrate has a strong influence on the anisotropy direction and other magnetic properties [45, 48].

Although high quality $L1_0$ Mn-Ga films can begrown on GaAs by MBE [26, 43, 50], growth of the $D0_{22}$ $Mn_{3-x}Ga$ phase on semiconductor substrates has been less successful [50]. For example films grown on GaSb-buffered Si [46] showed low coercivity and did not exhibit (001) texture. Nanoparticulate films with a perpendicular easy axis are obtained on Si/SiO_2 [51].

Sputtering is widely available in research laboratories and it is suitable for large-scale production. It is the method of choice for many metallic thin films and multilayers. The growth of complex materials such as Heusler compounds is a challenge, as the constituents may be materials with very different melting points, boiling points, vapor pressure, sputtering and ionization yields. The latter are affected by the configuration of the magnetron source and the sputtering voltage. An appropriate alloy sputtering target may be crucial to obtain the desired chemical composition of the thin film. Moreover, in order to guarantee a constant flux of atoms from a sputtered target, it should have a homogenous density of the constituent elements, which may be difficult to achieve for alloys with complex binary and ternary phase diagrams.

An alternative technique is confocal multi-target sputtering, where elemental or alloy targets are sputtered simultaneously towards a common focal point while the substrate rotates continuously. The sputtering rates are individually calibrated such that the desired stoichiometry is achieved across the film. This method provides the flexibility needed for research purposes.

When bombarded with an energetic beam of argon ions during sputtering, the target atoms may become ionized. These secondary ions can create disorder in the film growing film. It is best to prevent them from reaching the substrate. This is done by applying a repulsive dc bias, thereby forming the film only from neutral atoms.

The majority of Heusler alloy thin films are grown on lattice-matching crystal substrates. This is a limitation because most single-crystal substrates are expensive and unsuitable for large-scale production. A solution is to grow a sufficiently thick seed layer on Si or glass wafers that has lattice parameters within few percent of the material that is to be grown. This is not always possible, but there are promising examples where a crystalline material can be grown on an amorphous buffer layer, such as MgO(001) on CoFeB [52] and $SrTiO_3$(001) on SiO_x/Si(001) wafers [53, 54].

Many groups have successfully grown $D0_{22}$ and $L1_0$ Mn-Ga with perpendicular anisotropy by dc-sputtering [32, 43, 51, 55–65]. The first report, by Wu et al. [59], used Cr-buffered MgO substrates, but required post-annealing [58]. Initially, the aim was to make smooth textured films crystallized in the $D0_{22}$ structure, rather than obtain an exact 3:1 stoichiometry, which was reported to be difficult due to chemical instability [38, 66]. Growing very smooth and thin films in a multilayer stack is crucial to develop spintronic devices. So far, the best surface roughness has been achieved with Mn_3Ga films grown on $SrTiO_3(001)$ substrates [67] and with a Pt (001) seed layer [62, 64], which helps to maintain perpendicular anisotropy in ultra-thin films [65]. Pt offers the lowest crystallization temperature. Growth of well-crystallized Mn-Ga requires a heated substrate, but the surface roughness of the seed layer is degraded at elevated temperature. A compromise between crystallization and film roughness may be called for in thin film devices.

All the materials in a stack have to be compatible with the growth temperatures used and most importantly they should not alloy with each other to the point where their function in the stack is impaired. This is especially tricky to achieve in spin valve stacks where all layers are metallic. Magnetic tunnel junctions are easier, since the oxide or nitride tunnel barriers are very stable and they do not usually alloy with metals at the growth temperatures of most Heusler alloys. However, a report by Violbarbosa et al. [68] has shown that gallium in Mn-Ga can oxidize when an MgO tunnel barrier is grown on it. Therefore, an insertion layer, such as Mg, CoFeB, Fe or Co may be necessary before depositing the MgO.

The first results on sputtered $D0_{22}$ Mn_3Ge (001) epitaxial films were reported by Kurt et al. [69] on $SrTiO_3$ (001) single crystal substrates. Films grown at 450 °C are single-phase $D0_{22}$ with the c-axis perpendicular to the substrate. Subsequently Miyazaki [70, 71] demonstrated growth on MgO (001) substrates and on Cr(001) seed layers at 400–600 °C. In contrast, the $D0_{22}$ $Mn_{3-x}Ga$ series of alloys are well crystallized at a substrate temperature of 350 °C [59, 61, 62]. Therefore, $D0_{22}$ $Mn_{3-x}Ga$ could be integrated in a multilayered device structure more easily than Mn_3Ge, provided the adjacent layer does not alloy with Mn-Ga at 350 °C. The growth temperatures of $D0_{22}$ MnGe are higher than for Mn-Ga, limiting its use to a narrower set of materials in device stacks.

The films deposited on MgO tunnel barriers are usually sputtered at a temperature where the film is an atomically disordered polycrystalline state. This is a preventative measure to preserve the integrity of the very thin tunnel barrier. Subsequent postannealing after growing the full MTJ stack is required to achieve reasonable crystallographic order in the oxide barrier. Epitaxial growth of any crystalline material begins with the selection of a suitable seed layer, which must have a low surface roughness (ideally less than 0.5 nm rms). The same applies for the tunnel barrier that serves as a seed layer for the top ferromagnetic electrode in an MTJ. By choosing V(001) buffered MgO, Kurt et al. [72] have succeeded in producing cubic $D0_3/C1_b$ Mn_3Ga and Mn_2Ga films, a structure that is not accessible in the bulk. This is an example of how lattice-mismatch induced strain can be used to directly influence the crystal structure of Mn_2Ga grown on top of a seed layer. For example the lattice parameter of Cr is 2.88 Å and that of V is 3.02Å. There is considerable difference on the lattice strain of both seed layers. The variation of the crystal structure on both seed layers are shown in Fig. 7.5d below. The V (001) seed layer grown on MgO also shows a tetragonal distortion as seen from the V (004) peak position. This corresponds to an out of plane lattice constant of 3.27 Å. The bcc lattice of V shrinks in *ab* plane to fit crystallographically to the MgO lattice ($a = 4.212\,\text{Å}$, $a/\sqrt{2} = 2.98$ Å), and this expands the unit cell along c-axis.

Oriented hexagonal $D0_{19}Mn_3Ga$ films have be grown by sputtering onto an Si substrate with a Pt(111) [61] and Ru(0001) seed layers at 400 °C [37]. The first report of c-axis oriented $D0_{22}$ Mn_3Ge films was by Kurt et al. [69] on $SrTiO_3$ substrates.

Pulsed Laser Deposition (PLD) is a useful technique to make high quality oxide and nitride compound films. Very short but energetic pulses of focused laser radiation are used to ablate a small quantity of material from the irradiated area on the target, creating plasma of the material to be deposited. The film tends to reflect the stoichiometry of the target. Recent methods used for coating large wafers use target scanning and target and/or wafer tilting and rotation [73–75]. Relatively homogenous high- quality films of Heusler compounds have been obtained, such as Co_2FeSi [76] and Co_2(Fe,Mn)Si [77], and the magnetization of these PLD films is in agreement with the bulk values. So far, there are no reports of PLD growth of Mn-Ga or Mn-Ge films.

7.4 Structural Characterization

Thin films grown on a planar substrate are first characterized by X-ray diffraction (XRD). The substrate material is either single-crystal or polycrystalline with a specific texture to facilitate epitaxial or textured growth of the Heusler compound. The relative orientation, interplanar spacings and approximate out-of plane grain size can be determined. In an oriented film, only the c-axis interplanar distances are determined from the θ-2θ scans, but since they are different for cubic, tetragonal and hexagonal phases, one can usually distinguish which form is present in an oriented film. In order to determine the crystal structure one has to measure Bragg reflections from other planes, by orienting the specimen to generate a reciprocal space map (see for example Fig. 7.5c). The lattice parameters of the $D0_{22}$ and $D0_{19}$ Mn-Ga and Mn-Ge materials as well as cubic counterparts are included in Table 7.3.

Extended X-ray absorption spectroscopy (EXAFS) can be used to determine near-neighbour interatomic distances, and can help to determine the atomic site distribution. Perfect crystalline order is always difficult to obtain in Heuslers, since the X,Y and Z atoms can swap positions in the structure. The magnetic properties and electronic structure depend sensitively on the atomic order.

The interplanar distances and texture of the films inferred from the θ-2θ XRD scans are not by themselves enough to completely determine the crystal structure. They are indicators of whether the film has grown in a particular crystal phase and

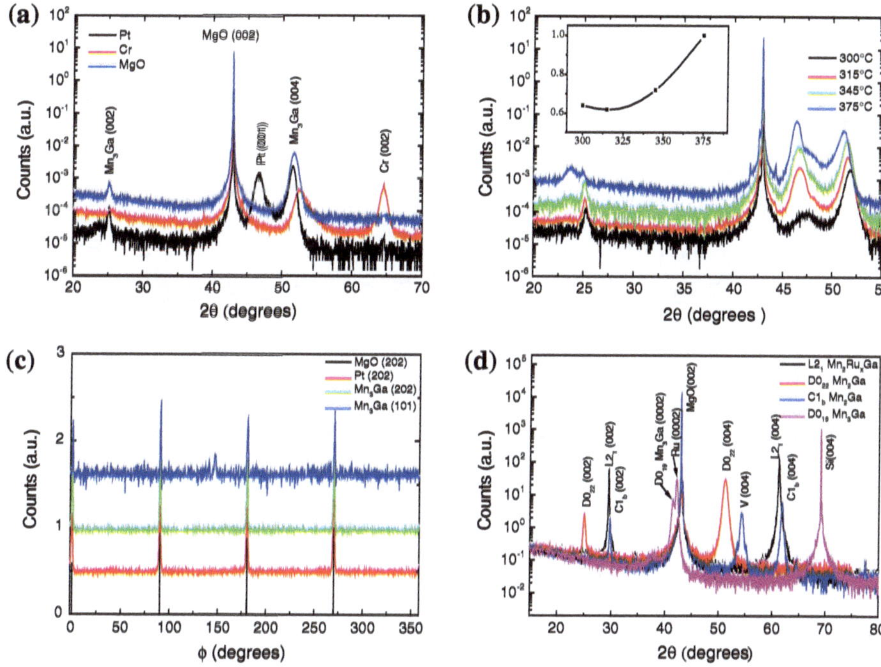

Fig. 7.5 **a** θ-2θ XRD scans of Mn_3Ga films grown at 350 °C on MgO (001) substrates, Cr (001) and Pt (001) seed layers **b** θ-2θ XRD scans of Mn_3Ga films grown on Pt (001) seed layers at various substrate temperatures. The inset shows the order parameter *S* as a function of growth temperature. **c** in-plane ϕ-scans of epitaxial $D0_{22}$ Mn_3Ga films grown on a Pt (001) seed layer that is grown on an MgO (001) substrate. **d** θ-2θ XRD scans of all phases of Mn-Ga thin films. All except the hexagonal Mn_3Ga were grown on MgO (001) single crystal substrates. Data from [62, 72]

texture, but the positioning of atoms in the unit cell also requires information about the diffraction peak intensities. A representative crystalline order parameter *S* can be obtained by comparing measured and calculated peak intensities from crystal planes which are not parallel. For example in the $D0_{22}$ Mn_3Ga structure shown in Fig. 7.6 the (101) and (204) planes are non-parallel and filled with different densities of Mn and Ga atoms; (101) planes contain Ga whereas (204) planes are filled with both Mn and Ga. Their intensity ratio gives a reliable indication of the order parameter *S* for the $D0_{22}$ structure; this is calculated as [59, 62].

$$S = \sqrt{}[I_{(101)}/I_{(204)}]_{\text{measured}}/\sqrt{}[I_{(101)}/I_{(204)}]_{\text{calculated}} \tag{7.3}$$

Films of non-stoichiometric compounds where both composition and atomic order depend on the growth temperature are difficult to characterize completely by X-ray diffraction. For example, the Mn_2Ga films grown on MgO, which may be considered as $Mn_{3-x}Ga$ with x = 1, show a significant variation of magnetization, anisotropy

Table 7.3 Substrate, deposition conditions and lattice constants of tetragonal, cubic and hexagonal Mn-Ga and Mn-Ge films

Substrate/ seedlayer	Film	T_s(°C)		Crystal structure	a (pm)	c (pm)	M_s (kA m^{-1})	K_u (MJ m^{-3})
MgO (001) [62]	Mn_3Ga	350	**S**	$D0_{22}$		707	150	1.91
MgO (001) [61]	Mn_2Ga	350	**S**	$D0_{22}$		716	470	2.35
MgO (001) [57]	Mn_2Ga	500	**S**	$D0_{22}$		721	360	1.10
Cr (001) [61]	Mn_3Ga	350	**S**	$D0_{22}$		696	140	0.40
Cr (001) [59]	$Mn_{2.5}Ga$	350	**S**	$D0_{22}$		707	250	1.20
SrTiO3 (001) [30]	Mn_3Ga	350	**S**	$D0_{22}$	392	708	155	1.35
SrTiO3 (001) [30]	$Mn_{2.77}Ga$	350	**S**	$D0_{22}$	391	712	315	1.37
SrTiO3 (001) [30]	Mn_2Ga	350	**S**	$D0_{22}$	391	722	430	1.50
SrTiO3 (001) [67]	$Mn_{2.9}Ga$	530	**S**	$D0_{22}$		722	270	
SrTiO3 (001) [67]	$Mn_{2.6}Ga$	530	**S**	$D0_{22}$		723	350	
SrTiO3 (001) [67]	$Mn_{2.3}Ga$	530	**S**	$D0_{22}$		725	350	
Pt (001) [62]	Mn_3Ga	250	**S**	$D0_{22}$		707	110	0.67
Pt (001) [62]	Mn_3Ga	250	**S**	$D0_{22}$		712	100	0.89
Pt (001) [30]	Mn_3Ga	350	**S**	$D0_{22}$	391	706	111	0.88
Cr(001)[32]	$Mn_{54}Ga_{46}$	400	**S**	$L1_0$	390	360	600	1.05
Cr(001)[32]	$Mn_{62}Ga_{38}$	400	**S**	$L1_0$	390	355	450	1.00
Cr(001)[32]	$Mn_{72}Ga_{28}$	400	**S**	$D0_{22}$	390	700	200	1.50
ScN (001) [49]	$Mn_{1.5}Ga$	250	**M**	$L1_0$	394.5	308	125	0.09
MgO(001) [56]	$Mn_{1.54}Ga$	450–500	**S**	$L1_0$	391	360	500	1.50
MgO(001) [56]	$Mn_{2.12}Ga$	450–500	**S**	$D0_{22}$	392	711	305	1.50
GaAs (001) [43]	$Mn_{1.5}Ga$	100–300	**M**	$L1_0$		347–351	29–270	0.20–2.17
GaAs (001) [25]	$Mn_{1.5}Ga$	100	**M**	$L1_0$	394.5	351.5	460	

(continued)

Table 7.3 (continued)

Substrate/ seedlayer	Film	T_s(°C)		Crystal structure	a (pm)	c (pm)	M_s (kA m^{-1})	K_u (MJ m^{-3})
GaN (0001) [81]	$Mn_{49}Ga_{51}$	250	**M**	$L1_0$	388.3	374.2	410	
GaN (0001) [81]	$Mn_{58}Ga_{42}$	250	**M**	$L1_0$	388.6	369.9	300	
GaN (0001) [81]	$Mn_{67}Ga_{33}$	250	**M**	$L1_0$	388.6	365.9	120	
Pt (111) [61]	Mn_3Ga	400	**S**	$D0_{19}$		437.7		
Pt (111) [61]	Mn_2Ga	400	**S**	$D0_{19}$		439		
Ru (0001) [37]	Mn_3Ga	400	**S**	$D0_{19}$	531	435.8	47	
Ru (0001)	Mn_2Ga	400	**S**	$D0_{19}$		438.5	160	
V (001) [72]	Mn_2Ga	350	**S**	$C1_b$		599	280[a]	
V(001) [72]	Mn_3Ga	350	**S**	$D0_3$		591	90	
MgO(001)[72]	Mn_2RuGa	350	**S**	$L2_1$		605.4	74	
MgO(001) [72]	$Mn_2Ru_{0.83}Ga$	350	**S**	$L2_1$		604.2	57	
MgO(001)[72]	$Mn_2Ru_{0.52}Ga$	350	**S**	$L2_1$		601.5	13	
MgO(001) [72]	$Mn_2Ru_{0.48}Ga$	350	**S**	$L2_1$		600.3	21	
MgO(001) [72]	$Mn_2Ru_{0.33}Ga$	350	**S**	$L2_1$		598.0	48	
SrTiO3 (001) [69]	Mn_3Ge	450	**S**	$D0_{22}$	385	718.5	73	0.91
MgO(001) [71]	$Mn_{3.55}Ge$	400–600	**S**	$D0_{22}$	381	722	130	1.00
Cr(001) [70]	Mn_3Ge	400	**S**	$D0_{22}$	380	722	100	1.18

Saturation magnetization M_s and uniaxial anisotropy constant K_u are measured at 300 K unless stated otherwise. For cubic films $a_\perp$ is measured and listed in the c column. **S** and **M** refers to 'sputtered' and 'MBE deposited' respectively

[a]at 100 K

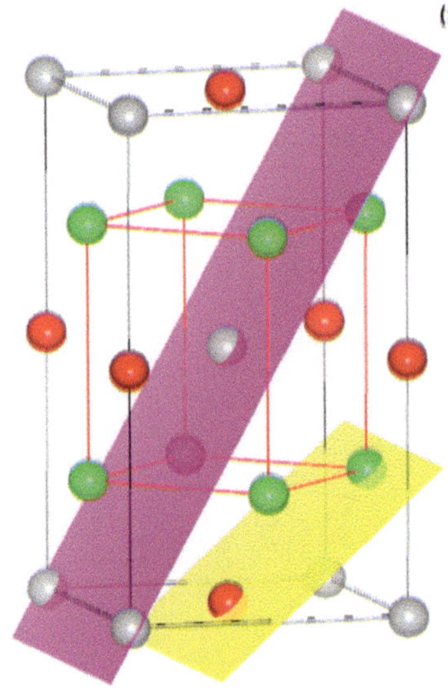

Fig. 7.6 (101) and (204) planes of the $D0_{22}$ Mn_3Ga

and coercivity [61], which is associated with the different atomic occupancy of Mn positions, and also antisite defects where Ga is replaced by Mn.

Alloy additions influence phase formation and the atomic site distribution. Addition of some cobalt in Mn_3Ga initially maintains the tetragonal $D0_{22}$ structure [78, 79], but the structure becomes cubic for $x \geq 0.5$ in $Mn_{3-x}Co_xGa$. When Mn_2Ga is alloyed with Fe and Cr the material maintains a tetragonal phase [80], but alloying with just a few percent of Ru leads it to crystallize in a cubic structure. Here extended X-ray absorption spectroscopy (EXAFS), which gives the interatomic distances between Mn,Ga and Ru atoms along different directions can be used to verify the cubic character. The X-ray absorption spectra on Mn and Ga K-edges with polarization in-plane and out of plane (see Fig. 7.7 below) shows that in the Mn_2Ru_xGa material, the interatomic spacings of Mn and Ga atoms are the same along both crystal axes, which confirms the cubic structure. A complete analysis of EXAFS data on Mn, Ga and Ru K-edges shown in Fig. 7.8 establishes that Mn_2Ru_xGa crystallizes in a variant of the $L2_1$ structure [72]. The Ru K edge also shows that there are no Ru-Ga nearest neighbors. The agreement between the fits and experimental values reveal that the structure is very close to that shown in Fig. 7.9.

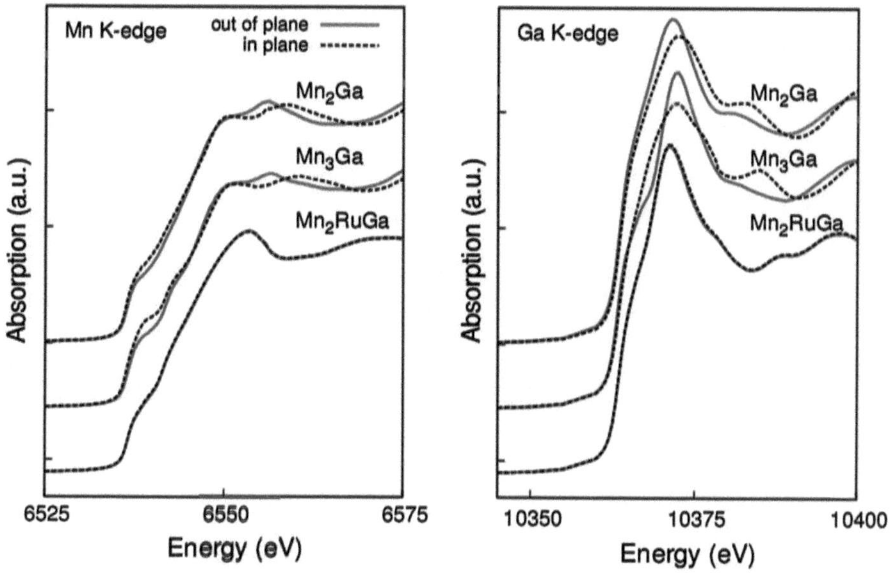

Fig. 7.7 X ray absorption spectra on Mn and K edges in $D0_{22}$ Mn_2Ga, $D0_{22}$ Mn_3Ga and $Mn_2Ru_{0.52}Ga$ thin films with polarization in-plane and out-of-plane

7.5 Magnetic Properties

Another way to distinguish a cubic magnetic material from a tetragonal or hexagonal one is by magnetization measurements on single crystals or oriented thin films. The uniaxial ferromagnetic materials often exhibit easy-axis magnetocrystalline anisotropy and they are magnetically hard. Ferromagnetic materials with cubic symmetry are magnetically soft, with low coercivity and in-plane anisotropy determined by thin film shape. Large magnetic fields, >10 T, are needed to saturate the magnetization of polycrystalline hard materials, but single crystals or c-axis oriented epitaxial or textured films can be saturated in fields that are a bit bigger than the coercivity.

Coercivity is not an intrinsic magnetic property of a material, since it is governed by reverse domain nucleation and domain wall pinning—features that depend on the microstructure [1]. However it is limited by the anisotropy field $\mu_0 H_a$, which does depends the intrinsic properties of saturation magnetization M_s and uniaxial anisotropy K_u

$$\mu_0 H_a = 2K_u/M_s \tag{7.4}$$

The saturation magnetization of a well-oriented c-axis film is deduced from the easy-axis magnetization curve, and the anisotropy field (and hence K_u) is deduced by extrapolating the hard-axis magnetization curve to saturation. The coercivity is rarely more than 10–20 % of the anisotropy field, and cannot in any case exceed it.

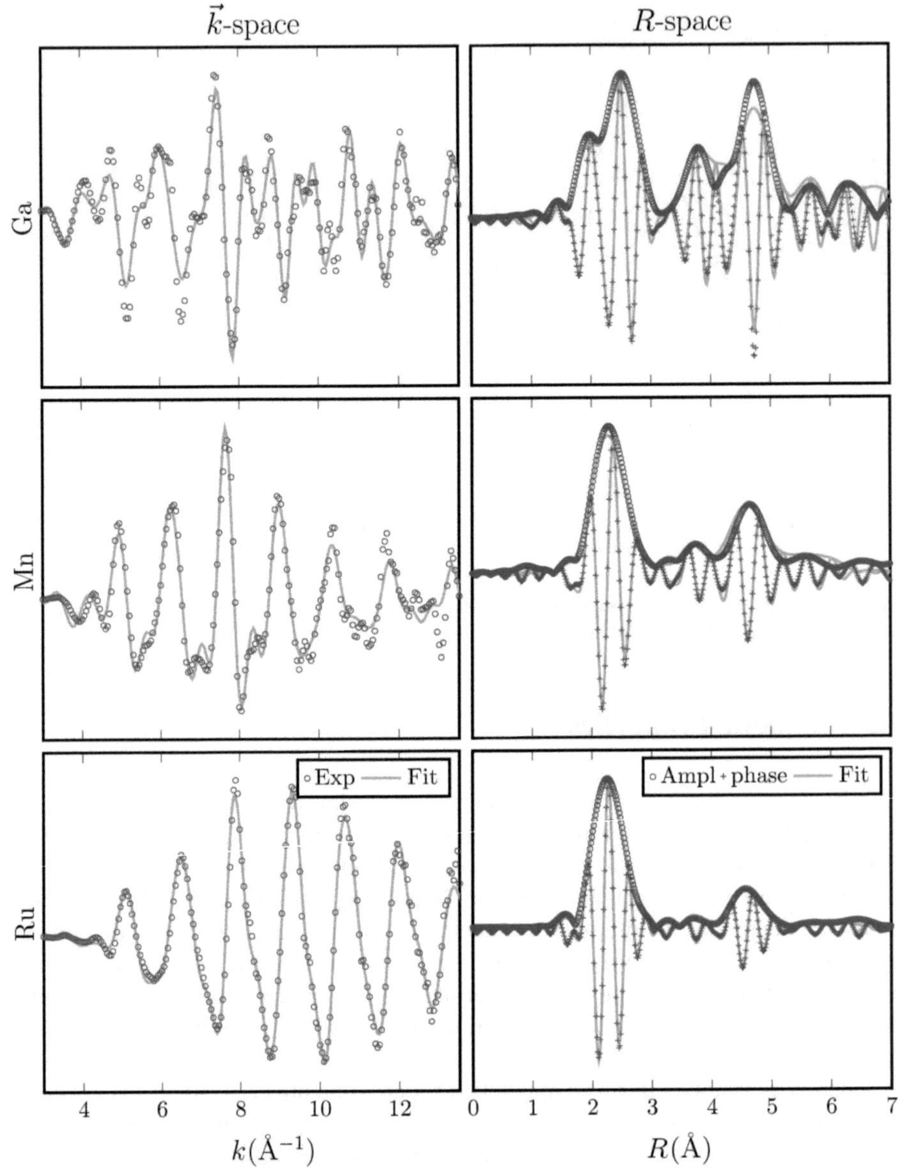

Fig. 7.8 Real and momentum space measurements and fits to the EXAFS data obtained on the K edges of Mn, Ga and Ru in $Mn_2Ru_{0.52}Ga$ thin film [30]

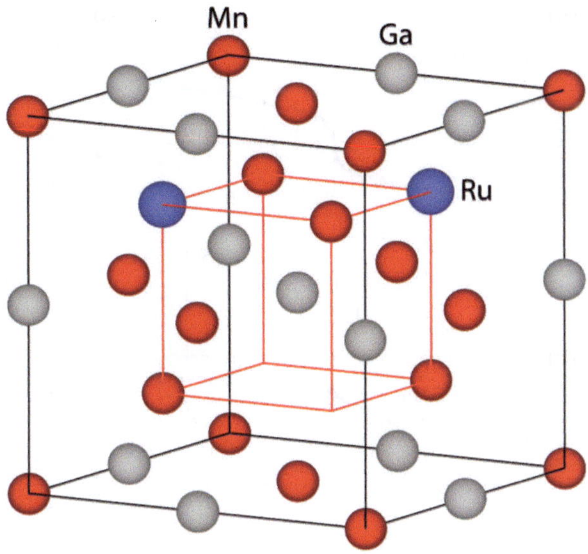

Fig. 7.9 Atomic structure of $Mn_2Ru_{0.5}Ga$. Ruthenium occupies half of the 4d sites and there are no Ru-Ga nearest neighbours

An example of relatively straightforward hard- and easy-axis magnetization behavior is shown for Mn_2Ga in Fig. 7.10b, where $M_s = 470$ kA m^{-1}, $\mu_0 H_a, = 10$ T and $\mu_0 H_c = 0.5$ T. The value of K_u calculated from 7.4 is 2.35 MJ m^{-3}. All these quantities depend on temperature. Note that for a given value of K_u, the anisotropy field diverges as M_s tends to zero. Large coercivity is to be expected in uniaxial materials with low magnetization.

The anisotropy and magnetization also depend on the atomic order. In uniaxial materials like tetragonal Mn-Ga, the coercivity usually decreases with increasing film thickness [61], and it also decreases in ultra-thin films [65]. Disorder increases the probability of antiferromagnetic coupling between Mn atoms in antisites, thereby decreasing the total magnetization and tending to increase coercivity. Thin films deposited at a temperature lower than the ideal crystallization temperature tend to be disordered. The crystal structure of the film follows the texture of the substrate but a low growth temperature favours disorder and imperfection.

The magnetic properties of thin films are small, but frequently measured using a SQUID magnetic property measurement system (MPMS) or a VSM magnetometer in physical property measurement system (PPMS) made by Quantum Design. The hysteresis loops of films with known thickness and lateral dimensions are measured with the magnetic field applied in-plane and out-of-plane, and data corrected for diamagnetic contributions from the substrate and geometrical factors are then used to determine the anisotropy field. The hysteresis curves in the perpendicular direction show that the easy directions of magnetization of $D0_{22}$ Mn-Ga and Mn-Ge are along the c-axis. The in-plane measurement of the $D0_{22}$ films reveals a rather soft magnetic component that does not show up in the magnetization measurements perpendicular to the plane (Fig. 7.10a, b). This indicates that the soft in-plane moment is not due to

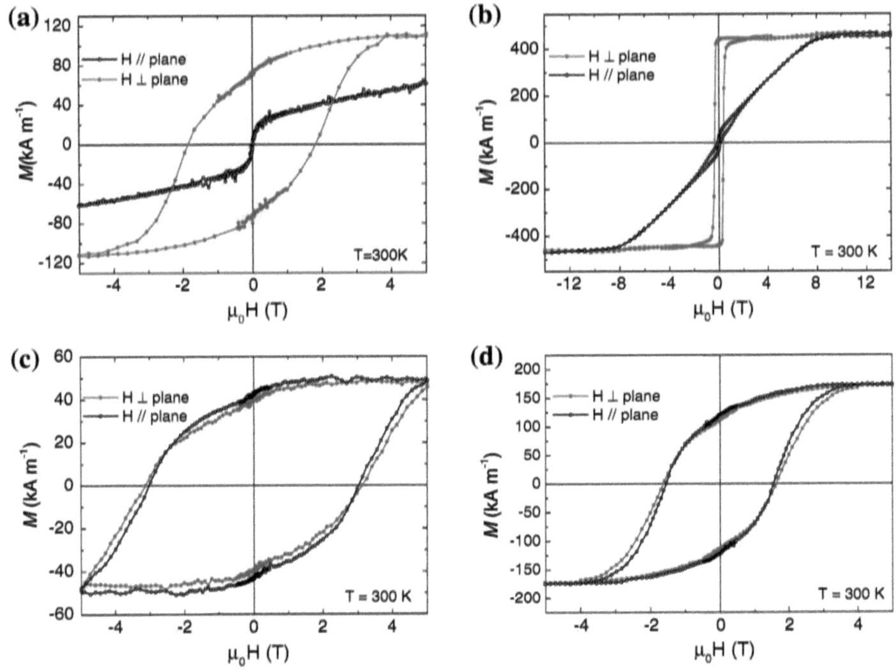

Fig. 7.10 In-plane and out of plane magnetization *curves* of **a** $D0_{22}$ Mn_3Ga (001) film, **b** $D0_{22}$ Mn_2Ga (001) film, **c** $D0_{19}$-Mn_3Ga (0001) film, **d** $D0_{19}$-Mn_2Ga (0001) film at room temperature

a secondary phase in the film, but that it is intrinsic to the structure, as demonstrated by the neutron diffraction measurements on $D0_{22}$ Mn_3Ga [30]; it is a unique and very interesting property of the $D0_{22}$ materials, because such a soft magnetic component coexisting in a highly anisotropic material may potentially be useful to generate radio-frequency signals, or facilitate spin torque switching. The magnitude of the in-plane moment differs for materials grown on different substrates, and it also depends on growth temperature, which changes the atomic order in the crystal. Here it is worth noting that the results of Mizukami et al. [56] showing 280 GHz oscillations in tetragonal Mn-Ga films could be due to the presence of such a soft moment.

Damping parameters of $L1_0$ and $D0_{22}$ phases have been measured by an optical pump-probe technique. The damping constant α is found to be rather low—0.008 for $L1_0$-$Mn_{1.54}Ga$ and 0.015 for $D0_{22}$ $Mn_{2.12}Ga$ films, both grown on MgO (001) substrates [56].

The magnetic properties of Mn-Ga and Mn-Ge thin films are included in Table 7.3. There is a tendency for the largest coercivity to be associated with the smallest magnetic moments. A series of hysteresis loops shown in Fig. 7.11, measured by Zhu et al. [43] on $L1_0$ $Mn_{1.5}Ga$ films deposited on substrates at different temperatures

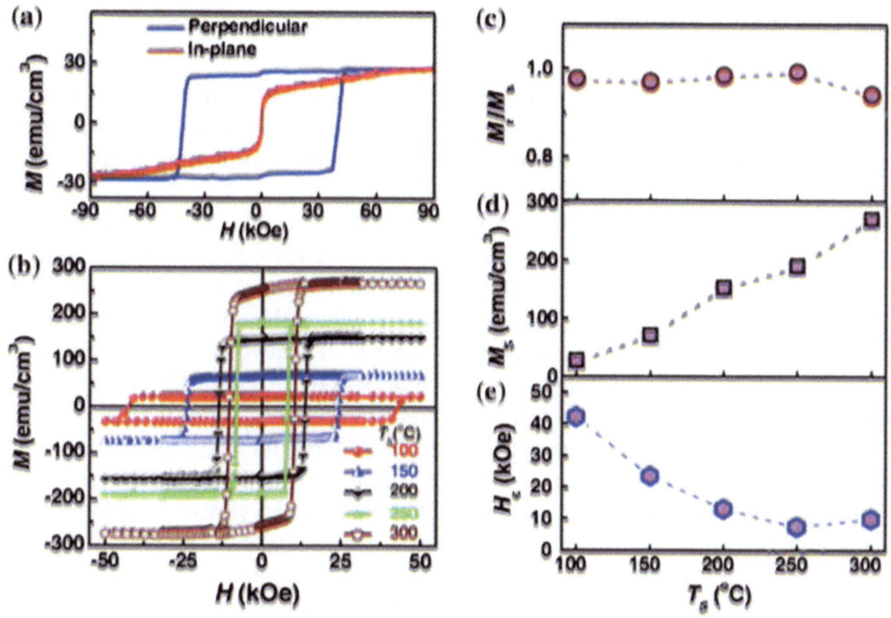

Fig. 7.11 **a** Perpendicular and in-plane hysteresis loops of the 100 °C-grown $L1_0$-$Mn_{1.5}Ga$ film measured by PPMS. **b** Perpendicular hysteresis loops of the $L1_0$-$Mn_{1.5}Ga$ films grown at different T_s measured by SQUID. **c** M_r/M_s, **d** M_s and **e** H_c plotted as a function of T_s. Figure taken from [43]

illustrates this tendency. The largest coercivity of 4.28 T is associated with the smallest magnetization of 29 kAm^{-1}. The cubic Mn_2Ga and Mn_3Ga films deposited on V-buffered MgO [72] exhibit negligible coercivity.

The magnetic moments of the epitaxial $D0_{19}$ Mn_3Ga and Mn_2Ga thin films are small, but significantly higher than those reported in the neutron diffraction experiments on bulk material [21, 37], and both types of material can exhibit high coercivity as shown in Fig. 7.10c, d. The disagreement between the magnetizations of the thin epitaxial films and the powder samples used for the neutron diffraction measurements is not surprising, as the data on the $L1_0$ material demonstrate the sensitivity of the moment to the atomic order. In view of the substantial intrinsic anisotropy values summarized in Table 7.3 (0.2–2.17 MJ m^{-3} for $Mn_{1.5}Ga$; 1.1 − 2.35 MJ m^{-3} for Mn_2Ga; 0.8 − 1.91 MJ m^{-3} for Mn_3Ga; and 0.91 − 1.18 $MJm^{-3}$for Mn_3Ge;) there are sometimes difficulties in completely saturating material with a high anisotropy field. However, the measurements shown in Fig. 7.10c, d on both $D0_{19}$ Mn_3Ga and Mn_2Ga films reveal that the net magnetization of these materials is almost isotropic.

The magnetic moments of the cubic $D0_3$ Mn_3Ga and $C1_b$ Mn_2Ga films are approximately 0.5 and 1.5 μ_B/formula, respectively [72]. A smaller moment of 0.1 μ_B/formula was found in nanocrystalline bulk $D0_3$ Mn_3Ga, which has a small tetragonal distortion [18]. In neither case is the zero-moment half-metallicity predicted for cubic Mn_3Ga [20] actually achieved.

7.6 The Zero-Moment Half-Metal

We now return to the question posed in the Introduction. Is it possible to find a zero-moment compound with a spin gap? This has long been an open question, and despite numerous theoretical proposal, no experimental confirmation of any of them had been forthcoming [82], untill recently [72].

In terms of cubic Heusler compounds, full Heuslers with 24 valence electrons [83], and half-Heuslers with 18 electrons are the prime candidates. For example, Cr_2CoGa has been suggested [84], but was found experimentally to have a small moment of 42 kA m^{-1} [18], possibly because the sample did not have the complete atomic order. The compound Mn_3Ga has the necessary number of valence electrons, but is not cubic in the bulk. A metastable nanocrystalline variant prepared by ball milling is nearly-cubic, and the magnetization is less than 20 kA m^{-1} [18] ($\sim 0.1 \mu_B$/formula). The cubic Mn_3Ga films we have stabilized on V-buffered MgO show a moment of 90 kA m^{-1} (0.53 μ_B/formula) [72]. The non-zero moments might be due to atomic disorder. Furthermore, the two inequivalent sublattices are expected to exhibit different temperature dependences [85], so measurements at 300 K may not reflect the situation at $T = 0$. Of course, it is also possible that counting valence electrons is simply too crude an indicator of the electronic density of states. The assumption of a perfect hybridization gap in the spin-polarized $3d_\downarrow$ band depends on perfect site occupancy of $4b$ positions by Ga or Ge, Spin-polarized density of states calculations can be of help here. Experimental measurements of the spin polarized density of states never yield 100 %, especially for Heusler half-metals, even at T $\sim$ 0 K.

It has not yet been possible to prepare cubic Mn_2Ge films, which are expected to be zero-moment half-metals. The material tends to grow in a triclinic or hexagonal phase [86, 87]. However, the compound Mn_2Ga is also stabilized in a cubic structure on V-buffered MgO, with a moment of (-) 1.6 μ_B/formula [72]. The minus sign reflects the idea that with 17 valence electrons, the Mn I sublattice is dominant. The EXAFS data shows that the interatomic spacings between Mn-Mn and Mn-Ga along a and c directions are the same, which confirms the cubic structure.

A small addition of Ru in Mn_2Ga films stabilizes the cubic structure, as discussed in Sect. 3.6.4. Density functional calculations [88] show that the Mn in an ideal cubic $C1_b$-Mn_2Ga structure occupies $4c$ and $4a$ sites and couples antiferromagnetically. This is also the case for Ru-doped $C1_b$-Mn_2Ga. Magnetization and anisotropy depend on the Ru concentration, and they change sign at $x = 0.5$, which corresponds to half occupancy of the Ru positions in the $L2_1$ unit cell (Fig. 7.9). Figure 7.12 shows magnetization, anisotropy and Curie temperature of Mn_2Ru_xGa cubic material as a function x.

The ruthenium addition in Mn_2Ru_xGa has the effect of adding both electrons and states to the density of states. The cubic end-member with $x = 1$ has a positive moment of 0.4 μ_B per formula. The crossover from negative to positive moments occurs at $x = 0.5$, and there is a remarkable change in the anisotropy and anomalous Hall coefficient [103] at this composition. Furthermore, the slope of the moment per formula dm/dx is exactly 2 at $x = 0.5$, where the valence electron number is

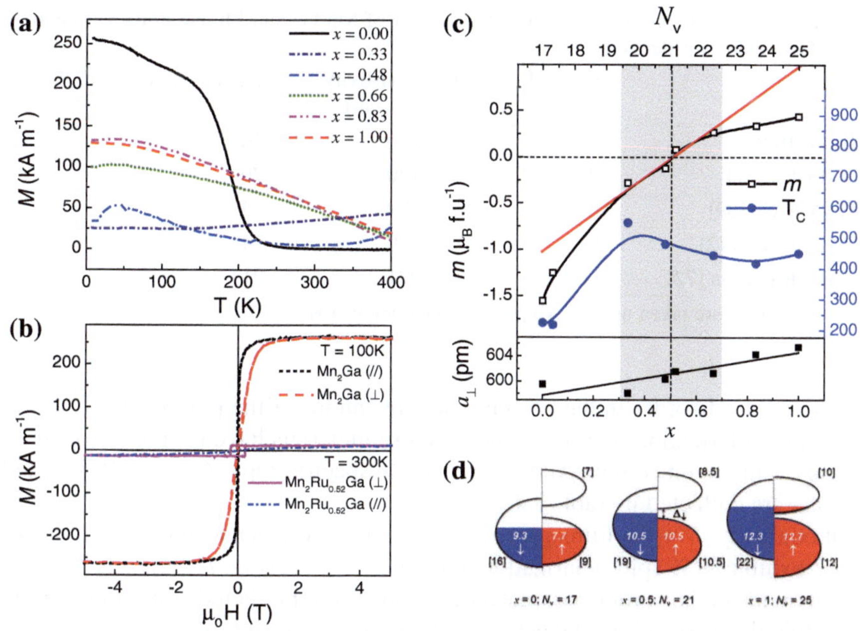

Fig. 7.12 **a** Thermal scans in 200 mT of the magnetization of $C1_b$-Mn_2Ru_xGa films with different Ru content. **b** In-plane and perpendicular magnetization of a cubic $C1_b$-Mn_2Ga film at 100 K and a $Mn_2Ru_{0.52}Ga$ film at 300 K. **c** Variation of Curie temperature and moment with Ru concentration x and number of valence electrons per formula unit. The magnetic anisotropy changes from in-plane to perpendicular when $x > 0.5$, and the moment is reversed in sign. Its variation close to $x = 0.5$ is indicated by a red line with slope 2. The half-metallic region is lightly shaded. The lower panel shows the variation of the lattice parameter $a_\perp$, normal to the substrate, as a function of x. **d** Schematic density of states showing the effect of Ru addition in Mn_2Ru_xGa, which illustrates the Fermi level crossing the spin gap. Figure taken from [72]

21 (Fig. 7.12c). This is the slope expected for a half metal. The effect of ruthenium addition in the cubic compound is to cross the spin gap, as illustrated schematically in in Fig. 7.12d. Electronic structure calculations by Archer et al. [89] confirm the half-metallic character near $x = 0.5$.

7.7 Electronic Properties

Mn-Ga and Mn-Ge films are metallic with high resistivities. They exhibit a high spin polarization at the Fermi level, which makes them potentially useful for perpendicular spin torque memory. Density functional theory calculations on several $Mn_{3-x}Ga$ compositions by Winterlik et al., show that stoichiometric $D0_{22}$ Mn_3Ga should exhibit 88 % spin polarization at the Fermi level [90]. However, non-stoichiometric compo-

Table 7.4 Fermi-level spin polarization and resistivity of Mn-Ga and Mn-Ge films measured by point contact Andreev reflection

Film stack	Spin polarization (%)	Resistivity (μΩcm)	Magnetization (kA m^{-1})
Pt/Mn_3Ga [62]	58	160	110
MgO/Mn_2Ga [61]	40	120	470
$SrTiO_3$/Mn_3Ge [69]	46	N/A	73
MgO/Mn_2RuGa [72]	41	280	74
MgO/$Mn_2Ru_{0.48}Ga$ [72]	54	221	21

The measurements are taken using a Nb point contact below 4 K

sition as well as disorder in the crystal structure can lower the polarization [24]. So far, the spin polarization measurements of these materials have been performed by Stamenov using point contact Andreev reflection spectroscopy [61, 62, 69, 72], and the results are collected in Table 7.4.

Epitaxial $D0_{22}$ Mn_3Ga films grown on Pt (001) seed layer with order parameter $S = 0.7$ exhibit 58 % spin polarization at the Fermi level (Fig. 7.13a). It is possible that films with higher order parameters may show the predicted spin polarization. In non-stoichiometric alloys the net magnetization and density of states can be very different depending on the position of the Mn vacancies. The measured spin polarization in a $D0_{22}$ Mn_2Ga thin film is 40 % [61] (Fig. 7.13b), which is lower than that of

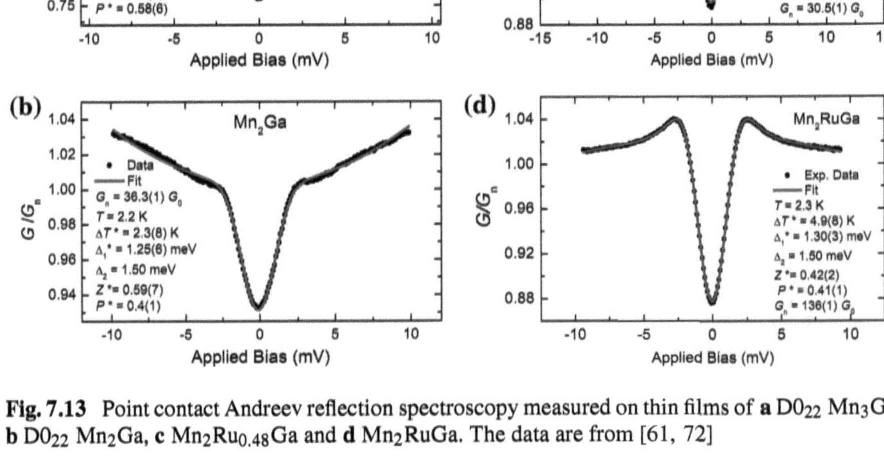

Fig. 7.13 Point contact Andreev reflection spectroscopy measured on thin films of **a** $D0_{22}$ Mn_3Ga, **b** $D0_{22}$ Mn_2Ga, **c** $Mn_2Ru_{0.48}Ga$ and **d** Mn_2RuGa. The data are from [61, 72]

Mn_3Ga [24, 62]. Ruthenium addition increases the spin polarization (see Fig. 7.13c, d) close to the composition where Ru occupies half of the 4*d* sites.

The electrical resistivities of $D0_{22}$-Mn_3Ga and Mn_2Ga films at room temperature are 160 μΩcm and 120 μΩcm respectively. These values are close to the minimum metallic conductivity.

Anomalous Hall effect measurements have been performed on samples grown on MgO and $SrTiO_3$ [57, 67, 72, 103]. The electronic properties of the thin films are correlated with their magnetic properties and depend both on stoichiometry and growth conditions. The maximum extraordinary Hall resistivity (ρ_{xy}) and angle (ρ_{xy}/ρ_{xx}) of $D0_{22}$ $Mn_{67}Ga_{33}$ films were 11.5 μΩcm and 5.7 % respectively. The Hall angle is found to be larger for thinner films grown on MgO substrate compared to the thicker films as shown below in Fig. 7.14. The substrate or seed layer creates either tensile or compressive stress on the material depending on the lattice mismatch. The MnGa

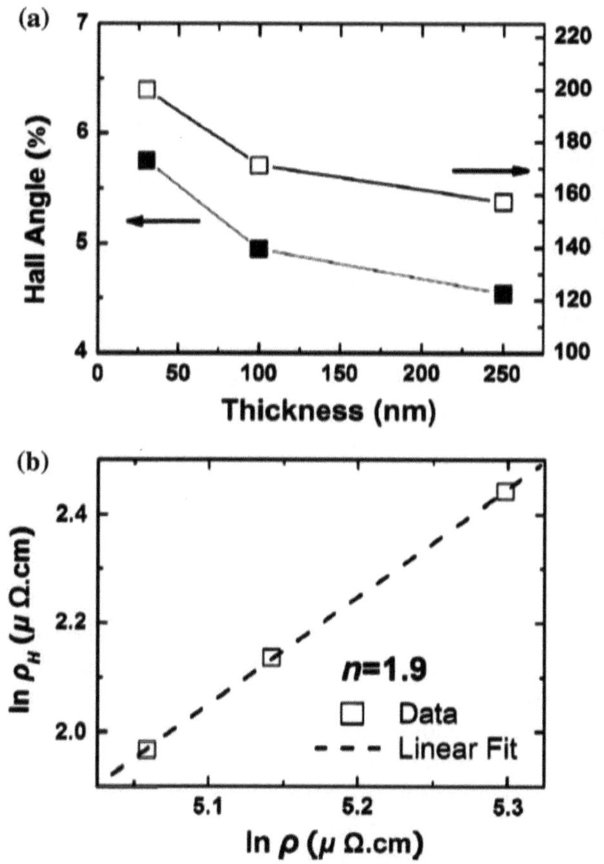

Fig. 7.14 **a** Hall angle and resistivity as a function of thickness of $Mn_{67}Ga_{33}$ films. **b** The slope of $\ln\rho_{xy}/\ln\rho_{xx}$ plot. Figure taken from [57]

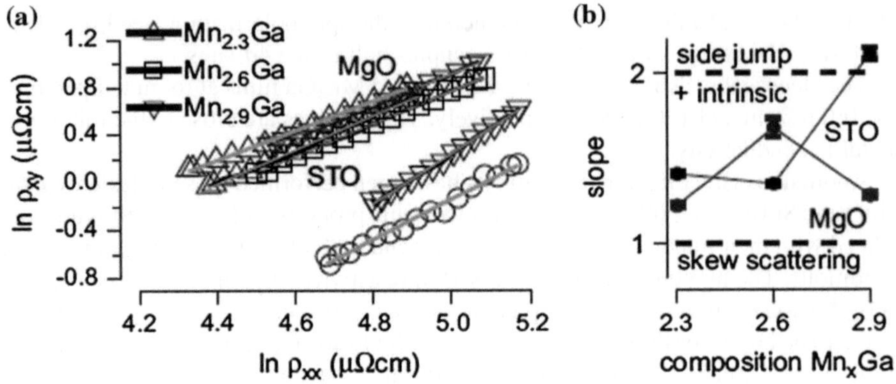

Fig. 7.15 **a** Extraordinary Hall effect measurements of various Mn-Ga films grown on MgO and STO substrates **b** The slope of $\ln\rho_{xy}/\ln\rho_{xx}$ plot of these thin films showing the dominant scattering mechanism. Figure taken from [67]

films have lattice parameters very close to that of STO, whereas the MgO creates a stress in the material, which relaxes with increasing film thickness. The relaxation leads to a reduction in the Hall angle, as seen in Fig. 7.14a. The EHE measurements of Glas et al. [67] show similar behavior, as can be seen in Fig. 7.15. The Hall angle for films grown on STO is smaller than those on MgO. In almost all films, except $Mn_{2.9}Ga$ on STO, the slope is between 1 and 2 showing that the anomalous Hall effect is dominated by intrinsic scattering [67].

Magnetoresistance and Hall measurements on Mn_2Ru_xGa are collected in Fig. 7.16. It shows that the high-field magnetoresistance is about two orders of magnitude lower in the half metal ($x = 0.33$) than in the normal ferromagnet ($x = 0.83$). There is a change of sign of the anomalous Hall effect, with the value in the half metal being an order of magnitude greater.

7.8 Device Structures

The combination of high spin polarization [24, 90], low magnetization, and high anisotropy makes $D0_{22}Mn_{3-x}Ga$ ($0 \leq x \leq 1$) and Mn_3Ge potentially attractive materials for high-density spin torque memory and future integrated logic circuits [62]. Initial attempts to fabricate perpendicular MTJs using Mn-Ga electrodes have resulted in magnetoresistance ratios well below those achieved using perpendicular CoFeB based structures [91]. It should be noted that while CoFeB system is attractive for the perpendicular MTJs, the CoFeB layer has to be very thin to maintain perpendicular anisotropy, which in turn limits the minimum dimensions consistent with data retention in nanoscale devices. Therefore perpendicular Mn-Ga materials, with their high uniaxial anisotropy constants K_u have the potential to fill a need where the minimum dimensions of the device have to be in the sub-10 nm regime.

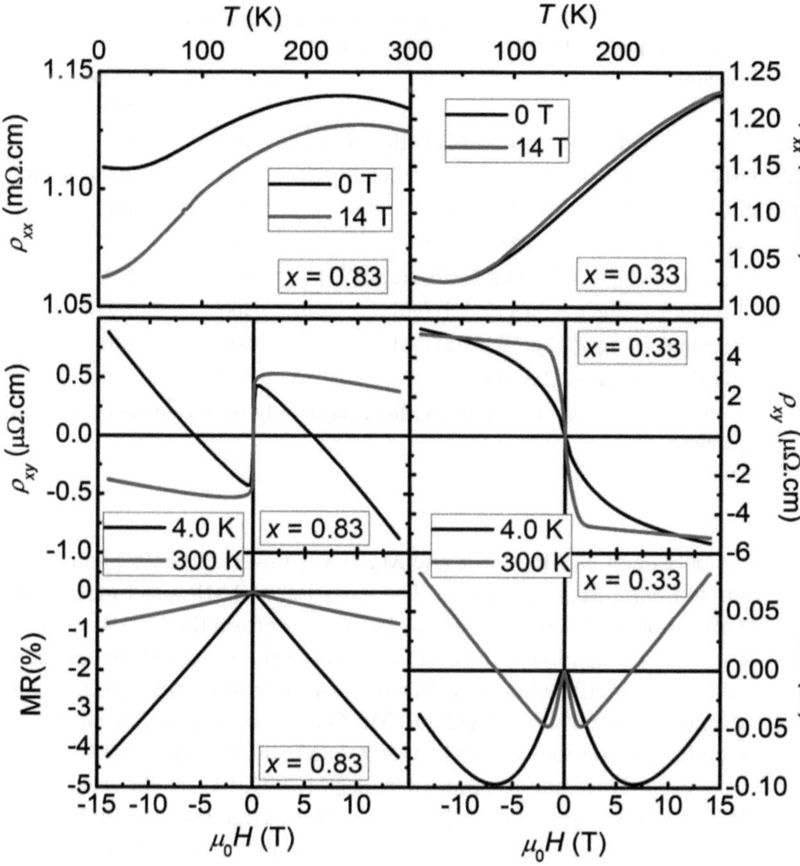

Fig. 7.16 Transport on half-metallic $Mn_2Ru_{0.33}Ga$ (*right*) and metallic $Mn_2Ru_{0.83}Ga$ (*left*) showing evidence of the change of magnetic and electronic structure on crossing the spin gap. Panels show resistivity (*top*), Hall effect (*middle*) and magnetoresistance (*bottom*). Figure taken from [72]

A full Mn-Ga/tunnel barrier/Mn-Ga perpendicular MTJ stack has not yet been reported but hybrid Mn-Ga/MgO/CoFe(B) or similar structures such as Mn-Ga/insertion layer/MgO/CoFe(B) have been prepared, but they all show significantly lower TMR compared to all-perpendicular CoFeB/MgO/CoFeB MTJs. Table 7.5 gives a summary of the Mn-Ga MTJs, preparation conditions and TMR ratios. The MgO tunnel barrier directly grown on $D0_{22}$ MnGa is not well crystallized, but this can be improved by inserting 0.4 nm Mg between $D0_{22}$ MnGa and MgO [92]. The TMR curves of these MTJs with 2.5 nm thick CoFe free layers are obtained at high field in order to pull the magnetization of the in-plane free layer out of plane. Recently, Ma et al. [93] fabricated MTJ stacks with thin Fe and Co insertion layers on top of several compositions of Mn-Ga alloys grown on Cr (001) seed layers. They also used 1.2 nm thick CoFeB as a perpendiular free layer. The device stack with 1.5 nm Fe and 1.5

Table 7.5 List of MTJs with Mn-Ga perpendicular bottom layers grown at different substrate temperatures and maximum TMR values at 300 and 10 K

MTJ core structure	T_s (°C)	TMR (%) at 300 K/ 10 K
$Mn_{62}Ga_{38}$(30)/Fe(1.5)/Mg(0.4)/MgO(2.2)/CoFeB(1.2) [93]	400	50/
$Mn_{62}Ga_{38}$(30)/Co(1.5)/Mg(0.4)/MgO(1.8)/CoFeB(1.2) [55]	400	40/80 at 5 K
$Mn_{2.4}Ga$(30)/Mg(0.4)/MgO(2.0)/CoFe(2.5) [92]	400	10/22
$Mn_{2.4}Ga$(30)/Mg(0.4)/MgO(2.0)/CoFe(2.5) [94]	400	9.8/22.1
$Mn_{62}Ga_{38}$(30)/Fe(1.1)/Mg(0.4)/MgO(2.2)/CoFe(2.5) [95]	400	24[a]/
$Mn_{62}Ga_{38}$(30)/Mg(0.4)/MgO(2.2)/CoFe(2.5) [96]	400	7/23
$Mn_{54}Ga_{46}$(30)/Mg(0.4)/MgO(2.2)/CoFe(2.5) [96]	500	6/18
$Mn_{71}Ga_{29}$(30)/Mg(0.4)/MgO(2.2)/CoFe(2.5) [96]	400	8/20

T_s shows the growth temperature of the bottom Mn-Ga layer only, the rest of the stack is grown at ambient temperature
[a]Not completely AP

nm Co layers between $L1_0$ $Mn_{62}Ga_{38}$ and Mg(0.4) showed 50 % [93] and 40 % [55] TMR at room temperature, respectively. The variation of TMR as a function of various Mn-Ga alloy compositions and various Fe (Co) insertion thicknesses is shown in Fig. 7.17. The same group also found that the direct exchange coupling between Mn-Ga and Fe is ferromagnetic, whereas with Co it is antiferromagnetic [93]. The use of a CoFeB interface layer between MgO and Mn-Ga is another possible solution [63].

First principles calculations for a Mn_3Ga(001)/MgO(001)/Mn_3Ga(001) structure predicted very high TMR due to spin filtering [94, 97], like in the case for MTJs with CoFeB electrodes. The difficulty here is to obtain a defect-free MgO tunnel barrier that can be grown on tetragonal Mn-Ga, with another Mn-Ga layer on top. The interfaces in these devices are critical. Unfortunately, the lattice mismatch between MgO (a = 4.21 Å) and $D0_{22}$ Mn-Ga (a = 3.91 Å) is too large to permit proper epitaxial growth. Atomic force microscopy measurements of very thin Mn_3Ga and Mn_2Ga grown on MgO confirm that a continuous $D0_{22}$ Mn-Ga film does not form on an MgO substrate [61, 62]. The thin films break up into nano-grains with pinholes. These discontinuities disappear for thicker films due to grain growth. However, the MgO thin films grown by radio frequency sputtering have a slight tetragonal distortion, [98] which may bring the *ab* plane of the MgO closer to the lattice parameter of the tetragonal Mn-Ga but this is not very helpful because this distortion so small. A plausible alternative tunnel barrier is $SrTiO_3$ (STO) which has a near-perfect lattice match with tetragonal Mn-Ga, and Mn-Ga films are known to grow epitaxially on STO substrates. However, the growth of insulating STO films is not straightforward. The sputtered films are oxygen deficient and become conducting when the oxygen deficiency is sufficiently high [99]. PLD is more suitable for growing this perovskite oxide, the growth temperature is high (~700 °C) and there are oxidation problems. In all deposition techniques, the STO loses a significantly amount of oxygen making the film slightly conducting. Better-quality STO films are obtained in the presence

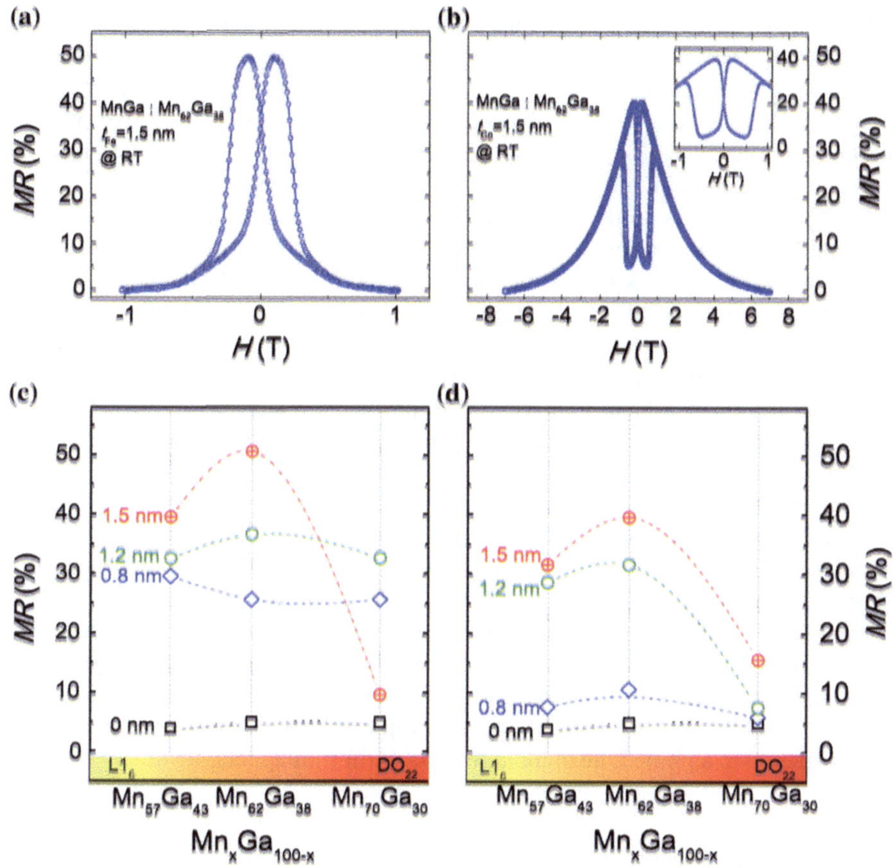

Fig. 7.17 **a** TMR curve for MTJs with $L1_0$-$Mn_{62}Ga_{38}$ electrode and Fe (1.5 nm) insert. **b** TMR curve for MTJs with $L1_0$-$Mn_{62}Ga_{38}$ electrode and Co(1.5nm) insert. **c** TMR response for Fe inserted MTJs. **d** TMR response for Co inserted MTJs. From [93]

of oxygen during deposition. Moreover, the growth temperature of high quality STO films is incompatible with the integrity of the $D0_{22}$ structure, which goes through a phase change at 700–770 K and becomes hexagonal. The epitaxial growth of a Mn-Ga/STO/Mn-Ga MTJ stack is therefore not straightforward. The $L1_0$ phase of Mn-Ga which does not exhibit a high-temperature phase change may be easier to combine with an STO tunnel barrier. Aside from the high temperature problem, the Mn oxidation may be prevented by depositing a very thin layer of Ti or Sr on Mn-Ga before depositing STO.

Tetragonal Mn_3Ge has not yet been used in devices, but this material is also promising for spin torque memories, as it exhibits high spin polarization. The difficulty is again related to finding a compatible tunnel barrier. A recent calculation also shows spin filter effect of the Δ_1 band of $D0_{22}$ Mn_3Ge with MgO barrier [71], but we have seen this is not easy to obtain due to lattice mismatch.

Spin valves with nonmagnetic metal spacers have been fabricated using a (112) textured $D0_{22}$ Mn-Ga thin film with high coercivity as the fixed layer, separated from a CoFe free layer by a thin Cu spacer [100]. This spin valve structure does not show very clear switching in the perpendicular direction since the free layer was not perpendicular and the $D0_{22}$ layer is not (001) textured. Such an approach has also been used for $L1_0$ FePt and $L1_0$ CoPt layers, but their magnetoresistance ratios are still low. $D0_{22}$ Mn_3Ga on the other hand has great promise for high TMR since the calculated spin polarization at the Fermi level is 88 % [90], and 58 % has already been measured in a film with an order parameter of $S = 0.7$ [62]. High quality films with smooth surfaces have so far been grown on MgO and STO substrates only, sometimes including metallic seed layers (Pt, Pd).

In summary, a fully perpendicular magnetic tunnel junction with $D0_{22}$ Mn-Ga or Mn-Ge has not yet been realized due to problems related to the growth of the crystalline tunnel barrier. This is mainly due to the large lattice mismatch between Mn-Ga and MgO, but possibly also to interface oxide formation. To counter this problem a very thin CoFeB, Fe or Co layer has been grown on the Mn-Ga and the MgO tunnel barrier is then grown directly on the CoFeB layer, which is a well-established technique to obtain polycrystalline MgO(001) barriers. However, the TMR values falls short of that expected for a fully epitaxial structure.

Another area of potential application of Mn-Ga thin films is for exchange bias. Large exchange bias that has been demonstrated using hexagonal $D0_{19}$ Mn_3Ga films is promising for spin electronic devices, since the bias can be achieved on cooling from only 100 °C [37]. Figure 7.18 shows a representative TMR curve of an MgO MTJ, which uses a pinned, hexagonal Mn_3Ga antiferromagnetic layer.

7.9 Future Prospects

Tetragonal Heusler alloys of the Mn-Ga and Mn-Ge families are attractive for spin transfer toque memory applications. However, for fully perpendicular Mn-Ga or Mn-Ge MTJ stacks a new tunnel barrier, such as $SrTiO_3$, needs to be developed and integrated into the stacks. The high magnetoresistance in such systems can enable scalable nonvolatile spin toque memories down to sub 10 nm lateral dimensions. For the case of GMR spin valves, a spacer material with long enough spin diffusion length is necessary to obtain sufficiently high magnetoresistance. The most suitable candidate with Mn-Ga electrodes is Pd [101], which can be epitaxially grown on $D0_{22}$ films due to the good lattice match.

$D0_{22}$ Mn_3Ga can be an ideal candidate for perpendicular spin torque memory and even very high frequency oscillators due to the presence of a the soft intrinsic in-plane magnetic moment coexisting in a highly anisotropic c-axis material [30]. This unique property of Mn_3Ga can foster new insights to realize very high frequency communication devices, reaching towards millimeter wave bands [102].

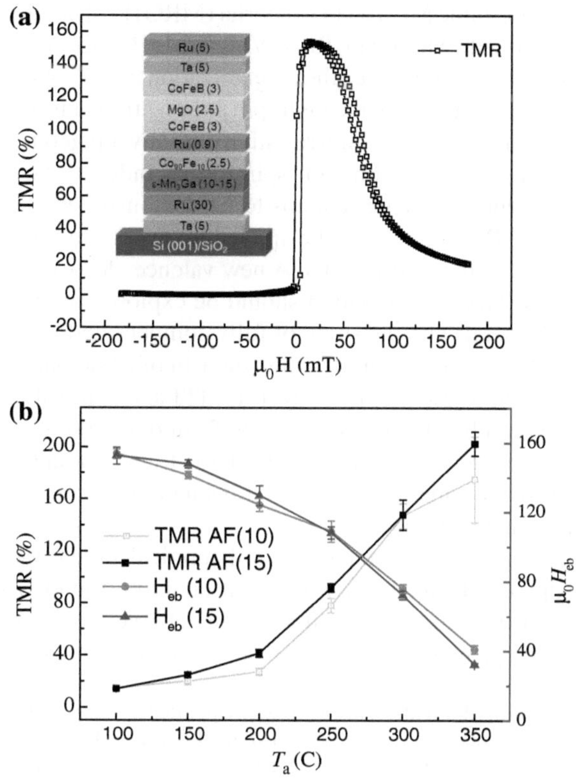

Fig. 7.18 **a** TMR curve of a representative MgO MTJ that is annealed at 300 °C. The inset shows a cartoon of the multilayer MTJ structure. **b** The variation of exchange bias field and TMR as a function of annealing temperature. Figure taken from [37]

The high magnetization in Mn_2Ga combined with the high anisotropy looks promising for future perpendicular bit-patterned recording media. The high anisotropy constant of 2.35 MJ m^{-3} can support thermally stable recording media up to 10 Tbit per square inch. Compared with other high anisotropy materials such as $L1_0$ FePt and $L1_0$ CoPt, $D0_{22}$ Mn_2Ga crystallizes at 350 °C, and the coercivity is smaller.

Both tetragonal Mn-Ga and Mn-Ge materials can be grown on suitable seed layers to obtain a smooth continuous granular structure. This could be a straightforward method to obtain isolated nanoscale magnetic regions on the surface. Heat-assisted magnetic recording (HAMR) is the next phase of magnetic recording technology where an intense laser pulse temporarily heats up a highly anisotropic material before cooling a small magnetic field to reverse the magnetization of the heated media, whether in a patch of nanoscale particles or in single bit-patterned media. Using this technique very high recording densities can be achieved. Current techniques involve subtractive lithography and ion milling to produce such nano-island structures on a surface. The ordered granular growth of these materials, if achieved with suitable substrates could eliminate the expensive nanoscale lithography and milling steps.

The cubic Mn_2Ru_xGa material (MRG) [72] has great potential because of its lattice match with MgO. $Mn_2Ru_xGa/MgO/Mn_2Ru_xGa$ MTJs could be useful for spin torque nano-oscillators since the magnetization is low. Prospects for obtaining high TMR are encouraging due to the high spin polarization of this material. The low magnetization, low anisotropy and high resistivity are advantageous for microwave oscillation driven by spin torque, although spin-torque induced oscillation or switching in a zero-moment half metal remains to be demonstrated. The results at hand show that the Slater-Pauling rule can be applied for average number of atoms in the formula which is 3.5, with 0.5 Ru atoms. A new valence electron rule with 21 valence electrons for an $X_2Y_{0.5}Z$ composition should be explored in the search for further zero-moment half metals and the study of fully compensated half metallic ferrimagnetism.

The triangular antiferromagnetism of hexagonal $Mn_{3-x}Ga$ ($0 \leq x \leq 1$) series can be used to pin the fixed layer in MTJ and spin valve stacks because good exchange bias can be obtained on cooling from only 100 °C. The isotropic magnetic moment due to the canting of the spins in the triangular antiferromagnet could be interesting for spin torque applications if this small canted moment is able to generate sufficiently high magnetoresistance.

Finally, there remains the theoretical prediction of a half-metallic single spin superconductor [19], which if it could be found would be one of the more exotic electronic states of condensed matter. Well-ordered cubic Mn_3Ga with Ru doping might provide be a first realization of such a material.

Acknowledgments This work was supported by Science Foundation Ireland, as part of the NISE project, contract 10/IN1.13006. JMDC also acknowledges support from the Humboldt Foundation. The authors thank K.Rode, P. Stamenov, M. Venkatesan and Y.-C. Lau for fruitful discussions.

References

1. J.M.D. Coey, *Magnetism and Magnetic Materials* (Cambridge University Press, Cambridge, 2010)
2. V. Sliwko, P. Mohn, K. Schwarz, J. Phys.: Condens. Matter **6**, 6557 (1994)
3. F. Heusler, Verh. DPG **5**, 219 (1903)
4. F. Heusler, W. Starck, E. Haupt, Verh. DPG **5**, 220 (1903)
5. J. Winterlik, G.H. Fecher, A. Thomas, C. Felser, Phys. Rev. B **79**, 064508 (2009)
6. T. Klimczuk et al., Phys. Rev. B **85**, 174505 (2012)
7. D.J. Singh, I.I. Mazin, Phys. Rev. B **57**, 14352 (1998)
8. S. Chadov, X. Qi, J. Kübler, G.H. Fecher, C. Felser, S.C. Zhang, Nat. Mater. **9**, 541 (2010)
9. H. Lin, L.A. Wray, Y. Xia, S. Xu, S. Jia, R.J. Cava, A. Bansil, M.Z. Hasan, Nat. Mater. **9**, 546 (2010)
10. R.A. de Groot, F.M. Mueller, P.G. van Engen, K.H.J. Buschow, Phys. Rev. Lett. **50**, 2024 (1983)
11. B. Behin-Aein, D. Datta, S. Salahuddin, S. Datta, Nat. Nanotechnol. **5**, 266 (2010)
12. G.H. Fecher, H.C. Kandpal, S. Wurmehl, C. Felser, G. Schonhense, J. Appl. Phys. **99**, 08J106 (2006)
13. R.J. Soulen et al., Science **282**, 85 (1998)
14. L. Ritchie et al., Phys. Rev. B **68**, 104430 (2003)

15. Y. Sakuraba, M. Hattori, M. Oogane, Y. Ando, H. Kato, A. Sakuma, T. Miyazaki, H. Kubota, Appl. Phys. Lett. **88**, 192508 (2006)
16. T. Graf, C. Felser, S.S.P. Parkin, Prog. Solid State Chem. **39**, 1 (2011)
17. H. van Leuken, R.A. de Groot, Phys. Rev. Lett. **74**, 1171 (1995)
18. M. Hakimi, M. Venkatesan, K. Rode, K. Ackland, J.M.D. Coey, J. Appl. Phys. **113**, 17B101 (2013)
19. W.E. Pickett, Phys. Rev. Lett. **77**, 3185 (1996)
20. S. Wurmehl, H.C. Kandpal, G.H. Fecher, C. Felser, J. Phys. Condens. Matter **18**, 6171 (2006)
21. E. Krén, G. Kádár, Solid State Commun. **8**, 1653 (1970)
22. K. Minakuchi, R.Y. Umetsu, K. Ishida, R. Kainuma, J. Alloy. Compd. **537**, 332 (2012)
23. E. Arras, D. Caliste, T. Deutsch, F. Lançon, P. Pochet, Phys. Rev. B **83**, 174103 (2011)
24. J. Winterlik, B. Balke, G.H. Fecher, C. Felser, M.C.M. Alves, F. Bernardi, J. Morais, Phys. Rev. B **77**, 054406 (2008)
25. K.M. Krishnan, Appl. Phys. Lett. **61**, 2365 (1992)
26. M. Tanaka, J.P. Harbison, J. DeBoeck, T. Sands, B. Philips, T.L. Cheeks, V.G. Keramidas, Appl. Phys. Lett. **62**, 1565 (1993)
27. H. Kono, J. Phys. Soc. Jpn. **13**, 1444 (1958)
28. G. Kádár, E. Krén, Int. J. Magn. **1**, 143 (1971)
29. N. Yamada, J. Phys. Soc. Jpn. **59**, 273 (1990)
30. K. Rode et al., Phys. Rev. B **87**, 184429 (2013)
31. N. Yamada, H. Sakai, H. Mori, T. Ohoyama, Phys. B+C **149**, 311 (1988)
32. S. Mizukami, T. Kubota, F. Wu, X. Zhang, T. Miyazaki, H. Naganuma, M. Oogane, A. Sakuma, Y. Ando, Phys. Rev. B **85**, 014416 (2012)
33. H. Liu, G.Y. Gao, L. Hu, Y. Ni, F. Zu, Zhu S, S. Wang, K.L. Yao, J. Appl. Phys. **115** (2014)
34. G.Y. Gao and K.L. Yao, Appl. Phys. Lett. **103** (2013)
35. S. Tomiyoshi, Y. Yamaguchi, T. Nagamiya, J. Magn. Magn. Mater. **31–34**, 629 (1983)
36. T. Nagamiya, S. Tomiyoshi, Y. Yamaguchi, Solid State Commun. **42**, 385 (1982)
37. H. Kurt, K. Rode, H. Tokuc, P. Stamenov, M. Venkatesan, J.M.D. Coey, Appl. Phys. Lett. **101**, 232402 (2012)
38. H. Niida, T. Hori, H. Onodera, Y. Yamaguchi, Y. Nakagawa, J. Appl. Phys. **79**, 5946 (1996)
39. T.A. Bither, W.H. Cloud, J. Appl. Phys. **36**, 1501 (1965)
40. T. Ohoyama, K. Yasukochi, K. Kanematsu, J. Phys. Soc. Jpn. **16**, 352 (1961)
41. L.-J. Zhu, S.-H. Nie, J.-H. Zhao, Chin. Phys. B **22**, 118505 (2013)
42. L. Zhu, J. Zhao, Appl. Phys. A **111**, 379 (2013)
43. L. Zhu, S. Nie, K. Meng, D. Pan, J. Zhao, H. Zheng, Adv. Mater. **24**, 4547 (2012)
44. A.W. Arins, H.F. Jurca, J. Zarpellon, J. Varalda, I.L. Graff, A.J.A. de Oliveira, W.H. Schreiner, D.H. Mosca, Appl. Phys. Lett. **102**, 102408 (2013)
45. W. Feng, D. Van Thiet, D.D. Dung, Y. Shin, S. Cho, J. Appl. Phys. **108**, 113903 (2010)
46. W. Feng, D.D. Dung, Y. Shin, D. Van Thiet, S. Cho, X. Hao, J. Korean Phys. Soc. **56**, 1382 (2010)
47. E. Lu, D.C. Ingram, A.R. Smith, J.W. Knepper, F.Y. Yang, Phys. Rev. Lett. **97**, 146101 (2006)
48. K. Wang, E. Lu, J.W. Knepper, F. Yang, A.R. Smith, Appl. Phys. Lett. **98**, 162507 (2011)
49. K. Wang, A. Chinchore, W. Lin, D.C. Ingram, A.R. Smith, A.J. Hauser, F. Yang, J. Cryst. Growth **311**, 2265 (2009)
50. L.J. Zhu, D. Pan, S.H. Nie, J. Lu, J.H. Zhao, Appl. Phys. Lett. **102**, 132403 (2013)
51. T.J. Nummy, S.P. Bennett, T. Cardinal, D. Heiman, Appl. Phys. Lett. **99**, 252506 (2011)
52. D.D. Djayaprawira, K. Tsunekawa, M. Nagai, H. Maehara, S. Yamagata, N. Watanabe, S. Yuasa, Y. Suzuki, K. Ando, Appl. Phys. Lett. **86**, 092502 (2005)
53. X. Gu, D. Lubyshev, J. Batzel, J.M. Fastenau, W.K. Liu, R. Pelzel, J.F. Magana, Q. Ma, V.R. Rao, J. Vac. Sci. Technol. B **28**, C3A12 (2010)
54. P. Tejedor, V.M. Fuenzalida, F. Briones, J. Appl. Phys. **80**, 2799 (1996)
55. Q.L. Ma, T. Kubota, S. Mizukami, X.M. Zhang, H. Naganuma, M. Oogane, Y. Ando, T. Miyazaki, Appl. Phys. Lett. **101**, 032402 (2012)
56. S. Mizukami et al., Phys. Rev. Lett. **106**, 117201 (2011)

57. F. Wu, E.P. Sajitha, S. Mizukami, D. Watanabe, T. Miyazaki, H. Naganuma, M. Oogane, Y. Ando, Appl. Phys. Lett. **96**, 042505 (2010)
58. F. Wu, S. Mizukami, D. Watanabe, H. Naganuma, M. Oogane, Y. Ando, T. Miyazaki, J. Phys.: Conf. Ser. **200**, 062037 (2010)
59. F. Wu, S. Mizukami, D. Watanabe, H. Naganuma, M. Oogane, Y. Ando, T. Miyazaki, Appl. Phys. Lett. **94**, 122503 (2009)
60. S. Ouardi, T. Kubota, G.H. Fecher, R. Stinshoff, S. Mizukami, T. Miyazaki, E. Ikenaga, C. Felser, Appl. Phys. Lett. **101**, 242406 (2012)
61. H. Kurt, K. Rode, M. Venkatesan, P. Stamenov, J.M.D. Coey, Phys. Status Solidi (b) **248**, 2338 (2011)
62. H. Kurt, K. Rode, M. Venkatesan, P. Stamenov, J.M.D. Coey, Phys. Rev. B **83**, 020405 (2011)
63. M. Glas, C. Sterwerf, J.M. Schmalhorst, D. Ebke, C. Jenkins, E. Arenholz, G. Reiss, J. Appl. Phys. **114**, 183910 (2013)
64. M. Li, X. Jiang, M.G. Samant, C. Felser, S.S.P. Parkin, Appl. Phys. Lett. **103**, 032410 (2013)
65. A. Köhler, I. Knez, D. Ebke, C. Felser, S.S.P. Parkin, Appl. Phys. Lett. **103**, 162406 (2013)
66. H.G. Meissner, K. Schubert, T.R. Anantharaman, Proc. Indian Acad. Sci. Sect. A **61**, 340 (1965)
67. M. Glas, D. Ebke, I.M. Imort, P. Thomas, G. Reiss, J. Magn. Magn. Mater. **333**, 134 (2013)
68. C. ViolBarbosa, S. Ouardi, S. Mizukami, G. Fecher, T. Kubota, E. Ikenaga, T. Miyazaki, C. Felser, arXiv preprint arXiv:1403.3556 (2014)
69. H. Kurt, N. Baadji, K. Rode, M. Venkatesan, P. Stamenov, S. Sanvito, J.M.D. Coey, Appl. Phys. Lett. **101**, 132410 (2012)
70. A. Sugihara, S. Mizukami, Y. Yamada, K. Koike, T. Miyazaki, Appl. Phys. Lett. **104**, 132404 (2014)
71. S. Mizukami, A. Sakuma, A. Sugihara, T. Kubota, Y. Kondo, H. Tsuchiura, T. Miyazaki, Appl. Phys. Express **6**, 123002 (2013)
72. H. Kurt, K. Rode, P. Stamenov, M. Venkatesan, Y.C. Lau, E. Fonda, J.M.D. Coey, Phys. Rev. Lett. **112**, 027201 (2014)
73. S. Sakai, M. Takahashi, K. Motohashi, Y. Yamaguchi, N. Yui, T. Kobayashi, J. Vac. Sci. Technol. A **25**, 903 (2007)
74. www.neocera.com
75. www.piezoflare.com
76. G. Oksana, H. Jaroslav, T. Simon, H. Burkard, S. Horst, J. Gerhard, J. Phys. D: Appl. Phys. **42**, 232001 (2009)
77. H. Schneider, E. Vilanova, B. Balke, C. Felser, G. Jakob, J. Phys. D: Appl. Phys. **42**, 084012 (2009)
78. P. Klaer, C.A. Jenkins, V. Alijani, J. Winterlik, B. Balke, C. Felser, H.J. Elmers, Appl. Phys. Lett. **98**, 212510 (2011)
79. V. Alijani, J. Winterlik, G.H. Fecher, C. Felser, Appl. Phys. Lett. **99**, 222510 (2011)
80. T. Gasi, A.K. Nayak, J. Winterlik, V. Ksenofontov, P. Adler, M. Nicklas, C. Felser, Appl. Phys. Lett. **102**, 202402 (2013)
81. A. Bedoya-Pinto, C. Zube, J. Malindretos, A. Urban, A. Rizzi, Phys. Rev. B **84**, 104424 (2011)
82. X. Hu, Adv. Mater. **24**, 294 (2012)
83. I. Galanakis, K. Özdoğan, E. Şaşıoğlu, B. Aktaş, Phys. Rev. B **75**, 172405 (2007)
84. I. Galanakis, E. Şaşıoğlu, Appl. Phys. Lett. **99**, 052509 (2011)
85. I. Galanakis, P.H. Dederichs, N. Papanikolaou, Phys. Rev. B **66**, 174429 (2002)
86. T. B. Massalski, Binary Alloy Phase Diagrams (ASM International, 1990)
87. E. Wachtel, E.T. Henig, Z. Metallkde, Aufbau des systems mangan-germanium **60**, 243 (1969)
88. N. Baadji and S. Sanvito, (Private communication)
89. T. Archer and S. Sanvito, (Private communication) (2014)
90. B. Balke, G.H. Fecher, J. Winterlik, C. Felser, Appl. Phys. Lett. **90**, 152504 (2007)
91. S. Ikeda et al., Nat. Mater. **9**, 721 (2010)
92. T. Kubota, S. Mizukami, D. Watanabe, F. Wu, X. Zhang, H. Naganuma, M. Oogane, Y. Ando, T. Miyazaki, J. Appl. Phys. **110**, 013915 (2011)

93. Q.L. Ma, T. Kubota, S. Mizukami, X.M. Zhang, H. Naganuma, M. Oogane, Y. Ando, T. Miyazaki, Phys. Rev. B **87**, 184426 (2013)
94. T. Kubota et al., Appl. Phys. Express **4**, 043002 (2011)
95. T. Kubota, Q. Ma, S. Mizukami, X. Zhang, H. Naganuma, M. Oogane, Y. Ando, T. Miyazaki, Appl. Phys. Express **5**, 043003 (2012)
96. T. Kubota et al., Appl. Phys. Lett. **99**, 192509 (2011)
97. Y. Miura, M. Shirai, Magnetics. IEEE Trans. **50**, 1 (2014)
98. H. Kurt, K. Oguz, T. Niizeki, J.M.D. Coey, J. Appl. Phys. **107**, 083920 (2010)
99. M. Bibes, J.E. Villegas, A. Barthélémy, Adv. Phys. **60**, 5 (2011)
100. C.L. Zha, R.K. Dumas, J. Persson, S.M. Mohseni, J. Nogues, J. Akerman, Magn. Lett. IEEE **1**, 2500104 (2010)
101. H. Kurt, R. Loloee, K. Eid, W.P. Pratt Jr, J. Bass, Appl. Phys. Lett. **81**, 4787 (2003)
102. A. Namai et al., Nat. Commun. **3**, 1035 (2012)
103. N. Thiyagarajah, Y.C. Lau, D. Betto, K. Borisov, P. Stamenov, J.M.D. Coey and K. Rode, Appl. Phys. Lett. **106**, 122402 (2015)

Chapter 8
Magnetic Phase Competition in Off-Stoichiometric Martensitic Heusler Alloys: The $Ni_{50-x}Co_xMn_{25+y}Sn_{25-y}$ System

Kanwal Preet Bhatti, Vijay Srivastava, Daniel P. Phelan, Sami El-Khatib, Richard D. James and Chris Leighton

Recently, motivated both by basic scientific interest and technological applications, and utilizing both experiment and theory, a number of investigators have independently identified a small group of off-stoichiometric Heusler alloys as having unusually interesting magnetic properties. These alloys take the form $Ni_{50-x}Co_xMn_{25+y}Z_{25-y}$ (Z = Sn, In, Ga, etc.), being Co-doped off-stoichiometric versions of the better-known Ni_2MnZ full Heusler compounds. In certain critical composition ranges these alloys are found to display unusually reversible martensitic phase transformations, multiferroicity (due to coexisting ferroelasticity and magnetic order), heightened sensitivity to compositional changes, and acute magnetic phase competition, leading to such exotic phenomena as spontaneous nanoscale magnetic inhomogeneity, collective cluster freezing, and intrinsic exchange bias. In terms of applications they can exhibit magnetic-field-induced phase transformations, magnetic shape memory behavior, magnetocaloric effects, and remarkably low thermal hysteresis, making them attractive for sensors and actuators, magnetic refrigeration, and energy conversion devices. In this chapter we briefly review the current state of knowledge on the magnetic properties of these alloys, before presenting new results on the prototypical $Ni_{50-x}Co_xMn_{40}Sn_{10}$ system in the critical composition range$0 \leq x \leq 14$.

K.P. Bhatti · D.P. Phelan · C. Leighton (✉)
Department of Chemical Engineering and Materials Science, University of Minnesota, Minneapolis, MN 55455, USA
e-mail: leighton@umn.edu

V. Srivastava · R.D. James
Department of Aerospace Engineering and Mechanics, University of Minnesota, Minneapolis, MN 55455, USA

S. El-Khatib
Department of Physics, American University of Sharjah, PO BOX 26666, Sharjah, United Arab Emirates

S. El-Khatib
NIST Center for Neutron Research, National Institute of Standards and Technology, MD 20899, Gaithersburg, USA

C. Felser and A. Hirohata (eds.), *Heusler Alloys*,
Springer Series in Materials Science 222, DOI 10.1007/978-3-319-21449-8_8

Combining comprehensive magnetometry, exchange bias studies, and both new and previously published neutron scattering data, we present a detailed picture of the magnetic phenomenology in this alloy system and construct a magnetic phase diagram. Most importantly, based on these results and the work of others, we discuss in detail potential origins of the unusual magnetic properties of these materials, most notably the magnetic phase competition and nanoscale inhomogeneity that dominate their low temperature magnetism.

8.1 Introduction

As aptly illustrated by the scope of contributions in this handbook, Heusler alloys are a class of materials that have attracted particularly broad attention, being of interest for their structural, mechanical, electronic, magnetic, and thermodynamic properties. The high interest in these materials is particularly obvious in the field of magnetism, perhaps the most noteworthy contemporary example being the pursuit of a half-metallic (or at least highly spin-polarized) ferromagnetic state [1]. Half Heuslers such as NiMnSb in fact played a key role in the development of the concept of half-metallic ferromagnetism [1–3], while full Heuslers such as Co_2MnSi and $Co_2Fe(Al,Si)$ (e.g. [4, 5]) are now intensively studied as options for highly spin-polarized electrodes in magnetic tunnel junctions, and in the current-perpendicular-to-plane giant magnetoresistance devices that could replace tunnel junctions in future hard disk drive read sensors.

Much, though certainly not all, of this contemporary work on magnetic Heusler alloys has focused on stoichiometric Heusler "compounds", the minimization of sources of non-stoichiometry and disorder, such as anti-site defects, often being an important theme (as with Heusler half-metals for example [3]). Quite recently however, significant research has been focused on the structural, magnetic, and thermodynamic properties of alloys that are deliberately synthesized in off-stoichiometric form. Perhaps the prime examples are the $Ni_{50}Mn_{25+y}Z_{25-y}$ alloys (Z = Sn, Ga, In, etc.) [6–9], derived from ordered full Heuslers such as Ni_2MnSn by deliberate substitution of Mn on the Z site. As an example, Ni_2MnSn is ferromagnetic (F) with a Curie temperature (T_C) around 340 K, testament to the strong ferromagnetic exchange between Mn, which is the predominant source of ordered moment [7–9]. When Mn is substituted on the Sn site however in $Ni_{50}Mn_{25+y}Sn_{25-y}$, Mn-Mn nearest neighbors result, with interatomic distances comparable to those occurring in Mn metal, favoring antiferromagnetic (AF) interactions [7–9]. Off-stoichiometric alloys such as $Ni_{50-x}Mn_{25+y}Sn_{25-y}$ [7–9] and $Ni_{50-x}Mn_{25+y}In_{25-y}$ [8, 9] thus have magnetic interactions and phases that are tunable by composition, leading to strong competition between the various possible magnetic ground states [6–9]. The second key ingredient generating the attention now being paid to these alloys is the common occurrence of martensitic phase transformations (MPTs) [6–9]. Such diffusionless symmetry-lowering transitions are common in these types of alloys, and, due to the sensitivity of exchange interactions to interatomic distances and local symmetry, these structural transformations typically drive accompanying first order magnetic phase transformations [6–9]. Common examples include transitions from a high temperature cubic austenite phase exhibiting long-range F, to a low temperature lower symmetry martensitic phase exhibiting no long-range magnetic order (i.e. paramagnetism (P)), long-range AF order, or long-range F order with distinct magnetization, anisotropy, etc. [6–9].

At least from the fundamental science perspective, the flurry of recent activity with alloys such as the $Ni_{50}Mn_{25+y}Z_{25-y}$ family is driven by the fact that this combination of magnetic phase competition and an MPT leads to a variety of unexpected magnetic phenomena. In particular, there is a growing realization that, on nanoscopic length scales, these and related materials can exhibit electronic and magnetic inhomogeneity (e.g. [10–13]), quite similar to the nanoscopic magneto-electronic phase separation now well-established in correlated electron systems such as complex oxides [14, 15]. In magnetic oxides such as the manganites for example, this inhomogeneity is widely believed to be responsible for some of their most unique properties (particularly colossal magnetoresistance), underscoring the importance of the effect [14, 15]. As discussed in more detail below, the magnetic inhomogeneity in these off-stoichiometric Heuslers is evident even from *macroscopic* properties, as nominally non-F martensitic phases exhibit phenomena such as superparamagnetic (SP)-like behavior [10–13], clearly indicating the formation of some form of magnetically nanostructured state at low temperatures. This is accompanied by "intrinsic" exchange bias in these nominally single-phase systems (e.g. [16–18]), possible due to interfacial exchange coupling between spontaneously formed nanoscopic F clusters and the magnetic matrix in which they form. Significant magnetic interaction amongst these clusters has even been claimed [10–12], and implicated as the origin of certain unusual exchange bias [18] and collective freezing phenomena [10–12]. Our recent work using small-angle neutron scattering (SANS) has provided the first direct observation, not only of the existence of these clusters, but also of their interactions [13]. Nevertheless, this advance has come only recently, and there remain a number of open questions with regard to the magnetic behavior of these complex systems, as reviewed later in this section.

The existence of these unusual magnetic properties has also generated high interest from the application perspective [19–26]. Multiferroic alloys of this type can, for example, exhibit field-induced magnetic phase transformations and magnetic shape memory effects, in addition to magneto- and baro-caloric effects [19–26]. Such broad functionality leads to potential application in sensors, actuators (in micro- or nano-electro-mechanical devices for instance), magnetic refrigeration systems, and a variety of energy conversion and harvesting devices [19–26]. A recent example of the latter from our own work uses cyclic phase transformation for the direct conversion of heat to electricity [23, 24]. In applications of this type, a common need in order to maximize efficiency is the minimization of thermal hysteresis associated with the MPT. The recent discovery that these off-stoichiometric Heusler alloys can also exhibit unusually reversible transformations, in addition to their already attractive magnetic properties, is thus very relevant [27]. The convergence of interest to a relatively small number of alloys based on the $Ni_{50}Mn_{25+y}Z_{25-y}$ materials was in fact guided to some extent by theoretical principles designed to identify martensitic systems with low thermal hysteresis. Specifically, and as discussed in more detail in [13] the theoretical framework of James et al. [28] predicts that the hysteresis at MPTs can be minimized in the situation in which an invariant plane occurs at the interface between the competing austenite and martensite phases. The condition for such an interface to occur can be simply stated as $\lambda_2 = 1$, where λ_1, λ_2, and λ_3 are the ordered eigenvalues of the transformation stretch matrix ($\boldsymbol{U}$) for the austenite to martensite phase transformation. In simple terms this is a condition for the lattice parameters of the phases under which an "exact" interface is possible during transformation. The validity of this approach was dramatically verified in the Ti-Ni-X system (X = Cu, Pd, Pt or Au), where the temperature hysteresis at the martensitic phase transformation was found to fall to values as low as a few K (60–70 K is typical) in the region where $\lambda_2 = 1.000 \pm 0.005$ [28–30]. The conjectured special interface between austenitic and martensitic phases was even observed directly in high-resolution transmission electron microscopy studies on appropriately-tuned Ti-Ni-Pd alloys [31].

Remarkably, the $Ni_{50}Mn_{25+y}Z_{25-y}$ alloys discussed above for their novel magnetic properties are already relatively close to satisfying the $\lambda_2 = 1$ criterion at certain compositions (at $y \approx 15$ for instance), and indeed exhibit rather modest temperature hysteresis in these parts of their phase diagrams [27]. Moreover, Substitution of Co for Ni drives changes in the lattice parameters of the two phases that brings λ_2 even closer to unity, minimizing thermal hysteresis at exceptionally low values of around 5 K, and demonstrating that the concepts discussed above are equally applicable to magnetic systems [13, 23, 24]. At the same time, this substitution has the additional advantage of increasing the saturation magnetization and T_C of the F austenite phase. The result is a group of alloys, such as $Ni_{50-x}Co_xMn_{25+y}Sn_{25-y}$ [10–13, 23, 24, 27, 33–36] and $Ni_{50-x}Co_xMn_{25+y}In_{25-y}$ [19, 22, 26, 32, 33] at $x \approx 5$ and $y \approx 15$, that exhibit a uniquely attractive set of properties, including high T_C (approaching 500 K), large magnetization in a low anisotropy F austenite phase (up to approximately 1200 emu/cm^3), a tunable MPT conveniently located just above ambient (e.g. 300–400 K), thermal hysteresis as low as 5 K, and a martensitic phase that is non-F, thus maximizing the magnetization change across the martensitic transformation. It is this unique set of properties that has generated such intense recent interest in these alloys, both from fundamental and applied perspectives.

In this chapter, we focus our attention on a representative member of this family of alloys, namely $Ni_{50-x}Co_xMn_{25+y}Sn_{25-y}$, at $y = 15$ and $0 \leq x \leq 14$, i.e. $Ni_{50-x}Co_xMn_{40}Sn_{10}$. This composition region spans the critical range over which an MPT occurs, in addition to the region in which the lattice parameters of the austenite and martensite phases dictate $\lambda_2 \approx 1$. Since 2006 or so, alloys in this composition range have been the focus of rapidly expanding interest, being studied for their magnetic shape memory effect [10–12], magnetocaloric effect [34–36], anomalous SP-like freezing [10–13], enhanced saturation magnetization [13, 27, 33, 35], minimal thermal hysteresis [13, 23, 24, 27], and glassy magnetism [10, 12], culminating in 2012 with the establishment of a detailed magnetic phase diagram [12]. Of most relevance to the current work, the strong magnetic phase competition in the vicinity of the MPT, and the apparently related formation of a magnetically inhomogeneous ground state at low temperatures are now well established [10–13]. Key factors in this regard include:

(i) Detection of clear SP-like behavior in the nominally non-F low temperature martensitic phase [10–13]. This is evident from the existence of an obvious SP-like freezing temperature (at 50–100 K, dependent on exact composition), with bifurcation of field cooled (FC) and zero-field-cooled (ZFC) magnetization curves [10–13], a significant frequency-shift in the freezing temperature from AC susceptibility measurements [10, 12], and field- and temperature-dependent magnetization above the freezing point that is well described by a modified Langevin function [10–13].

(ii) Suggestions of *collective* freezing phenomena such as the formation of a super-spin-glass state [10, 12]. Such claims have been made on the basis of specific frequency-scaling of the freezing temperature, as well as the demonstration of aging, rejuvenation, and memory effects, which are thought to distinguish such a collective state from a weakly-interacting or non-interacting cluster ensemble [10, 12]. This clearly implicates the existence of non-negligible inter-cluster interactions, a phenomenon that in fact forms the basis for the explanation of the surprising zero-magnetization-cooled exchange bias recently found in the related $Ni_{50}Mn_{25+y}In_{25-y}$ system [18].

(iii) A recent SANS study of our own on $Ni_{50-x}Co_xMn_{40}Sn_{10}$ at $x = 6$ and 8 [13], providing important new information on the nature of the magnetic ordering across the MPT. These measurements confirmed long-range F order in the austenite phase, in addition to elucidating the non-trivial magnetism in the martensitic phase. This martensite was concluded to be P or long-range AF in nature, the latter being favored (albeit without direct verification) on the basis of a number of related observations. Most importantly, the inhomogeneity of this low temperature magnetic phase was directly verified for the first time, *via* the clear observation of scattering from a liquid-like distribution of nanoscopic F clusters. The mean center-to-center spacing of these clusters was determined to be 12 nm, while the cluster diameter was estimated to lie in the region of 2–6 nm. These SANS data also yielded a direct measure of the magnetic correlation length, which was found to amount to approximately 5 inter-cluster spacings, directly verifying the existence of significant magnetic interactions [13].

(iv) The advancement of a qualitative model for the origin of the magnetically clustered state, raising the possibility of nm-scale variations in the nature of the magnetic order, due to local compositional fluctuations [13].

While these findings advance our grasp of the essential materials science and physics of these alloys, there remain large gaps in our fundamental understanding. For example, while the theory of James et al. [28] clarifies the interplay between thermal hysteresis and composition, the acute sensitivity of the magnetic properties to minor chemical substitutions is not yet fully understood. Indeed, given this heightened sensitivity to composition [12, 27, 33, 35], and the suggestion of the importance of local compositional fluctuations [13], it is unclear to what extent the phase behavior and macroscopic properties of these systems can be expected to be robust with respect to disorder, defects, variations in synthesis and processing conditions [37], etc. Further work on phase behavior, the details of the magnetic phase diagrams, etc., is thus clearly important. Additionally, while the local compositional fluctuation picture has been advanced [13], and appears qualitatively consistent with some currently available observations, much additional work is required to test this picture, or to develop alternative models if it is found lacking. In particular, we currently have little understanding of the interplay between compositional fluctuations and the complex microstructure anticipated in the martensitic phase, or how this is expected to influence magnetic properties. In terms of gathering additional clues to the essential physics behind the unique magnetism in these alloys, two obvious areas of exploration are further elucidation of the nature of the link between the MPT and the low temperature magnetic inhomogeneity, and a more complete appreciation of what can be learned by using the exchange bias effect as an indirect probe of nanoscale spin clusters, the matrix in which they form, and their interfacial interactions. Each of these issues, in addition to various others, is addressed in Sect. 8.3 of this chapter, following a brief discussion of experimental details.

8.2 Sample Preparation, Structural Characterization, and Experimental Details

Polycrystalline $Ni_{50-x}Co_xMn_{40}Sn_{10}$ ingots with $0 \leq x \leq 14$ were prepared by arc melting high purity Ni (99.999 %), Co (99.99 %), Mn (99.98 %), and Sn (99.99 %) under a positive Ar pressure [13]. After melting and turning multiple times the ingots were then annealed in vacuum at 900 °C for 24 h and water quenched to further promote homogeneity. The mass loss upon arc melting was verified to be $< 1\,\%$, and the exact composition was determined *via* carefully calibrated energy dispersive spectroscopy. Differential Scanning Calorimetry (DSC) was performed in a commercial instrument at heating/cooling rates of 10 K/min and was used to determine m_s, m_f, a_s, a_f, the martensite start, martensite finish, austenite start, and austenite finish temperatures, respectively. (The minimum temperature for such DSC measurements was 225 K). The extent of thermal hysteresis was then defined as $\Delta T_m = a_f - m_s$. As an example, at $x = 6$ we determined m_s= 392 K, $m_f = 382$ K, $a_s = 388$ K and $a_f = 398$ K. In addition to the wide temperature range neutron diffraction measurements presented in Sect. 8.3.2 below, temperature-dependent wide-angle x-ray diffraction (WAXRD) characterization was performed on multiple samples, and used for subsequent structural refinement, as in our prior work at $x = 6$ and 8 [13]. In all cases the high temperature austenite phase was determined to be in the $Fm\bar{3}m$ space group, analogous to the $L2_1$ base Heusler Ni_2MnSn. At the compositions that exhibit a martensitic phase transformation, the low temperature phase was found to be a $5M$-modulated martensite with space group $P2_1$. As a specific example, at $x = 6$ [13] we deduced a cubic structure with a lattice parameter of 5.987 Å at 410 K, transforming to the $5M$-modulated martensite with $a = 4.407$ Å, $b = 5.643$ Å, $c = 21.69$ Å and

$\beta = 87.05°$ at 300 K. These lattice parameters result in $\lambda_2 = 1.0051$, consistent with the low ΔT_m of 6 K. In fact, in the range $6 \leq x \leq 8$, λ_2 reaches values very close to unity, varying from 1.0051 to 1.0057. At these compositions we find $\Delta T_m = 6-7$ K (i.e. very low thermal hysteresis), consistent with these λ_2 values.

Specimens cut *via* electric discharge were used for DC magnetometry performed in a Quantum Design SQUID magnetometer and a Quantum Design PPMS vibrating sample magnetometer. Such measurements were performed at temperature (T) between 5 and 600 K, in applied magnetic field (H) up to 90 kOe. For very low H measurements the remnant field profile in the superconducting magnet was measured and the field at the sample position nulled to $H << 1$ Oe. Neutron diffraction data were taken on samples that were powdered by simple grinding, to avoid preferred orientation effects. The ground powder was annealed in vacuum to relieve any strain introduced during the grinding process. Samples were sealed in a V can with a Pb gasket and measured in a closed cycle refrigerator (between 40 and 600 K) on the BT-1 neutron diffractometer at the NIST Center for Neutron Research. A wavelength of 2.079 Å was selected by a Ge(311) monochromator, and an in-pile collimation of 60' was employed.

8.3 Results and Analysis

We first present a detailed experimental magnetic phase diagram for $Ni_{50-x}Co_xMn_{40}Sn_{10}$ (Sect. 8.3.1), in addition to a basic one for $Ni_{50}Mn_{25+y}Sn_{25-y}$ in order to provide context. We then present and discuss the magnetic properties in each of the three important regions of the $Ni_{50-x}Co_xMn_{40}Sn_{10}$ phase diagram, showing neutron diffraction measurements at the representative $x = 6$ composition in Sect. 8.3.2, before progressing to magnetometry characterization and analysis in Sect. 8.3.3. In Sect. 8.3.4 we focus on the low temperature exchange bias effect. Throughout, we place these results in the context of prior work on this and other related alloy systems, in addition to our prior SANS study at $x = 6$ and 8 [13].

8.3.1 Magnetic Phase Diagram

In order to provide context for experiments and discussion on $Ni_{50-x}Co_xMn_{40}Sn_{10}$, we first present in Fig. 8.1a a basic magnetic phase diagram for the $Ni_{50}Mn_{25+y}Sn_{25-y}$ system from which $Ni_{50-x}Co_xMn_{25+y}Sn_{25-y}$ is derived. Note the top axis, which shows the conversion to the valence electron per atom ratio, *e*/*a*, which facilitates comparison with other systems. The data used to construct Fig. 8.1a come from our own samples and measurements but are quantitatively similar to earlier work by Krenke et al. [7]. We show here only the basic phases and phase transitions (i.e., T_C and the average martensitic phase transformation temperature, $T_m = (a_f - m_f)/2)$, although, as discussed above, complex magnetic phase behavior, magnetic phase

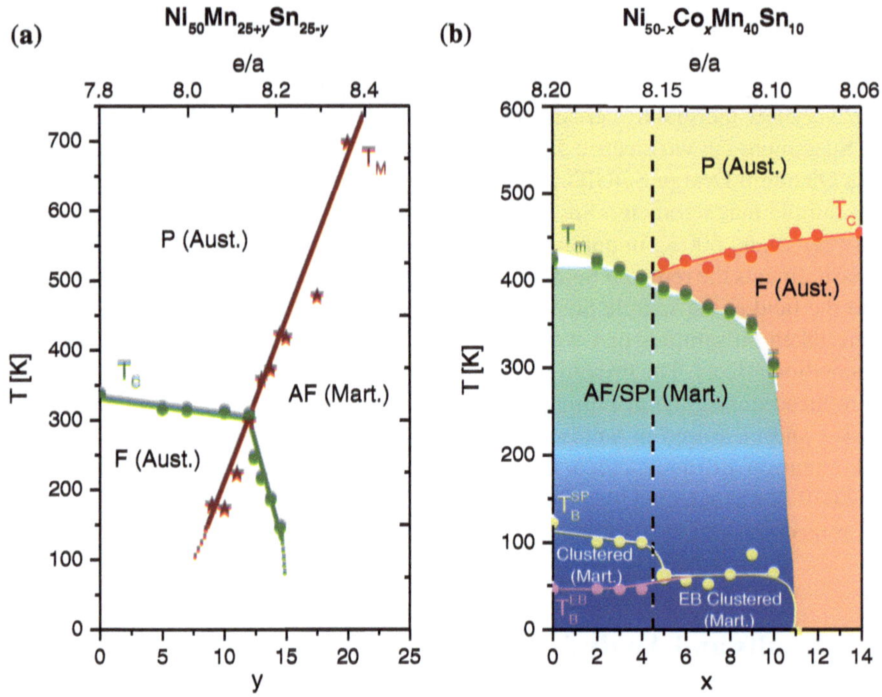

Fig. 8.1 **a** Magnetic phase diagram of $Ni_{50}Mn_{25+y}Sn_{25-y}$. The top axis shows the conversion to the e/a ratio defined in the text. The temperatures T_m and T_C are the martensitic phase transformation temperature (average) and Curie temperature, respectively. "Aust." and "Mart." refer to austenite and martensite, while P, F and AF refer to paramagnetic, ferromagnetic, and antiferromagnetic, respectively. **b** Magnetic phase diagram of $Ni_{50-x}Co_xMn_{40}Sn_{10}$. The top axis shows the conversion to the e/a ratio defined in the text. The temperatures T_m, T_C, T_B^{SP} and T_B^{EB} are the martensitic phase transformation temperature (average), Curie temperature, superparamagnetic blocking temperature, and exchange bias blocking temperature, respectively. "Aust." and "Mart." refer to austenite and martensite, while P, F, AF and SP refer to paramagnetic, ferromagnetic, antiferromagnetic, and superparamagnetic, respectively. The error bar on T_m, and the accompanying white region, represent the start and finish temperatures of the transformation. At low temperatures the exchange-biased clustered phase ("EB clustered") is distinguished from the unbiased clustered phase ("clustered"), as discussed in the text. The vertical dotted line marks the critical composition, $x = 4.5$, where ferromagnetism emerges

competition, and magnetic inhomogeneity are known to occur in certain regions [7, 9, 16]. Starting at $y = 0$, i.e. the ordered full Heusler alloy Ni_2MnSn, the obvious important feature is the conventional second order magnetic phase transition (at $T_C = 340\,K$) from P austenite to F austenite, with no accompanying MPT. Such behavior persists out to $y \approx 8$, with T_C decreasing slowly, reaching approximately 310 K at $y \approx 8$. Around this composition a MPT is detected for the first time (at $T_m \approx 180\,K$), increasing approximately linearly with y at a rate, dT_m/dy, of approximately 55 K/at. % out to $y = 20$. In the region $8 \leq y \leq 12$ the system is thus

F both above and below T_m, but with distinctly different F properties in the high T and low T phases. Above $y \approx 12$ on the other hand, T_C begins to drop rapidly with y. This reflects the fact that the first-order phase transformation from P austenite to martensite now occurs at a T_m that exceeds T_C, resulting in weakening of the F interactions and strengthening of the AF interactions [7, 9]. In the region above $y \approx 12$ this leads to the complex behavior related to AF/F competition, as discussed in the introduction. Finally, as $y \rightarrow 25$ (i.e. $Ni_{50}Mn_{50}$), the first order T_m separating P austenite from martensite is stabilized to far above room temperature [18]. We note that a number of features of this and related phase diagrams have been quantitatively understood using a combination of first principles electronic structure and Monte Carlo methods [9, 38, 39].

One particularly noteworthy issue here is the nature of the magnetism in the martensitic phase for $y > 12$, i.e. the large phase field on the right side of Fig. 8.1a. Unambiguous determination of the magnetic state in this portion of this, and related phase diagrams, has proven difficult. While $Ni_{50}Mn_{50}$ ($y = 25$) is understood to be long-range AF [18], tracking the evolution of the AF order as y is decreased is non-trivial, as this order is thought to set in simultaneously with T_m [18]. One example of the complications this introduces is the difficulty of reliably determining magnetic structures and ordering types from neutron powder diffraction (NPD) data (see Sect. 8.3.2 below) when the AF ordering occurs simultaneously with a structural phase transformation to a low symmetry phase with a large number of atoms per unit cell. This difficulty has motivated neutron polarization analysis experiments for instance in $Ni_{50}Mn_{37}Sn_{13}$ and $Ni_{50}Mn_{40}Sb_{10}$ [40], which indeed confirmed AF spin correlations below the MPT, as well as DSC-based studies of $Ni_{50}Mn_{25+y}In_{25-y}$ based on the presumption that AF ordering is simultaneous with T_m [18]. The latter suggest that AF order is maintained down to $y = 10$. We thus tentatively label the large phase field below T_m on the right side of Fig. 8.1a as “AF”. This assignment is strengthened by the significant number of features of our recently reported SANS data that are suggestive of AF order [13], some indirect observations from the NPD data presented here (see Sect. 8.3.2 below), and the ease with which certain aspects of the exchange bias phenomena can be interpreted based on an AF matrix (see [13] as well as additional data in Sect. 8.3.4 below). We do not speculate as to whether this “AF” order is long-ranged.

The discussion above provides context for the introduction of Fig. 8.1b, where we present our experimentally-determined magnetic phase diagram for $Ni_{50-x}Co_xMn_{40}Sn_{10}$ ($0 \leq x \leq 14$), the primary focus of the current work. As in Fig. 8.1a we provide on the top axis the conversion to *e/a*, which *decreases* with the addition of Co. The measurements upon which this phase diagram are based will be discussed in detail in Sects. 8.3.2–8.3.4 below, but we provide now some initial discussion of the basic features. The $Ni_{50-x}Co_xMn_{40}Sn_{10}$ series is derived from $Ni_{50}Mn_{40}Sn_{10}$, i.e. the $y = 15$ composition on the phase diagram shown in Fig. 8.1a, where $T_m \approx 430$ K and long-range F ordering has been suppressed. The $Ni_{50-x}Co_xMn_{40}Sn_{10}$ phase diagram shown in Fig. 8.1b thus starts at $x = 0$ ($e/a = 8.20$) with a system that undergoes a MPT from a P austenite phase to a non-F martensite phase around 430 K. As just discussed, we tentatively assign some type of AF character to this phase. As x is

increased we find that the phase diagram can be conveniently divided into three distinct regions, namely Regions I ($0 \leq x < 4.5$), II ($4.5 < x < 11$), and III ($x \geq 11$). These are discussed in turn below.
Region I ($0 \leq x < 4.5$): In this region T_m decreases with x (as expected based on the decrease in e/a (see Fig. 8.1a), reaching about 410 K at $x = 4$. A central observation is that although the martensitic phase is clearly non-F in character (see discussion above), significant magnetization (M) nevertheless develops at low T. The behavior of $M(T,H)$ down to $T \approx 120$–100 K in this region is in fact consistent with an unblocked (i.e. thermally unstable) SP, strongly suggestive of the spontaneous formation of some type of magnetically nanostructured state, as discussed in the introduction (Sect. 8.1). As T is lowered below 120–100 K, SP blocking occurs (at a temperature T_B^{SP}), indicating freezing of the magnetizations of the nanoscale clusters responsible for the SP-like behavior. As will be shown in detail in Sect. 8.3.4 below, a second important T scale emerges around 50 K in Region I, at which point the nanoclusters become exchange biased (at a temperature T_B^{EB}). Interestingly, this separation of exchange bias and SP blocking temperatures is unique to Region I, and was in fact not detected in prior work on $Ni_{50-x}Co_xMn_{40}Sn_{10}$ [12].
Region II ($4.5 < x < 11$): F behavior emerges in this region (at $e/a = 8.155$), T_C being stabilized above T_m around $x = 5$ then increasing with increasing x up to 450 K at $x = 14$, the highest Co content studied. In this region a second order magnetic transition from P austenite to F austenite on cooling is thus followed by a first order MPT to a non-F martensitic phase, which again exhibits SP-like character. It is in this region that our prior SANS work at $x = 6$ and 8 directly detected a liquid-like distribution of nanoscopic F clusters (with a center-to-center spacing of 12 nm and a diameter of 2–6 nm) [13], which we thus presume to be the origin of the similar SP-type behavior seen in Region I. This presumption is substantiated below, in Sect. 8.3.3. At the lowest temperatures we again find SP blocking, but, in contrast to Region I, this is coincident with the exchange bias blocking temperature, i.e. the onset of exchange biasing is coincident with the onset of thermal stability of the F clusters ($T_B^{SP} = T_B^{EB}$). Another noteworthy feature in this region is the decrease in T_m with x, which is relatively slow out to $x = 9$, but is followed, between $x = 10$ and 11, by a remarkable 300 K decrease in T_m over only 1 % in Co content. As a final comment on Region II we note that it is in this range that λ_2 becomes closest to unity, resulting in extremely low thermal hysteresis, as indicated by the error bars on T_m and the white region in Fig. 8.1b.
Region III ($x \geq 11$): Following the dramatic suppression of the MPT between $x = 10$ and $x = 11$, we find in Region III a phase field in which F austenite dominates below a T_C that increases weakly with x (again consistent with the decreasing e/a). Notably, the suppression of the MPT is found to stabilize apparently conventional F behavior, the SP-like character at low T being completely eliminated. We thus find in this portion of the phase diagram what appears to be a conventional F exhibiting only a standard second order phase transition to a P state at high T.

The measurements providing the physical basis for the construction of this phase diagram are presented in detail in Sects. 8.3.2–8.3.4 below.

8.3.2 Neutron Powder Diffraction (NPD)

As a precursor to detailed magnetometry experiments on $Ni_{50-x}Co_xMn_{40}Sn_{10}$ ($0 \leq x \leq 14$), NPD measurements were taken to supplement our previously reported WAXRD characterization [13]. As an example, we show in Fig. 8.2 NPD data taken at $x = 6$, representative of the particularly interesting Region II. Data are shown in the P austenite phase (500 K, Fig. 8.2a), and in the martensite phase at both 295 K (Fig. 8.2b) and 40 K (Fig. 8.2c). Rietveld refinement of the 500 K data were carried out with the GSAS package [41] using the EXPGUI interface [42], resulting in the fit and residual shown in the figure (green and blue lines, respectively), with $\chi^2 = 1.706$ and $R_{wp} = 4.33\,\%$. The refinement confirms the expected $Fm\bar{3}m$ space group, corresponding to the $L2_1$ structure with lattice parameter, $a = 5.9852(4)$ Å. The only potentially anomalous factor from this refinement is the unusually large displacement parameter detected on the Mn/Sn site, which could potentially implicate significant local displacements. It must be noted however that the uncertainty on this parameter is significant, due to the less than ideal scattering wave vector range studied, which was optimized for magnetic scattering rather than structure determination.

As can be seen from Fig. 8.2b, the 295 K pattern clearly reveals the much lower symmetry below the MPT, which was previously [13] identified (*via* x-ray diffraction) with the $P2_1$ space group, i.e. a monoclinic $5M$-modulated martensite. It must be noted however that the atomic displacements that give rise to this martensitic structure have not yet been determined. As touched upon above in Sect. 8.3.1, this uncertainty in the precise structure, the very low symmetry, and the potential coincidence of magnetic and structural phase transformations combine to make identification of any magnetic ordering type and spin structure in the martensite very difficult. Some insight into this issue is provided by Fig. 8.2c however, where the 40 K NPD pattern is presented. Close inspection reveals some differences between the 40 K and 295 K data in the low 2θ range (where the magnetic form factor is non-negligible), as illustrated by the difference plot in the inset to Fig. 8.2c. Specifically, a small increase in intensity is observed in some locations (e.g. at 31.5 and 35.7°), which could indicate magnetic order. It is worth noting that this signal would be *expected* to be weak if the magnetic ordering were already well-developed at 295 K, i.e. if the natural magnetic ordering temperature of the system was well in excess of the T_m at which it is cut off. Future work with polarized neutrons could potentially clarify this situation by directly determining which reflections have a magnetic component. Nevertheless, the current observations, specifically the potential magnetic reflections identified in the inset to Fig. 8.2c, further add to the growing evidence for long-range AF order in the martensitic phase of this alloy series.

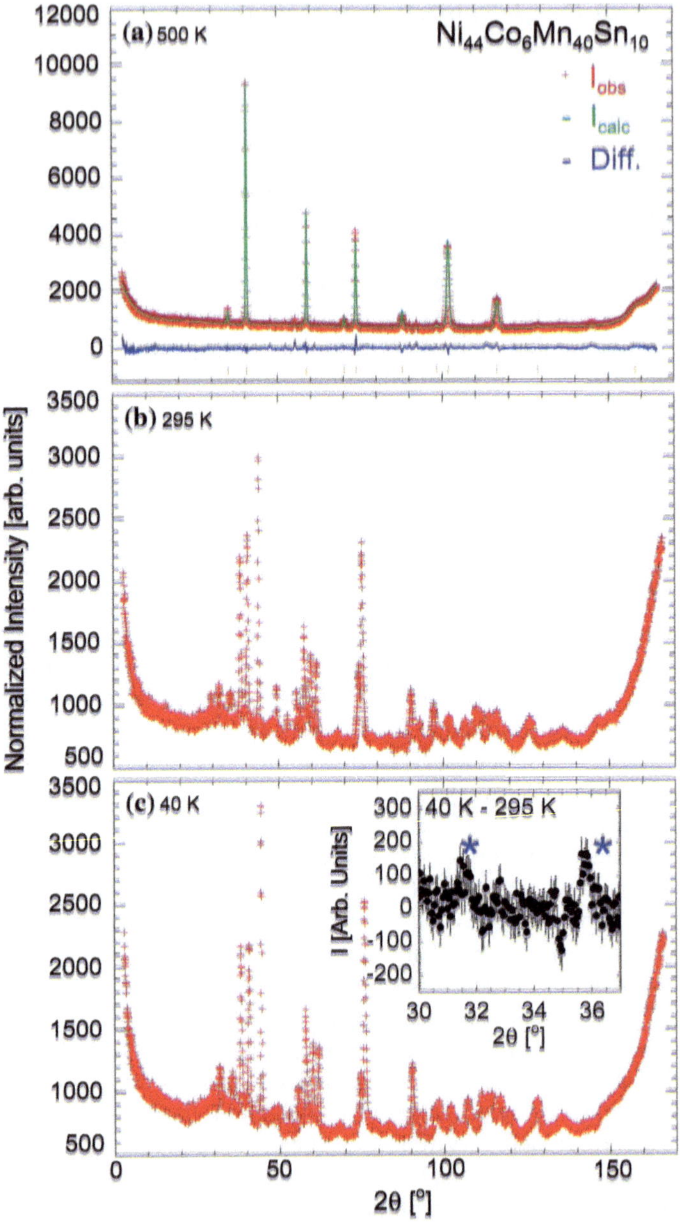

Fig. 8.2 Neutron powder diffraction patterns at (**a**) 500 K, (**b**) 295 K, and (**c**) 40 K for $x = 6$. The data are indicated by the *red* points. In (**a**) the *green solid line* through the points is the result of a Rietveld refinement as described in the text, with the difference between the data and refinement shown at the *bottom* of the panel (in *blue*). The *inset* to (**c**) is a difference pattern between 40 and 295 K, illustrating the possible existence of two low scattering wave-vector magnetic reflections (labeled with asterisks)

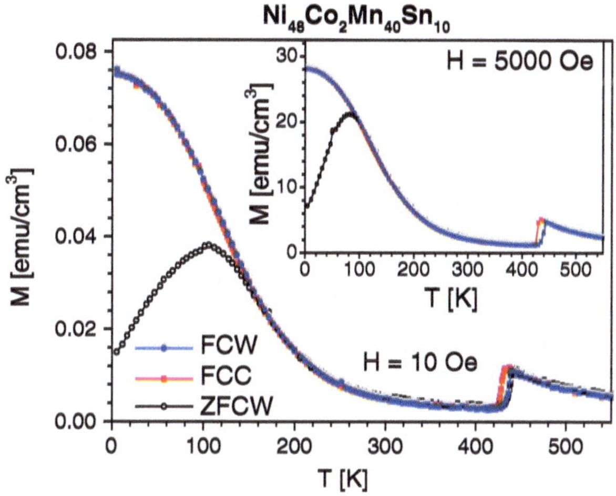

Fig. 8.3 Temperature dependence of the magnetization (in an applied field of 10 Oe) at $x = 2$. Data are shown for FCW (field-cooled warming), FCC (field-cooled cooling), and ZFCW (zero-field-cooled warming). The *inset* shows the same quantities in a field of 5000 Oe

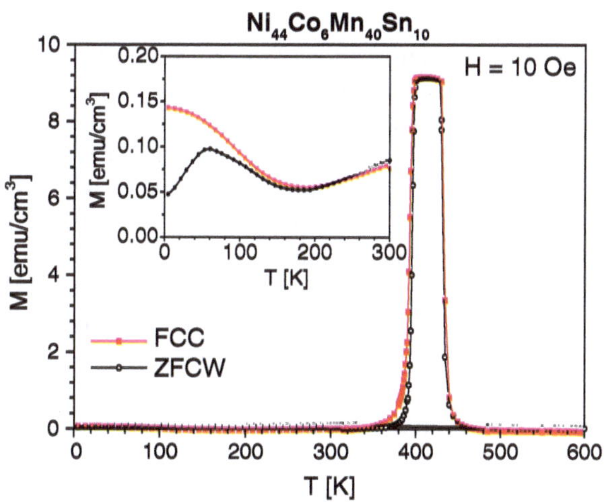

Fig. 8.4 Temperature dependence of the magnetization (in an applied field of 10 Oe) at $x = 6$. Data are shown for FCC (field-cooled cooling), and ZFCW (zero-field-cooled warming). The *inset* shows an expanded view from 0–300 K. Adapted from [13]

8.3.3 Magnetometry

Example characterization of $M(T,H)$ at $x = 2$, 6 and 14 (representative of Regions I, II and III) is shown in Figs. 8.3, 8.4 and 8.5, respectively. $M(T)$ data are shown for field-cooled cooling (FCC), field-cooled warming (FCW), and zero-field-cooled warming (ZFCW) conditions, the main panels of Figs. 8.3, 8.4 and 8.5 being in measuring and cooling fields of 10 Oe. As shown in Fig. 8.3, at the $x = 2$ composition (in Region I) the T_m of approximately 415 K is marked by an abrupt thermally hysteretic decrease in M on cooling, an observation that is again consistent with a MPT from P austenite to AF martensite. As expected from the discussion in Sect. 8.1 above, the magnetic properties in the non-F martensitic phase are non-trivial however,

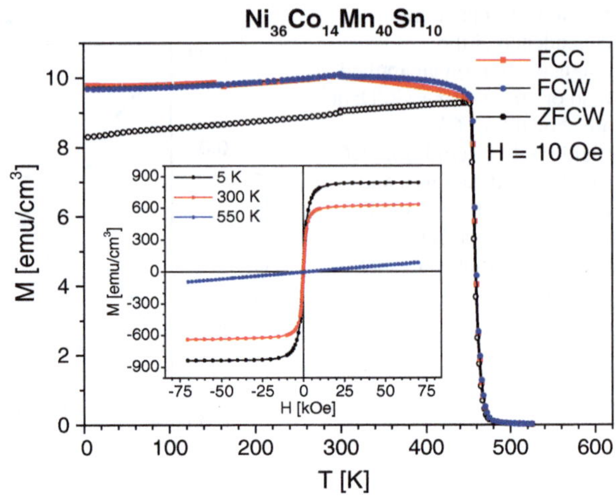

Fig. 8.5 Temperature dependence of the magnetization (in an applied field of 10 Oe) at $x = 14$. Data are shown for FCW (field-cooled warming), FCC (field-cooled cooling), and ZFCW (zero-field-cooled warming). The *inset* shows isothermal magnetization hysteresis loops at 5, 300 and 550 K, taken after zero-field-cooling from above the Curie temperature

$M(T)$ showing bifurcation of the FC and ZFC curves below 100–150 K, followed at 100 K by a clear peak in the ZFCW data. This is consistent with prior classification as SP-like in this composition range, indeed with a qualitatively similar blocking temperature to other work ($T_B^{SP} \approx 100\,\mathrm{K}$) [12]. This SP-like response is further verified below *via* measurement and analysis of $M(H)$ at various T (Fig. 8.6). In addition, the data shown in the inset to Fig. 8.3 confirm that this SP-like $M(T)$ is preserved even at higher fields (5 kOe in this case), with little change other than the expected increase in M.

Qualitatively different behavior is found at $x = 6$ (in Region II), where $M(T)$ in $H = 10\,\mathrm{Oe}$ (Fig. 8.4) clearly reflects the onset of F ordering at $T_C = 440\,\mathrm{K}$, followed at $T_m \approx 390\,\mathrm{K}$ by an abrupt thermally hysteretic decrease in M due to the MPT. As in Region I the martensitic phase is clearly not a long-range F but the low T behavior nevertheless reveals non-trivial magnetic response, as is more clearly illustrated by the blow-up of the low T region in the inset to Fig. 8.4; bifurcation of ZFC and FC curves and a peak in the ZFCW $M(T)$ again indicate SP-like behavior. Although this SP behavior is *qualitatively* similar to that seen at $x = 2$ (characteristic of Region I), it must be emphasized that T_B^{SP} in this case is around 60 K (from the peak in the ZFC $M(T)$, as well as the onset of finite coercivity (see Fig. 8.8 below)), significantly reduced in comparison to the 120 K found in Region I. This is a point returned to a number of times below.

In Fig. 8.5, $M(T)$ data are plotted at $x = 14$, representative of Region III of the $Ni_{50-x}Co_xMn_{40}Sn_{10}$ phase diagram (Fig. 8.1b). In the absence of a MPT, apparently simple long-range F order is observed below $T_C = 460\,\mathrm{K}$, with no indication of any form of the complex low T freezing phenomena that are so evident in Regions II and III. Note that some minor FC/ZFC splitting is observed below T_C but that this can be understood upon consideration of the relative magnitudes of the measuring and coercive fields; higher measuring fields (data not shown) progressively suppress this splitting to lower T. The simple F response deduced from these $M(T)$ data

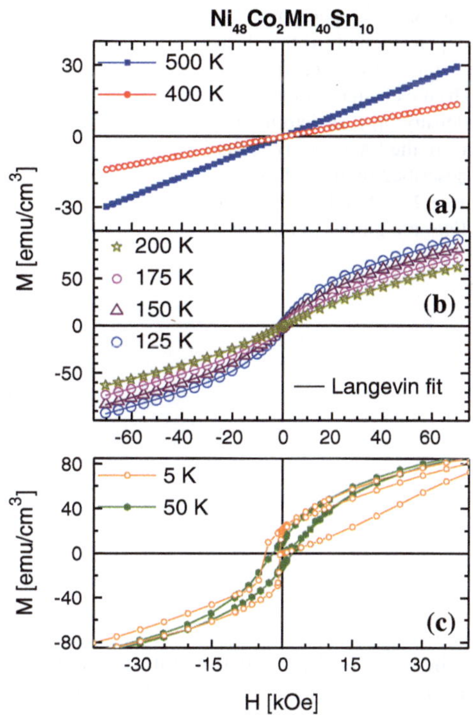

Fig. 8.6 Isothermal magnetization hysteresis loops for $x = 2$ at (**a**) 500 and 400 K, (**b**) 200, 175, 150 and 125 K, and (**c**) 50 and 5 K. All data were taken after zero-field-cooling from above the martensitic phase transformation temperature. In (**b**) the solid lines are a fit to a modified Langevin form, as described in the text. In (**c**) virgin (magnetizing) curves are shown in addition to major loops. Note that in (**c**) a magnified region from −40 to 40 kOe is shown, but the data were recorded to ± 70 kOe

is further verified by $M(H)$ measurements at $T = 5, 300$ and 550 K, as shown in the inset to Fig. 8.5. The linear paramagnetic response at 550 K converts to simple F behavior below T_C, eventually yielding (at 5 K), a soft ferromagnet with a saturation magnetization of 820 emu cm^{-3}, similar to prior reports in the F phase of $Ni_{50-x}Co_xMn_{40}Sn_{10}$. For context note that the saturation magnetization of the Full Heusler Ni_2MnSn is around 600 emu cm^{-3} [7].

Following similar analyses previously performed at other compositions [13], we present in Fig. 8.6 $M(H)$ measurements on an $x = 2$ sample (Region I) at temperatures around T_m (500 and 400 K, Fig. 8.6a), $T_B^{SP} < T < T_m$ (200, 175, 150 and 125 K, Fig. 8.6b), and $T < T_B^{SP}$ (50 and 5 K, Fig. 8.6c). Figure 8.6a shows that the decrease in low field $M(T)$ that occurs on cooling through the MPT at this composition (see Fig. 8.3) is preserved even at higher H, $M(H)$ being linear at both 500 and 400 K, but with distinctly smaller susceptibility at 400 cf. 500 K. As might be anticipated from the SP-like behavior deduced from Fig. 8.3, this situation changes on cooling (Fig. 8.6b), $M(H)$ becoming progressively non-linear from 200 down to 125 K. As in prior work at other compositions [10–13], $M(H)$ in this regime can be fit to a modified Langevin form:

$$M(H,T) = n_C(T)\mu_C(T)\left[\coth\left(\frac{\mu_C(T)H}{k_BT}\right) - \frac{k_BT}{\mu_C(T)H}\right] + \chi_{BG}(T)H, \quad (8.1)$$

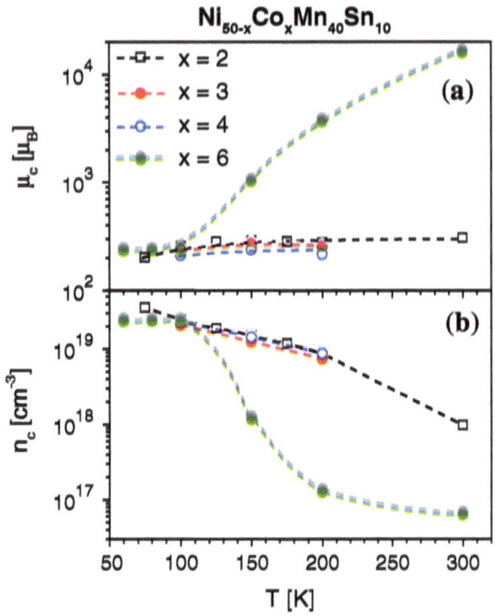

Fig. 8.7 Temperature dependence of (**a**) the cluster magnetic moment (μ_c) and (**b**) the cluster volume density (n_c), as determined from the Langevin-type fits described in the text, for $x = 2, 3, 4$ and 6. The *dotted lines* are guides to the eye

where n_c and μ_c are the volume density and individual magnetic moment of the F clusters, and χ_{BG} is a "background" or "matrix" susceptibility of the martensitic phase [13]. As can be seen from the solid lines in Fig. 8.6b this form, which describes a thermally fluctuating (i.e. unblocked) assembly of isotropic classical macrospins (as in a simple SP), provides a good fit to the data down to $T \approx T_B^{SP}$. The resulting fitting parameters are shown as a function of T in Fig. 8.7, which also includes the results of similar experiments and fitting procedures on samples with x = 2, 3, 4, and 6. In addition to revealing remarkably systematic trends, the results provide significant insight into the differences between Regions I and II. Specifically, the data for $x = 2$, 3 and 4, i.e. those compositions in Region I, are essentially identical, showing a T-independent μ_c of 200–300 μ_B, and an n_c that grows only relatively slowly on cooling, from 10^{18} cm^{-3} at 300 K to a T-independent value of 2–4 × 10^{19} cm^{-3} as $T \rightarrow T_B^{SP}$. These μ_c values result in simple estimates [13] of ferromagnetic cluster diameters and center-to-center spacings (assuming spherical non-overlapping clusters and reasonable values of the F magnetization) of 2–6 nm. While the low T limiting behavior is similar at $x = 6$ (in Region II), the higher T behavior is remarkably different, n_c decreasing by a factor of more than 400 from 100 to 300 K, while μ_c increases by a factor of 70. In prior work [13] we interpreted this behavior at $x = 6$, supported by careful analysis of $M(T)$, in terms of *microscopic* clusters of retained austenite with F character. These *microscopic* F clusters are gradually extinguished on cooling far below T_m, eventually leaving only the SP signature of the *nanoscopic* F clusters below approximately 125 K. The data of Fig. 8.7 provide considerable support for this picture, as the evidence for this retained F austenite, specifically the strong T-dependence of n_c and μ_c at higher T, vanishes on entry to

Region I ($x < 4.5$), where long-range F behavior in the austenite is extinguished due to the increased T_m (Fig. 8.1b). The other noteworthy feature of the data shown in Fig. 8.7 is the very weak x dependence of the low T limiting values of n_c and μ_c. This is significant as it implies that the nanoscopic F cluster density and diameter vary only weakly with x, a point that will be returned to below.

Completing our discussion of Fig. 8.6, we return now to the behavior of $M(H)$ for $x = 2$ at temperatures *below* the SP blocking point, i.e. $T < T_B^{SP}$. Data are shown at 50 and 5 K in Fig. 8.6c, including not only the hysteresis loops but also virgin curves measured after zero-field-cooling to the particular measurement temperature. As expected, the $M(H)$ hysteresis loops gradually open below T_B^{SP} as the system develops finite coercivity and remnance. As is the case in a number of related off-stoichiometric Heusler alloy systems, and as we previously reported at $x = 6$, exchange bias also develops as the system is cooled, i.e. the hysteresis loop becomes systematically displaced from $H = 0$ [13, 16–18]. Of particular interest in this case, and in contrast to observations at $x = 6$ in Region II, the exchange bias can be induced simply by "field poling" the system after zero-field-cooling. This is seen in the 5 K data plotted in Fig. 8.6c where the act of measuring the virgin $M(H)$ curve out to large positive H (i.e. poling in the positive H direction) was sufficient to induce a significant shift of the subsequent $M(H)$ loop in the negative H direction (i.e. negative exchange bias). This is the unconventional exchange bias phenomenon after zero-field-cooling to an unmagnetized state, as reported by Wang et al. in $Ni_{50}Mn_{25+y}In_{25-y}$ [18]. This has been ascribed to a field-induced growth in the F cluster size, leading to a transition from super-spin-glass to super-ferromagnetic behavior, thus creating unidirectional anisotropy. As in [18] we find that this exchange bias after zero-field-cooling is accompanied by a virgin $M(H)$ curve that extends *outside* the subsequent hysteresis loop, consistent with the notion of H-induced alteration of the clustered state.

In addition to this observation of zero-field-cooled exchange bias, which we again emphasize does not occur in Region II, additional measurements of the conventional (field cooled) exchange bias reveal further differences between Region I and II, as discussed in the next section.

8.3.4 Exchange Bias Effects

The T dependence of the coercivity (H_C) and exchange bias (H_E) using the conventional measurement protocol, i.e. after field cooling to the measurement temperature (in this case in 70 kOe) are shown in Fig. 8.8. (Note that the full H_{FC} dependence was explored in prior work [13], at least for $x = 6$). The behavior at $x = 6$ (Region II) is similar to previous reports; finite H_C sets in on cooling below $T_B^{SP} = 60$ K (Fig. 8.8a, consistent with the $M(T)$ shown in Fig. 8.4), coincident with the onset of H_E at $T_B^{EB} = 60$ K (Fig. 8.8b). SP-like blocking and exchange bias blocking are thus simultaneous, a conclusion that we find to hold throughout Region II. The T dependence of H_E and H_C are both monotonic in this composition range, H_E reaching

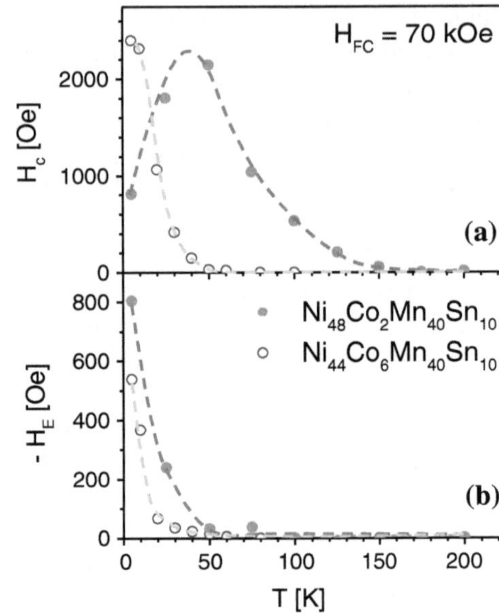

Fig. 8.8 Temperature dependence of (**a**) the coercivity (H_C) and (**b**) the (negative) exchange bias (H_E) at $x = 2$ and 6. The data were taken after field cooling from above the blocking temperature in 70 kOe. The *dotted lines* are guides to the eye

about −550 Oe at 5 K, at which point H_C attains a value around 2400 Oe. At the $x = 2$ composition on the other hand, representative of Region I, distinctly different exchange bias behavior is found. Finite H_C sets in now at $T_B^{SP} \approx 125$ K (Fig. 8.8a), in reasonable agreement with $M(T)$ (Fig. 8.3), confirming the distinctly larger T_B^{SP} in Region I cf. Region II. An increase in H_C then occurs on initial cooling, followed at about 50 K by a clear peak in $H_C(T)$, again in contrast with the behavior seen in Region II. Simultaneous with this H_C peak is the onset of H_E (Fig. 8.8b), with a magnitude that increases monotonically on cooling to about −800 Oe at 5 K, substantially higher than that seen at $x = 2$. As returned to in the discussion section below (Sect. 8.4), Regions I and II thus differ in terms of their T_B^{SP} values, abilities to support zero-magnetization-cooled exchange bias, overall magnitude of H_E, separation of T_B^{SP} and T_B^{EB}, and the existence, or otherwise, of a peak in $H_C(T)$ at T_B^{EB}.

As a final analysis of the exchange bias phenomena in these alloys we present in Fig. 8.9 a comparison of the behavior of $H_E(T)$ and $H_C(T)$ for $x = 2$ (Region I) after conventional field cooling (in 70 kOe) and after establishing the bias by field poling (to 70 kOe) after zero-field-cooling to an unmagnetized state. The first observation from these data is that the conclusion of distinct T_B^{SP} and T_B^{EB} values in Region I is independent of the manner in which the exchange bias is set; T_B^{EB} remains around 60 K, regardless of whether bias is established by field cooling or by field poling after zero-magnetization-cooling. The second observation is that the peak in $H_C(T)$ is similarly robust with respect to the mechanism by which the bias is established, clear peaks occurring in both cases shown in Fig. 8.9. Finally we note, as discussed previously by Wang et al. [18] for the $Ni_{50}Mn_{25+y}Sn_{25-y}$ system, that the exchange bias after zero field cooling in fact significantly exceeds that obtained by conventional field cooling, in our case by a factor of almost 3 at 5 K.

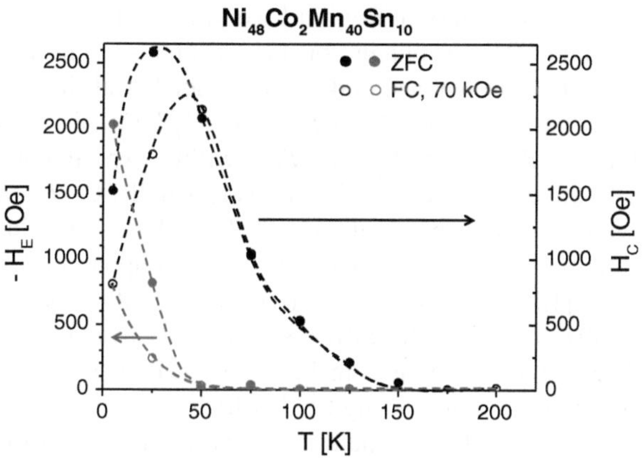

Fig. 8.9 Temperature dependence of the (negative) exchange bias (H_E, *left axis*) and coercivity (H_C, *right axis*) for $x = 2$, after zero-field-cooling (*solid points*) and 70 kOe field-cooling (*open points*). The *dotted lines* are guides to the eye

8.4 Discussion

Further to the discussion already presented, we consider three main points from the data and analysis above that warrant additional discussion: A recap of the issue of the nature of the magnetism in the martensite matrix, the distinctly different low T behavior in Regions I and II of the $Ni_{50-x}Co_xMn_{40}Sn_{10}$ phase diagram, and the implications of the behavior seen in Region III for the origin of the nanoscale magnetic inhomogeneity. These are now discussed in turn.

As discussed in Sect. 8.3.1 a full appreciation of the nature of the magnetism in the martensitic state in this and related systems remains elusive. We wish to recap however the gathering indirect evidence now accumulated for some form of AF order in this phase. This includes the AF ordering temperature deduced from T_m in $Ni_{50}Mn_{25+y}In_{25-y}$ from $y = 25$ down to $y = 10$ [18], and our prior SANS data at $x = 6$ and 8 showing that the Porod scattering intensity and exponent in the martensitic phase are both distinctly different from the P austenite phase, and T-dependent, suggestive of (non-F) long-range order [13]. From the current work we may add the identification of some low scattering wave vector ND reflections with weak T dependence below T_m (Fig. 8.2), and the observed decrease in susceptibility on cooling below T_m in Region I (Fig. 8.3). Additionally, and as alluded to above, it should be reiterated that the existence of AF order in the martensite provides one simple interpretation (though certainly not the only one [43]) for the multitude of exchange bias phenomena observed here, and in several prior works, when F clusters form in the martensitic phase. Nevertheless, it should be emphasized that none of these pieces of evidence is direct and that additional work with other techniques will likely be required to completely resolve this issue. Polarized ND and zero-field NMR (on Mn [37] and/or Co sites) are promising in this regard and will hopefully be pursued in the compounds studied here.

The second issue that warrants additional discussion is the distinctly different low T magnetic behavior in Regions I and II of the phase diagram (Fig. 8.1b). As a succinct summary in Region I we find: Higher T_B^{SP} (100–120 K) distinct from T_B^{EB} (around 50 K), exchange bias even after ZFC in an unmagnetized state, relatively large FC H_E values, and a pronounced peak in H_C at T_B^{EB}. In Region II on the other hand, we find: Distinctly lower T_B^{SP} (50–80 K), simultaneous SP and exchange bias blocking (i.e. $T_B^{SP} = T_B^{EB}$ within experimental uncertainty), no measurable exchange bias after ZFC, relatively small FC H_E values, and monotonic $H_C(T)$ with no observable peak around T_B^{EB}. The first point to emphasize is that this distinction between Regions I and II is not apparent from the phase diagram of Cong et al. [12], where T_B^{SP} decreases only weakly with x, from just over 100 K at $x = 0$, to 80 K at $x = 8$. This differs significantly from the abrupt decrease in T_B^{SP} (by about 50 K) at $x \approx 4.5$ seen in Fig. 8.1b. This discrepancy cannot be ascribed to small differences in measuring field (which we have experimentally verified to have little impact), or different methods of extracting T_B^{SP} from $M(T)$, as [12] shows $M(T)$ data at $x = 5$ for instance, with $T_B^{SP} \approx 100$ K, quite different from Fig. 8.1b. A more likely explanation lies with variations in local chemical composition and structure arising from (potentially even small) differences in synthesis and thermal treatment. Such local chemical and structural changes are probable given the manifest chemical and structural complexity of alloys such as $Ni_{50-x}Co_xMn_{25+y}Sn_{25-y}$ [38]. The unusual sensitivity of magnetic properties to chemical composition [12, 27, 33, 35], the proximity to multiple structural, electronic and magnetic phase transitions [12, 13, 33, 35], and the acute phase competition [12, 13], could then be anticipated to lead to local variations in magnetic properties, as has been implicated as the source of the nanoscale magnetic inhomogeneity in these materials [13]. Significantly, the recent work of Belesi et al. [37] has directly observed sensitivity to preparation conditions using Mn NMR, even in the Ni_2MnGa system which one could naively expect to be somewhat simpler than $Ni_{50-x}Co_xMn_{25+y}Sn_{25-y}$. Clearly, additional future work with a variety of local probes is warranted in order to more fully understand the interplay between preparation conditions, and short-range chemical, structural, electronic, and magnetic variations.

Regardless of the exact origin of the apparent differences between the samples used in the current work and those discussed in [12], it is useful to consider how the numerous distinguishing features of the magnetic properties in Regions I and II of our phase diagram (Fig. 8.1b) can be rationalized. Some source of increased thermal stability for the spin clusters in Region I cf. Region II is a factor that would seem to account for many of the key observations. This would trivially explain the increased T_B^{SP} in Region I for instance, although it must be noted that the analysis shown in Fig. 8.7, particularly the x independence of μ_c and n_c, strongly suggest that this does not arise from an increase in cluster volume in Region I versus Region II. (We note parenthetically that future SANS studies to directly verify this are clearly worthwhile). In the absence of a cluster volume increase the most obvious sources of improvement in thermal stability in Region I are increases in cluster anisotropy, cluster-matrix interactions, or inter-cluster interactions. Exchange bias observations (Figs. 8.6, 8.7, 8.8 and 8.9) are important in this context, as the generally larger H_E values and peak in $H_C(T)$ at T_B^{EB} in Region I (Fig. 8.8) are also broadly consistent with the concept of improved cluster thermal stability. Adapting arguments from the literature on thin film F/AF bilayers (e.g. [44]), improved thermal stability of the clusters (from higher F cluster anisotropy for instance), would dictate that magnetization reversal of the F could not proceed without significant perturbation of the AF spin structure. This rearrangement of the AF spin structure would thus provide a substantial contribution to the measured H_C, leading to a peak in $H_C(T)$ at T_B^{EB} as the AF loses thermal stability. Such phenomena are notably absent in Region II, where $H_C(T)$ is monotonic (Fig. 8.8).

While this notion of improved thermal stability of the F spin clusters in Region I can thus qualitatively rationalize several observed differences between the low T magnetic behavior in Regions I and II, the question remains as to how this thermal stability improvement arises. The most notable observation in this regard is the abrupt change in behavior (consider T_B^{SP} (x) for instance in Fig. 8.1b) at $x \approx 4.5$, i.e. the point where T_C exceeds T_m. This indicates, quite intriguingly, that the sequence of phase transitions that takes place on cooling plays a vital role in distinguishing the behavior of the low T magnetic inhomogeneity in Regions I and II. Apparently, when $T_C > T_m$ (Region II) and long-range F order sets in prior to the MPT on cooling, this leads to decreased thermal stability of the low T spin clusters in comparison to Region I where the MPT suppresses F order. Further work, perhaps combining local experimental probes of magnetism, structure and chemistry with modeling [9, 38, 43], will likely be required to understand this. As a final comment on the differences in behavior seen in Regions I and II we note that the existence of zero-magnetization-cooled exchange bias in Region I only again points to fundamental differences in the magnetically clustered ground state in the two regions, this time highlighting that they likely respond very differently to large applied magnetic fields.

As a third item, we consider the implications of the behavior seen in Region III of the phase diagram for the origins of the nanoscale magnetic inhomogeneity in this complex alloy system. The primary observation here is that when T_m is suppressed to zero by increased Co alloying (or alternatively decreased *e/a*), the resulting F state is apparently quite conventional, with no evidence found for complex low T magnetism or overt short-range magnetic inhomogeneity. This important observation vividly illustrates the essential role of the MPT in establishing nanoscale magnetic inhomogeneity, likely *via* the role the MPT plays in suppressing F and promoting AF interactions. Further work in this regime with neutron scattering for instance is nevertheless worthwhile in order to probe local magnetic fluctuations that could be concealed by long-range F order.

Finally, we wish to identify a further item that our results point to a beginning understanding of, but which would clearly benefit from additional work. This is the issue of statistics and collective behavior of nucleation sites at first-order phase transformations of the type seen in $Ni_{50-x}Co_xMn_{25+y}Sn_{25-y}$. The explanation for the importance of the $\lambda_2 = 1$ criterion is explained in [28] as being related to the free energy barrier seen by an austenite-twinned martensite nucleus as temperature is increased or decreased. Essentially, this barrier is dramatically lowered as the transition layer between the austenite and martensite disappears, together with the total twin interfacial energy, when $\lambda_2 \rightarrow 1$. This leads to low thermal hysteresis around the phase transformation. At certain compositions of $Ni_{50-x}Co_xMn_{25+y}Sn_{25-y}$, specifically $5 \leq x < 10$ for the $y = 15$ studied here, the F character of the austenite phase, and thus the microscopic clusters of retained austenite (see Fig. 8.7 and related discussion in Sect. 8.3.3), render magnetometry (and potentially other magnetic techniques) a probe of the statistical distributions of sizes, spacings, and interactions of these clusters. More detailed studies could therefore provide important data to inform the development of future models of nucleation and phase transformation. Of particular interest are collective effects that may underlie an understanding of the observed reproducibility of the hysteresis loops at these transformations.

8.5 Summary

We have presented a brief recap of the current understanding of the fascinating magnetic properties of the off-stoichiometric full Heusler alloys, followed by a study of the specific system $Ni_{50-x}Co_xMn_{25+y}Sn_{25-y}$ in the critical composition range $y = 15$, $0 \leq x \leq 14$. Neutron powder diffraction data, magnetometry data, and studies of exchange bias phenomena have been presented, leading to the construction of a detailed magnetic phase diagram. While some of the major features of this phase diagram are in agreement with prior work, and can be rationalized based on recent computational work, several aspects differ from prior experimental findings and continue to pose significant challenges to our understanding. The primary example of the latter lies with the inhomogeneous nanoscale magnetic structure deduced to occur at low temperatures in the range $0 \leq x < 11$. This magnetically inhomogeneous ground state results from strong magnetic phase competition, with short-range structural and chemical disorder likely playing vital, but as yet not fully understood, roles. We have highlighted a number of open questions in this context (focusing on the interplay between preparation conditions and local chemical, crystal and magnetic structure), arguing that careful application of additional local probes from neutron scattering, x-ray scattering and spectroscopy, NMR, and electron microscopy/spectroscopy (among others) could well yield substantial advances. These advances would not only generate improved fundamental understanding of the magnetic properties of these fascinating materials, but could also potentially impact numerous potential applications of these alloys.

Acknowledgments Work at UMN supported primarily by DOE under award DE-FG02-06ER 46275. Additional support from the UMN Institute for Renewable Energy and the Environment (IREE) is acknowledged. VS and RDJ also acknowledge support from MURI (W911NF-07-1-0410) and NSF/PIRE (OISE-0967). We acknowledge the support of the National Institute of Standards and Technology, US Department of Commerce, in providing the neutron research facilities used in this work. We also acknowledge the use of computing resources at the University of Minnesota Supercomputing Institute.

References

1. C.Y. Fong, J.E. Pask, L.H. Yang, *Half-metallic Materials and Their Properties* (Imperial College Press, World Scientific Publishing, London, 2013)
2. R.A. de Groot, F. Mueller, P. Engen, K. Buschow, Phys. Rev. Lett. **50**, 2024 (1983)
3. D. Orgassa, H. Fujiwara, T.C. Schulthess, W.H. Butler, Phys. Rev. B **60**, 13237 (1999)
4. R. Shan, H. Sukegawa, W.H. Wang, M. Kodzuka, T. Furubayashi, T. Ohkubo, S. Mitani, K. Inomata, K. Hono, Phys. Rev. Lett. **102**, 246601 (2009)
5. Y. Sakaruba, K. Izumi, T. Iwase, S. Bosu, K. Saito, K. Takanashi, Y. Miura, K. Futatsukawa, K. Abe, M. Shirai, Phys. Rev. B **82**, 094444 (2010)
6. A. Planes, L. Mañosa, A. Saxena (eds.), *Magnetism and Structure in Functional Materials*, Springer Series in Materials Science, vol. 79 (Springer, New York, 2005)
7. T. Krenke, M. Acet, E.F. Wassermann, X. Moya, L. Manosa, A. Planes, Phys. Rev. B **72**, 014412 (2005)
8. T. Krenke, M. Acet, E.F. Wassermann, X. Moya, L. Manosa, A. Planes, Phys. Rev B **73**, 174413 (2006)
9. P. Entel, M. Siewart, M.E. Gruner, H.C. Herper, D. Comtesse, R. Arroyave, N. Singh, A. Talapatra, V.V. Sokolovskiy, V.D. Buchelnikov, F. Albertini, L. Righi, V.A. Chernenko, Eur. Phys. J. B **86**, 65 (2013)
10. D.Y. Cong, S. Roth, J. Liu, Q. Luo, M. Potschke, C. Hurrich, L. Schultz, Appl. Phys. Lett. **96**, 112504 (2010)
11. D.Y. Cong, S. Roth, M. Potschke, C. Hurrich, L. Schultz, Appl. Phys. Lett. **97**, 021908 (2010)
12. D.Y. Cong, S. Roth, L. Schultz, Acta Mater. **60**, 5335 (2012)
13. K.P. Bhatti, S. El-Khatib, V. Srivastava, R.D. James, C. Leighton, Phys. Rev. B **85**, 134450 (2012)
14. E. Dagotto, T. Hotta, A. Moreo, Phys. Rep. **344**, 1 (2001)
15. E. Dagotto, *Nanoscale Phase Separation and Colossal Magnetoresistance* (Springer, New York, 2002)
16. Z. Li, C. Jing, J. Chen, S. Yuan, S. Cao, J. Zhang, Appl. Phys. Lett. **91**, 112505 (2007)
17. M. Khan, I. Dubenko, S. Stadler, N. Ali, Appl. Phys. Lett. **91**, 072510 (2007)
18. B.M. Wang, Y. Liu, P. Ren, B. Xia, K.B. Ruan, J.B. Yi, J. Ding, X.G. Li, L. Wang, Phys. Rev. Lett. **106**, 077203 (2011)
19. H.E. Karaca, I. Karaman, B. Basaran, Y. Ren, Y.I. Chumlyakov, H.J. Maier, Adv. Funct. Mater. **19**, 983 (2009)
20. L. Manosa, D. Gonzalez-Alonso, A. Planes, E. Bonnot, M. Barrio, J.-L. Tamarit, S. Aksoy, M. Acet, Nat. Mater. **9**, 478 (2010)
21. T. Krenke, E. Duman, M. Acet, E.F. Wassermann, X. Moya, L. Manosa, A. Planes, Nat. Mater. **4**, 450 (2005)
22. R. Kainuma, Y. Imano, W. Ito, Y. Sutou, H. Morito, S. Okamoto, O. Kitakami, K. Oikawa, A. Fujita, T. Kanomata, K. Ishida, Nature **439**, 957 (2006)
23. V. Srivastava, Y. Song, K.P. Bhatti, R.D. James, Adv. Energy Mater. **1**, 97 (2011)
24. Y. Song, K.P. Bhatti, V. Srivastava, C. Leighton, R.D. James, Energy Environ. Sci. **6**, 1315 (2013)
25. A. Planes, L. Manosa, M. Acet, J. Phys. Condens. Matter **21**, 233201 (2009)
26. J. Liu, T. Gottschall, K.P. Skokov, J.D. Moore, O. Gutfleisch, Nat. Mater. **11**, 620 (2012)
27. V. Srivastava, X. Chen, R.D. James, Appl. Phys. Lett. **97**, 014101 (2010)
28. Z. Zhang, R.D. James, S. Muller, Acta Mater. **57**, 4332 (2009)
29. J. Cui, Y.S. Chu, O.O. Famodu, Y. Furuya, J. Hattrick-Simpers, R.D. James, A. Ludwig, S. Thienhaus, M. Wuttig, Z. Zhang, I. Takeuchi, Nat. Mater. **5**, 286 (2006)
30. R. Zarnetta, R. Takahashi, M.L. Young, A. Savan, Y. Furuya, S. Thienhaus, B. Maass, M. Rahim, J. Frenzel, H. Brunken, Y.S. Chu, V. Srivastava, R.D. James, I. Takeuchi, G. Eggeler, A. Ludwig, Adv. Funct. Mater. **20**, 1917 (2010)
31. R. Delville, D. Schryvers, Z. Zhang, R.D. James, Scr. Mater. **60**, 293 (2009)

32. P. Klaer, H.C. Herper, P. Entel, R. Nieman, L. Schultz, S. Fahler, H.J. Elmers, Phys. Rev. B **88**, 174414 (2013)
33. W. Ito, X. Xu, R.Y. Umestu, T. Kanomata, K. Ishida, R. Kainuma, Appl. Phys. Lett. **97**, 242512 (2010)
34. T. Krenke, E. Duman, M. Acet, X. Moya, L. Manosa, A. Planes, J. Appl. Phys. **102**, 033903 (2007)
35. R.Y. Umetsu, A. Sheikh, W. Ito, B. Ouladdiaf, K.R.A. Ziebeck, T. Kanomata, R. Kainuma, Appl. Phys. Lett. **98**, 042507 (2011)
36. V. Basso, C.P. Sasso, K.P. Skokov, O. Gutfleisch, V.V. Khovaylo, Phys. Rev. B **85**, 014430 (2012)
37. M. Belesi, L. Giebeler, C.G.F. Blum, B. Buchner and S. Wurmehl, arXiv:1311.7414v1 28 Nov 2013
38. V.V. Sokolovskiy, V.D. Buchelnikov, M.A. Zagrebin, P. Entel, S. Sahoo, M. Ogura, Phys. Rev. B **86**, 134418 (2012)
39. B. Himmetoglu, V.M. Katukuri, M. Cococcioni, J. Phys. Condens. Matter **24**, 185501 (2012)
40. S. Aksoy, M. Acet, P.P. Deen, L. Manosa, A. Planes, Phys. Rev B **79**, 212401 (2009)
41. A.C. Larson, R.B.V. Dreele, Los Alamos National Laboratory Report No. LAUR 86–748 (2004)
42. B.H. Toby, J. Appl. Crystallogr. **34**, 210 (2001)
43. V. Buchelnikov, V. Sokolovskiy, I. Taranenko, S. Taskaev, P. Entel, J. Phys. Conf. Ser. **303**, 012084 (2011)
44. C. Leighton, M.R. Fitzsimmons, A. Hoffmann, J. Dura, C.F. Majkrzak, M.S. Lund, I.K. Schuller, Phys. Rev. B **65**, 064403 (2002)

Part III
Applications

Chapter 9
Heusler Alloy Films for Spintronic Devices

Atsufumi Hirohata, James Sagar, Luke R. Fleet and Stuart S.P. Parkin

Abstract This chapter reviews the requirements for the Heusler-alloy films to be used in spintronic devices. Four key requirements are identified to be large giant magnetoresistance (GMR), large tunnelling magnetoresistance (TMR), large spin-transfer torque and fast spin resonance. These requirements can be achieved by utilising the fundamental properties of the Heusler alloys, such as atomic substitution, generalised Slater-Pauling behaviour, crystalline ordering, half-metallicity, low damping constant, high Curie temperature, good lattice matching and large activation volume. To date the main obstacles for the Heusler-alloy films to be used in spintronic devices are their (i) high crystallisation temperature, (ii) interfacial atomic disordering and (iii) small activation volume. Here, we have investigated these properties

A. Hirohata (✉)
Department of Electronics, University of York, Heslington, York YO10 5DD, UK
e-mail: atsufumi.hirohata@york.ac.uk

A. Hirohata
PRESTO, Japan Science and Technology Agency, Kawaguchi 332-0012, Japan

J. Sagar · L.R. Fleet
Department of Physics, University of York, Heslington, York YO10 5DD, UK
e-mail: james.sagar@ucl.ac.uk

J. Sagar
Present address: London Centre for Nanotechnology, University College London, Gordon Street, London WC1E 6BT, UK

L.R. Fleet
Present address: Department of Materials, Imperial College London, South Kensington Campus, London SW7 2AZ, UK
e-mail: l.fleet@imperial.ac.uk

S.S.P. Parkin
Almaden Research Center, IBM, San Jose, CA 95120-6099, USA
e-mail: stuart.parkin@mpi-halle.mpg.de

S.S.P. Parkin
Max Planck Institute, Weinberg 2, 06120 Halle, Germany

C. Felser and A. Hirohata (eds.), *Heusler Alloys*,
Springer Series in Materials Science 222, DOI 10.1007/978-3-319-21449-8_9

for both epitaxial and polycrystalline films and have found a favourable crystallisation orientation to lower the ordering temperature by inducing a two-dimensional growth. We have demonstrated the effect of interfacial dusting to maintain the crystalline ordering from atomic diffusion by annealing. We have also established that the above requirements can be controlled by the competition between the structural and magnetic volume, the latter of which can be defined as activation volume. In all cases, the polycrystalline films have found to be advantageous over the epitaxial ones due to their strain-free growth with controlled grain size. We anticipate that the optimised polycrystalline films can be used in the next generation hard disk read heads and magnetic random access memory cells.

9.1 Spintronics

9.1.1 Concept

Spintronics is an emerging field based on a combination of three conventional information carriers; electron charges, electron spins and photons shown schematically in Fig. 9.1 [1]. These carriers represent three major fields in information and communications technology (ICT); data processing with electron transport, data storage with an assembly of spins and data transfer via optical connections. Recent developments of ICT require larger capacity data storage and faster data processing. The performance of both data storage and processing devices has been improving at a very

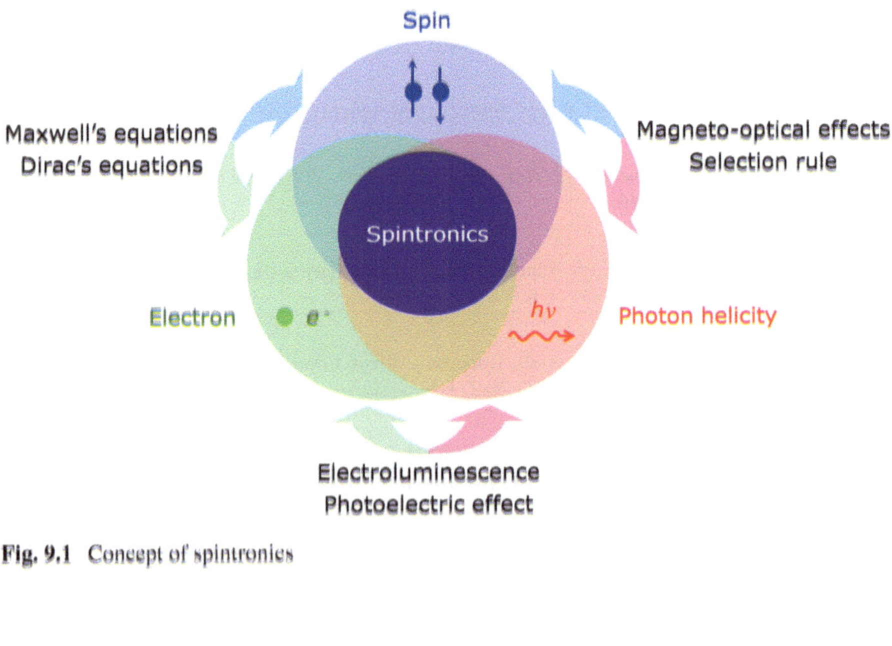

Fig. 9.1 Concept of spintronics

high rate (approximately 60 % per year in data storage, and 100 % in 18 months for Si-based processors, known as Moore's law) but such progress cannot continue using conventional technologies with miniaturisation. Since the possibility of reaching the physical limits in both ferromagnet based memories and conventional semiconductor based electronic processors is known, an interdisciplinary field between magnetism and electronics is required. Spintronics, based on spin-polarised electron transport, is believed to meet the requirements for future technologies, namely low power consumption due to the non-volatility of recorded data and resulting normally power-off operation without data refreshing. Furthermore, the electron spin can be connected to optics via photon helicity, which is expected to allow for much faster data transfer.

Studies of spintronics started in 1988 after the discovery of giant magnetoresistance (GMR), which is a change in resistance dependent upon the magnetisation orientation of magnetic mulitlayers separated by a non-metallic spacer. Within 10 years this discovery was implemented into hard disk drives (HDD), allowing for a drastic increase in the areal density due to the increased sensitivity of the GMR read heads. This is the most common information storage media and the dramatic impact of the technology saw the discovery recognised via the Nobel Prize for Physics in 2007 [2]. Since then, significant efforts have been devoted to advancing spin memories, including the recent development of a racetrack memory [3], as listed in Fig. 9.2. In parallel to the developments in spin memoires, similar efforts have been made to realise spin transistors. These devices can be categorised into three generations in terms of their operation: (i) spin transport, (ii) spin dynamics and (iii) 3-dimension (3D) and quantum physics. Initially, these devices were operated using dc currents or ac currents at low frequency, below ~kHz, with successful demonstrations

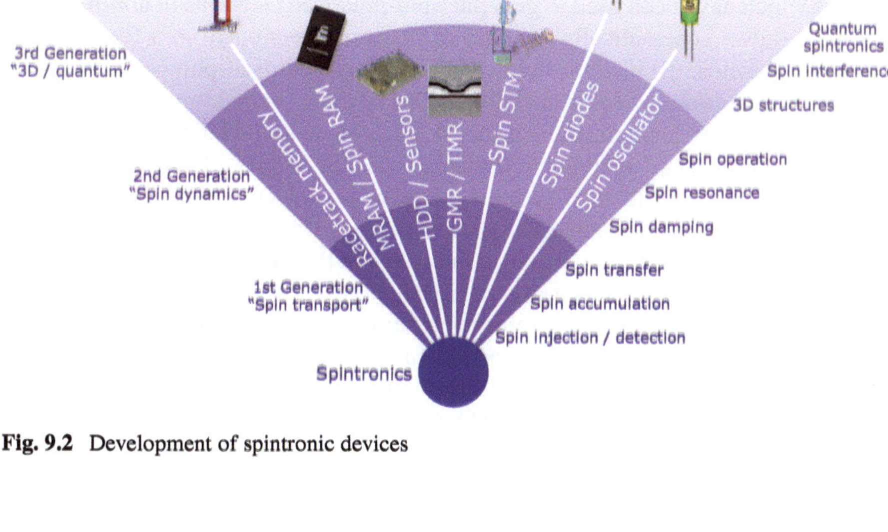

Fig. 9.2 Development of spintronic devices

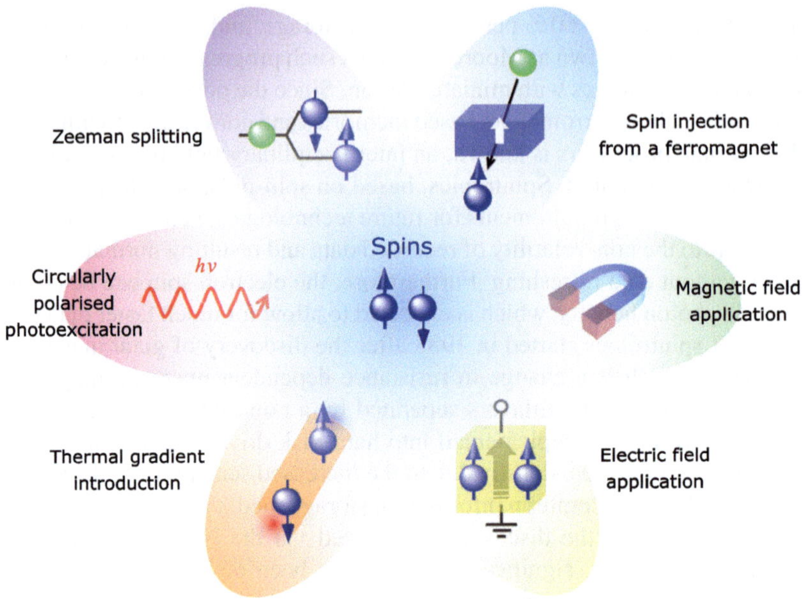

Fig. 9.3 Spin-polarised current generation

highlighting spin injection, detection, accumulation and transfer in spin-polarised electron (or hole) current transport. By increasing the operation frequency spin damping, resonance and manipulation becomes prominent in the devices, leading to the study of spin dynamics. In recent years the trend has been shifting towards 3D architectures and additional functionalities based on quantum physics, such as spin interference effects.

In spintronic devices, efficient generation of a spin-polarised current is an essential issue for which a practical solution has yet to be found. To date, there are six methods which have been used to generate spin-polarised electrons in non-magnetic materials, shown schematically in Fig. 9.3: (1) spin injection from a ferromagnet, (2) magnetic fields, (3) electric fields, (4) circularly polarised photoexcitation, (5) thermal gradients and (6) Zeeman splitting. One of the most common methods is (1) spin injection from a ferromagnetic material, e.g., conventional ferromagnetic metals (Fe, Co, Ni and Gd), half-metallic ferromagnets (HMF) [4] and dilute magnetic semiconductors (DMS) [5]. In such systems the spins are injected into a non-magnetic metal or semiconductor through an ohmic contact or tunnel barrier, although it is important to minimise the conductance mismatch at the interface [6]. Conventional ferromagnets typically possess a spin polarisation of $<50\,\%$ and hence cannot inject electrons with a large degree of spin polarisation into the non-magnetic material. The injected spin polarisation could be increased by employing a HMF or a DMS, both of which have $100\,\%$ spin polarisation in theory, although this has currently not been achieved at room temperature (RT). In a DMS, (6) Zeeman splitting induces ferromagnetic behaviour and its conductance mismatch with conventional semiconductors is not an issue, However, Curie temperatures are typically below 80 K [5]. For a conventional ferromagnets, in order to maintain the injected spin polarisation into a semiconductor from the ferromagnetic source, when injecting into semiconductors which have a much greater resistance, a tunnel barrier can to be employed to avoid the polarisation drop at the interface due to the conductance mismatch [7]. So far, however, only around 30 % spin polarisation into a semiconductor has been achieved in such structures [8].

As already mentioned, although injection from ferromagnets remains one of the most sought after methods for spin generation, there are several alternatives. (2) Stray fields at the edge of a ferromagnet can also be used to induce a population difference in spin-polarised electrons in a non-magnetic material. The magnitudes of the stray fields, however, are difficult to control due to the difficulties in controlling the edge-shape definition and the resulting field distributions [9]. (3) An electric field can also be utilised to create spin currents as it can provoke motion of spin-polarised carriers in a non-magnet towards a favourable direction based on spin-dependent Halleffects. Electric fields can also be utilised to induce magnetic phase transition, demonstrated in an InMnAs layer sandwiched within a FET [10]. A positive bias on the gate created an electric field which repels holes, causing the Mn magnetic moments to rotate randomly (paramagnetism). A negative voltage generated an electric field which attracted holes, causing the Mn magnetic moments to align along the field direction (ferromagnetism). Since this effect has only been observed in a DMS, these phenomena were only seen at 25 K with a bias voltage of ± 125 V, which is not suitable for device applications at this stage. A similar effect has, however, recently been observed in conventional ferromagnets at RT, with electric fields used to control the ferromagnetic phase transition through tuning the Curie temperature [11]. (4) Circularly-polarised light can also be used to excite spin-polarised electrons in a direct-bandgap semiconductor, dependent upon optical selection rules. This is the reverse effect of circularly polarised light emission generated by a spin-polarised electron current in a spin-polarised light emitting diode (LED). By measuring electron transport across a NiFe/GaAs Schottky junction, clear spin-dependent currents have been reported [12]. In such junctions, light is typically scattered at the surface of the semiconductor and it is subsequently difficult to introduce spin-polarised electrons into bulk regions of the semiconductor. In addition, (5) thermal gradients have been found to produce a spin-polarised carrier flow due to a spin-dependent Seebeck effect. When a temperature gradient is applied across a NiFe strip, spin-polarised electrons with the opposite sign can be collected at both ends, dependent upon the magnetisation direction of the strip and the temperature gradient. This can be measured as a voltage through the spin Seebeck effect [13]. This effect converts thermal energy into a spin voltage, which is another method to generate spin-polarised electrons. In order to raise a measurable spin current, the temperature gradient needs to be large, which results in devices typically having contacts separated by greater than 7 mm , held at different temperatures [13].

9.1.2 Spintronic Devices

Almost 80 % of man-made data is currently stored magnetically onto HDD, leading to a market size of USD 100B [14]. In 1956 the first HDD was introduced by IBM, which consisted of 50, 24-inch disks giving a total capacity of 5 MB at a density of 2 kbit/in^2 [15]. Since then the areal density of the HDD has increased at an annual rate of up to 100 % leading to a current density of >600 Gbit/in^2. In a HDD the media typically consists of CoPtCr/SiO_2 thin films deposited onto a metallic (Al) or glass platter [15]. Read/write heads, attached to an arm, are moved precisely over the disks using linear motor actuators. A stray magnetic field from an inductive write head is used to write data into the disk. The magnetic easy axis of current media now lies out of plane, providing much greater stability than the previous generations of in-plane media. The fidelity of the recording depends on the small grain size of the media. These grains, which constitute a written bit, are read by an MR head which detects a change of the direction of the stray field from the media. A sharp change in the magnetisation orientation induces a change in the MR value, detected electrically. The energy stored in one grain is defined as $K_u V$ (K_u: anisotropy constant and V: volume). To store data for 10 years, thermal stability $K_u V / k_B T$ (k_B: Boltzmann constant and T: temperature) therefore needs to exceed 60. In 2005 perpendicular media was introduced by Seagate, Hitachi and Toshiba and has subsequently led to densities over 600 Gb/in^2, corresponding to $\sim$20 × 100 nm^2 bit size.

Although switching to perpendicular media enhanced the stability of the grains, this came at a price as this stability makes it much more difficult to write bits. In order to overcome the difficulty of writing into a highly anisotropic medium, heat assisted magnetic recording (HAMR) has been proposed [16]. In HAMR media, a laser beam is employed to locally heat the media and reduce the thermal stability of the data bit to be written. Successful demonstrations of HAMR have already been reported, with a scanning near-field optical microscope implemented into a write head to create regions of localised heating [16]. Such heat assisted recording is expected to lead to recording densities of up to 10 Tb/in^2.

Other methods for increasing the areal density are also being explored with patterned media, fabricated using top-down lithography, an attractive alternative. In patterned media, each bit could be made to consist of a single domain, so small grains would no longer be required. Patterned bits can therefore be made to be larger than conventional grains, which helps with the thermal stability whilst still leading to an increase in recording density. In order to achieve an areal density >1 Tbit/in^2), a recording bit with a width of $\sim$12.5 nm, separated by trenches of a similar size, needs to be realised [15]. In addition, both large magnetoresistance (MR) ratios and small resistance-area products (*RA*) are required, as shown in Fig. 9.4(a) [17]. Namely, the requirements for next-generation HDDs above 2 Tbit/in^2 is MR > 50 % and $RA < 0.1\ \Omega \cdot \mu m^2$ [18].

As well as exploring different methods to increase the storage capacity of HDDs there is great interest in technologies which could replace them. One of the major ongoing device studies is in magnetic random access memory (MRAM) [19]. MRAM

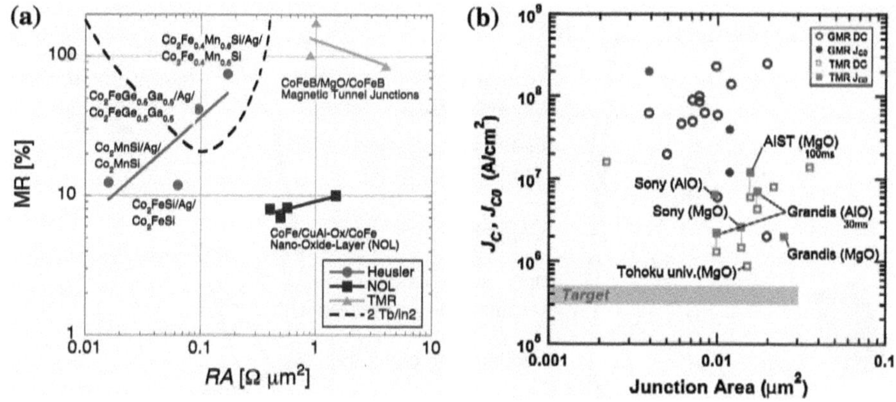

Fig. 9.4 Current and target values of **a** MR ratios and *RA* for a HDD read head after [1] and **b** J_c and *R* for a MRAM cell [2]

can achieve the following features: (1) nonvolatility similar to HDD without using a mechanical head, (2) read and write times of the order of a nano-second, (3) high density and (4) low power consumption. Hence, MRAM is a good candidate for a universal memory. MRAM can also avoid the delay and power consumption arising from circuit leads as a high-density circuit can be achieved by converting passive current leads into memory cells. In 2010, EverSpin Technologies started to ship a 16-Mbit MRAM which operated at 3.3 V with a current of 60 mA (110 mA) for read-out (write-in) at a speed of 35 ns. The initial architecture of such MRAM chips consisted of memory cells located at the crossing points of read and write lines. The current flow in the write line creates an Ampère field, reversing the magnetisation in the free-layer of the cell. Recent studies, however, have shown that spin-transfer torque (STT) can be utilised to reverse the magnetisation of the free layer of a MR nano-pillar using less current when the junction of the cell is below 0.1 μm^2. This current-induced magnetisation switching has already been demonstrated by Toshiba, in collaboration with Hynix, who successfully showed a 64-Mbit MRAM [14]. Improving on the density of MRAM was initially hampered by the fact that an additional transistor was required for each cell for operation but this structural limitation can now be overcome due to an improved architecture, with one transistor for each 4MTJs, recently proposed [20].

It should be noted that a cell size below 0.1 μm^2 is the minimum requirement for a Gbit/in^2 recording density for MRAM. In order to achieve such high recording density, both small switching current density J_c and small junction area A are required as shown in Fig. 9.4(b) [18]. Namely, the requirements for next-generation STT-MRAM cells are $J_c < 5 \times 10^5$ A/cm^2 and $A < 0.03$ μm^2 in addition to a large MR ratio above 100 %. A Gbit MRAM demonstration is expected in 2020 with Grandis-Samsung a likely candidate for such demonstration, with great efforts being employed based on their nearly 200 patents on MRAM technologies.

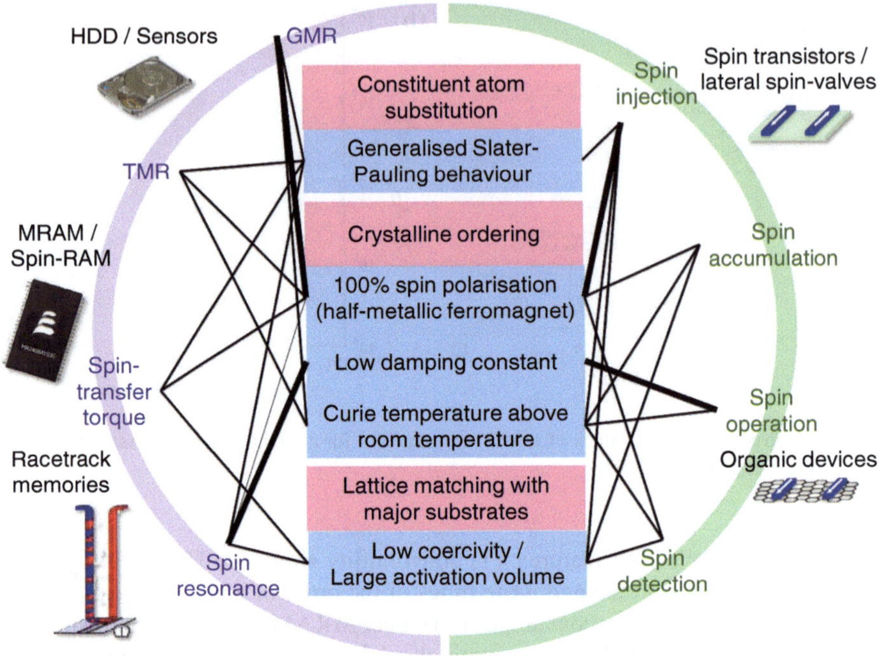

Fig. 9.5 Schematic relationships between spintronic devices and their requirements to use half-metallic Heusler-alloy films

9.1.3 Half-Metallic Heusler-Alloy Ferromagnets

In order to satisfy the above requirements, half-metallic ferromagnets (HMFs) have been rigorously investigated to exploit 100 % spin polarisation induced by spontaneous magnetisation [21]. HMFs possess a bandgap (δ) at the Fermi level (E_F) for only one spin sub-band, thus achieving 100 % spin polarisation at E_F. However, to date there has been no experimental reports demonstrating half-metallicity at RT, crucial for device applications. Among the many theoretically proposed HMFs, Heusler alloys hold the greatest potential to realise half-metallicity at RT due to their high Curie temperatures (above RT), lattice constant matching with major substrates (such as III–V semiconductors and MgO), and generally large δ at E_F [22]. Heusler alloys are categorised into half- and full-Heusler alloys with the form of XYZ and X_2YZ (X and Y atoms: transition metals, and Z: either semiconducting or non-magnetic metal), respectively [22].

The utilisation of Heusler-alloy films as highly-spin-polarised ferromagnetic layers requires several conditions to be met, as listed in Fig. 9.5. First of all, there are a great number of Heusler-alloys, approximately 2500 combinations, which can be formed. Changing the alloy through the substitution of constituent atoms has a significant effect on many of its properties, offering the potential for tuneability of the

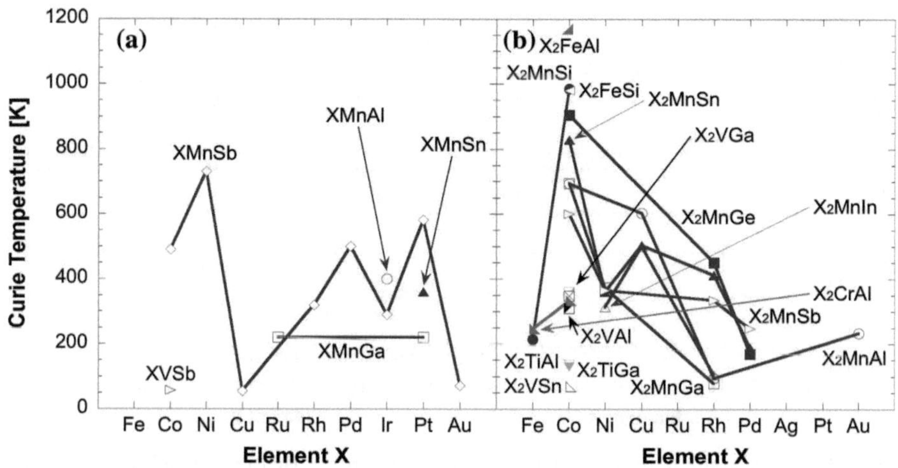

Fig. 9.6 Curie temperature distribution of both **a** half and **b** full Heusler bulk alloys with respect to the element X [22]

spin density of states (DOS). Of the many Heusler alloys that have currently been explored, some show more promise for half-metallicity than others. Block et al. have presented a large negative MR at RT in a quarternary full Heusler $Co_2Cr_{0.6}Fe_{0.4}Al$ alloy due to its high degree of spin polarisation [23]. As a result, a great number of subsequent attempts have been made to utilise this alloy to achieve even larger MR ratios, arising from the half-metallic nature [24–26]. We have previously grown stoichiometric epitaxial $Co_2Cr_{1-x}Fe_xAl$ films directly onto GaAs(001) substrates, for the first time, by using three-source co-evaporation with an ultrahigh vacuum (UHV) molecular beam epitaxy (MBE) technique [25]. We successfully demonstrated $L2_1$ ordering in bulk regions of these films, resulting in 98 % of the saturation magnetisation (M_S) predicted from the generalised Slater-Pauling curve [27]. The minor discrepancy in M_S from the predicted value is attributed to atomically disordered phases at the interfaces and surfaces. The controllability of Heusler alloys and the many potential systems enables researchers to engineer M_S, following the generalised Slater-Pauling curve, as detailed in Chap. 1.3.

Secondly, and as already been eluded to, crystalline ordering is critical to achieve half-metallicity, or very high spin polarisations. Such a high degree of ordering also effects the damping constant, as discussed in Sect. 7.6, as well as the Curie temperature, as shown in Fig. 9.6. The influence of atomic disorder on the half-metallicity of $Co_2Cr_{1-x}Fe_xAl$ full Heusler alloys has been systematically studied by Shirai et al. by using first-principles calculations [28]. In the Co_2CrAl alloys atomic disorder between Cr and Al, which eventually deforms the crystalline structure from $L2_1$ into $B2$ at a disorder level of 50 %, maintains a very high spin polarisation (P) of 97 % for $L2_1$ and even 93 % for $B2$ [29]. The Co-Cr type disorder, however, destroys the half-metallicity rapidly: P falls to zero at a disorder level of 40 %, with the magnetic moment falling to 2.0 μ_B/f.u. at full disorder. These results help to

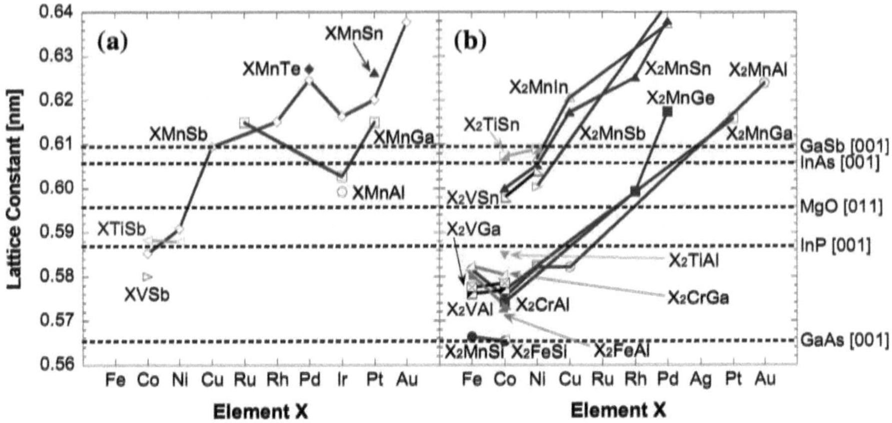

Fig. 9.7 Lattice constant distribution of both **a** half and **b** full Heusler bulk alloys with respect to the element X [22]. Lattice constants of representative III–V semiconductors are also shown as references

explain the decrease in the spin polarisation near interfaces, as atomic disorder and reconstruction occurs. It should be noted that these properties can also be engineered by atomic substitution in the alloy, as shown in Fig. 9.6. As already mentioned, the Curie temperature can be controlled by substituting one of the constituent atoms. For instance, the Curie temperature of X_2CrAl is about 300 K, however, it can be increased by substituting the Cr atoms with Fe atoms (X_2FeAl) to almost 1200 K. Such controllability provides flexibility to design Heusler alloys to match its lattice constant to the under-layer, with the Curie temperature well above RT.

Finally, lattice matching with a substrate or neighbouring layer is crucial to maintain the subsequent structural and magnetic properties (see Fig. 9.7). In addition, the coercivity and activation volume are predominantly dependent upon the lattice matching. Accordingly, the correct combination of substrate, Heusler alloy and non-magnetic spacer is needed to maximise the spin polarisation and corresponding MR ratios. Recently, magnetic tunnel junctions (MTJ) with an epitaxial $L2_1$ Co_2MnSi film have been reported to show very large TMR ratios of 70 % at RT and 159 % at 2 K [29]. These are the largest TMR ratios obtained in a MTJ employing a Heusler alloy film and AlO_x barrier. This is purely induced by the intrinsic spin polarisation of the Heusler electrodes, different from MTJs with an oriented MgO barrier for which TMR ratios of 386 % have been achieved at RT (832 % at 9 K) when using $Co_2FeAl_{0.5}Si_{0.5}$ [30]. Such rapid decrease in the TMR ratio with increasing temperature does not follow the temperature dependence of the magnetisation $T^{3/2}$, suggesting that a small fraction of atomically disordered phases cannot be ignored in the spin-polarised electron transport at finite temperatures. The elimination of such disordered phases, especially near the barrier interface, could improve the TMR ratios further and allow for the realisation of half-metallicity at RT. It should also be noted that a large GMR ratio of 42 % has been reported using a $Co_2FeGe_{0.5}Ga_{0.5}$/Ag/$Co_2FeGe_{0.5}Ga_{0.5}$ junctions [31]. These junctions clearly have the capability of being used in the next generation of read heads.

In this chapter we identify three major issues for the implementation of the Heusler-alloy films into the spintronic devices: crystalline ordering, coercivity and activation volume. A high degree of crystalline ordering has been achieved in films annealed at high temperature (~500 °C) [29–31] but in order to become compatible with Si-based device fabrication processes, annealing temperature between 200 and 250 °C are required [32]. It is therefore essential to establish a crystallisation process with lower annealing temperatures below 250 °C, discussed in Sect. 9.3. Fast magnetisation reversal and a high spin polarisation at the Heusler-alloy/non-magnet interfaces are also required. Fast reversal is crucial for the device operation in a GHz regime and can be achieved by minimising the number of domain-wall pinning centres in the MR nano-pillar of these devices as the magnetisation reversal in Heusler alloys is governed by domain wall motion [46]. Such pinning centres can be estimated by calculating the activation volume, as listed in Sect. 9.4. These two issues also correlate to the lattice matching. In addition, the coercivity needs to be precisely controlled, especially in a magnetically pinned layer, to ensure reproducible magnetisation reversal in a counterpart free layer. This can be achieved by inducing an exchange bias at an interface between the Heusler alloy and neighbouring antiferromagnetic layers, as detailed in Sect. 9.5. Addressing these three issues we propose a promising candidate, with relevant properties, for the future magnetic memories.

9.2 Film Growth

In this chapter, we prepared two sets of Heusler-alloy films, epitaxial and polycrystalline, for the following characterisations.

9.2.1 Epitaxial Films

A 20nm thick film of Co_2FeSi was deposited onto MgO(001) substrates using UHV magnetron sputtering from an alloyed target of optimised stoichiometric composition $Co_{43.7}Fe_{27.95}Si_{28.35}$ at an average rate of 0.03 Å/s [33]. Before deposition of the Co_2FeSi layer, the MgO(001) substrate was annealed at 800 °C for 30 min to clean the substrate surface. A 10 nm MgO buffer layer was then deposited using MBE and annealed for a further 30 min at 500 °C. Post deposition *ex-situ* annealing was carried out for 30 min at temperatures of 500 and 600°C at a pressure of 5×10^{-5} Pa to achieve $L2_1$ ordering, confirmed using X-ray diffraction (XRD) with a Rigaku (Rigaku SmartLab).

9.2.2 Polycrystalline Films

Polycrystalline films were prepared using a high target utilisation sputtering (HiTUS) sputtering system [34] with full control of the mean grain size [35]. The target bias was adjusted to deposit Co_2MnSi and Co_2FeSi films with a series of initial grain sizes. MgO(001) and Si/SiO_2 substrates were cleaned with acetone and isopropanol then annealed at 300 °C for 20 min at a base pressure of 3×10^{-5} Pa. The plasma was generated by an RF field at 3×10^{-1} Pa Ar pressure and steered onto a target with a DC bias (V_B) ranging from −250 to −990 V, controlling the deposition rate. This resulted in the change of initial average grain size. The smaller grains are expected to represent grain-boundary properties, while the larger ones hold bulk characteristics. The Co_2MnSi and Co_2FeSi films were grown onto the substrates, as well as Si_3N_4 electron microscopy grids, at 300 °C, with the thickness maintained at 23 nm with a 2 nm thick Ru capping layer. The chemical compositions of the samples were confirmed by energy dispersive X-ray spectroscopy (EDX). These films were systematically annealed at 300 °C for 3–9 h. Again, the crystalline structures were characterised by glancing-angle XRD.

9.3 Crystallisation Energy

9.3.1 Crystallisation Temperature

To date the half-metallicity has been demonstrated at low temperature [36–39] and in the bulk form [40]. In addition, these films and bulk samples have been annealed typically around 500 and 1000 °C, respectively (see Fig. 9.8). Hence a question remains why the Heusler-alloys do not show the half-metallic behaviour at RT in the film form and whether these alloys can be crystallised at lower annealing temperature to be compatible with the chip and memory processes. In order to answer these questions, the details of the crystallisation processes are required to be observed and analysed, which can be beneficial for the optimisation of the crystallisation in general.

Current models to describe the crystallisation processes have been developed phenomenologically [42]. These models suggest that the crystallisation favours particular axes, which have great potential to reduce the crystallisation energy. Such reduction is essential for the implementation of any functional films into the current chip and memory industry due to their compatibility with the annealing processes typically as high as 200 or 250 °C [32].

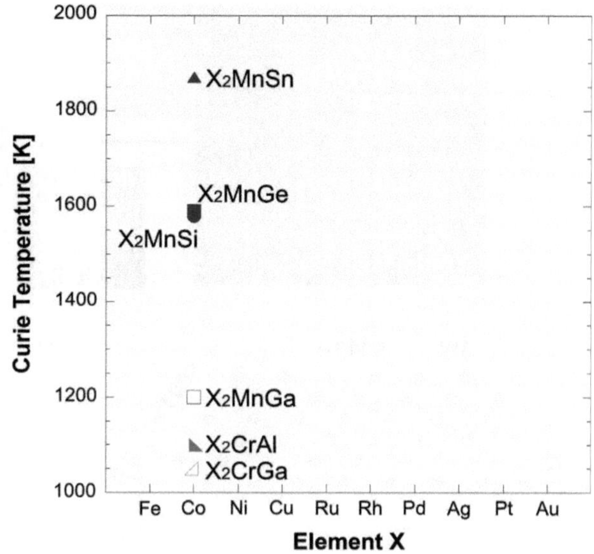

Fig. 9.8 Phase transformation temperatures between the $L2_1$ and $B2$ phases for the full-Heusler alloy bulk [41]

9.3.2 Low-Temperature Crystallisation

We developed a purpose-built double spherical aberration corrected (Cs) (scanning) transmission electron microscope [(S)TEM, JEOL JEM-2200FS] with widened sample space to observe in situ reactions without sacrificing the atomic resolution [43, 44]. A heating stage was used to mount a Si_3N_4 TEM grid to allowing for in-situ annealing to be performed up to 600 °C.

Figure 9.9a shows an initially crystallised Co_2FeSi grain nucleated in Co_2FeSi amorphous-like matrix after annealing at 235 °C for 5 min The grain is orientated with a (110) type surface plane, as can be seen from the digital diffractogram Fig. 9.9d, which is found to be typical for grains annealed for just a few minutes. The atomic structure of the [110] zone axis is illustrated schematically in Fig. 9.9e, calculated for a cubic (*Fm-3m*, $L2_1$ ordering) crystallite with Co_2FeSi unit cells (3 cells along a, b and c axes, i.e., $3 \times 3 \times 3$).

Following nucleation the grains are expected to grow three-dimensionally (3D). However, when the grain height reaches the thickness of the film, i.e., 20 nm that 3rd dimension is constrained. It is thought that a layer-by-layer growth process occurs in these films as suggested in the previous report [46]. Here, those grains with the (110) surface plane parallel to the film surface require the minimum crystallisation energy in the layer-by-layer growth mode. As the crystallites are hexagonal, with the (110) surface facing upwards, the surface facets lie along <111> and <001>. The layer-by-layer growth is therefore expected to occur along these directions.

To explore the growth of the grains in details in situ crystallisation process was observed using both aberration corrected TEM and also selected area electron diffraction. Figure 9.10a shows the grain size evolution with annealing time for five

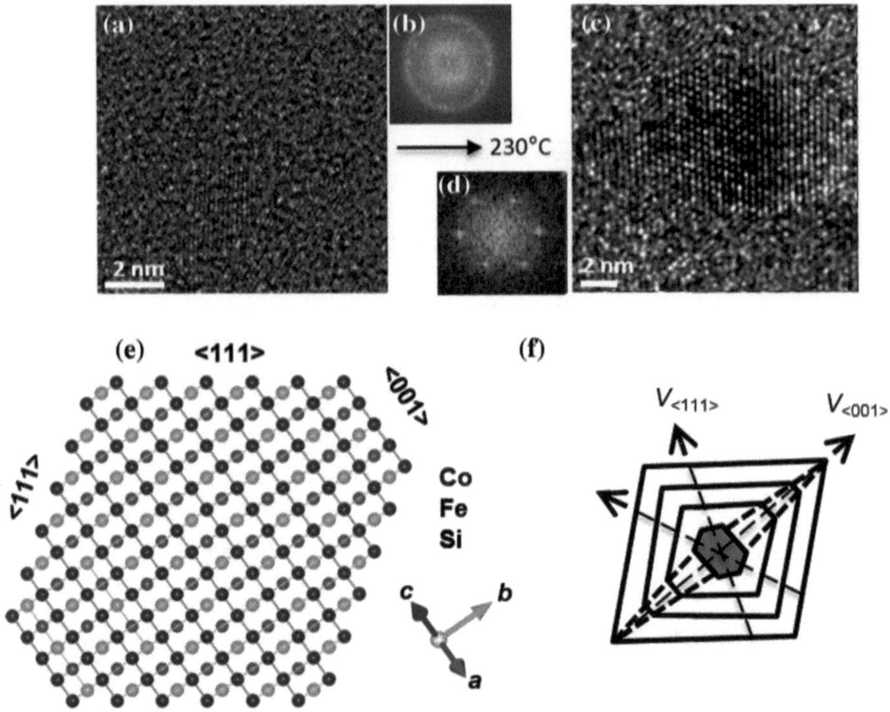

Fig. 9.9 High-resolution (HR)-TEM images of (**a**) the as deposited and (**c**) an initially nucleated Co_2FeSi grain with (**b**) and (**d**) showing the corresponding fast Fourier transformation (FFT). **e** Schematic atomic structure with (111) surface facet, consisting of $6 \times 6 \times 6$ unit cells. **f** Schematic crystallisation process [44]

representative crystallites (D1–D5). The initial crystallisation speed was found to be almost the same for all crystallites, approximately (2.6 ± 0.1) nm/min It should also be noted that all the initial nucleation of crystallites occurs within the first 5 min of annealing, suggesting that the nucleation is governed by an intrinsic parameter of the film.

To investigate the structural ordering of the samples, X-ray diffraction (XRD) scans were used to determine explore the evolution of the (220) peaks, known to be related to $B2$ ordering [47], shown in Fig. 9.10d. It can be seen that the films contain a large degree of, at least $B2$ ordering, which itself has already been shown to lead to a large spin polarisation in Co_2FeSi [48]. It should be empathised, however, that the films may in fact be $L2_1$ ordered but could not easily be determined in these samples due to their grain size.

To explore the magnetic properties of the films, the ex situ annealed samples were measured using an alternating gradient force magnetometer. The evolution of the magnetic moment as a function of time can be seen in Fig. 9.10e. It can be seen that the majority of the moment of the sample arises within 2 h with a steady increase

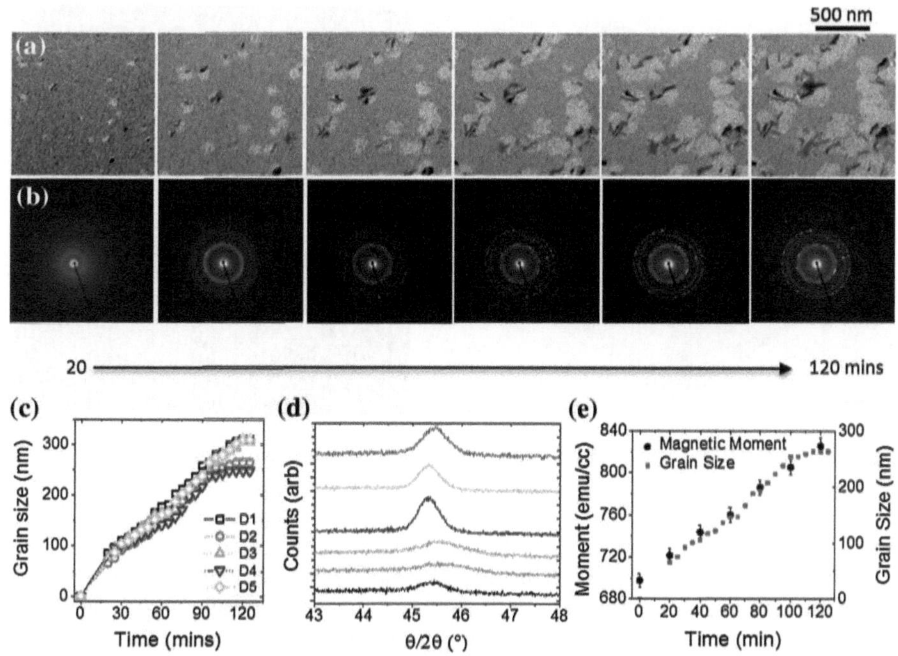

Fig. 9.10 **a** TEM images during the crystallisation from 20 to 120 min with 20 min steps at 235 °C with (**b**) corresponding selected area electron diffraction (SAED) patterns. **c** Grain-size evolution during the crystallisation of 5 representative grains along with (**d**) evolution of (111) and (220) peaks and (**e**) magnetic moment [44]

occurring from the as-deposited state due to the steady increase in grain size. The increase which occurs after 100 min occurs due to a restructuring of the grains, as no further growth occurs. The maximum value corresponds to 72 % of the theoretical value predicted from the generalised Slater-Pauling curve [21] and is similar to those annealed at 500 °C for 6 h [48].

After between 100 min (for D4 and D5) and 120 min (for D1–D3), the crystallisation speed decreases down to 0.5 nm/min on average. As shown in the TEM images in Fig. 9.10a, the crystallites start to merge around this annealing time. This means that the re-aligning of the crystalline axes and planes require almost 5 times more energy than crystallite growth. Since the (110) surface plane is the common plane for the crystallites, the merging process for the crystallites occurs 2-dimensionally (2D) as shown in Fig. 9.11. The stripe patterns seen in Fig. 9.12 suggest that such a grains contain mis-aligned surface planes and stress. Indexing of the fourier transform shown in Fig. 9.11b indicates that grain coalescence occurs via the alignment of {111} planes of the two crystals. In order to facilitate coalensence the crystallites rotate to adjust the orientation of the (110) plane first and then relax the interfacial stress. These merging grains then cover approximately 90 % of the surface of the film, after annealing at 235 °C for 3 h. By dividing the coverage by the density of

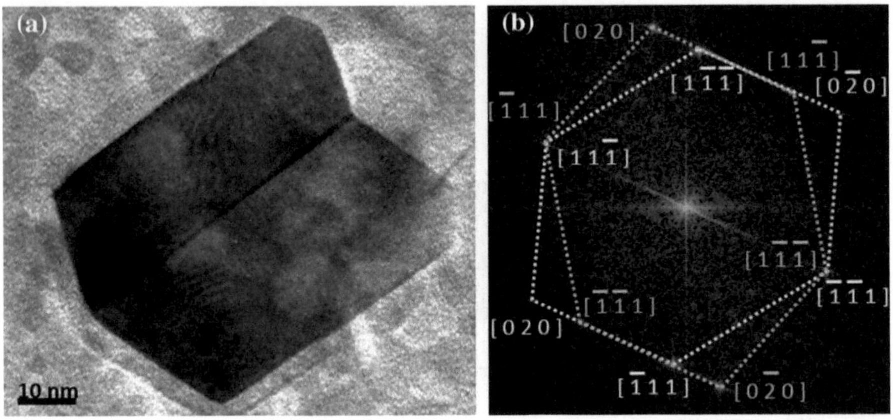

Fig. 9.11 **a** HR-TEM image and **b** corresponding digital difftractogram of Co_2FeSi grains merging along the (110) planes [44]

nucleated crystallites in a HR-TEM image, the average grain size is estimated to be 290 nm in diameter assuming a cylindrical grain shape. This value agrees well with that measured by the grain-size analysis from the films grown on a TEM grid.

Here, what is clear is that through utilising in situ aberration corrected microscopy, X-ray diffraction and alternating gradient force magnetometry we have shown that it is possible to prepare highly spin-polarised Heusler alloys films using annealing temperatures compatible with many current technologies. These samples are therefore ideal candidates for use as spin injectors and/or detectors in spintronic devices.

Although we have shown that these films have desirable structural and magnetic properties through low temperature annealing, we have not yet discussed the details of how this is achieved. It is well known that crystallisation occurs at an interface between a crystallite and matrix, where their chemical potentials differ [49]. The crystallisation speed v can be defined using the difference between the chemical potentials for the interface μ_I and for the matrix μ_M:

$$v = K\left(\mu_I - \mu_M\right), \tag{9.1}$$

where K is a kinetic growth coefficient dependent upon the crystalline orientation and temperature. This crystallisation process leads to an equilibrium shape of a crystal, which minimises the surface free energy density (Wulff's figure).

As shown in Fig. 9.9a, the initial crystallites were found to be hexagonal in the matrix with the (101) zone axis normal to the film surface leading to facets along <111> and <001>. By comparing this schematic crystallite with HR-TEM images we find that, in general, for films annealed at low temperatures the crystallisation speed along <111> $v_{<111>}$ is greater than that along <001> $v_{<001>}$ whilst remaining constant relative to one another thus maintaining four <111> faces to each crystal. For high annealing temperatures, however, the speed along <111> $v_{<111>}$ can

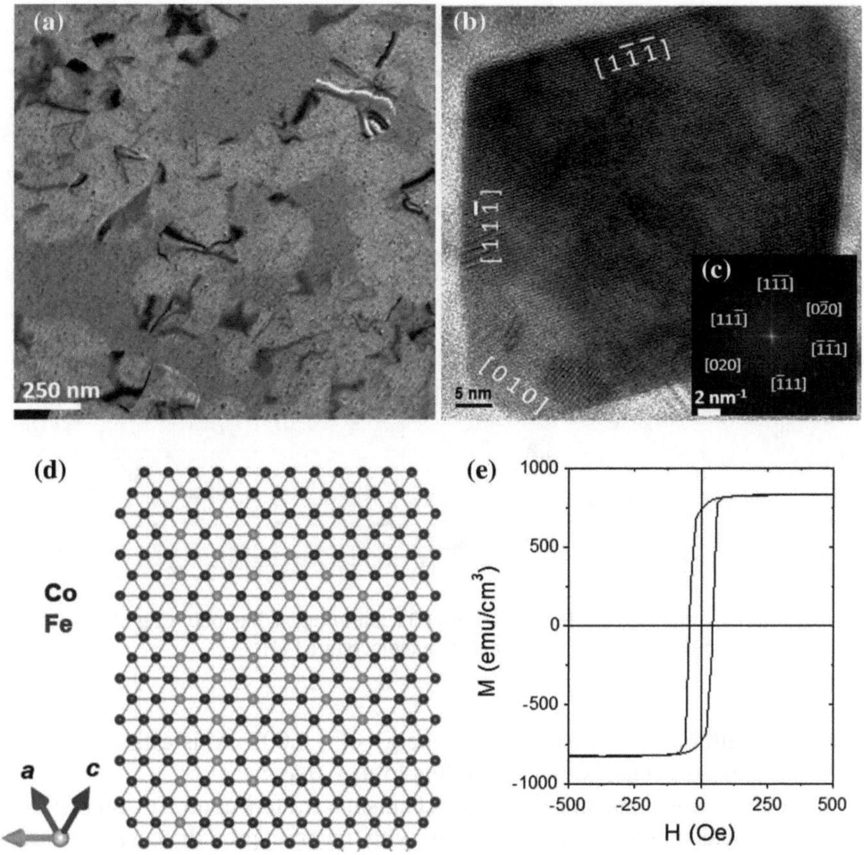

Fig. 9.12 **a** TEM image Co_2FeSi film after 3 h annealing at 235 °C with (**b**) HRTEM image and (**c**) corresponding digital diffractogram of typical grain. **d** schematic of the layer-by-layer grain structure. **e** Hysteresis loop at RT [44]

become much greater, leading to grains whose shape is more rhombic than square [34, 46]. This is particularly significant for the case of high-temperature annealing (500 °C), shown in Figs. 9.13a–f, as excess thermal energy can lead to a 3D grain rotation, causing the (112) plane to face upwards [46].

In our thin-film samples, the surface planes of the two merging grains are perfectly aligned but mirrored along (111) face, as can be seen in Fig. 9.11. This means that the crystallites rotate to adjust the (110) plane first and then relax the interfacial stress. The re-aligning of the crystalline axes and planes therefore requires almost 5 times more energy than crystallite growth as discussed above. The rotation of the crystalline axes therefore consumes approximately 85 % of the thermal energy introduced by annealing, estimated from the change in the crystallisation speed. The films were annealed at 235 °C, which corresponds to the thermal energy of $k_B T \sim 0.044$ eV. Hence ~0.037 eV, 85 % of $k_B T$ at 235 °C, is consumed for the 2D in-plane rotation

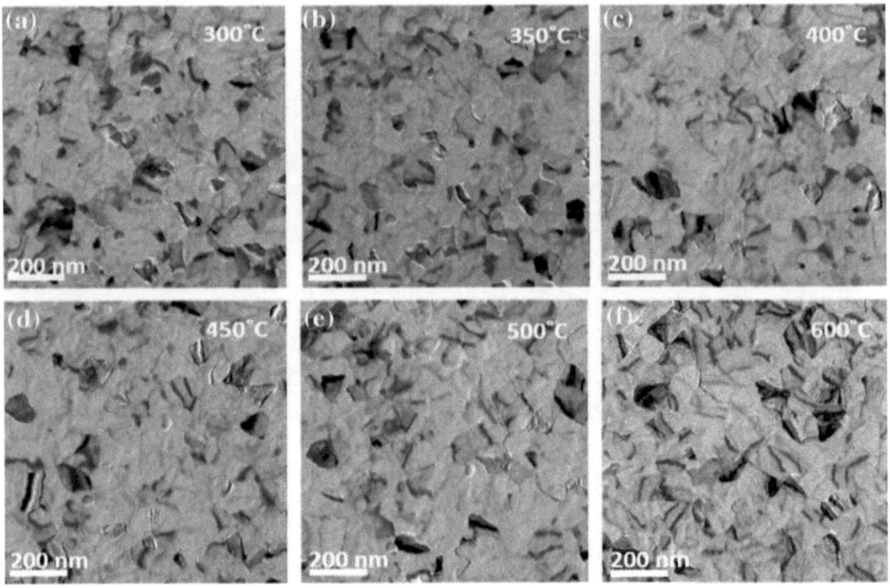

Fig. 9.13 **a–f** Crystallised Co_2FeSi grains after 3 h annealing at 300–600 °C. **g** Magnetic moment and **h** structural order as a function of annealing temperature [44]

of the crystallite axes. The remaining energy of ~0.007 eV is estimated to be the pure crystallisation energy, corresponding to $\mu_I - \mu_M$.

It should be noted that this growth mechanism is in stark contrast with that for bulk samples annealed at higher temperatures, with the grain growth occurring through a 3D crystallisation process [45]. In 3D crystallisation, with fully randomly oriented grains, much larger interfacial area S per grain needs to be considered for $\mu_I - \mu_M$.

$$\int (\mu_I - \mu_M)\, dS. \tag{9.2}$$

This results in the increase in the crystallisation energy by 0.051 eV, which is much larger than that for 2D in-plane rotation. This significant reduction in the crystallisation energy by over 50 % when compared to bulk samples, therefore arises from the fact that the crystallisation in these films occurs through a 2D layer-by-layer growth mode. This leads to a significant reduction of the annealing temperature required from around 830 °C down to just 235 °C. Even for pseudo-2D thin films, with an epitaxial plane orientation of (001), an energy of ~0.055 eV is required to crystallise along different orientations from the favoured <111> axis. This leads to the annealing as high as 500 °C used in the conventional crystallisation process in a film form [37–39].

To explore the crystallisation in more detail we also annealed the films at higher temperatures, up to 600 °C. As already mentioned, it was found that at higher temperatures, above 500 °C, some of the grains rotated, leading to the (112) surface facing upwards. This grain rotation most likely occurs due to the excess amount of thermal energy. This means that the grain shapes are much more randomly distributed for the high-temperature annealed films, as can be seen in Figs. 9.13a–f. This finding also supports our assumption that the excess energy introduced during annealing induces 3D crystallite rotation, resulting in a random orientation of grains with a wide distribution in facets and rough surfaces.

In summary, by restricting the crystallisation processes to 2D, the formation energy is significantly reduced. We show that the crystallisation energy of Co_2FeSi films can be reduced by over 50 % as compared with bulk samples, to allow for highly ordered, highly spin-polarised films to be prepared at annealing temperatures of just 235 °C. This crystallisation process maintains a (110) surface facet, allowing independent crystallites to merge through a 2D in-plane rotation. Such low-temperature annealing with smooth surfaces makes these films ideal candidates for the use in current Si-based chip and memory technology and could allow for the development of many future spintronic materials.

9.4 Grain Volume and Magnetic Properties

9.4.1 Grain Volume and Crystallisation

Figure 9.14 shows the mean grain-volume evolution for a series of V_B, which determines the growth rate of polycrystalline films using a high-target utilisation sputtering system (PlasmaQuest, HiTUS). As can be seen, the grain volume increases with increasing annealing time t_a at 400–500 °C up to 6 h and decreases with t_a. This decrease can be attributed to Si segregation as previously reported [34]. The maximum mean grain size, which is estimated by assuming a cylindrical shape with the film thickness to be the height, is obtained to be 220 nm. This value corresponds to the mean grain volume of 7.6×10^{-22} m^3. This is approximately an order of magnitude larger than the other Co-base Heusler-alloy polycrystalline films, e.g., Co_2MnSi [34]. Hence we focus on Co_2FeSi films throughout this review. It should be noted that the crystalline structures of these grains are found to be almost predominantly $L2_1$ as confirmed by selected area electron diffraction (SAED) patterns in Fig. 9.11, showing that the films consist of the highly ordered grains with some degrees of disordered phases at the grain boundaries and the matrices.

9.4.2 Magnetic Properties with Grain Volume

As discussed in the previous chapter, the coercivity of the Heusler-alloy films provides a measure of their crystallisation. Figure 9.15 shows representative magneti-

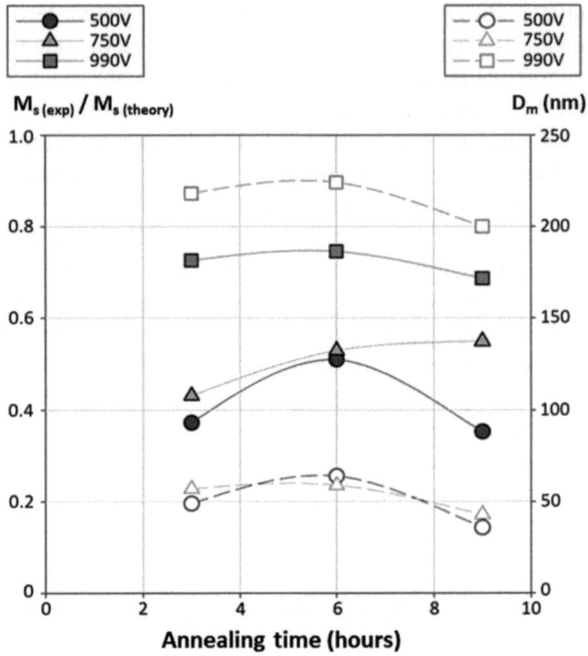

Fig. 9.14 Grain-size distributions and saturation magnetisation of Co_2FeSi polycrystalline films sputtered at various bias voltages V_B

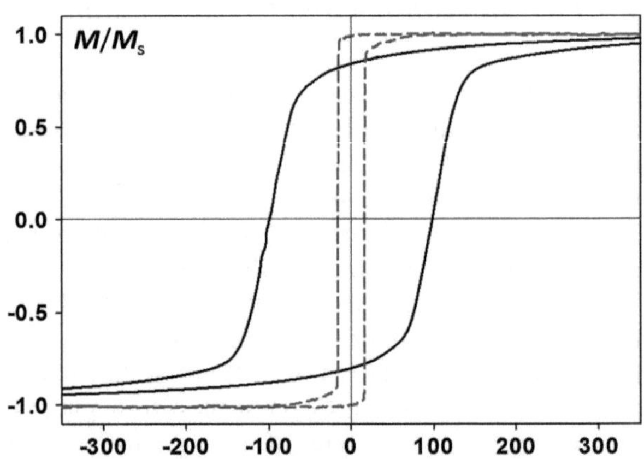

Fig. 9.15 Magnetisation curves of epitaxial (*dashed line*) and polycrystalline (*solid line*) Co_2FeSi film after annealing at 500 °C for 3 h

sation curves of both epitaxial and polycrystalline Co_2FeSi films after annealing at 500 °C for 3 h. The epitaxial film shows strong cubic anisotropy with the easy axis along [100] and the hard axis along [110]. The coercivity H_c of the film shows the largest value of ~10 Oe along the easy axis and the minimum (~5 Oe) along the diagonal direction between the easy and hard axes. Since this film is confirmed to form the $L2_1$ ordering by XRD [33], this anisotropy is dominated by the crystalline anisotropy as detailed in Sect. 3.1. This indicates that the magnetisation reversal is controlled by such strong anisotropy.

To the contrary, the polycrystalline shows almost isotropic behaviour as the ordered grains are orientated randomly in the two dimensions of the film giving values of H_c between 50 and 500 Oe dependent upon the crystallite size in the film. This grain-evolutional behaviour agrees with the corresponding magnetic properties, i.e., saturation magnetisation M_S, as shown in Fig. 9.14. The maximum M_S obtained is 760 emu/cm^3, which is almost 75 % of the theoretically predicted value from the generalised Slater-Pauling curve [21]. This indicates that the grain-size increase results in the increase in M_S, proving the bulk-like regions in the Heusler-alloy films possess highly crystalline ordering as stated above.

9.4.3 Exchange Bias

The ability to create highly ordered grains is not unique to growing on conventional substrates but can also be achieved when depositing onto other suitable layers. A high degree of ordering has been shown for a NiCr/IrMn/(Mn)/Co_2FeSi/Ta stack grown onto a thermally oxidised Si substrate [50]. As shown by the cross-sectional HR-TEM image in Fig. 9.19, the IrMn layer is highly ordered along the (111) direction, as previously reported [51]. This ordering is found to provide a template to crystallise the above Heusler-alloy layer, clearly visible when analysing the SAED patterns in Fig. 9.16. The Co_2FeSi films are predominantly found to form the $L2_1$ ordered phase after annealing at 500 °C for 10 min based on their SAED patterns inserted in Fig. 9.19. This annealing condition also provides an exchange bias of ~400 Oe, which could be utilised in many device applications. Furthermore, this system is by no means optimised and could be improved through greater control of the interfacial atomic ordering. These findings provide a way to improve the structural properties of polycrystalline Heusler alloy films to achieve half-metallicity at RT.

The effects of annealing were also investigated, shown in Fig. 9.17. It can be seen that low temperature annealing does not affect the magnetic properties of the Heusler-alloy layers, despite the probably diffusion of Mn, as reported previously [51]. Even so, these Heusler-alloy layers maintain ~80 % of the bulk M_S. H_c, on the other hand, shows large distributions except for layers grown at $V_B = 400$ V. This is because in order for exchange bias to be observed IrMn grains above 7 nm are required, something which can only be achieved when using $V_B > 400$ V [50]. As shown in Fig. 9.17a, H_{ex} is a minimum when the Heusler alloy is deposited at $V_B = 400$ V, reaching a maximum of 240 Oe when the deposition is carried out at $V_B = 800$ V. This suggest that the Heusler alloy grain size is well matched to that of the IrMn, as can be seen in the insert in Fig. 9.16. This optimises both the grain size distribution of both films and also maximises the interfacial exchange coupling between the two layers. It has been shown that by changing the thickness of the both the Heusler alloy and IrMn layers that the grain volume distribution can also be

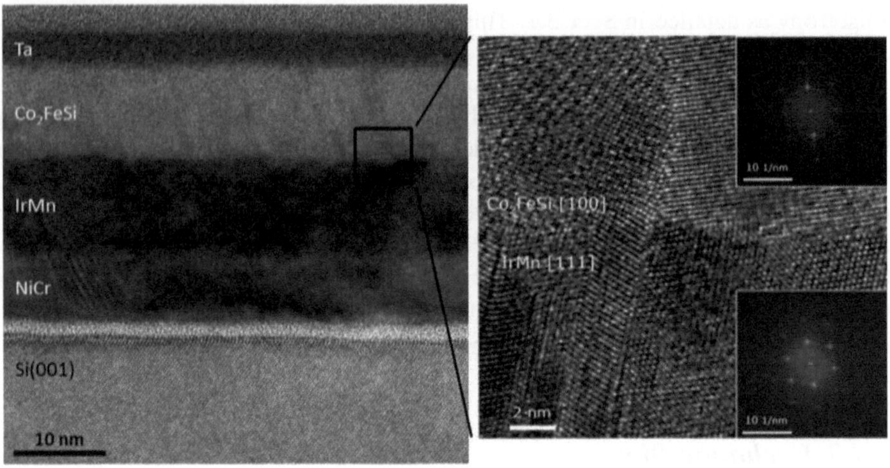

Fig. 9.16 Cross-sectional TEM images of a polycrystalline NiCr/IrMn/(Mn)/Co_2FeSi/Ta stack with the blow-up showing the IrMn/Co_2FeSi interface

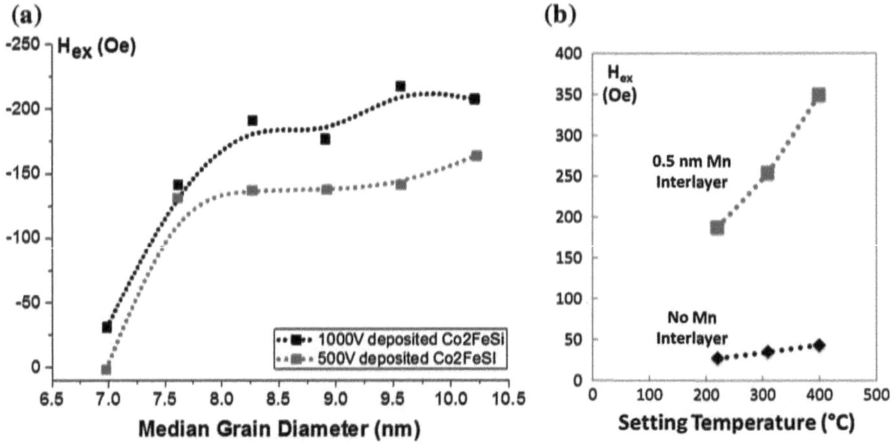

Fig. 9.17 **a** Median grain diameter D_m dependences of exchange bias H_{ex} of the Ir_3Mn/Co_2FeSi stacks after annealing at 310 °C for 30 min under a magnetic field of 1 kOe. **b** the effect of an interfacial Mn dusting layer on H_{ex}

optimised to further improve upon this value of H_{ex} with the a maximum occurring at film thickness of 8.5 nm for the IrMn and 7.6 nm for the Heusler alloy [50].

In summary, saturation magnetisation, coercivity, exchange bias and the resulting magnetoresistance is predominantly controlled by the grain size. In order to maximise these properties, the larger grains or ideally epitaxial films are required. Namely, the exchange bias has been further increased up to 350 Oe by inserting 0.5 nm thick Mn layer at the interface which helps to reduce the effects of Mn diffusion Fig. 9.17b [51]. These results show that Heusler alloys can be utilised to create exchange biased junctions offering tuneable characteristics which make of great interest for many future device applications.

9.5 Activation Volume

9.5.1 *Definition*

The activation volume is defined as the smallest volume which reverses in a single step during magnetisation reversal. In an elongated particle, this corresponds to the sub-volume of the particle that reverses coherently during incoherent reversal. In a thin film this can reveal cooperative reversal by exchange-coupled grains and regions, so the study of activation volumes is of great interest, particularly for magnetic storage media [52]. In a system where the magnetisation reversal is dominated by domain-wall motion, the activation volume is the volume swept out by the wall in a single step during the reversal process. This is analogous to the volume involved in a single Baukhausen jump.

In order to estimate the activation volume, both magnetisation and DC demagnetised (DCD) remanence curves were measured using an alternating gradient force magnetometer (AGFM, Princeton Measurement Micromag Model 2900) [53]. This system has a noise level better than 2×10^{-8} emu, allowing for time-dependent measurements to be taken. The samples were initially saturated, followed by time-decay measurements of their magnetisation, between positive and negative saturation, with incremental field steps as schematically shown in Fig. 9.18. The time decay was observed to be non-linear in ln$t$ for all the samples and fields (H), where t is the time over which the magnetisation (M) is allowed to decay. From these measurements, a magnetic viscosity (S) can be determined by [54]

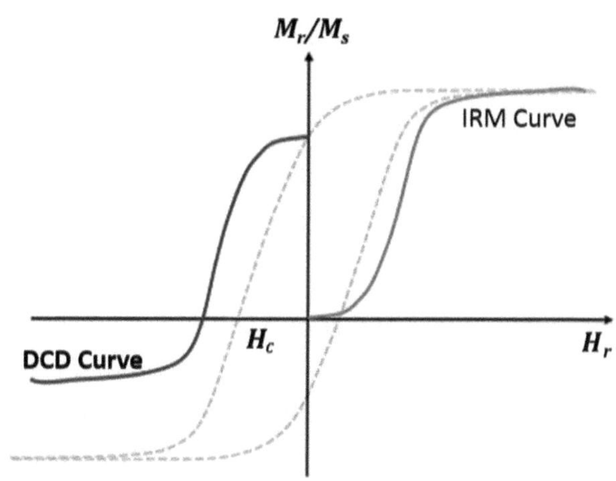

Fig. 9.18 Schematic diagram of DC demagnetised (DCD) remanence and isothermal remanence (IRM) curves with reference to a hysteresis curve

$$dM(H) = d\ln t = S_0 + S_1(H)\ln t + \{S_2(H)\ln t\}^2 + \cdots .$$

For a film with distributed energy barriers for magnetisation reversal, S has a magnetic field dependence leading to a distribution of values [55]:

$$S(H) = 2M_S k_B T f\,[\Delta E(H)],$$

where $\Delta \mathrm{E}(H)$ represents the energy barrier distribution at a magnetic field H. $S(H)$ becomes minimum at the coercivity H_c.

From the DCD remanence curves (see Fig. 9.14), the irreversible susceptibility (χ_{irr}) is given by

$$\chi_{irr}(H) = d\,\{\mathrm{DCD}\,(H)\}/dH.$$

From the magnetic viscosity S and irreversible susceptibility χ_{irr}, a fluctuation field (H_f) can be deduced:

$$H_f = S(H)/\chi_{irr}(H).$$

The activation volume (V_{act}) is finally given by [54]

$$V_{act} = k_B T / M_S H_f,$$

where T is the temperature at which the measurements were carried out, M_S the saturation magnetisation of the sample and k_B the Boltzmann constant. In our analysis, the activation volume is modelled to consist of a cylinder, with a circular surface activation diameter (D_{act}), with the height assumed to be the same as the film thickness (t).

An example of a time-dependent measurement for a Co_2FeSi polycrystalline film, deposited at $V_B = -800$ V followed by post-annealing at 500 °C for 6 h, is shown in Fig. 9.19a, measured with 20 Oe steps. Here, the magnetisation decay was measured for 600 s for each value of magnetic field providing 150 data points for each scan. By taking the differential of the linear time dependence measurements in lnt values of $S(H)$ can be found following equation (number). By also taking the differential of the DCD curve for the same sample curves for S(H) and $\chi_{irr}(H)$ can be found. The peak values of each of these curves are then used to find the value for H_f and subsequently D_{act}. It must be noted however that if the time dependence measurements are not linear in ln$t$ then the quadratic term in series expansion of dM(H) must be considered and as a result determination of H_f become non-trivial.

9.5.2 Epitaxial Films

The structure of Co_2FeSi films, grown on MgO(001) substrates, were observed by HR-TEM (JEOL JEM-2011) and double aberration-corrected (S)TEM (JEOL JEM-2200FS). Cross-section TEM samples were prepared by mechanical grinding

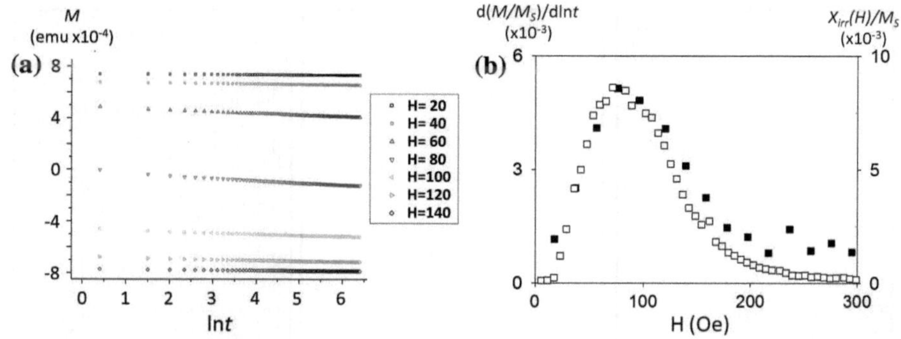

Fig. 9.19 **a** Time-dependence curves and **b** the corresponding magnetic viscosities (*closed squares*) and irreversible susceptibilities (*open squares*) for the Co_2FeSi polycrystalline film ($V_B = -800$ V and annealed at 500 °C for 6 h)

and Ar-ion thinning to electron transparency using the method described in [56]. The substrate/film epitaxial relationship was found to be MgO(001)//Co_2FeSi(001) and MgO(100)//Co_2FeSi(110) which corresponds to a mismatch of approximately 4.5 %. Dark-field images, obtained using the (202) diffraction spots, highlights strain within the Co_2FeSi film up to 5 nm from the interface with the MgO buffer layer (see Fig. 9.20). To explore further the effect of strain, HAADF images were analysed using Geometric Phase Analysis (GPA) [57] using a Digital Micrograph-script [58]. The strain analysis shows the lattice constant of Co_2FeSi is −4.4 % smaller than the MgO buffer layer, which is in a good agreement with the expected value for fully relaxed films based on the bulk lattice constants. This lattice mismatch is compensated by misfit dislocations, as shown in the superimposed strain map in Fig. 9.20c. The periodicity of dislocations, the missing of a single atomic plane of MgO, occurs every 20–25 planes of Co_2FeSi, found using both GPA and analysis of HAADF STEM images.

The activation diameter D_{act} for such films, estimated from the DCD magnetisation measurements, can be seen in Fig. 9.21a. Whilst V_{act} is not an absolute measurement to determine microscopic magnetic structures, the fact that it is constant with grain size implies that there is an array of regularly spaced pins, such as the defect array seen in Fig. 9.16. This implies that the nucleation field (H_n) occurs as a single grain event, related to the grain diameter, whereas domain wall pinning is an interface effect related to the spacing of misfit dislocations (D_{defect}). The number of these misfit dislocations can be minimised by selecting a buffer layer which matches the Heusler alloy lattice spacing. It should be noted that the spacing of the misfit dislocations D_{defect} is on the same order as the grain size estimated from the XRD profiles by Scherrer's equation (D_{XRD}) [59]. It should be noted that a full elimination of dislocations is not currently possible. Hence, epitaxial films are liable to suffer from potentially small activation volumes, preventing fast magnetisation switching.

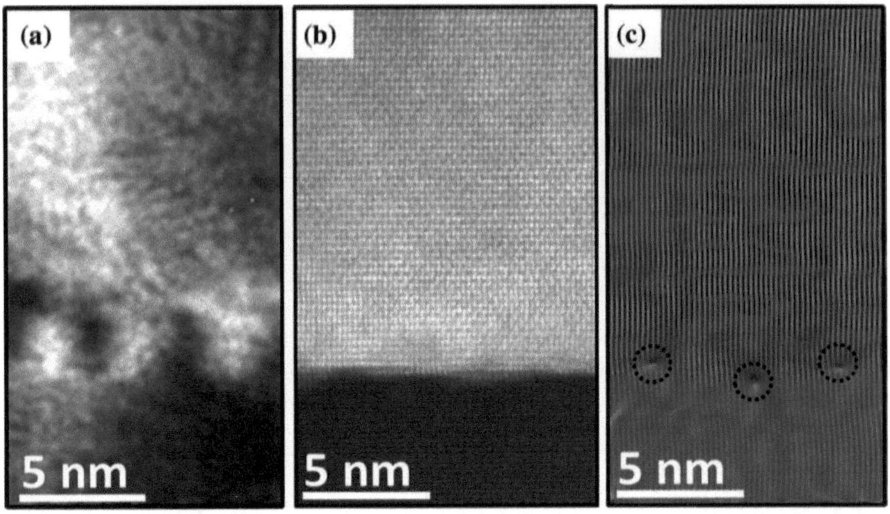

Fig. 9.20 **a** Dark-field weak beam TEM image of the Co_2FeSi/MgO interface showing periodic contrast variation due to stress. **b** High resolution HAADF STEM image of the Co_2FeSi/MgO interface. **c** HAADF STEM image masked to show only (010) planes superimposed with GPA image highlighting the misfit dislocations at the interface

9.5.3 Polycrystalline Films

The activation diameters D_{act} were similarly estimated for polycrystalline Co_2FeSi films and compared with the grain diameters measured from films grown onto Si_3N_4 TEM grids using TEM (JEOL JEM-2011). In general, in polycrystalline sputtered films the grain size follows a log-normal distribution [60]. In order to measure the grain-diameter distribution accurately, a circular light with a variable diameter (Carl Zeiss Particle Analyser) has been used to measure a grain diameter D_m, providing the equivalent area of the grain, for between 600–1000 grains per sample.

As shown in Fig. 9.21b, the activation diameter D_{act} is estimated to be $\sim$40 nm and is almost independent of the sputtering bias voltage. This means that the maximum activation volume measured (5.7×10^{-23} m^3) represents the intrinsic magnetic volumes for the Heusler alloy and fits well within the structural volume as measured in Sect. 9.3 (7.6×10^{-22} m^3). The measured activation volume is more than one order of magnitude larger than that for an Fe particle for instance [61]. This finding proves the advantage for the Heusler alloys in high-frequency operation, including a low damping constant [62].

The grain diameters D_{XRD}, on the other hand, are maximum for the largest bias voltage. It should be noted that the sample grown at the bias voltage of 250 V initially has $D_{act} \sim D_{XRD}$. For these films, once a reverse field is applied, a domain nucleates in each grain and then propagates through the grain at high speed, allowing for a sharp reversal with a low coercivity. In contrast, the films grown at -750 V show a

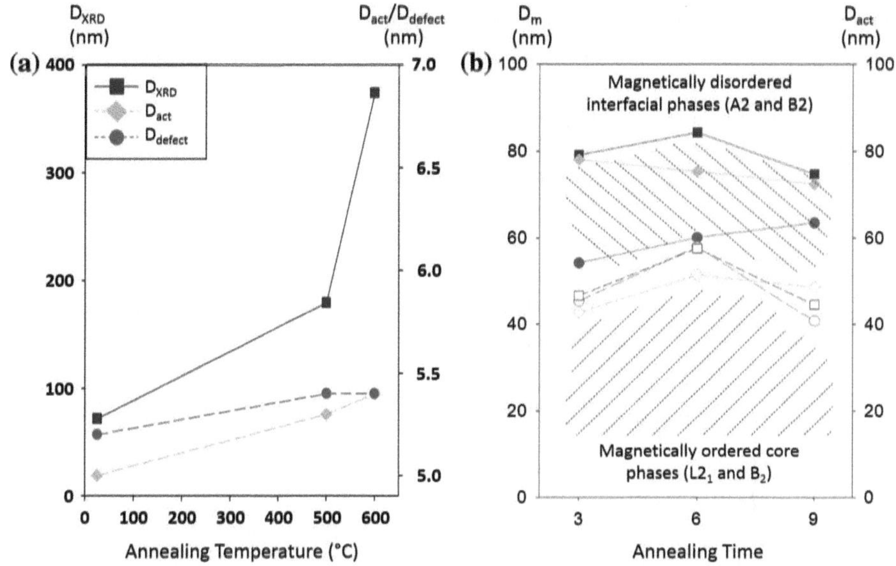

Fig. 9.21 (a) D_{XRD} (*squares*), D_{act} (*rhombuses*) and D_{defect} (*circles*) displayed as a function of annealing temperature for epitaxial Co_2FeSi films. **b** D_m (*closed symbols*) and D_{act} (*open symbols*) displayed as a function of annealing time for polycrystalline Co_2FeSi films grown with bias voltages V_B of −250 (*circles*), −500 (*rhombus*) and −750 V (*squares*)

multistage reversal process so once a reverse field is applied, a domain nucleates in the ordered centre of the grain then sweeps through the rest of the grain.

In summary, these results unambiguously indicate that the polycrystalline films have great advantages over epitaxial films for implementation into GMR or TMR nano-pillar junctions for next-generation HDD read heads or STT-RAM. For device implementation further improvements can be made through elimination of additional domain-wall pinning sites, such as those arising from edge-roughness.

9.6 Concluding Remarks

In this chapter, we discuss two types of volume to characterise structural and magnetic properties of these films; (i) grain and (ii) activation volumes as summarised in Table 9.1. The grain volume is an averaged physical size of the polycrystalline grains, which determines the magnetic behaviour induced at their interfaces against their neighbouring layers. For the Heusler-alloy junctions, such behaviour includes the MR effects against an insulator and non-magnet as well as exchange bias against a antiferromagnet. The activation volume is an averaged magnetic size, which can be reversed at once during a magnetisation reversal process. For the Heusler-alloy

Table 9.1 Comparison between the mean grain volume and activation volume

	Structural volume (mean grain volume)	Magnetic volume (activation volume)
Corresponding structural properties	Crystallisation speed	
	Crystallisation process	N/A
	Atomic segregation	
	Surface/interface area	
Induced magnetic phenomena	Saturation magnetisation	Magnetisation reversal
	GMR/TMR	Number of pinning site
	Exchange bias	Magnetic dynamics
	Spin injection	Damping constant
	Interfacial spin polarisation	Domain-wall pinning

junctions, the activation volume corresponds to magnetic domain size surrounded by domain walls and determines their magnetic dynamics.

As is commonly understood, epitaxial films have great advantages over polycrystalline films for achieving $L2_1$ ordering with abrupt and smooth interfaces. However, due to the lattice mismatch, epitaxial films always suffer from small activation volumes induced by misfit dislocations at the Heusler-alloy/(buffer)/substrate interface. Polycrystalline films, on the other hand, reveal their activation diameter, which defines the average diameter of a cylindrical volume with domain-wall-free magnetisation switching, to be much larger, on the order of 40 nm. By controlling the sputtering conditions, the grain size of polycrystalline films can be tuned to be of the same size as the activation diameter, thus allowing for extremely fast switching. This is ideal if Heusler alloys are to be utilised in MR nano-pillars in the next-generation HDD read heads and the STT-MRAM cells.

Acknowledgments The authors would like to thank Professor Kevin O'Grady for his support on the magnetic measurements and for fruitful discussion. We are also grateful for Professors Hiraku Endo and Tadachika Nakayama for their support on the film growth. This work was partially supported by EPSRC (EP/H026126/1 and EP/K03278X/1), Royal Society Research Grant and European Commission (NMP3-SL-2013-604398).

References

1. T. Dietl, D.D. Awschalom, M. Kaminska, H. Ohno (eds.), *Spintronics* (Elsevier, Amsterdam, 2008)
2. S.M. Thompson, J. Phys. D Appl. Phys. **41**, 093001 (2008)
3. S.S.P. Parkin, M. Hayashi, L. Thomas, Science **320**, 190 (2008)
4. R.A. de Groot, F.M. Mueller, P.G. van Engen, K.H.J. Buschow, Phys. Rev. Lett. **50**, 2024 (1983)
5. R. Fiederling, M. Keim, G. Reuscher, W. Ossau, G. Schmidt, A. Waag, L.W. Molenkamp, Nature **402**, 787 (1999); Y. Ohno et al. ibid. **402**, 790 (1999)

6. G. Schmidt, D. Ferrand, L.W. MolenKamp, A.T. Filip, B.J. van Wees, Phys. Rev. B **62**, R4790 (2000)
7. E.I. Rashba, Phys. Rev. B **62**, 16267 (2000)
8. S.A. Crooker, M. Furis, X. Lou, C. Adelmann, D.L. Smith, C.J. Palmstrøm, P.A. Crowell, Science **309**, 2191 (2005)
9. M. Johnson, Science **260**, 324 (1993)
10. H. Ohno, D. Chiba, F. Matsukura, T. Omiya, E. Abe, T. Deitl, Y. Ohno, K. Ohtani, Nature **408**, 944 (2000)
11. D. Chiba, S. Fukami, K. Shimamura, N. Ishiwata, K. Kobayashi, T. Ono, Nat. Mater. **10**, 853 (2011)
12. A. Hirohata, Y.B. Xu, C.M. Guertler, J.A.C. Bland, S.N. Holmes, Phys. Rev. B **63**, 104425 (2001); ibid. **66**, 035330 (2002)
13. K. Uchida, S. Takahashi, K. Harii, J. Ieda, W. Koshibae, K. Ando, S. Maekawa, E. Saitoh, Nature **455**, 778 (2008)
14. http://www.toshiba.co.jp
15. A. Moser, K. Takano, D.T. Margulies, M. Albrecht, Y. Sonobe, Y. Ikeda, S. Sun, E.E. Fullerton, J. Phys. D Appl. Phys. **35**, R157 (2002)
16. H. Katayama, S. Sawamura, Y. Ogimoto, J. Nakajima, K. Kojima, K. Ohta, J. Magn. Soc. Jpn. **23**(S1), 233 (1999)
17. M. Takagishi, K. Yamada, H. Iwasaki, H.N. Fuke, S. Hashimoto, IEEE Trans. Magn. **46**, 2086 (2010)
18. M. Oogane, T. Miyazaki, Magnetic random access memory. in *Epitaxial Ferromagnetic Films and Spintronic Applications*, ed. by A. Hirohata, Y. Otani (Research Signpost, Trivandrum, 2009), pp. 335–361
19. J. de Boeck, G. Borghs, Phys. World **12**(4), 27 (1999)
20. H. Tanizaki, T. Tsuji, J. Otani, Y. Yamaguchi, Y. Murai, H. Furuta, S. Ueno, T. Oishi, M. Hayashikoshi, H. Hidaka, Asian Solid-State Circuits Conf. Dig. Tech. Pap. 303 (2006)
21. I. Galanakis, P.H. Dederichs (eds.), *Half-Metallic Alloys* (Springer, Berlin, 2005)
22. A. Hirohata, M. Kikuchi, N. Tezuka, K. Inomata, J.S. Claydon, Y.B. Xu, G. van der Laan, Curr. Opin. Solid State Mater. Sci. **10**, 93–107 (2006)
23. T. Block, C. Felser, G. Jakob, J. Ensling, B. Mühling, P. Gütlich, R.J. Cava, J. Solid State Chem. **176**, 646 (2003)
24. K. Inomata, S. Okamura, R. Goto, N. Tezuka, Jpn. J. Appl. Phys. **42**, L419 (2003)
25. A. Hirohata, H. Kurebayashi, S. Okamura, M. Kikuchi, T. Masaki, T. Nozaki, N. Tezuka, K. Inomata, J. Appl. Phys. **97**, 103714 (2005)
26. A. Hirohata, H. Kurebayashi, S. Okamura, T. Masaki, T. Nozaki, M. Kikuchi, N. Tezuka, K. Inomata, J.S. Claydon, Y.B. Xu, J. Appl. Phys. **97**, 10C308 (2005)
27. I. Galanakis, P.H. Dederichs, N. Papanikolaou, Phys. Rev. B **66**, 174429 (2002)
28. Y. Miura, K. Nagao, M. Shirai, Phys. Rev. B **69**, 144413 (2004)
29. Y. Sakuraba, J. Nakata, M. Oogane, H. Kubota, Y. Ando, A. Sakuma, T. Miyazaki, Jpn. J. Appl. Phys. **44**, L1100 (2005)
30. N. Tezuka, N. Ikeda, F. Mitsuhashi, S. Sugimoto, Appl. Phys. Lett. **94**, 162504 (2009)
31. Y.K. Takahashi, A. Srinivasan, B. Varaprasad, A. Rajanikanth, N. Hase, T.M. Nakatani, S. Kasai, T. Furubayashi, K. Hono, Appl. Phys. Lett **98**, 152501 (2011)
32. K. Suzuki, S. Matsui, Y. Ochiai, *Sub-Half-Micron Lithography for ULSIs* (Cambridge University Press, Cambridge, 2000)
33. J. Sagar, H. Sukegawa, L. Lari, V.K. Lazarov, S. Mitani, K. O'Grady, A. Hirohata, Appl. Phys. Lett. **101**, 102410 (2012)
34. A. Hirohata, S. Ladak, N.P. Aley, G.B. Hix, Appl. Phys. Lett. **95**, 252506 (2009)
35. M. Vopsaroiu, M.J. Thwaites, S. Rand, P.J. Grundy, K. O'Grady, IEEE Trans. Magn. **40**, 2443 (2004)
36. Y. Sakuraba, M. Ueda, Y. Miura, K. Sato, S. Bosu, K. Saito, M. Shirai, T.J. Konno, K. Takanashi, Appl. Phys. Lett. **101**, 252408 (2012)
37. J. Sato, M. Oogane, H. Naganuma, Y. Ando, Appl. Phys. Express **4**, 113005 (2011)

38. H. Sukegawa, Z. Wen, K. Kondou, S. Kasai, S. Mitani, K. Inomata, Appl. Phys. Lett. **100**, 182403 (2012)
39. H.-X. Liu, Y. Honda, T. Taira, K.-I. Matsuda, M. Arita, T. Uemura, M. Yamamoto, Appl. Phys. Lett. **101**, 132418 (2012)
40. P. Klaer, M. Kallmayer, H.-J. Elmers, L. Basit, J. Thöne, S. Chadov, C. Felser, J. Phys. D Appl. Phys. **42**, 084001 (2009)
41. R. Y. Umetsu, A. Okubo, M. Nagasako, M. Ohtsuka, R. Kainuma, K. Ishida, in *Spin* **4**, 1440018 (2014)
42. I.V. Markov, *Crystal Growth for Beginners*, 2nd edn. (World Scientific, Singapore, 2002)
43. P.L. Gai, E.D. Boyes, Microsc. Res. Tech. **72**, 153–164 (2009)
44. P.L. Gai, E.D. Boyes, *Handbook on Nanoscopy*, ed. by G. van Tendeloo, D. van Dyke, S. Pennycook (Wiley, New York, 2012), pp. 375-404
45. J. Sagar, L.R. Fleet, M.Walsh, L. Lari, E.D. Boyes, O. Whear, T. Huminiuc, A. Vick, A. Hirohata, Appl. Phys. Lett. **105**, 032401 (2014)
46. L.R. Fleet, G. Cheglakov, K. Yoshida, V.K. Lazarov, T. Nakayama, A. Hirohata, J. Phys. D. Appl. Phys. **45**, 032001(FTC) (2012)
47. P.J. Webster, K.R.A. Ziebeck, J. Phys. Chem. Solids **34**, 1647 (1973)
48. P. Bruski, S.C. Erwin, M. Ramsteiner, O. Brandt, K.-J. Friedland, R. Farshchi, J. Herfort, H. Riechert, Phys. Rev. B **83**, 140409(R) (2011)
49. I. V. Markov, *Crystal Growth for Beginners: Fundamentals of Nucleation, Crystal Growth and Epitaxy*, 2nd edn. (World Scientific, Singapore, 2003)
50. H. Endo, A. Hirohata, T. Nakayama, K. O'Grady, J. Phys. D. Appl. Phys. **44**, 145003 (2011)
51. N.P. Aley, R. Kroeger, B. Lafferty, J. Agnew, Y. Lu, K. O'Grady, IEEE Trans. Magn. **45**, 3869 (2009)
52. M. El-Hilo, K. O'Grady, R.W. Chantrell, J. Magn. Magn. Mater. **120**, 244 (1993)
53. J. Sagar, L.R. Fleet, A. Hirohata, K. O'Grady, IEEE Trans. Magn. **47**, 2440 (2011)
54. E.P. Wohlfarth, J. Phys. F Met. Phys. **14**, 155 (1984)
55. P. Gaunt, J. Appl. Phys. **59**, 4129 (1986)
56. L. Lari, S. Lea, C. Feeser, B.W. Wessels, V.K. Lazarov, J. Appl. Phys. **111**, 07C311 (2012)
57. M. Hytch, F. Snoeck, R. Kilaas, Ultramicroscopy **74**, 131 (1998)
58. http://elim.physik.uni-ulm.de/?page_id=564
59. B.D. Cullity, S.R. Stock, *Elements of X-Ray Diffraction*, 3rd edn. (Prentice Hall, Upper Saddle River, 2001)
60. K. O'Grady, L. Fernandex-Outon, G. Vallejo-Fernandez, J. Magn. Magn. Mater. **322**, 883 (2010)
61. R.L. Stamps, A. Stollo, M. Madami, S. Tacchi, C. Carlotti, G. Gubbiotti, M. Fabrizioli, J. Fujii, Phys. Rev. B **74**, 134401 (2006)
62. T. Kubota, S. Tsunegi, M. Oogane, S. Mizukami, T. Miyazaki, H. Naganuma, Y. Ando, Appl. Phys. Lett. **94**, 122504 (2009)

Chapter 10
Thermoeletric Heusler Compounds

Julia Krez and Benjamin Balke

Abstract Thermoelectric converters for power generation aim at reducing CO_2 emission via the conversion of a part of the low-grade waste heat generated by engines, industrial furnaces, gas pipes, etc. to electricity. The recovery of waste heat from the exhaust of an automotive engine, in particular, is an attractive, albeit not very efficient way for reduction of fuel consumption. Thermoelectric converters with high overall efficiency convert heat directly into electricity without moving parts and, thus, not only decrease our reliance on fossil fuels but also actively counteract global warming. State-of-the-art converters are simply too inefficient to be economic, partly due to expensive elementary constituents (Te, Ge, etc.). On this background, Heusler compounds with $C1_b$ structure stand out on account of their relatively low cost components and have been extensively studied as potential thermoelectric materials for high temperature power generation up to 1000 K during the last years.

10.1 Introduction

The search for alternative energy technologies has taken an accelerated pace in recent years as climate change has become more noticeable and the use of nuclear energy introduces political controversy for many countries. The quest for sustainable energy sources has piqued interest in different research fields to find new energy conversion techniques to satisfy the worlds rising demand for energy. In 2013 the Lawrence Livermore National Laboratory has suggested that more than the half (i.e. 61 %) of the energy that flows through our economy is ultimately wasted [1]. The most of the energy is dissipated as waste heat and only 25 % is used as mechanical power [2, 3]. Some of this thermal energy could be converted directly into electrical energy by a thermoelectric generator (TEG) [4, 5]. Presently, thermoelectric (TE) devices are

J. Krez · B. Balke (✉)
Institut für Anorganische und Analytische Chemie,
Johannes Gutenberg - Universität, 55099 Mainz, Germany
e-mail: balke@uni-mainz.de

J. Krez
e-mail: krez@uni-mainz.de

C. Felser and A. Hirohata (eds.), *Heusler Alloys*,
Springer Series in Materials Science 222, DOI 10.1007/978-3-319-21449-8_10

actively considered as a clean energy source for waste heat recovery in automobiles, since they operate silently and do not have any moving parts or environmentally harmful fluids. The captured energy can be used for a vehicle's electrical components such as air conditioning, lights and windows without additional engine load [6]. This in turn can make a crucial contribution to an improvement in fuel economy, and thus a reduction of CO_2 emissions. Beside waste heat recovery, thermoelectric (TE) devices have been investigated for their use in TE-solar hybrid systems [4], TE-refrigeration [7] and as radioisotope TEGs for deep-space application of NASAs Voyager and Cassini missions [8].

The pioneer in introducing semiconductors as promising materials for TE application was Ioffe [9, 10]. His work led to a very active period in TE research in the 1950 s and early 1960s, when many new TE materials were discovered and investigated. At this time, Goldsmid [11] showed the high potential of Bi_2Te_3 as a TE material, and this material remains the basis for the TE industry up to the present time. Ioffes proposal to employ semiconductor alloys rather than simple binary compounds in order to lower the thermal conductivity lead to todays huge variety of material classes such as skutterudites, clathrates, Half-Heusler compounds and complex chalocogenides as promising candidates for TE application. An extensive overview of all material classes and their TE properties can be found in [12, 13]. Guided by the figure of merit zT, the most promising materials with high zT seems to be narrow band gap semiconductors with heavy elements and complex unit cells.

The theoretical predictions by Hicks and Dresselhaus [14, 15] and later by Slack [16] have stimulated a new wave of interest in complex TE materials, where the realization of new ideas mainly emerges from interdisciplinary collaborations between chemistry, physics and material science. Thus, a materials engineering approach must be pursued to either enhance more established materials or to discover completely new classes of materials. In fact, little enhancement in highly efficient TE materials has been made in the past six decades. The figure of merit of the TE compounds for industrial applications has remained static with $zT \approx 1$ for a temperature difference $\Delta T = 300\,K$. State-of-the-art converters are simply too inefficient to be economic, partly due to expensive elementary constituents (Te, Ge, etc.). Currently Bi_2Te_3-based TE materials are used for applications from room temperature to 200 °C, for higher temperatures PbTe-based materials are appropriate material choice. Although Bi_2Te_3 and PbTe are high performance TE materials, a general application becomes more difficult, because Te is a rare and the environmentally friendliness of Pb and Bi is questionable.

A commercially available TE module for power generation or cooling, as shown in Fig. 10.1a, consists of many *n*- and *p*-doped semiconductors, so called TE elements. Each TE element is connected by metal contacts and is sandwiched between electrically insulating ceramic plates, which are high thermal conductive [5]. It is important that the contact material is rigid and exhibits similar thermal expansion coefficients to avoid contact cracks. The TE elements are connected electrically in series but thermally in parallel, to optimize the heat flow from the hot to the cold end [5]. Considering the Seebeck effect, the temperature gradient drives charge carriers from the hot end to the cold end in the material, where the rejected heat must

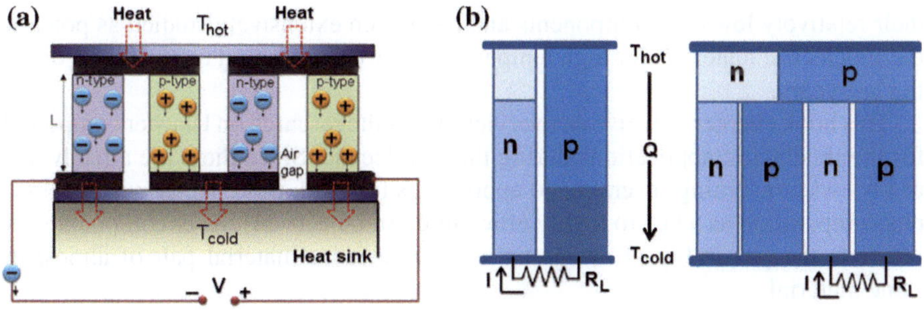

Fig. 10.1 **a** Illustration of a TE module. **b** Illustration of a segmented TEG (*left*) and a cascaded TEG with a cascading ratio of three (*right*)

be removed by a heat sink. Thus, electrons diffuse from the hot to the cold end in the n-type material, and holes in the p-type material, respectively.

To break out of the limited applicability of TE devices and realize adoption in other application fields, it is necessary to increase the TE efficiency of a TEG. The theoretical assumption is that doubling of the device ZT up to 2 could lead to a tenfold increase in the number of TE purposes [17]. Therefore, the main efforts in TEG optimization are focused on the materials optimization but also due to the device engineering, e.g. cascaded or segmented generators (see Fig. 10.1b), where the n- and p-type TE elements (or TE legs) are combined with different materials and joined in series [18]. Ideally, the segmentation helps to improve the TE efficiency, if both materials are operating in their most efficient temperature range. For any TE material to become viable for future applications, it needs to fulfill a set of criteria (see Fig. 10.2), beyond those of being an efficient TE material. A commercial and nationwide application a of TE material has to meet the requirements of being non-toxic, cheap and earth abundant, mechanically and thermally stable and furthermore provide the possibility of processable and reproducible high volume manufacturing. On this background, Heusler compounds with $C1_b$ structure stand out on account of

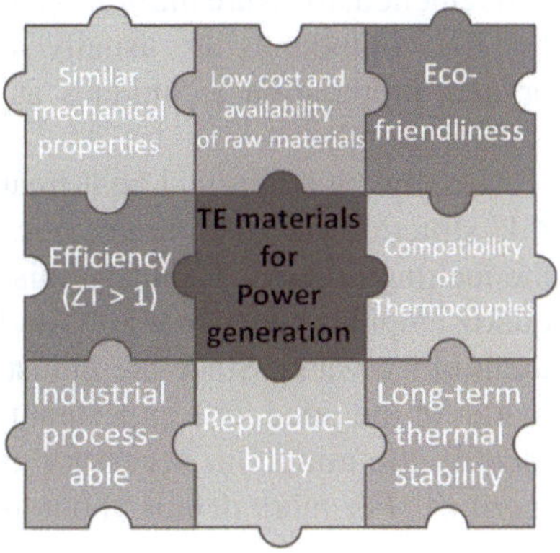

Fig. 10.2 Illustration of the criteria for TE materials required for large-scale power generation

their relatively low cost components and have been extensively studied as potential thermoelectric materials for high temperature power generation up to 1000 K during the last years.

This book chapter will give an overview about the research on Heusler compounds for thermoelectric applications during the last decade. Since there are already a lot of reviews concerning several nano approaches [19–22], this chapter will focuss on phase separation as a key to highly efficient thermoelectric Heusler compounds and illustrate the possibility of designing a thermoelectric material pair of almost the same material.

10.2 Structure and Production of Thermoelectric Heusler Compounds

Half-Heusler (HH) compounds with the general formula XYZ (X, Y = transition metal, Z = main group element) crystallize in a non-centrosymmetric cubic $C1_b$ structure ($F\overline{4}3m$, spacegroup 216), [23, 24] shown in Fig. 10.3, are in focus of the present thermoelectric research, since they fulfill almost all requirements for a commercial applications [19, 25–45]. Their mechanical and thermal stability is exceptional in comparison to the commonly used thermoelectric materials [46]. The possibility to substitute small amounts of elements from the parent compound without destructing the lattice structure allows tuning the electronic properties [47, 48]. This tunability also allows to avoid the use of toxic and expensive elements. The reported thermoelectric Heusler compounds exhibit high electrical conductivity and moderate values of the Seebeck coefficients, which lead to a high powerfactor [37].

HH compounds contain high melting point elements such as Hf: 2231 °C, Zr: 1855 °C, Ti: 1668 °C, Ni: 1455 °C, and Co: 1495 °C as well as elements with relatively low melting point such as Sn: 232 °C and Sb: 631 °C. Therefore, a high temperature alloying method is necessary. Usually, this is done by stoichiometric weighting the constituting elements, followed by arc-melting in an Ar atmosphere for at least three times using a home-made or a commercial system. After this the samples are crushed and remelted, to insure the homogeneity of the samples. Other preparation techniques for HH compounds are usually induction heating [31], solid state reaction [45] or new preparation techniques as melt spinning [49] or spark plasma sintering (SPS) [34, 50].

Usually, an additional heat treatment is necessary for all samples to obtain the $C1_b$ structure and to eliminate possible secondary phases which could occur during the melting process. Typically, this annealing of the ingots is done under vacuum in quartz ampules between 900 °C and 1050 °C for 3 to 7 days, followed by quenching them in ice water. Since the industrial application of a product is always as well a matter of the price, one important task for the ongoing thermoelectric research is to identify the minimum necessary annealing time while preserving a high figure of merit zT. For more details about all the possibilities, opportunities, and challanges

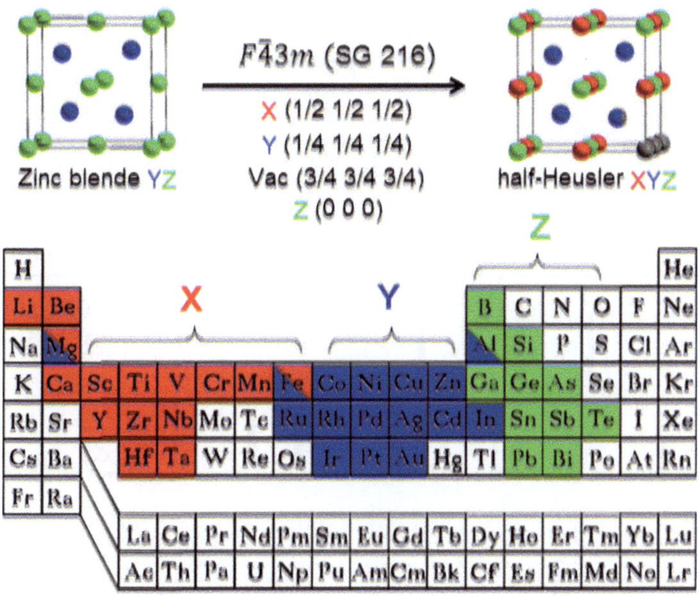

Fig. 10.3 Periodic table of elements including the crystal structure $F\bar{4}3m$ of zinc blende and HH compound. The various number of HH compounds can be formed by combination of the different elements according the color scheme

concerning the production of HH compounds see [27]. A relatively new approach is using an energy and time-efficient process involving solid-state preparation in a commercial microwave oven [38, 51].

Even though one can read quite often that mechanical alloying (MA) of HH compounds of the elements is not successful due to limited energy input and the high melting points, usually the [52] is cited, this might be true for NiTiSn-based compounds but for sure is not the case for CoTiSb-based compounds. Using a high energy ball mill (Fritsch *Planeten-Mikromühle PULVERISETTE 7 premium line*) we succeed with the mechanical alloying of several state-of-the-art *p*-type HH compounds such as TiCoSb, HfCoSb, ZrCoSb, $TiCo_{0.85}Fe_{0.15}Sb$, $TiCo_{0.7}Fe_{0.3}Sb$, $Zr_{0.5}Hf_{0.5}CoSb_{0.8}Sn_{0.2}$, and $ZrCoSb_{0.9}Sn_{0.1}$. Exemplary for all those HH compounds, the results on the MA of $TiCo_{0.85}Fe_{0.15}Sb$ are described in more details. Figure 10.4d shows the x-ray diffraction (XRD) patterns of the as-milled $TiCo_{0.85}Fe_{0.15}Sb$ HH nanoparticles as a function of milling time. After milling for more than 6 h, the HH phase appears. Longer milling, however, results in some impurity phase (for example, as the case of 20 h). Transmission Electron Microscope (TEM) investigations reveal the size of Heusler nanoparticles is around 15 nm (see Figs. 10.4a,b). Mechanical ball milling method is a top-down based method and is facile to scale up. Heusler nanoparticles produced by this approach, however, are easily agglomerated, severely strained, presented with numerous intermediate phase. For a very recent review about Heusler nanoparticles see [53]. Bartholomé et al. built thermoelectric modules based

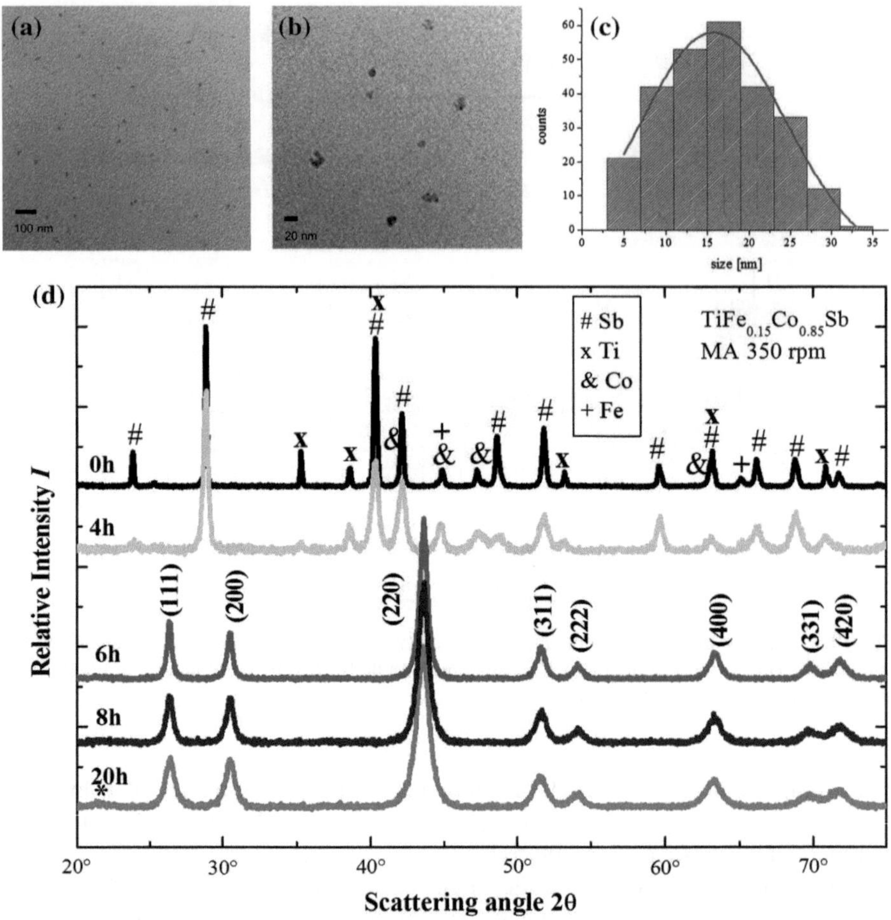

Fig. 10.4 Exemplary TEM micrographs **a b**, and the corresponding histogram **c** for a $TiCo_{0.85}Fe_{0.15}Sb$ nanoparticle sample milled for 8 h with 700 rpm. **d** XRD patterns as a function of milling time for as-milled (350 rpm) $TiCo_{0.85}Fe_{0.15}Sb$ nanoparticles. For comparision the XRD result of the mixture of the elemental powders before milling is shown

on HH compounds from material synthesized in kg-batches [31]. The material performance is in line with the published values for comparable material compositions and exhibits peak ZT-values of 0.7 for the n-type and 0.5 for the p-type samples. The modules built from these materials have a maximum power output of 2.8 W with a total module area of $16 \times 16\,mm^2$, resulting in the highest values for the power density of $3.2\,W/cm^2$ and a Z-value of $3.1 \times 10^{-4}\,K^{-1}$ for HH modules published so far. The long-term stability and reproducibility of these modules could be verified by the authors.

10.3 Electronic Structure of Thermoelectric Heusler Compounds

According to the literature, both n-type compounds with the elemental formula MNiSn (M = Ti, Zr, Hf) and p-type MCoSb display a high potential for exceptional ZT values. Research has, therefore, focussed mainly on improvement of these ternary intermetallic compounds. Figure 10.5 shows the calculated band structure (a) and the density of states (b) of NiTiSn. The compound is a semiconductor with an indirect gap. The valence band maximum appears at Γ and the conduction band minimum at X. The band gap has a size of $\Delta E_{\text{gap}} = 0.45\,\text{eV}$. The optical gap at Γ is considerably larger ($\Delta E_{\Gamma} = 1.38\,\text{eV}$).

The electronic structure exhibits at 5 eV to 8 eV below the Fermi energy the typical sp hybridization gap that separates the low lying a_1 (s) from the t_1 (p) bands. The high density of states at about −2 eV emerges mainly from Ni d states. The high density of states at −0.7 eV arises mainly from Ni d states with e symmetry. Most important for the transport properties, the states at both band edges are due to Ti d states with t_2 symmetry. From the band structure shown in Fig. 10.5 it is obvious that electron (n) or hole (p)doping will have rather different results. It is easily seen that p-type doping creates holes in the triply degenerate valence band at Γ whereas the situation is completely different for n-type doping that fills electrons into the single conduction band above the indirect gap at X.

For comparision, Fig. 10.5c,d shows the calculated band structure and the density of states of CoTiSb. The topmost valence band at ϵ_F of the semiconducting pure CoTiSb has t_2 character at the Γ point that has T_d symmetry. It is followed by d bands of e (−1.56 eV) and bands of t_2 (−2.39 eV) character. The low lying a_1 states are found below −9.4 eV and are separated from the higher lying bands by the sp hybridization gap that is typical for Heusler compounds. The pure CoTiSb exhibits an indirect $\Gamma - X$ band gap of about 1.06 eV, whereas the optical gap at Γ is considerably larger (1.83 eV). Both, topmost valence band and lowest conduction band appear very low dispersing with a bandwidth of ≈ 150 meV in the $\Gamma - L$ direction and

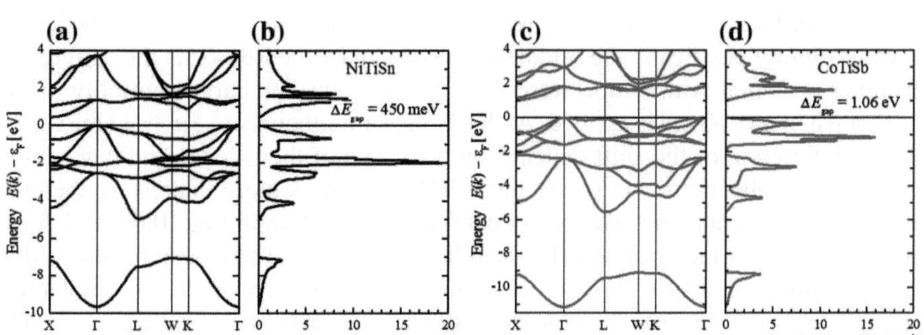

Fig. 10.5 Electronic structure of NiTiSn **a b** and CoTiSb **c d**

thus allow easily for direct optical transitions at energies of about 1.8 eV. For more details about this electronic band structure calculations see [54, 55]. Miyamoto and coworkers studied the electronic structures of the Heusler compounds MNiSn by means of photoelectron spectroscopy [56]. They observed "in gap" states close to the Fermi edge and suggested that these electronic states are mainly created by chemical disorder, which could be the key to control the thermoelectric properties. For more details about this "in gap" states see [54, 55, 57].

In order to achieve the best performance of thermoelectric module the $n-$ and $p-$ type materials to be used should be designed to exhibit similar chemical and physical properties [58]. This can be easily realized when starting from the same material, e.g. the Heusler compound NiTiSn. Yang and coworker evaluated theoretically the thermoelectric-related electrical transport properties of several Heusler compounds, they calculated the maximum power factors and the corresponding optimal $n-$ or $p-$type doping levels [59], which can provide guidance to experimental work. Horyn and coworkers investigated the effect of a partial substitution of Ti and Zr by Sc on the thermoelectric properties of MNiSn-based compounds and obtained at room temperature a fairly high positive Seebeck coefficient of about 121 μV/K with 5 % Sc substitution of Zr [60].

In the commonly used one-parabolic-band approaches $n-$ or $p-$type doping lead to rather similar results, just with opposite signs for the Seebeck coefficient. The situation in practical materials is more difficult. Depleting the valence band or filling the conduction band acts on electronic states with rather different characters. Starting from the calculated electronic structure, the transport properties are calculated using Boltzmann transport theory [61]. Doping the semiconducting materials by electrons or holes will change the transport properties. The doping will cause the chemical potential to change its position. At high doping levels it will shift into the valence band (hole doping) or conduction band (electron doping). Figure 10.6a, b show the calculated Seebeck and power coefficients as function of the position of the chemical potential, respectively. The calculations were performed for NiTiSn and for $T = 300$ K. It is assumed that the shift $\delta\mu = 0$ corresponds to the middle of the band gap at $T = 0$. The Seebeck coefficient exhibits the typical semiconductor behavior under doping, it is positive for hole and negative for electron doping. It is largest for small shifts of the chemical potential from the original position. At 300 K it is already slightly positive in the middle of the band gap. the Reason is the shift of the chemical potential with temperature $\mu = \mu(T)$ to ensure charge neutrality of the system when no voltage is applied. At 300 K the shift amounts to $\delta\mu \approx 13$ meV. The size and direction of the shift depends on the shape of the valence and conduction bands. For applications, the power factor $PF = S^2\sigma$ is more interesting than the Seebeck coefficient alone. The power coefficient, as shown in Fig. 10.6b, is defined here by the power factor divided by the relaxation time τ. The reason for this is that the calculations deliver σ/τ rather than directly the pure conductivity. The largest power coefficient appears for hole doping when the chemical potential is already shifted slightly outside of the band gap into the valence band. The maximum for electron doping is smaller and also appears when μ is slightly shifted outside of the gap. The reason for this is in both cases the high conductivity when the compound

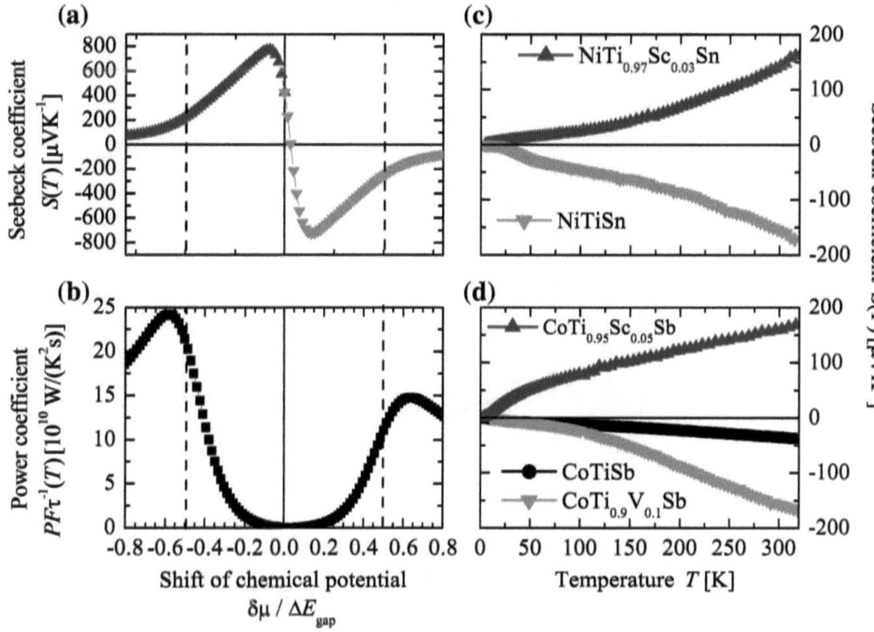

Fig. 10.6 Calculated Seebeck **a** and power coefficients **b** of NiTiSn for $T = 300\,\mathrm{K}$. The shift of the chemical potential is given with respect to the size of the gap. The valence and conduction band extrema are marked by *dashed lines*. Temperature dependence of the Seebeck coefficient $S(T)$ of NiTiSn and $NiTi_{0.97}Sc_{0.03}Sn$ in **c** and of CoTiSb, $CoTi_{0.95}Sc_{0.05}Sb$ and $CoTi_{0.9}V_{0.1}Sb$ in **d**

goes over into the completely metallic state. From the integrated density of states it is estimated that the maxima of the power coefficient are reached at an electron or hole doping of about 1.1 % or 1.4 %, respectively.

Substituting Ti by V or Sc will act as electron or hole doping, respectively. The difference in the number of valence electrons is in both cases one, such that a substitution by an amount x changes the electron concentration also by $\pm x$. Exemplary, Fig. 10.6c, d show—in full agreement with the calculations—the temperature dependence of the Seebeck coefficient $S(T)$ of pure NiTiSn compared to the substituted compound $NiTi_{0.97}Sc_{0.03}Sn$ in (c) and of pure CoTiSb compared to the substituted compounds $CoTi_{0.95}Sc_{0.05}Sb$ and $CoTi_{0.9}V_{0.1}Sb$ in (d). This shows the possibility to create $n-$ and $p-$type thermoelectric materials based on the same basic compound and thus to produce suitable, well matched pairs for thermoelectric devices out of HH materials. Thus, the probability of cracks and material distortion in the TE device can be reduced and better TE efficiencies can be attained under operating conditions.

The site substitution is mainly used to reduce the thermal conductivity due to an enhanced disorder scattering of phonons [62, 63]. In addition to the reduction of the thermal conductivity, site substitution with non-isoelectronic elements significantly affects the transport properties in HH materials, owing to a shift of the chemical potential. The size and direction of the shift depends on the shape of the valence and

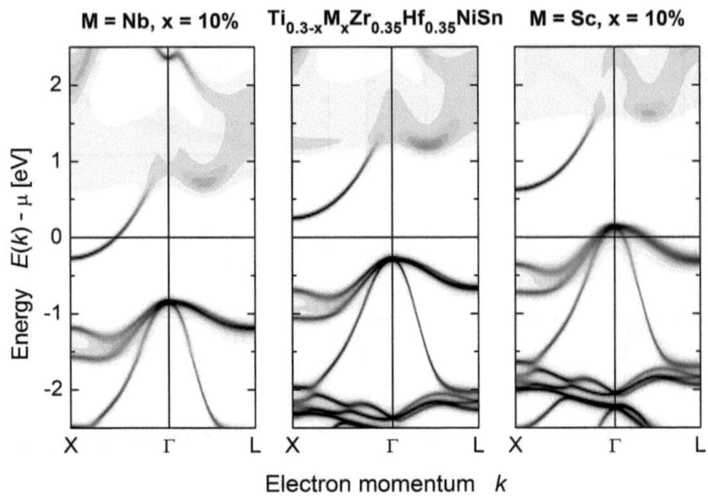

Fig. 10.7 Electronic structure of the *n*-type $Ti_{0.30}Zr_{0.35}Hf_{0.35}NiSn$ from ab initio calculations performed by Dr. G.H. Fecher at Max-Planck-Institute for Chemical Physics of Solids (Dresden). The Fermi energy is shifted into the conduction band with 10 % Nb substitution and into the valence band with 10 % Sc substitution

conduction bands and the origin of the substituted elements (see Fig. 10.7). Elements with donor dopant properties (elements with more valence electrons) introduce additional electrons into the HH system, leading to a shift of the Fermi energy towards the conduction band (i.e. higher energies) and increasing of the band gap size. Therefore, minority carriers are suppressed and cause a shift of the Seebeck coefficient to higher temperatures. Acceptor dopants (elements with less valence electrons) introduce holes into the system, shift the Fermi energy towards the valence band (to lower energies) and cause a change of the electronic transport from *n*- to *p*-type conductivity. Thus, the best *n*-type HH materials based on the MNiSn system and many of the fairly good *p*-type HH compounds are based on the MCoSb HH system [32, 64].

Ouardi et al. have shown the impact of acceptor doping in the *n*-type $Ti_{0.30}Zr_{0.35}Hf_{0.35}NiSn$ HH system [57]. The substitution of 4 % Sc on the Ti-position induced a change in transport properties from *n*- to *p*-type conduction, where the Seebeck coefficient α reveals a maximum for both *n*- and *p*-type compound at $|\alpha| \approx 210\,\mu VK^{-1}$ at 600 K. Figure 10.7 illustrates the influence of site substitution in the *n*-type $Ti_{0.30}Zr_{0.35}Hf_{0.35}NiSn$ compound with Nb as a donor dopant and Sc as an acceptor dopant. It can be seen, that electron doping changes the position of the Fermi energy with 10 % Nb substitution into the conduction band, and with 10 % Sc substitution into the valence band. Tobola et al. [65] have shown that the impact on the transport properties via electronic doping in HH materials depends on the sublattice where the substitution takes place. Thus, adding one electron on the Z-site (e.g. Sn replaced by Sb) does not have the same effect as increasing the number of electrons at the X-site by replacing Ti by Nb, due to the different origin of the electrons and their localization in the DOS. It can be assumed, that a change in the number of electrons on the X-site (Ti replaced by a non-isoelectronic TM) has the greatest influence on the electronic properties, since these elements are responsible for the appearance of the electronic states near the Fermi level [39, 57]. Krez et al. investigated the Hall carrier concentration of the $Ti_{0.3-x}Nb_xZr_{0.35}Hf_{0.35}NiSn$ system [66]. For samples with high Nb content, it could be observed that the carrier concentration remained constant and no intrinsic excitation occurred at high temperatures. The Hall mobility, μ_H decreased with increasing Nb concentration, which may be caused by the scattering of the charge carriers as a result of the additional metal inclusions. The variation in mobility with temperature can be described by a power law equation: $\mu \propto T^{-\nu}$, where the value of ν is an indicator of the most dominant scattering mechanism in the system. For phase-separated HH compounds, Xie et al. were able to show that the major mechanism is alloy scattering [67].

10.4 Phase Separation as a Key to a Thermoelectric High Efficiency

A general challenge in improving HH compounds for thermoelectric applications is the comparatively high thermal conductivity of the order of 10 $WK^{-1}m^{-1}$. In the late 1990 s a common approach to reduce the thermal conductivity was to increase the phonon scattering. Hohl et al. reduced the thermal conductivity by a factor of three for different temperatures by introducing disorder on the X-sites of $X_{0.5}X'_{0.5}NiSn$ (X, X′ = Ti, Zr, Hf) [68, 69]. Substitution of tin by antimony increases both the thermal and electrical conductivities [70]. Substitution at the Ni position also decreases the thermal conductivity [71]. The composition $Zr_{0.5}Hf_{0.5}Ni_{0.8}Pd_{0.2}Sn_{0.99}Sb_{0.01}$ possesses a figure of merit $ZT = 0.7$ at 527 °C [72]. For a long time the TE applicability of HH compounds was limited by their high thermal conductivities $\kappa > 7$–10 $WK^{-1}m^{-1}$ [73].

But owing to the development in the TE community, the introduction of mass disorder due to site substitution (alloying) [32, 64] and nanostructures [29, 45] is an effective way to produce additional phonon scattering and with it to decrease the thermal conductivity. Thus, the substitution of non- and isoelectronic elements leads to a drastic decline in the thermal conductivity ($\kappa < 4$ W/Km) in HH materials [63]. The resulting higher disorder due to higher mass and strain fluctuations and an intrinsic phase separation in multi-component HH materials are responsible for the strong reduction in κ. Another approach is to implement a nano- or microstructure in the thermoelectric material. This can be achieved by phase separation, composite materials, pulverization with additional spark plasma sintering or by a complex lattice structure [74, 75]. The experimental efforts of site substitution in HH compounds significantly improved the TE performances of HH materials. Hereby, zT values of 1.5 at 427 °C in the n-type $Zr_{0.25}Hf_{0.25}Ti_{0.5}NiSn_{0.998}Sb_{0.002}$ [63, 76] and zT $\approx$ 1 in the p-type $Ti_{0.12}Zr_{0.44}Hf_{0.44}CoSn_{0.8}Sb_{0.2}$ material [64] could be attained.

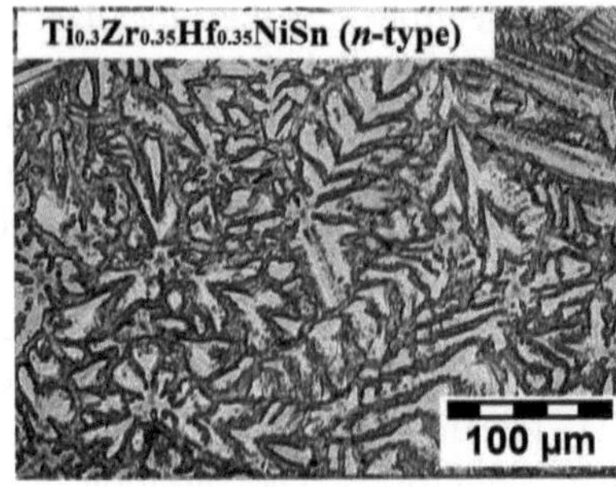

Fig. 10.8 Light microscope image of n-type $Ti_{0.30}Zr_{0.35}Hf_{0.35}NiSn$ compound. The sample was etched with a $HCl : HNO_3 : HF : H_2O$ solution to emphasize the intrinsic phase separation

A remarkable characteristic of multi-component HH compounds is their intrinsic phase separation, shown in Fig. 10.8, when the compounds are solidified by rapid cooling, leading to a significant reduction of κ, and thus to improved TE efficiencies [30]. Sakurada et al. [63] have studied the effect of Ti substitution on the TE properties in $Ti_x(Zr,Hf)_{1-x}NiSn$. Their results revealed that the thermal conductivity can be reduced significantly with $x = 0.3$ Ti content to $\kappa \approx 3\,WK^{-1}m^{-1}$ at room temperature. This strong decline in κ has two origins. First, the high mass strain and disorder within the $C1_b$ structure due to the Ti substitution at the X-site leads to an effective phonon scattering. Secondly, this multi-component HH system reveals an intrinsic phase separation into two HH phases, which provides a further scattering of phonons at the interfaces.

The eutectic microstructure consists of a main Ti-poor HH 1 phase (bright phase), which solidifies first and a Ti-rich HH 2 phase (dark phase), which is dendritically interlaced through the microstructure (see Fig. 10.8). Due to the rapid solidification method, small constitutional variations occur in the composition, leading to the precipitation of small Sn impurities, which are homogenously distributed throughout the dendritic microstructure with an average size of about 5–10 μm. The composition of the two HH phases carried out by electron microscope phase analysis (EMPA) was found to be $Ti_{0.18}Zr_{0.40}Hf_{0.41}NiSn$ for the Ti-poor HH 1 phase and $Ti_{0.65}Zr_{0.20}Hf_{0.17}NiSn$ for the Ti-rich HH 2 phase. It seems that Ti is the driving force of the intrinsic phase separation since the two HH phases differs the most in the Ti content. The formation of these HH 1 and HH 2 phases is strongly kinetically favored. Efforts to attain these two HH phases as single phase materials revealed also phase separations in the sample, resulting from constitutional changes under the solidification conditions.

The thermal analysis of $Ti_{0.30}Zr_{0.35}Hf_{0.35}NiSn$ carried out by differential scanning calorimetrie (DSC) up to 2000 K in argon gas atmosphere with a heating rate of $10\,Kmin^{-1}$ is shown in Fig. 10.9a. The DSC curve shows two endothermic reactions at 1680 K and 1720 K, indicating the eutectic and the melting point of $Ti_{0.30}Zr_{0.35}Hf_{0.35}NiSn$, respectively. This incongruently melted compound reveals a strong exothermic recrystallization peak at 1620 K. The supercooling behavior is a result of the decelerated nucleation in the compound. The melting and recrystallization peaks at about 503 K are caused by Sn impurities in the sample. Specific heat c_p

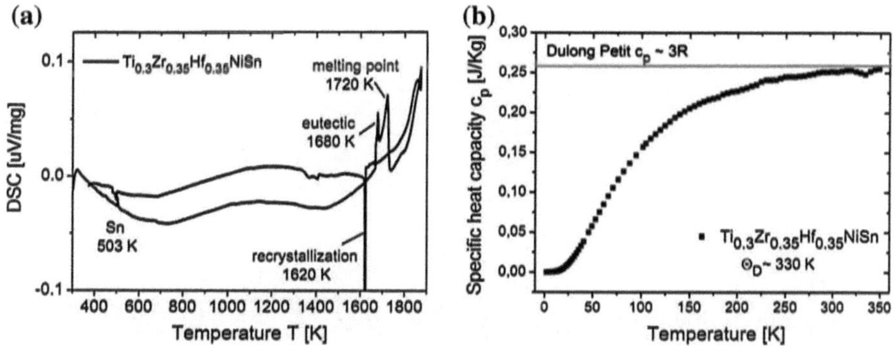

Fig. 10.9 **a** DSC measurement of $Ti_{0.30}Zr_{0.35}Hf_{0.35}NiSn$ exemplifies the eutectic temperature at 1680 K and the melting point at 1780 K. The supercooled melt recrystallized at 1680 K. The peaks at 503 K indicate Sn impurities. **b** Specific heat capacity c_p measurement of $Ti_{0.30}Zr_{0.35}Hf_{0.35}NiSn$ compound, indicating the Dulong-Petit limit of $3R$ for Θ_D

measurements of the parent $Ti_{0.30}Zr_{0.35}Hf_{0.35}NiSn$ compound were performed from 10 K to 350 K using a PPMS (see Fig. 10.9b). Since this HH compound has a large number of atoms in the unit cell, many phonon modes appear above room temperature. The Deby temperature was determined to be $\Theta_D = 330\,K$. For $T \gg \Theta_D$ the specific heat capacity c_p reached its Dulong-Petit limit of $3R$, corresponding to the value of $c_p \approx 0.26\,Jg^{-1}K^{-1}$.

Owing to the high melting points of HH materials the intrinsic phase separation is extraordinary temperature stable and can be considered as the key to high TE efficiencies in HH materials. This leads to the suggestion that temperature stable eutectic microstructures can establish an entirely new area in the TE community, where the phase separation microstructure can be independently engineered to fulfill desired TE properties. For commercial applications, the thermal stability of the phaseseparated microstructure under operating conditions ($\approx 873\,K$ in automotive applications) needs to be guaranteed. Therefore, the long-term stability of n- and p-type HH materials based on the $Ti_{0.30}Zr_{0.35}Hf_{0.35}NiSn$-system after 500 cycles (1700 h) from 373 K to 873 K was investigated [77]. The SEM images, shown in Fig. 10.10, revealed the intrinsic phase separation in n-type HH compounds. Surprisingly, the dendritic microstructure showed no obvious change in n-type materials under temperature cycling conditions. The XRD and EMPA results emphasized the extraordinary temperature stability of these HH materials in a moderate temperature range. Due to the high melting point($T \approx 1720\,K$) of this (Ti,Zr,Hf)NiSn HH system the resulting phase separation is stable under the given conditions, which is of utmost importance to maintain low thermal conductivities in these HH materials.

The transport properties of the semiconducting n-type $Ti_{0.30}Zr_{0.35}Hf_{0.35}NiSn$ HH system, shown in Fig. 10.11, were measured up to 873 K after 500 cycles (1700 h). The electrical conductivity σ improved after 50 cycles due to an enhancement of the structural order (see Fig. 10.11a). The Seebeck coefficient α is negative at all temperatures, indicating electrons as the majority charge carriers, with a peak value of

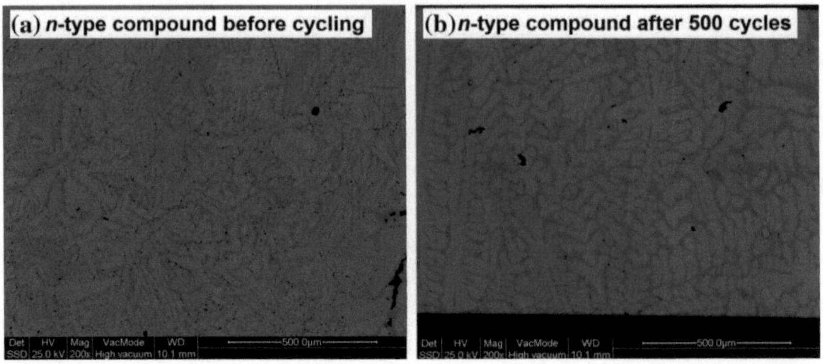

Fig. 10.10 SEM images of the n-type $Ti_{0.30}Zr_{0.35}Hf_{0.35}NiSn$ compound before cycling **a** and after 500 cycles (1700 h) from 373 K to 873 K **b**

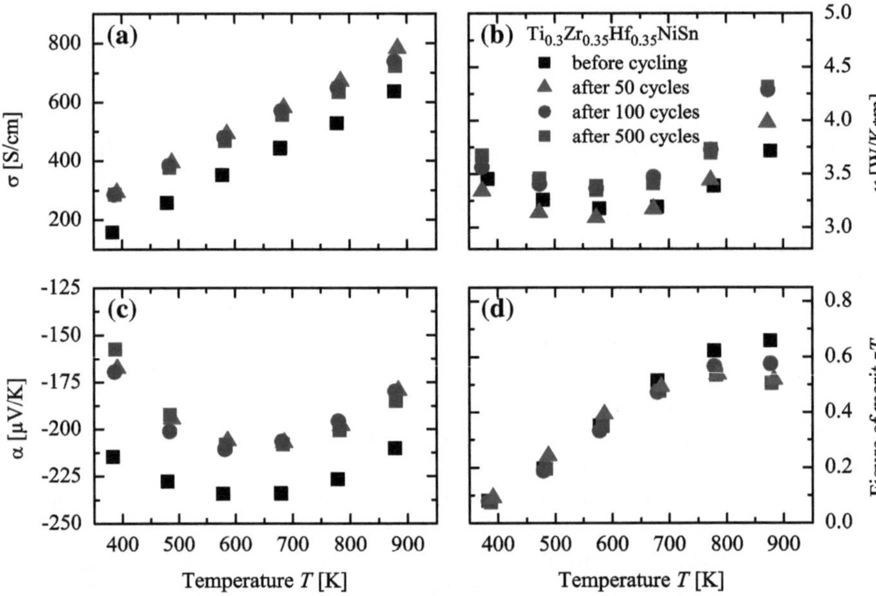

Fig. 10.11 Thermoelectric properties as a function of temperature of the n-type $Ti_{0.30}Zr_{0.35}Hf_{0.35}NiSn$ compound under the long-term treatment of 500 cycles. **a** Electrical conductivity σ(T). **b** Thermal conductivity κ(T). **c** Seebeck coefficient α(T). **d** Figure of merit zT(T)

$\alpha \approx$ -210 μVK^{-1} at 600 K (see Fig. 10.11b). The large Seebeck coefficient emerges from the high density of states, which is caused by the d-states of the transition metals near the Fermi leve [57]. The decrease in α above 600 K is caused by the thermal excitation of intrinsic carries. The power factor PF shows a peak value of 2.5×10^{-3}WK^{-2}m^{-1} at 773 K after 50 cycles, which is comparable to values of n-type Bi_2Te_3, i.e., $PF \approx 2.6\times10^{-3}$WK^{-2}m^{-1} at 423 K [78]. A recent investigation of the long-term efficiency of a commercial available bulk-Bi_2Te_3 TEG with 31 thermocouples showed a reduction in α and σ caused by material deterioration [79]. The thermal conductivity κ of (Ti,Zr,Hf)NiSn compounds is much than the values of ternary HH alloys, where $\kappa >$ 7-10 WK^{-1}m^{-1} [73]. The low κ values of the phase-separated HH system occur due to different phonon scattering agents in the structure. Owing to the improved structural order of the n-type compound, κ increases slightly after 500 cycles (see Fig. 10.11c). The increase in κ above 600 K is caused by the excitation of intrinsic carriers. The figure of merit zT shows no drastic change after 500 cycles since the values lie within the acceptable error range. This results strongly demonstrate the suitability of phase-separated HH materials, which also comply with requirements such as reproducibility and environmental friendliness via mechanical and thermal stability, for a commercial TE application at moderate temperature.

As already mentoined above, Toshiba (Japan) reported a maximum ZT of 1.5 for $Zr_{0.25}Hf_{0.25}Ti_{0.5}NiSn_{0.998}Sb_{0.002}$ at 427 °C [63, 76]. These high ZT values were never reproduced by any other group since the original publication in 2005. In 2013, Schwall and Balke showed that the high Figure of Merit values reported by Shutoh and Sakurada [63, 76] could almost be reproduced (see Fig. 10.12g) [30]. The origin of the exceptional low thermal conductivity is the phase decomposition (see Fig. 10.12a–f), which does not influence the electrical conductivity significantly because of semi-coherent interfaces existing between the three co-existing Heusler phases. These intrinsic properties of the $Ti_{0.5}Zr_{0.25}Hf_{0.25}NiSn_{(1-x)}M_x$ | M= Sb, Bi, Te; x= 0-0.006 system show again that the Heusler compounds in matters of the transport properties are competitive thermoelectric materials.

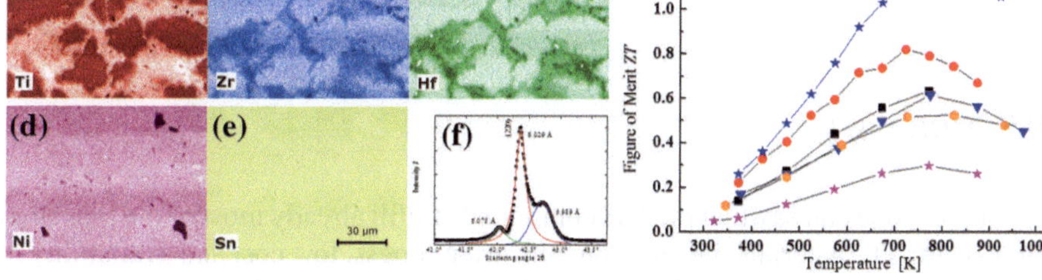

Fig. 10.12 Element-specific EDX mappings of the five constituents of the phase separated $Ti_{0.5}Zr_{0.25}Hf_{0.25}NiSn$ compound with brightness proportional to the concentration **a-e**. (220) reflection of $Ti_{0.5}Zr_{0.25}Hf_{0.25}NiSn$ measured by synchrotron radiation **f**. Temperature dependence of the Figure of Merit of $Ti_{0.5}Zr_{0.25}Hf_{0.25}NiSn_{(1-x)}M_x$ compounds **g**

Most of the examined HH materials are n-type semiconductors. As already mentoined above, HH materials can also be p-type semiconductors; XCoSb compounds are the most thoroughly examined p-type systems due to their comparatively high ZT values. Wu et al. reported an iron-doped TiCoSb with the elemental formula $TiFe_{0.3}Co_{0.7}Sb$, that reaches a maximum in ZT of 0.42 at 470 °C. The p-type semiconductors have a thermal conductivity that is 2–3 times higher than the n-type materials and therefore the reduction of the thermal conductivities of the p-type material to the level of the n-type semiconductors, which is nowadays in the range of 3 $WK^{-1}m^{-1}$, seems a very attractive research target with a high technical impact.

Atomic-scale substitution on the various lattice sites of the HH compounds is not the only means to attain increased phonon scattering. Nano-structures at an appropriate scale can also be generated and fine-tuned by suitable thermal treatments, based on a thorough knowledge of the relevant phase diagrams. Disproportionation reactions, as recently shown in multicomponent chalcogenide compounds, while retaining the initial crystal structure of the components, can give rise to nano-structured composites with a reduced thermal conductivity and hence enhanced ZT [80]. Recently, Yan et al. were able to enhance the ZT of p-Type HH compound via a nanostructuring approach [34]. They succeeded in achieving grain sizes smaller than 200 nm in p-type HH samples with a composition of $Zr_{0.5}Hf_{0.5}CoSb_{0.8}Sn_{0.2}$ by ball milling the alloyed ingot into nanopowders and then hot pressing them into dense bulk samples, resulting in a simultaneous increase in Seebeck coefficient and a significant decrease in thermal conductivity, which led to a 60 % improvement in peak ZT from 0.5 to 0.8 at 700 °C.

Using the approach of introducing stronger phonon scattering by larger differences in atomic mass and size in p-type HH compounds $Hf_{1-x}Ti_xCoSb_{0.8}Sn_{0.2}$ Yan et al. were able to increase the ZT to 0.9 at 700 °C [35]. Very recently, it has been shown, that it is possible to introduce an intrinsic phase separation as well in the p-type Heusler system (Ti/Zr/Hf)$CoSb_{0.8}Sn_{0.2}$ and (Ti/Zr/Hf)$CoSb_{0.85}Sn_{0.15}$ analogue to the n-type system [81]. Mandatory is the isoelectronic alloying of Ti with Hf on the Y position, which results in a microstructuring of the material consisting of at least two Heusler phases - one rich in Ti and Sn and another rich in Hf and Sb. The subsequent reduction of the thermal conductivity leads to a maximum ZT of 1.15 at 710 °C [82]. For recent reviews about nanocomposite HH materials see [19, 27, 29].

10.5 Summary

In summary, one can say that the HH compounds fulfill already most of the industrial demands for TE materials, i.e. environment-friendliness, low cost and availability of raw materials, long term stability, processible in industrial production and chemical and mechanical resistance toward high temperatures. The most recent results strongly underline the importance of phase separations as an important tool in the design of highly thermoelectric efficient materials which fulfill the industrial demands for a thermoelectric converter. For achieving the goal of a greater fundamental understanding and later on for the sophisticated design of phase separated HH compounds with thermoelectric properties beyond the state of the art one has to clearly separate general effects of nanostructuring from material specific influences and work on a very detailed fundamental investigation of the crystallographic, mechanical, and thermoelectric properties of HH compounds.

Acknowledgments This work was financially supported by the thermoHEUSLER Project (Project No. 0327876D) of the German Federal Ministry of Economics and Technology (BMWi) and the TEG 2020 project of the German Federal Ministry of Education and Research (BMBF). Additonal financial support by the Deutsche Forschungsgemeinschaft (projects BA4171/2-2 and FE633/8-1 of the DFG Priority Programm SPP 1386 and project INST247/897-1 FUGG) and Stiftung Innovation Rheinland-Pfalz (No. 961-386261/931) is gratefully acknowledged. The authors thank G.H. Fecher, M. Schwall, E. Rausch, S. Ouardi, T. Graf, J. Barth, S. Beccard, M. Cambaz, J. Schmitt, R. Stinshoff, W. Schnelle, S. Kostmann, M. Eckert, M.P. Schmidt, J.M. Stahlhofen and all the student assistants during the last 5 years for their help with theory and experiments, and for fruitful discussions. Especially, the authors thank C. Felser for all the support during the last ten years. BB thanks the whole team of the thermoHEUSLER project for fruitful discussions and especially the team of the Fraunhofer IPM (Freiburg, Germany) for help with experiments and the verification of experimental results. J. Krez and M. Schwall were recipient of a fellowship of the Graduate School of Excellence "MAterials Science IN MainZ" MAINZ through the Excellence Initiative (DFG/GSC 266).

References

1. https://flowcharts.llnl.gov/. 2014
2. J. Yang, F.R. Stabler, J. Electron. Mater. **38**, 1245 (2009)
3. K. Schierle-Arndt, W. Hermes, Chemie in unserer Zeit **47**, 92 (2013)
4. T.M. Tritt, Annu. Rev. Mater. Res. **41**, 433 (2011)
5. G.J. Snyder, E.S. Toberer, Nat. Mater. **7**, 105 (2008)
6. P. Ball, T. Caillat, *MRS Bulletin Energy Quarterly, June* Energy Quarterly Thermoelectric Heat Recovery could boots Auto Fule Economy (2011)
7. B.C. Sales, Science **295**, 1248 (2002)
8. J. Yang, T. Caillat, MRS Bull. **31**, 224 (2006)
9. A. Ioffe, *Semiconductor Thermoelements and Thermoelectric Cooling* (Infosearch, London, 1957)
10. A. Ioffe, Sci. Am. **199**, 31 (1958)
11. H. Goldsmid, *Applications of Thermoelectricity*, Methuen's Monographs on Physical Subjects (Methuen, London, 1960)
12. G.S. Nolas, J. Sharp, H.J. Goldsmid, *Thermoelectrics: Basic Principles and New Materials Developments* (Springer, Berlin, 2001)
13. J.R. Sootsman, D.Y. Chung, M. Kanatzidis, Angew. Chem. Int. Ed. **48**, 8616 (2009)
14. L.D. Hicks, M.S. Dresselhaus, Phys. Rev. B **47**, 12727 (1993)
15. L.D. Hicks, M.S. Dresselhaus, Phys. Rev. B **47**, 16631 (1993)
16. G.A. Slack, in *CRC Handbook of Thermoelectrics*, ed. by D.M. Rowe (CRC, 1995), pp. 407–440
17. K. Nielsch, J. Bachmann, J. Kimling, H. Böttner, Adv. Energy Mater. **1**, 713 (2011)
18. G.J. Snyder, Appl. Phys. Lett. **84**, 2436 (2004)
19. W. Liu, X. Yan, G. Chen, Z. Ren, Nano Energy **1**, 42 (2012)
20. Z.-G. Chen, G. Han, L. Yang, L. Cheng, J. Zou, Prog. Nat. Sci.: Mater. Int. **22**, 535 (2012)
21. P. Pichanusakorn, P. Bandaru, Mater. Sci. Eng.: R: Rep. **67**, 19 (2010)
22. H. Alam, S. Ramakrishna, Nano Energy **2**, 190 (2013)

23. W. Jeischko, Metall. Mater. Trans. B **1**, 3159 (1970)
24. H.C. Kandpal, C. Felser, R. Seshadri, J. Phys. D: Appl. Phys **39**, 776 (2006)
25. F. Heusler, Verh. d. DPG **5**, 219 (1903)
26. L.O. Grondahl, S. Karrer, Phys. Rev. Ser. I **33**, 531 (1911)
27. S. Chen, Z. Ren, Mater. Today **16**, 387 (2013)
28. F. Casper, T. Graf, S. Chadov, B. Balke, C. Felser, *Semicond. Sci. Technol.***27** (2012)
29. S.J. Poon, D. Wu, S. Zhu, W. Xie, T.M. Tritt, P. Thomas, R. Venkatasubramanian, J. Mater. Res. **26**, 2795 (2011)
30. M. Schwall, B. Balke, Phys. Chem. Chem. Phys. **15**, 1868 (2013)
31. K. Bartholomé, B. Balke, D. Zuckermann, M. Köhne, M. Müller, K. Tarantik, Jan König, J. Electron. Mater. **43**, 1775 (2014)
32. G. Joshi, T. Dahal, S. Chen, H.Z. Wang, J. Shiomi, G. Chen, Z.F. Ren, Nano Energy **2**, 82 (2013)
33. T. Jäger, C. Mix, M. Schwall, X. Kozina, J. Barth, B. Balke, M. Finsterbusch, Y.U. Idzerda, C. Felser, G. Jakob, Thin Solid Films **520**, 1010 (2011)
34. X. Yan, G. Joshi, W. Liu, Y. Lan, H. Wang, S. Lee, J.W. Simonson, S.J. Poon, T.M. Tritt, G. Chen et al., Nano Lett. **11**, 556 (2011)
35. X. Yan, W. Liu, H. Wang, S. Chen, J. Shiomi, K. Esfarjani, H. Wang, D. Wang, G. Chen, Z. Ren, Energy Environ. Sci. **5**, 7543 (2012)
36. P.H. Ngan, D.V. Christensen, G.J. Snyder, L.T. Hung, S. Linderoth, N.V. Nong, N. Pryds, phys. Status Solidi (a) **211**, 9 (2014)
37. M. Schwall, B. Balke, Appl. Phys. Lett. **98**, 042106 (2011)
38. C.S. Birkel, W.G. Zeier, J.E. Douglas, B.R. Lettiere, C.E. Mills, G. Seward, A. Birkel, M.L. Snedaker, Y. Zhang, G.J. Snyder et al., Chem. Mater. **24**, 2558 (2012)
39. J.W. Simonson, D. Wu, W.J. Xie, T.M. Tritt, S.J. Poon, Phys. Rev. B **83**, 235211 (2011)
40. H. Hazama, M. Matsubara, R. Asahi, J. Electron. Mater. **41**, 1730 (2012)
41. H.-H. Xie, C. Yu, T.-J. Zhu, C.-G. Fu, G.J. Snyder, X.-B. Zhao, Appl. Phys. Lett. **100**, 254104 (2012)
42. M.-S. Lee, F.P. Poudeu, S.D. Mahanti, Phys. Rev. B **83**, 085204 (2011)
43. Y. Gelbstein, N.T.A. Yarmek, Y. Rosenberg, M.P. Dariel, S. Ouardi, B. Balke, C. Felser, M.Köhne, J. Mater. Res. **26** (2011)
44. M. Mikami, Y. Kinemuchi, K. Ozaki, Y. Terazawa, T. Takeuchi, J. Appl. Phys. **111**, 093710 (2012)
45. J.P.A. Makongo, D.K. Misra, X. Zhou, A. Pant, M.R. Shabetai, X. Su, C. Uher, K.L. Stokes, and Pierre F.P. Poudeu, J. Am. Chem. Soc. **133**, 18843 (2011)
46. M.A. Verges, P.J. Schilling, P. Upadhyay, W.K. Miller, R. Yaqub, K.L. Stokes, P.F.P. Poudeu, Sci. Adv. Mater. **3**, 659 (2011)
47. P. Klaer, M. Kallmayer, C.G.F. Blum, T. Graf, J. Barth, B. Balke, G.H. Fecher, C. Felser, H.J. Elmers, Phys. Rev. B **80**, 144405 (2009)
48. P. Klaer, T. Bos, M. Kallmayer, C.G.F. Blum, T. Graf, J. Barth, B. Balke, G.H. Fecher, C. Felser, H.J. Elmers, Phys. Rev. B **82**, 104410 (2010)
49. C. Yu, T. Zhu, K. Xiao, J. Shen, X. Zhao, Funct. Mater. Lett. **03**, 227 (2010)
50. J.E. Garay, Annu. Rev. Mater. Res. **40**, 445 (2010)
51. C.S. Birkel, J.E. Douglas, B.R. Lettiere, G. Seward, N. Verma, Y. Zhang, T.M. Pollock, R. Seshadri, G.D. Stucky, Phys. Chem. Chem. Phys. **15**, 6990 (2013)
52. M. Zou, J.-F. Li, B. Du, D. Liu, T. Kita, J. Solid State Chem. **182**, 3138 (2009)
53. C. Wang, J. Meyer, N. Teichert, A. Auge, E. Rausch, B. Balke, A. Hütten, G.H. Fecher, C. Felser, J. Vac. Sci. Technol. B **32**, 020802 (2014)
54. S. Ouardi, G.H. Fecher, B. Balke, X. Kozina, G. Stryganyuk, C. Felser, S. Lowitzer, D. Ködderitzsch, H. Ebert, E. Ikenaga, Phys. Rev. B **82**, 085108 (2010)
55. S. Ouardi, G.H. Fecher, C. Felser, M. Schwall, S.S. Naghavi, A. Gloskovskii, B. Balke, J. Hamrle, K. Postava, J. Pištora et al., Phys. Rev. B **86**, 045116 (2012)
56. K. Miyamoto, K. Kimura, K. Sakamoto, M. Ye, Y. Cui, K. Shimada, H. Namatame, M. Taniguchi, S.I. Fujimori, Y. Saitoh, E. Ikenaga, K. Kobayashi, J. Tadano, T. Kanomata, Appl. Phys. Express **1**, 081901 (2008)

57. S. Ouardi, G.H. Fecher, B. Balke, M. Schwall, X. Kozina, G. Stryganyuk, C. Felser, E. Ikenaga, Y. Yamashita, S. Ueda et al., Appl. Phys. Lett. **97**, 252113 (2010)
58. D.M. Rowe, *Thermoelectrics Handbook: Macro to Nano* (CRC Taylor & Francis, Boca Raton, 2006)
59. J. Yang, Adv. Funct. Mater. **18**, 2880 (2008)
60. A. Horyn', O. Bodak, L. Romaka, Y. Gorelenko, A. Tkachuk, V. Davydov, Y. Stadnyk, J. Alloy. Compd. **363**, 10 (2004)
61. J.M. Ziman, *Electrons and Phonons* (Oxford University Press, Oxford, 1960)
62. S.R. Culp, J.W. Simonson, S.J. Poon, V. Ponnambalam, J. Edwards, T.M. Tritt, Appl. Phys. Lett. **93**, 022105 (2008)
63. S. Sakurada, N. Shutoh, Appl. Phys. Lett. **86**, 2105 (2005)
64. X. Yan, W. Liu, S. Chen, H. Wang, Q. Zhang, G. Chen, Z. Ren, Adv. Energy Mater. **3**, 1195 (2013)
65. J. Tobola, J. Pierre, S. Kaprzyk, R.V. Skolozdra, M.A. Kouacou, J. Phys. Condens. Matter **10**, 1013 (1998)
66. J. Krez, J. Schmitt, G.J. Snyder, C. Felser, W. Hermes, M. Schwind, *J. Mater. Chem. A***2**,13513 (2014)
67. H. Xie, H. Wang, Y. Pei, C. Fu, X. Liu, G.J. Snyder, X. Zhao, T. Zhu, Adv. Funct. Mater. **23**, 5123 (2013)
68. H. Hohl, A. Ramirez, W. Kaefer, K. Fess, Ch. Thurner, Ch. Kloc, E. Bucher, Mater. Res. Soc. Symp. Proc. **478**, 109 (1997)
69. H. Hohl, A. Ramirez, C. Goldmann, G. Ernst, B. Wolfing, E. Bucher, J. Phys.: Condens. Matter **11**, 1697 (1999)
70. S. Öğüt, K.M. Rabe, Phys. Rev. B **51**, 10443 (1995)
71. Q. Shen, L.M. Zhang, L.D. Chen, T. Goto, T. Hirai, J. Mater. Sci. Lett. **20**, 2197 (2001)
72. Q. Shen, L. Chen, T. Goto, T. Hirai, J. Yang, G.P. Meisner, C. Uher, Appl. Phys. Lett. **79**, 4165 (2001)
73. S.R. Clup, S.J. Poon, N. Hickman, T.M. Tritt, J. Blumm, Appl. Phys. Lett. **88**, 042106 (2006)
74. T. Graf, J. Barth, C.G.F. Blum, B. Balke, C. Felser, P. Klaer, H.J. Elmers, Phys. Rev. B **82**, 194420 (2010)
75. T. Graf, P. Klaer, J. Barth, B. Balke, H.J. Elmers, C. Felser, Scr. Mater. **63**, 1216 (2010)
76. S. Sakurada, N. Shutoh, J. Alloy. Compd. **389**, 204 (2005)
77. J. Krez, B. Balke, W. Hermes, M. Schwind, C. Felser, arXiv:1502.01828 (2015)
78. L. Zhao, B.-P. Zhang, W. Liu, H. Zhang, J.-F. Li, J. Alloy. Compd. **467**, 91 (2009)
79. E. Hatzikraniotis, K.T. Zorbas, I. Samaras, Th Kyratsi, K. Paraskevopoulos, J. Electron. Mater. **39**, 2112 (2010)
80. Y. Gelbstein, B. Dado, O. Ben-Yehuda, Y. Sadia, Z. Dashevsky, M.P. Dariel, Chem. Mater. **22**, 1054 (2010)
81. E. Rausch, B. Balke, S. Ouardi, C. Felser, Phys. Chem. Chem. Phys. **16**, 25258 (2014)
82. E. Rausch, S. Ouardi, U. Burkhardt, C. Felser, J.M. Stahlhofen, B. Balke, arXiv:1502.03336 (2015)

Chapter 11
Thermodynamics and Energy Conversion in Heusler Alloys

Yintao Song, Chris Leighton and Richard D. James

Abstract Heusler alloys, particularly in the family Ni_2MnZ ($Z = Ga$, In, Sn and Sb) and nearby compositions, often exhibit martensitic phase transformation from a high-temperature cubic ($L2_1$) structure to a low-temperature, low-symmetry martensitic phase. These transformations are commonly accompanied by a change in magnetic ordering, due to the sensitivity of spin interactions to the change in interatomic distances and local symmetry. Various forms of energy can be made to interconvert during this multiferroic phase transformation. A particularly interesting family of alloys for this purpose is $Ni_{50-x}Co_xMn_{25-y}Z_y$ (Z = Ga, In, Sn and Sb). Over a small compositional range, the phase transformation in these alloys is accompanied by a large change in magnetization—up to 1.2 MA/m (1200 emu/cm^3) in $Ni_{45}Co_5Mn_{40}Sn_{10}$. In such materials it is possible to design an energy conversion device that directly converts heat to electricity using Faraday's law of induction and cyclic phase transformation. Both the efficiency and work output per volume of such an energy conversion device are significantly affected by the size of the hysteresis, which is however tunable in these systems. A thermodynamic theory for such a energy conversion method is presented based on a free energy function that includes contributions from phase transformation and magnetism. Material constants in the free energy functions are determined by magnetic and calorimetric measurements. We give the estimates of efficiency and power output in terms of material constants, design parameters and working conditions for this energy conversion method. The

Y. Song · R.D. James (✉)
Department of Aerospace Engineering and Mechanics,
University of Minnesota, Minneapolis, MN 55455, USA
e-mail: james@umn.edu

Y. Song
e-mail: ytsong@umn.edu

C. Leighton
Department of Chemical Engineering and Materials Science,
University of Minnesota, Minneapolis, MN 55455, USA
e-mail: leighton@umn.edu

C. Felser and A. Hirohata (eds.), *Heusler Alloys*,
Springer Series in Materials Science 222, DOI 10.1007/978-3-319-21449-8_11

first-order nature of the phase transformation, leading to an effect of magnetic field on transformation temperature and a mixed-phase region, play a critical role for the effectiveness of these methods.

11.1 Introduction

With increasing demand for electric power and unabated, approximately linear increase of fossil fuel consumption, energy conversion from sustainable sources to electricity by environmentally friendly methods has become an imperative. Waste heat from the production of electricity, and from the devices that use this electricity, is ubiquitous, but is limited to a regime of relatively small temperature difference from ambient, too small to utilize steam generation, the main worldwide source of electricity. In addition, the deserts and arctic regions of the world provide natural sources in this small temperature difference regime. First order solid-solid phase transformations with an abrupt change in magnetic or electric polarization are well suited to do energy conversion in this regime. The basic idea is to periodically induce the phase transformation using the heat source, bias the material by an applied field or stress, and use processes of induction or charge separation to generate electricity. Essentially, a portion of the latent heat absorbed is converted to electricity. These can be termed *direct* methods for the conversion of heat to electricity [1] because they do not need a separate electrical generator. In principle, the device can be quite compact and, therefore, distributed generation becomes a possibility. The critical aspect that enables the possibility of large efficiency and power output, consistent with the given temperature difference, is the same as that of a steam generator: unlike in single phase materials, the first order nature of the phase transformation implies the existence of a mixed phase region that supports Carnot and related high efficiency cycles.

Heusler alloys play into this energy conversion method because many of them have the following properties: (1) a first order phase transformation emitting/absorbing a large latent heat; (2) an abrupt change in ferromagnetic or ferroelectric ordering near the transition temperature; (3) the tunability to achieve conditions of compatibility that imply low thermal hysteresis and high phase reversibility that, in turn, enable numerous conversion cycles. These specifications provide a reliable alloy development strategy to look for the best candidates for energy conversion applications. Starting with the most popular Ni_2MnGa ferromagnetic shape memory alloy [2, 3], a large number of its nearby Heusler alloys with reversible martensitic phase transformations have been studied for mechanical caloric and magnetocaloric effects [4–6], for actuation, and for solid state refrigeration [7–9]. Similarly, the periodic oscillations in magnetization between phases can be used for electromagnetic induction. The family of Ni_2MnX and its slightly off-stoichiometric compositions where $X =$ In, Sn, and Sb is of particular interest as an applicable waste heat energy conversion medium. Many in this family of Heusler alloys have phase transformations from a weakly ferromagnetic (including paramagnetic and anti-ferromagnetic) martensite to a strongly ferromagnetic austenite. While the study of these methods is just beginning, there are now at least two demonstrations [1, 10] of energy conversion devices in the family Ni_2MnX (X = In, Sn, Sb) of Heusler alloys with appropriate chemical substitution.

To use these materials as the working medium for a large number of conversion cycles, an important prerequisite is to control the phase reversibility and thermal hysteresis while the material changes its crystal structure back and forth. The change of crystal structure results in formation of microstructure. If the phase interface is elastically incompatible [11, 12], stress accumulates in the material at the interface. The creation of these stressed transition layers has been identified as the main cause of thermal hysteresis [12, 13], via the mechanism of the formation of an energy barrier [14]. Under cyclic operation, dislocations and other defects may also form in these transition layers and cause degradation and eventual failure of the material [15, 16]. In Sect. 11.2, we describe a theoretical approach to minimize the thermal hysteresis and enhance the transformation reversibility in martensitic materials that is apparently widely applicable to known Heusler alloys having a martensitic phase transformation. The simplest result of this theory is the alloy-development criterion $\lambda_2 = 1$, where λ_2 is the middle eigenvalue of the transformation stretch tensor. The deviation of λ_2 from 1 represents microstructural incompatibility of a single variant of martensite with austenite. Experimentally, by systematically changing the atomic composition using doping elements with similar chemical properties, the condition $\lambda_2 = 1$ can often be satisfied. An example of the use of this strategy occurs in the off-stoichiometric Heusler system $(Ni, Co)_2MnSn$ with small amount of cobalt doping from 3 to 9 at.% [6]. This system not only shows low thermal hysteresis (only 6°C) during the phase transformation, but the maximum magnetization under modest applied field (1 T) also jumps up to 1.2 MA/m (1200 emu/cm^3), which is ideal for energy conversion applications.

A typical specific concept of an energy conversion device is the following. A plate or rod of an alloy, say $Ni_{45}Co_5Mn_{40}Sn_{10}$, is surrounded by a pick-up coil, and is placed near a permanent magnet. By heating the sample through the transformation, the magnetization $\mathbf{M}$ in the material is suddenly increased to its maximum value, biased by the permanent magnet. The dipolar relationship $\mathbf{B} = \mu_0(\mathbf{H} + \mathbf{M})$ implies that the large $\partial\mathbf{M}/\partial t$ is partitioned between $\partial\mathbf{B}/\partial t$ and $-\partial\mathbf{H}/\partial t$. This partitioning is mainly governed by micromagnetic principles, and it is well-known that the demagnetization energy (and the associated shape effect) plays an important role. With a suitable shape and field, the large $\partial\mathbf{M}/\partial t$ produces a large $\partial\mathbf{B}/\partial t$. Owing to Faraday's law, $\partial\mathbf{B}/\partial t = -\mathrm{curl}\,\mathbf{E}$, i.e., an electric field is created in the coil by electromagnetic induction. When cooling the sample to pass through the martensite temperature, the magnetization in the material is decreased to its smallest value while the electric field is generated in an opposite direction. Oscillating the phase transformation by a periodic temperature field, an alternating current can be generated directly from heat through the periodic change of magnetization in such a material. A demonstration based on these ideas is given by Song et al. [17].

Thermodynamic models for energy conversion can also be used to analyze magnetocaloric refrigeration [8, 9, 18]. In fact, the Heusler alloys Ni-Mn-Sn have an "inverse magnetocaloric effect" [19]. Since the thermodynamic cycles of a refrigerator and a heat engine working at the same temperature difference are identical except for the signs of the net work done and heat absorbed, the explicit free energy and our analysis presented below of energy conversion cycles can be easily adopted to magnetic refrigerators that use the same materials.

One difficulty for the prediction of the power output is the quasistatic nature of the ferroic-caloric energy conversion. The simplest generally accepted finite-time thermodynamics is the endoreversible thermodynamics proposed by Curzon and Ahlborn [20]. The terminology "endoreversibility" means that the dissipation only comes from the heat exchange between the working material and heat reservoirs, while the working material internally still performs reversible processes [21, 22]. The losses due to the finite rate of processes are located only in the interaction between the reversible subsystem and its environment. Van den Broeck et al. [23] proved that the efficiency formula derived by Curzon and Ahlborn [20] (the C–A efficiency) agrees with, without approximation, the theory of linear irreversible thermodynamics, universally up to the quadratic order in the deviation from equilibrium [24]. By adopting the assumption of endoreversibility, Song [25] established a finite-rate thermodynamic model for energy conversion by phase-transforming Heusler alloys that is capable of predicting the power output of the energy conversion method reviewed here.

11.2 Multiferroic Heusler Alloys with Low Hysteresis at Phase Transformation

To achieve low hysteresis of phase transformation for the sustainability of cyclic applications, the lattice parameters in the two phases should satisfy the *compatibility condition* that $\lambda_2 = 1$ where λ_2 is the middle eigenvalue of the transformation stretch matrix $\mathbf{U}$. The stretch matrix $\mathbf{U}$ is the pure stretch part of the deformation gradient mapping the unit cell of austenite to one of martensite. The deformation gradient $\mathbf{F}$ for a martensitic transformation is calculated by the Cauchy-Born rule that

$$\mathbf{e}_i^{\mathrm{m}} = \mathbf{F}\mathbf{e}_i^{\mathrm{a}}, \tag{11.1}$$

where $\mathbf{e}_i^{\mathrm{a}}$ and $\mathbf{e}_i^{\mathrm{m}}$, $i = 1, 2, 3$, are three linearly independent lattice vectors of austenite and martensite as shown in Fig. 11.1. The transformation stretch matrix $\mathbf{U} = \sqrt{\mathbf{F}^T\mathbf{F}}$. Physically, $\lambda_2 = 1$ is the necessary and sufficient condition that there exists an undistorted interface between austenite and a single variant of martensite [11, 12].

Multiferroic Heusler alloys often undergo cubic ($L2_1$) to 4 periodic layer modulated orthorhombic (4O) or 7, 10 periodic layer modulated monoclinic (7M, 10M) structural transformations. In these kinds of complex lattices, the deformation gradient is still given by (11.1) (according to the Cauchy-Born rule), but the lattice vectors

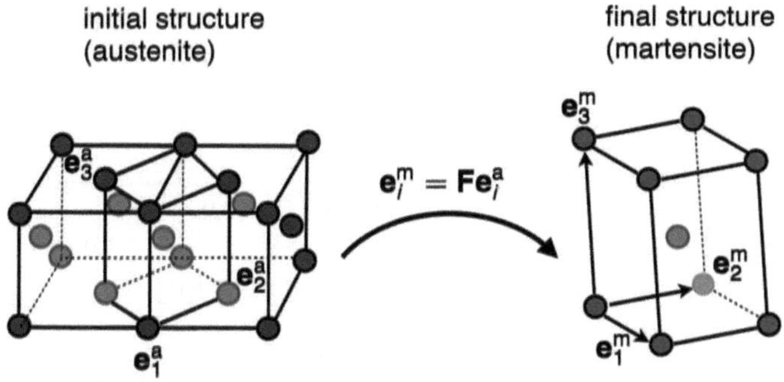

Fig. 11.1 Illustration of the Cauchy-Born rule for the face-centered cubic to monoclinic phase transformation

$\mathbf{e}_i^a$ and $\mathbf{e}_i^m$, $i = 1, 2, 3$, are interpreted as suitable sublattice vectors of austenite and primitive lattice vectors of martensite, respectively. Particularly, the Heusler alloy family $Ni_{50}Mn_{50-x}Sn_x$ undergoes an $L2_1$ to (7M, 10M) transformation, and the form of the transformation stretch matrix is

$$\mathbf{U} = \begin{bmatrix} \alpha & \beta & 0 \\ \beta & \delta & 0 \\ 0 & 0 & \gamma \end{bmatrix}, \tag{11.2}$$

in the orthonormal cubic basis. Here, the components α, β, δ and γ depend on the lattice parameters of the austenite and martensite phases. The lattice parameters measured by X-ray diffraction are used to determine the transformation stretch matrix and its middle eigenvalue λ_2. The hysteresis of the transformation is calculated as $(A_s + A_f - M_s - M_f)/2$ where A_s, A_f, M_s and M_f are austenite start/finish and martensite start/finish temperatures respectively. These transformation temperatures can be characterized by Differential Scanning Calorimetry (DSC). Table 11.1 shows the transformation stretch matrices, the corresponding middle eigenvalues and the thermal hysteresis upon martensitic transformation in the Heusler alloy system $Ni_{50}Mn_{50-x}X_x$ (X = In, Sb, Sn). Slight changes in lattice parameters results in significant variation of the condition of compatibility between the two phases. A direct benefit of the satisfaction of compatibility is to drop the transformation hysteresis and increase phase reversibility [12]. Using $NiMn_{40}Sn_{10}$ as a starting point for a further compositional tuning process, the thermal hysteresis decreases to 6°C by substitution of Co for Ni.

In the alloy system $Ni_{50-x}Co_xMn_{40}Sn_{10}$, $x = 3, 5, 7, 9$ have been investigated and it turns out that $x = 5, 6$ satisfy the condition $\lambda_2 = 1$ closely [6]. Due to the Co substitution, the saturation magnetization of the austenite phase in this

Table 11.1 Lattice parameters, transformation stretch matrices and λ_2 for the family of Heusler alloy Ni-Mn-X (X = In, Sb, Sn) undergoing martensitic transformation

Alloy	Latt. param. (Å)		Transformation stretch tensor	λ_2	Hys. (°C)
	Aus.	Mart.			
$Ni_{50}Mn_{37}Sb_{13}$ [26]	5.971	4.305 2.885 8.407	$\begin{bmatrix} 0.8515 & 0.1682 & 0 \\ 0.1682 & 0.8515 & 0 \\ 0 & 0 & 1.4080 \end{bmatrix}$	1.01963	9
$Ni_{50}Mn_{35}In_{15}$ [5]	5.998	4.377 5.654 21.594 91.93	$\begin{bmatrix} 1.0077 & 0.0069 & 0 \\ 0.0069 & 1.0423 & 0 \\ 0 & 0 & 0.9426 \end{bmatrix}$	1.00643	12
$Ni_{50}Mn_{36}Sn_{14}$ [27]	5.988	8.617 5.702 4.359	$\begin{bmatrix} 1.0235 & 0.006 & 0 \\ 0.006 & 1.0235 & 0 \\ 0 & 0 & 0.9522 \end{bmatrix}$	1.01756	12
$Ni_{50}Mn_{37}Sn_{13}$ [26]	5.973	4.313 2.870 8.401	$\begin{bmatrix} 0.8504 & 0.1708 & 0 \\ 0.1708 & 0.8504 & 0 \\ 0 & 0 & 1.4065 \end{bmatrix}$	1.02118	15
$Ni_{50}Mn_{40}Sn_{10}$ [28]	5.965	4.333 5.57 29.971 93.84	$\begin{bmatrix} 0.9864 & 0.0061 & 0 \\ 0.0061 & 1.0548 & 0 \\ 0 & 0 & 0.9338 \end{bmatrix}$	0.9859	10
$Ni_{45}Co_5Mn_{40}Sn_{10}$ [6]	5.986	4.405 5.642 21.68 87.03	$\begin{bmatrix} 1.0590 & 0.0082 & 0 \\ 0.0082 & 1.0054 & 0 \\ 0 & 0 & 0.9425 \end{bmatrix}$	1.0047	6
$Ni_{44}Co_6Mn_{40}Sn_{10}$ [29]	5.987	4.407 5.643 21.69 87.05	$\begin{bmatrix} 1.0591 & 0.0082 & 0 \\ 0.0082 & 1.0060 & 0 \\ 0 & 0 & 0.9425 \end{bmatrix}$	1.0042	6

system increases up to 1.2 MA/m (1200 emu/cm^3) (at $x = 5$) in sharp contrast to those moderate magnetization values in the NiMnSn system. Figure 11.2 shows the calorimetry (DSC) and magnetometry (VSM) measurements for $Ni_{45}Co_5Mn_{40}Sn_{10}$ with $\lambda_2 = 1.0047$.

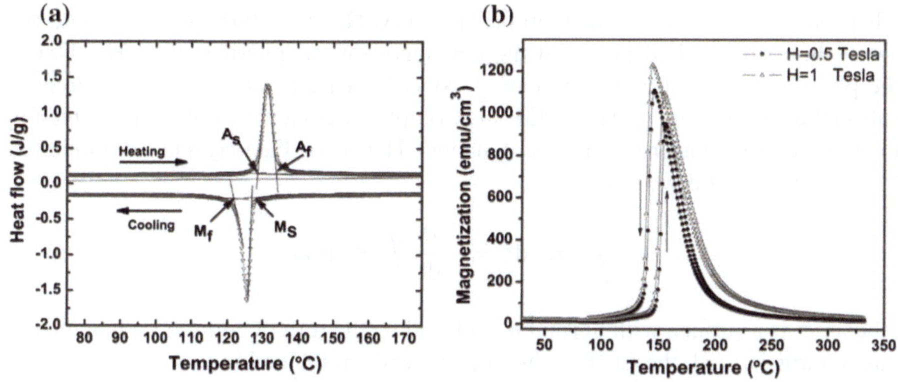

Fig. 11.2 **a** Calorimetry and **b** magnetometry measurements on $Ni_{45}Co_5Mn_{40}Sn_{10}$. Reproduced from [6] with permission © American Institute of Physics

11.3 Energy Conversion Devices

The fact that the Heusler alloy $Ni_{50-x}Co_xMn_{25-y}Z_y$ (Z = Ga, In, Sn and Sb) has a low thermal hysteresis and a big jump of magnetization at the phase transformation has inspired various designs of energy conversion devices [1, 10]. The main idea is first to convert thermal energy into magnetostatic and/or kinetic energy by driving the phase transformation cyclically, then further to convert the latter to electric energy by Faraday's law of induction.

Figure 11.3 is the schematic view of the prototype built by Srivastava et al. [1]. According to the aether relation, the magnetization **M**, the magnetic induction **B** and magnetic field **H** satisfy

$$\mathbf{B} = \mu_0(\mathbf{H} + \mathbf{M}), \tag{11.3}$$

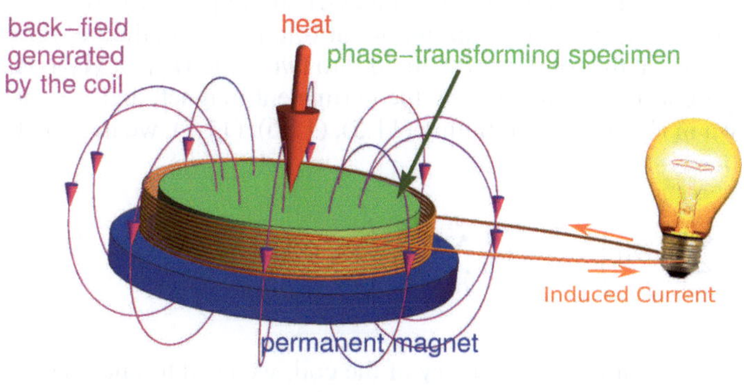

Fig. 11.3 A design based on oscillatory temperature field and fixed specimen. Reproduced from [17] with permission © The Royal Society of Chemistry

where $\mu_0 = 4\pi \times 10^{-7}$ is vacuum permeability. $\mathbf{H}$ can be further divided into two parts: the external field $\mathbf{H}_{\text{ext}}$ and the demagnetization (internal) field $\mathbf{H}_{\text{d}}$. During the phase transformation upon heating, $\mathbf{M}$ is increased in the specimen, giving a substantial component of $\dot{\mathbf{M}} := \partial\mathbf{M}/\partial t$. Complete demagnetization is prevented by the presence of the nearby permanent magnet. Hence, by Faraday's law of induction, for each turn of coil we have

$$\oint_{\partial\mathscr{O}} \mathbf{E} \cdot \mathrm{d}\boldsymbol{\ell} = -\frac{\mathrm{d}}{\mathrm{d}t} \int_{\mathscr{O}} \mathbf{B} \cdot \mathbf{n} \mathrm{d}a. \tag{11.4}$$

Above, $\mathscr{O}$ is a surface with normal $\mathbf{n}$ enclosed by the turn of coil $\partial\mathscr{O}$. When the coil has N turns in total, the EMF across the two ends of the coil is

$$V = \sum_{i=1}^{N} \oint_{\partial\mathscr{O}_i} \mathbf{E} \cdot \mathrm{d}\boldsymbol{\ell} = -\frac{\mathrm{d}}{\mathrm{d}t} \sum_{i=1}^{N} \int_{\mathscr{O}_i} \mathbf{B} \cdot \mathbf{n} \mathrm{d}a. \tag{11.5}$$

Assuming the load can be modeled as a resistor with resistance R, and the resistivity of coil is ρ, by Ohm's law, we obtain the current

$$I = \frac{V}{R + (\rho/a) \sum_{i=1}^{N} |\partial\mathscr{O}_i|}, \tag{11.6}$$

where a is the cross sectional area of the wire.

Notice that this current in the coil will generate a secondary external field, called the back field $\mathbf{H}_{\text{b}}$, which satisfies the Biot-Savart law

$$\mathbf{H}_{\text{b}} = \frac{1}{4\pi} \sum_{i=1}^{N} \oint_{\partial\mathscr{O}_i} \frac{I \mathrm{d}\boldsymbol{\ell} \times \mathbf{r}}{|\mathbf{r}|^3}, \tag{11.7}$$

where $\mathbf{r}$ is the vector from the wire element $\mathrm{d}\boldsymbol{\ell}$ to the point where the back field is calculated. The back-field plays an important role in this method of energy conversion. We can divide the total external field into two parts $\mathbf{H}_{\text{ext}} = \mathbf{H}_0 + \mathbf{H}_{\text{b}}$ where $\mathbf{H}_0$ is the background field generated by the permanent magnet, and is constant for the design shown in Fig. 11.3. Combining (11.5), (11.6), (11.7), we have an ODE for the current I:

$$I\left(R + \frac{\rho}{a}\sum_{i=1}^{N} |\partial\mathscr{O}_i|\right) = -\mu_0 \frac{\mathrm{d}}{\mathrm{d}t} \sum_{i=1}^{N} \int_{\mathscr{O}_i} \left(\mathbf{H}_{\text{d}} + \mathbf{M} + \frac{1}{4\pi} \sum_{j=1}^{N} \oint_{\partial\mathscr{O}_j} \frac{I \mathrm{d}\boldsymbol{\ell} \times \mathbf{r}}{|\mathbf{r}|^3}\right) \cdot \mathbf{n} \mathrm{d}a. \tag{11.8}$$

To solve (11.8), besides the geometry of the coil, we need to know $\dot{\mathbf{H}}_{\text{d}} + \dot{\mathbf{M}}$, i.e. the dynamics of phase transformation. That requires heat transfer and thermodynamic analysis, which will be discussed in later sections.

The main driving force of the system is $\dot{\mathbf{M}}$, i.e., I is almost proportional to $-\dot{\mathbf{M}} \cdot \mathbf{n}$. Besides, $(\mathrm{d}\boldsymbol{\ell} \times \mathbf{r}) \cdot \mathbf{n} > 0$. That is, $\dot{\mathbf{H}}_\mathrm{b}$ weakens the driving power of $\dot{\mathbf{M}}$. This is easy to understand by realizing that the back field is a recoil effect caused by the changing of $\mathbf{M}$. But the back field $\mathbf{H}_\mathrm{b}$ is not a malicious factor of the system at all. In fact, it is the source of energy output, in this particular design. This can be seen by writing down the first law of thermodynamics in two different approaches [17]. In the first approach, the thermodynamic system is chosen to be the specimen only. Denoting the region occupied by the specimen Ω, the first law reads

$$\frac{\mathrm{d}}{\mathrm{d}t}\int_\Omega u\mathrm{d}x = -\oint_{\partial\Omega} \mathbf{q} \cdot \mathbf{n}\mathrm{d}a + \mu_0 \int_\Omega (\mathbf{H}_0 + \mathbf{H}_\mathrm{b}) \cdot \dot{\mathbf{M}}\mathrm{d}x, \tag{11.9}$$

where u is the internal energy density, $\mathbf{q}$ is the heat flux per unit area, and $\mathbf{H}_0$ is the background field coming from the permanent magnet. Integrating (11.9) over a cyclic process $\mathscr{C} \in [0, \infty)$, history independent terms all vanish, and we have

$$-\oint_\mathscr{C}\oint_{\partial\Omega} \mathbf{q} \cdot \mathbf{n}\mathrm{d}a\mathrm{d}t = -\mu_0 \int_\Omega \mathbf{H}_\mathrm{b} \cdot \dot{\mathbf{M}}\mathrm{d}x. \tag{11.10}$$

That is that the net heat absorbed equals to the net work done by the back field. Then we need to show that the net heat absorbed is the same as the total electricity delivered to the load. To show this, let us introduce the second approach, where the system is chosen to include all the components, i.e. specimen, permanent magnet, coil and load resistor. Now the system only exchanges energy with the environment by heat transfer through the specimen, the coil and the resistor. The former is the same as the left hand side of (11.10), the latter must be a net cooling that equals the accumulated Joule heating over a cycle in order to bring the system to the initial state after a cycle. Finally, we have the *equivalence of thermal-magnetic and thermal-electric energy conversion*:

$$-\mu_0 \int_\Omega \mathbf{H}_\mathrm{b} \cdot \dot{\mathbf{M}}\mathrm{d}x = \oint_\mathscr{C} I^2 \left(R + \frac{\rho}{a} \sum_{i=1}^{N} |\partial \mathscr{O}_i| \right) \mathrm{d}t. \tag{11.11}$$

Equation (11.11) states that all the electricity comes from the magnetostatic work done by the back field, and is distributed on the load and the coil.

The design discussed so far is not the only possible design. For example, Gueltig et al. proposed an alternative design utilizing a thin film specimen, as shown in Fig. 11.4. Compared to the previous design, this one uses a fixed temperature gradient (difference) and an oscillating specimen, instead of an alternating temperature field and a fixed specimen, to drive cyclic phase transformation. This mechanically oscillatory design has been used in pyromagnetic energy conversion devices [30, 31], but its application to first order phase transformation in Heusler alloys is rather recent [10, 32].

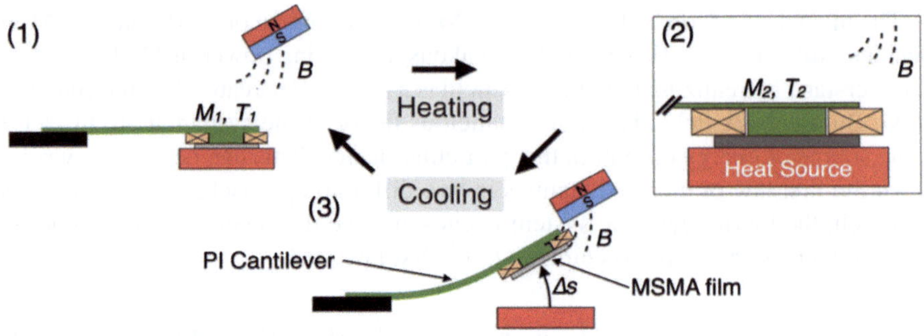

Fig. 11.4 A design based on fixed temperature gradient and oscillatory specimen. Reproduced from [10] with permission © John Wiley & Sons, Inc

11.4 Quasistatic Thermodynamic Analysis

11.4.1 Gibbs Free Energy and Clausius-Clapeyron Relation

As discussed in the previous section, the overall behavior of a whole cycle can be understood by studying only the thermal-magnetic energy interaction. For simplicity, from this section, we assume all variables are spatially uniform in the specimen. Let us define a Gibbs free energy ψ as a function of *external* magnetic field H and temperature T, which satisfies the following two Maxwell relations:

$$\frac{\partial \psi(H,T)}{\partial H} = -\mu_0 M, \quad \frac{\partial \psi(H,T)}{\partial T} = -S. \tag{11.12}$$

Here we choose

$$M = \left| \frac{1}{|\Omega|} \int_\Omega \mathbf{M} \mathrm{d}x \right| \tag{11.13}$$

and H such that

$$HM = \frac{1}{|\Omega|} \int_\Omega \mathbf{M} \cdot \mathbf{H} \mathrm{d}x. \tag{11.14}$$

Integrating the first relation of (11.12), we have

$$\psi(H,T) = -\mu_0 \int_0^H M(h,T)\mathrm{d}h + f(T). \tag{11.15}$$

where $f(T)$ is the field-independent component of the free energy, and $M(H,T)$ is the magnetization as a function of external field and temperature. The latter can be obtained from single-phase M-H and M-T measurements reported in the following sections, where the exact method of interpolation is also provided.

Recall the definition of heat capacity

$$C(H,T) = T\frac{\partial S(H,T)}{\partial T} = -T\frac{\partial^2 \psi(0,T)}{\partial T^2}. \tag{11.16}$$

Then the zero field heat capacity is

$$C(0,T) = -Tf''(T). \tag{11.17}$$

$f(T)$ can be obtained by integrating twice (11.17). Equation (11.17) only holds for single variant specimens. If a mixed-phase specimen is considered, extra terms related to phase distribution (e.g. volume fraction) must be included. For a single phase specimen, the general solution for the field-independent free energy is

$$f(T) = f(T_0) + \int_{T_0}^{T} \left[f'(T_0) - \int_{T_0}^{t_2} \frac{C(0,t_1)}{T} \mathrm{d}t_1 \right] \mathrm{d}t_2. \tag{11.18}$$

T_0 is some reference temperature, which is usually chosen to be the transformation temperature under zero-field. $f(T_0)$ and $f'(T_0)$ are two integral constants. For example, for the Heusler alloy $Ni_{44}Co_6Mn_{40}Sn_{10}$, $T_0 = 390\,\mathrm{K}$.

Combining (11.15) and (11.18) gives the full Gibbs free energy function, and hence an important thermodynamic function: entropy.

$$S(H,T) = \mu_0 \int_0^H \frac{\partial M(h,T)}{\partial T} \mathrm{d}h + \int_{T_0}^{T} \frac{C(0,t)}{t} \mathrm{d}t - f'(T_0). \tag{11.19}$$

Clearly $-f'(T_0)$ is an integral constant representing the field-independent entropy at the temperature T_0.

The Gibbs free energy function (11.15) defines a function on the H-T plane for each phase, as shown in Fig. 11.5a. Austenite (red surface in Fig. 11.5a) has lower energy than martensite (blue surface) does at high temperature. Therefore austenite is more stable at high temperature. The fundamental rule of quasistatic thermodynamics is that phase transformation occurs when the Gibbs free energies of the two phases become equal. In Fig. 11.5, that is represented by the intersection (green curve) of the two Gibbs free energy surfaces.

Using subscripts a and m to denote the austenite and martensite respectively, at every point on the curve we have

$$\psi_{\mathrm{m}}(H,T) = \psi_{\mathrm{a}}(H,T). \tag{11.20}$$

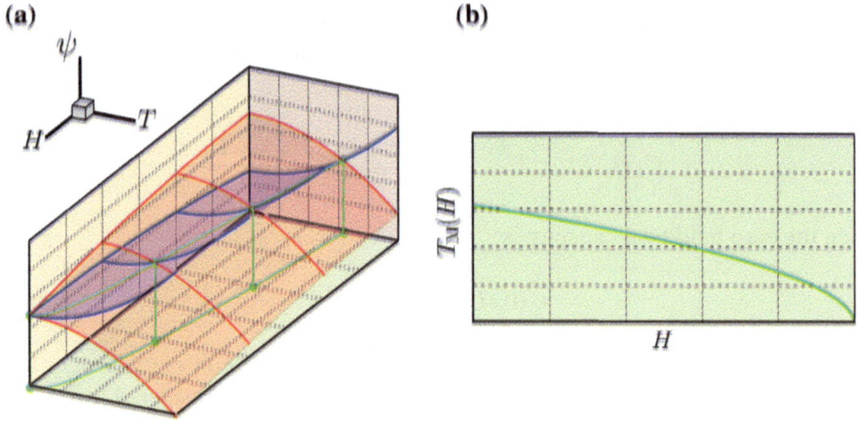

Fig. 11.5 **a** Gibbs free energy and **b** Clausius-Clapeyron relation. *Red* and *blue* surfaces represent the Gibbs free energy of austenite and martensite respectively. *Green curve* is where the phase transformation could occur. (**b**) is the projection of the *green curve* in (**a**) on the H-T plane

The solution to (11.20) gives the transformation temperature as a function of external field $T_M(H)$, which satisfies

$$T'_M(H) = -\mu_0 \frac{M_a(H, T_M) - M_m(H, T_M)}{S_a(H, T_M) - S_m(H, T_M)}, \tag{11.21a}$$

$$T_M(0) = T_0. \tag{11.21b}$$

Equation (11.21a) is also called the Clausius-Clapeyron relation. The denominator must be positive because transforming from martensite to austenite absorbs latent heat. So if $M_a > M_m$, $T_M(H)$ curves has a negative slope. It is equivalent to say that when $M_a > M_m$, external magnetic field stabilizes austenite.

Substituting (11.19) into (11.21a), the denominator becomes

$$\Delta S(H, T_M) = \mu_0 \int_0^H \frac{\partial[M_a(h, T_M) - M_m(h, T_M)]}{\partial T} dh + \int_{T_0}^{T_M} \frac{C_a(0, t) - C_m(0, t)}{t} dt + \frac{L}{T_0}, \tag{11.22}$$

where L is the latent heat of the transformation under zero field, and the last term comes from the definition of latent heat

$$f'_a(T_0) - f'_m(T_0) = -L/T_0. \tag{11.23}$$

Without loss of generality, we can fix the four integral constants in the two Gibbs free energy functions as

$$f_m(T_0) = 0, \quad f'_m(T_0) = 0, \quad f_a(T_0) = 0, \quad f'_a(T_0) = -L/T_0. \tag{11.24}$$

To sum up, if we know $M_{a,m}(H, T)$, $C_{a,m}(0, T)$, L and T_0, we can completely determine the Gibbs free energy functions and the Clausius-Clapeyron relation. For a given material these functions and quantities can be determined through calorimetry and magnetometry. Song et al. [17] derived these functions and quantities for the Heusler alloy $Ni_{44}Co_6Mn_{40}Sn_{10}$.

11.4.2 Thermodynamic Cycles

Engines work by repeating cyclic thermodynamic processes. Let the system be the phase-transforming specimen in an external field; consider the following cycle that consists of subprocesses:

1. Start with the state $\mathscr{S}_1 = (T_-, S_m(H_-, T_-))$, i.e. martensite under the external field H_- and at the temperature $T_- = T_M(H_-)$.
2. Change the field to H_+ isothermally. The state is now $\mathscr{S}_2 = (T_-, S_m(H_+, T_-))$.
3. Heat up under the constant field H_+ to the transformation temperature $T_+ = T_M(H_+)$.[1] The state is now $\mathscr{S}_3 = (T_+, S_m(H_+, T_+))$.
4. Transform from martensite to austenite under the field H_+ and the temperature T_+. The end state is $\mathscr{S}_4 = (T_+, S_a(H_+, T_+))$.
5. Change the field back to H_-, isothermally. The state becomes $\mathscr{S}_5 = (T_+, S_a(H_-, T_+))$.
6. Cool down under the constant field H_- to $T_- = T_M(H_-)$, i.e. $\mathscr{S}_6 = (T_-, S_a(H_-, T_-))$.
7. Transform from austenite to martensite under the field H_- and the temperature T_-. The system comes back to the initial state $\mathscr{S}_1$.

Denote the rate of heat absorption $q(t)$. For a quasistatic process from time t_a to t_b, the net heat absorbed during the process Q_{ab} is

$$Q_{ab} = \int_{t_a}^{t_b} q(t)\mathrm{d}t = \int_{S(t_a)}^{S(t_b)} T\mathrm{d}S = T(t_b)S(t_b) - T(t_a)S(t_a) - \int_{t_a}^{t_b} S[H(t), T(t)]\dot{T}(t)\mathrm{d}t \tag{11.25}$$

Hence, in the cycle defined above,

$$Q_{12} = T_-[S_m(H_+, T_-) - S_m(H_-, T_-)], \tag{11.26a}$$

$$Q_{23} = T_+S_m(H_+, T_+) - T_-S_m(H_+, T_-) - \int_{T_-}^{T_+} S_m(H_+, T)\mathrm{d}T, \tag{11.26b}$$

$$Q_{34} = T_+[S_a(H_+, T_+) - S_m(H_+, T_+)], \tag{11.26c}$$

[1] H_+ is not necessarily higher than H_-.

$$Q_{45} = T_+[S_a(H_-, T_+) - S_a(H_+, T_+)], \tag{11.26d}$$

$$Q_{56} = T_- S_a(H_-, T_-) - T_+ S_a(H_-, T_+) - \int_{T_+}^{T_-} S_a(H_-, T)\mathrm{d}T, \tag{11.26e}$$

$$Q_{61} = T_-[S_m(H_-, T_-) - S_a(H_-, T_-)]. \tag{11.26f}$$

Following Carnot's argument, the efficiency of a (quasistatic) cycle $\mathscr{C} \in [0, \infty)$ is

$$\eta = \frac{2\int_{\mathscr{C}} q\mathrm{d}t}{\int_{\mathscr{C}} q + |q|\mathrm{d}t}. \tag{11.27}$$

By the physical meaning of latent heat, $S_a[H, T_M(H)] > S_m[H, T_M(H)]$ for all H. By the positivity of heat capacity, $\partial S_{a,m}(H, T)/T > 0$ for all H and T. Then using (11.26), we have

$$\int_{\mathscr{C}} q\mathrm{d}t = \int_{T_-}^{T_+} S_a(H_-, T) - S_m(H_+, T)\mathrm{d}T, \tag{11.28a}$$

$$\frac{1}{2}\int_{\mathscr{C}} q + |q|\mathrm{d}t = Q_{23} + Q_{34} + \frac{Q_{12} + |Q_{12}|}{2} + \frac{Q_{45} + |Q_{45}|}{2}. \tag{11.28b}$$

Using (11.27) and (11.28), we can calculate the efficiency of the aforementioned energy conversion cycle. After some tedious algebra, the efficiency can be expressed in the form

$$\eta = \frac{\mathbb{A}T_0 + L(T_+ - T_-)}{\mathbb{B}T_0 + LT_+}. \tag{11.29}$$

$\mathbb{A}$ and $\mathbb{B}$ have complicated dependence on $M_{a,m}(H, T)$ and $C_{a,m}(0, T)$.

This is not the only valid, or the best, cycle for this kind of energy conversion. Song et al. [17] discussed another two types of thermodynamic cycles, namely the thermomagnetic Rankine cycle and the thermomagnetic Ericsson cycle. Moreover, the cycle proposed here is only a simplification of the true cycles used by the devices [1, 10] mentioned in the previous section. For the device of Srivastava et al. [1], the difference between H_+ and H_- should be considered as arising solely from the back field, i.e. $H_\pm \approx H_0 + H_{b\pm}$. For the device built by Gueltig et al. [10], one must add elastic potential energy into the Gibbs free energy and modify the analysis above. As discussed in the equivalence of thermal-magnetic and thermal-electric energy conversion, the result will be similar to the analysis above.

11.4.3 A Simplified Model

Here, we propose an idealized model for the phase-transforming material to simplify the Gibbs free energy functions and the efficiency formula. The same simplified model will also be used in later sections.

First, we make the magnetization functions M_{a} and M_{m} constants, and denote the difference $\Delta M = M_{\mathrm{a}} - M_{\mathrm{m}}$. Then

$$\int_0^H M_{\mathrm{a,m}}(h,T)\mathrm{d}h = M_{\mathrm{a,m}}H, \quad \text{and} \quad \int_0^H \frac{\partial M_{\mathrm{a,m}}(h,T)}{\partial T}\mathrm{d}h = 0. \tag{11.30}$$

Second, let the zero-field heat capacities C_{a} and C_{m} be constants. Equation (11.18) gives

$$f_{\mathrm{m}}(T) = -C_{\mathrm{m}}\left[T_0 - T + T\ln\left(\frac{T}{T_0}\right)\right], \tag{11.31a}$$

$$f_{\mathrm{a}}(T) = -C_{\mathrm{a}}\left[T_0 - T + T\ln\left(\frac{T}{T_0}\right)\right] - \frac{L(T-T_0)}{T_0}. \tag{11.31b}$$

And (11.19) gives

$$S_{\mathrm{m}}(H,T) = S_{\mathrm{m}}(T) = C_{\mathrm{m}}\ln\left(\frac{T}{T_0}\right), \tag{11.32a}$$

$$S_{\mathrm{a}}(H,T) = S_{\mathrm{a}}(T) = C_{\mathrm{a}}\ln\left(\frac{T}{T_0}\right) + \frac{L}{T_0}. \tag{11.32b}$$

So (11.28) becomes

$$\int_{\mathscr{C}} q\mathrm{d}t = (C_{\mathrm{a}} - C_{\mathrm{m}})\left[T_+\ln\left(\frac{T_+}{T_0}\right) - T_-\ln\left(\frac{T_-}{T_0}\right) - (T_+ - T_-)\right] + \frac{L(T_+ - T_-)}{T_0}, \tag{11.33a}$$

$$\frac{1}{2}\int_{\mathscr{C}} q + |q|\mathrm{d}t = (C_{\mathrm{a}} - C_{\mathrm{m}})T_+\ln\left(\frac{T_+}{T_0}\right) + C_{\mathrm{m}}(T_+ - T_-) + \frac{LT_+}{T_0}. \tag{11.33b}$$

Since both phases are similar metals, terms involving $C_{\mathrm{a}} - C_{\mathrm{m}}$ are expected to be small compared to other terms. So the efficiency as defined by (11.27) is now

$$\eta = \frac{L(T_+ - T_-)}{C_{\mathrm{m}}T_0(T_+ - T_-) + LT_+}. \tag{11.34}$$

Clearly, $\eta \leqslant (T_+ - T_-)/T_+$, the Carnot efficiency. In other words, when $(C_{\mathrm{a}} - C_{\mathrm{m}})T_0 \ll L$ and $C_{\mathrm{m}}T_0 \ll L$, the quasistatic cycle defined earlier gives exactly the Carnot efficiency in terms of $T_\pm$. We will continue adopting these two assumptions in the remaining part of this chapter. Note that this pair of assumptions is equivalent to $C_a T_0 \ll L$ and $C_m T_0 \ll L$.

Finally, let $H_\pm = H_0 + H_{b\pm}$, for constant H_0 and $H_{b\pm}$, and try to rewrite the efficiency (11.34) in terms of $H_{b\pm}$. Start by reducing (11.22) to

$$\Delta S(T_M) = (C_a - C_m) \ln\left(\frac{T_M}{T_0}\right) + \frac{L}{T_0}. \tag{11.35}$$

Then the Clausius-Clapeyron relation (11.21a) is reduced to a constraint between $T_M(H)$ and H,

$$(C_a - C_m)\left[T_M \ln\left(\frac{T_M}{T_0}\right) - (T_M - T_0)\right] + \frac{L(T_M - T_0)}{T_0} = -\mu_0 \Delta M H \tag{11.36}$$

Again, assuming $(C_a - C_m)T_0 \ll L$,

$$T_M(H) = T_0 - \frac{\mu_0 \Delta M H T_0}{L}. \tag{11.37}$$

Thus

$$T_\pm = T_0 - \frac{\mu_0 \Delta M (H_0 + H_{b\pm}) T_0}{L}, \tag{11.38}$$

and the Carnot efficiency can be expressed as

$$\eta = \frac{T_+ - T_-}{T_+} = \frac{-2\mathbb{C}(H_{b+} - H_{b-})}{\Delta M - \mathbb{C}(H_0 + H_{b+})}, \tag{11.39}$$

where the dimensionless number $\mathbb{C}$ is the Clausius-Clapeyron coefficient

$$\mathbb{C} = \frac{\mu_0 (\Delta M)^2}{L}. \tag{11.40}$$

Since $\eta > 0$ and $-H_0 < H_{b+} < 0 < H_{b-} < H_0$, $d\eta/d\mathbb{C} > 0$, i.e. when the cycle is driven by $H_{b\pm}$, the efficiency increases as $\mathbb{C}$ increases. $\mathbb{C}$ is a material constant. For the Heusler alloy $Ni_{44}Co_6Mn_{40}Sn_{10}$, $\Delta M \approx 10^6$ A/m and $L \approx 10^8$ J/m^3, so $\mathbb{C} \approx 0.012$.

11.5 Finite-Time Performance Analysis

The quasistatic analysis above cannot be applied to the discussion of power output, because the process is assumed infinitely slow. The thermodynamics of fully non-equilibrium processes is still a difficult problem nowadays. Often, various assumptions are adopted to study finite-rate processes so that analysis analogous to that used in equilibrium thermodynamics, e.g. Gibbs' theory, can be used. In this section, we validate Gibbs' picture of first order phase transformation by adopting the assumption of endoreversibility [20, 23, 24]. Hence, we are able to model finite-rate—but not far from equilibrium—processes for energy conversion using phase-changing multiferroic materials. We should mention that the assumption of endoreversibility also gives results in broad agreement with many kinds of energy conversion devices in use today [20, 33].

The basic idea of an endoreversible process is that the irreversibility only comes from the energy exchange between the system and the environment, e.g. heat transfer. The processes in the system, called the inner processes (inner cycles), are still quasistatic. So the work output for a process is calculated in the same way as for a quasistatic process. To get power output, we just need to know the time duration of the process. That is, for a cyclic process the power output is

$$P = \frac{1}{|\mathscr{C}|} \int_{\mathscr{C}} q \mathrm{d}t. \tag{11.41}$$

For example, the system undergoes a Carnot cycle with working temperature T_+ and T_-, called the inner cycle. The isothermal heating and cooling are modeled by Newton's law of cooling

$$q_{\pm} = \alpha h_{\pm}(\tau_{\pm} - T_{\pm}), \tag{11.42}$$

where α is the ratio between the area for heat exchange and the volume of the specimen, $h_{\pm}$ are respectively the heat transfer coefficient in the unit W m^{-2} K^{-1}, and $\tau_{\pm}$ are the ambient temperature (temperature of the heat reservoir) during the two isothermal processes. Clearly, we have the constraint $\tau_- < T_- < T_+ < \tau_+$. Curzon and Alborn [20] discussed these kind of cycles in detail. Here, we focus on those driven by a first-order phase transformation.

Recalling the assumptions $C_a T_0 \ll L$ and $C_m T_0 \ll L$, we note that the two adiabatic processes in the endoreversible Carnot cycle take negligible time. Using (11.33), the power output is therefore

$$P = \frac{L(T_+ - T_-)}{T_0(t_+ + t_-)}, \tag{11.43}$$

where $t_{\pm}$ are the time cost by the two isothermal processes. The efficiency is still the Carnot efficiency, but based on the environmental temperatures. However, since our control parameters are in fact $\tau_{\pm}$ instead of $T_{\pm}$ in this case, we need to rewrite efficiency and power output in terms of the former.

The relation between $T_{\pm}$ and $\tau_{\pm}$, and the values of $t_{\pm}$ originate from the heat transfer law (11.42) and the Clausius-Clapeyron relation (11.38). Taking the heating at T_+ as an example, the total amount of heat that needs to be supplied in order to finish the phase transformation is, according to (11.35),

$$L_+ = T_+ \Delta S(T_+) = \frac{LT_+}{T_0}. \tag{11.44}$$

The time t_+ is therefore

$$t_+ = \frac{L_+}{q_+} = \frac{LT_+}{\alpha h_+ T_0(\tau_+ - T_+)}. \tag{11.45}$$

To further eliminate T_+, we need to use the fact that T_+ is the transformation temperature that deviates from T_0 due the presence of the back field H_b via the Clausius-Clapeyron relation (11.38).

Recalling the previous discussion in Sect. 11.3, in particular (11.8), the back field is related to the rate of change of magnetization. A full simulation of (11.8) is quite complicated and requires a lot of additional information about the device. A reasonable estimate of the rate of change in magnetization is the linear interpolation $\dot{M}_\pm = \pm \Delta M / t_\pm$. Then the back field can be obtained by first solving (11.8) for the current I. As proposed by Song [25], a simple upper bound of the current can serve as an estimate, which leads to a linear dependence of the back field on $\dot{M}_\pm$. Hence the total external fields during heating and cooling are respectively $H_0 - \Pi \dot{M}_\pm$. Then (11.38) reads

$$\frac{T_+}{T_0} = 1 - \frac{\mu_0 \Delta M (H_0 - \Pi \dot{M}_+)}{L}. \tag{11.46}$$

Applying the same argument to T_- leads to the new expressions of efficiency and power output

$$\eta = \frac{\mathbb{C}\Pi(\dot{M}_+ - \dot{M}_-)}{\Delta M - \mathbb{C}(H_0 - \Pi \dot{M}_+)}, \tag{11.47a}$$

$$P = -\frac{L\mathbb{C}\Pi \dot{M}_+ \dot{M}_-}{(\Delta M)^2}. \tag{11.47b}$$

And the corresponding frequency is

$$f = -\frac{\dot{M}_+ \dot{M}_-}{\Delta M (\dot{M}_+ - \dot{M}_-)}. \tag{11.48}$$

Note that this frequency is a result of enforcing $\dot{M}_\pm$ and the assumption that the inner cycle is a Carnot cycle, so it is not a control parameter of the cycle.

Substituting $\dot{M}_\pm = \pm\Delta M/t_\pm$ into (11.45) yields

$$\dot{M}_\pm = \frac{g(zz'_\pm, \hat{\tau}_\pm)(\Delta M - \mathbb{C}H_0)}{2\mathbb{C}\Pi}, \tag{11.49}$$

where $z'_+ = 1$, $z'_- = h_-/h_+$, and $\hat{\tau}_\pm = \tau_\pm \Delta M\big/(\Delta M - \mathbb{C}H_0)T_0$ are the ambient temperatures normalized by the transformation temperature under H_0. The dimensionless function $g(x, y)$ is defined as

$$g(x, y) = -x - 1 + \sqrt{(x+1)^2 + 4x(y-1)}. \tag{11.50}$$

Finally, we obtain as a *figure of merit* the dimensionless parameter

$$z = \frac{\mathbb{C}\Pi\alpha h_+ T_0 \Delta M}{(\Delta M - \mathbb{C}H_0)L}. \tag{11.51}$$

Two time scales are involved in the figure of merit: (i) the electromagnetic induction time scale Π and (ii) the heat transfer time scale $L/\alpha h_+ T_0$. A larger ratio of the former to the latter leads to a larger z. A larger $\mathbb{C}$ gives a larger z, too.

Efficiency, power output and frequency can all be expressed in terms of z and the function $g(x, y)$:

$$\eta = \frac{g(zz'_+, \hat{\tau}_+) - g(zz'_-, \hat{\tau}_-)}{2 + g(zz'_+, \hat{\tau}_+)}, \tag{11.52a}$$

$$\hat{P} = P\frac{\Delta M}{(\Delta M - \mathbb{C}H_0)\alpha h_+ T_0} = -\frac{g(zz'_+, \hat{\tau}_+)g(zz'_-, \hat{\tau}_-)}{4z}, \tag{11.52b}$$

$$\hat{f} = f\frac{2\mathbb{C}\Pi\Delta M}{\Delta M - \mathbb{C}H_0} = -\frac{g(zz'_+, \hat{\tau}_+)g(zz'_-, \hat{\tau}_-)}{g(zz'_+, \hat{\tau}_+) - g(zz'_-, \hat{\tau}_-)}. \tag{11.52c}$$

Since $\lim_{x\to\infty} g(x, y) = 2(y-1)$, we have

$$\lim_{z\to\infty} \eta = \frac{\hat{\tau}_+ - \hat{\tau}_-}{\hat{\tau}_+}, \qquad \lim_{z\to\infty} P = 0, \qquad \lim_{z\to\infty} f = \frac{\Delta M - \mathbb{C}H_0}{\mathbb{C}\Pi\Delta M}\frac{(\hat{\tau}_+ - 1)(1 - \hat{\tau}_-)}{\hat{\tau}_+ - \hat{\tau}_-}. \tag{11.53}$$

That is as $z \to \infty$, the Carnot efficiency is recovered, but the power output vanishes. When x is in the vicinity of zero, $g(x, y) = 2x(y-1) + O(x^2)$. Thus, we have the simplified formulas for small z, up to the order $O(z)$,

$$\eta = z \cdot \left[(\hat{\tau}_+ -) + h'(1 - \hat{\tau}_-)\right], \tag{11.54a}$$

$$P = z \cdot \alpha h_- T_0 \left(1 - \frac{\mathbb{C}H_0}{\Delta M}\right)(\hat{\tau}_+ - 1)(1 - \hat{\tau}_-), \tag{11.54b}$$

$$f = \frac{\alpha h_+ T_0}{L}\frac{h_-(\hat{\tau}_+ - 1)(1 - \hat{\tau}_-)}{h_+(\hat{\tau}_+ - 1) + h_-(1 - \hat{\tau}_-)}. \tag{11.54c}$$

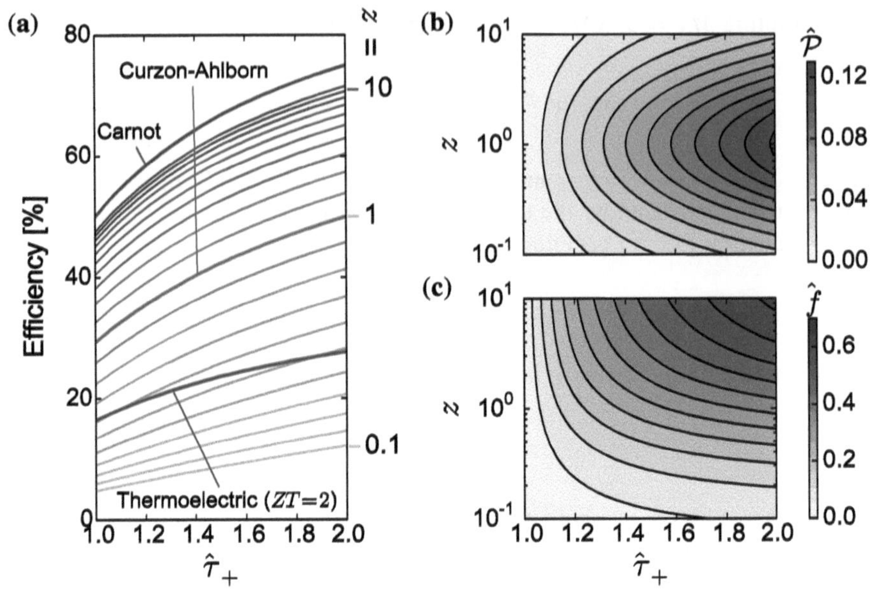

Fig. 11.6 **a** Efficiency, (**b**) effective power output, and (**c**) effective frequency of the proposed device with the figure of merit varying from 0.1 to 10. In **a**, *green thin solid lines* are the efficiency for the thermal-magnetic generators with different z values. The z value increases from 0.1 to 10, as the efficiency curve moves from *bottom* to *top*. The *purple* and *red thick solid curves* represent the Carnot and Curzon-Ahlborn [20] (C-A) efficiencies, respectively. The *blue thick solid curve* corresponds to thermoelectric generators with the ZT value of 2. **b** and **c** are contour plots. In all of these plots, we choose $z'_{\pm} = h' = 1$ and $\hat{\tau}_- = 0.5$. Reproduced from [25] with permission © The PCCP Owner Societies (Color figure online)

So when z is small, η and P grow in the order of z.

Song [25] discussed in detail this endoreversible model, and Fig. 11.6 is the calculation for $z \in [0.1, 10]$. In Fig. 11.6, the efficiency of thermoelectric is calculated using the formula

$$\eta_{\mathrm{TE}} = \frac{\tau_+ - \tau_-}{\tau_+} \frac{\sqrt{1 + ZT} - 1}{\sqrt{1 + ZT} + \tau_+/\tau_-}, \tag{11.55}$$

where $T = (\tau_+ + \tau_-)/2$, and ZT is the so-called figure of merit for thermoelectric materials. The value of ZT depends on the Seebeck coefficient, the electric and thermal conductivity of the material, in addition to the working temperatures. For most of the existing thermoelectric materials, ZT is less than 2 over a wide range of temperature. In Fig. 11.6, ZT is chosen to be 2 for the comparison. An important observation from Fig. 11.6 is that the power output reaches a maximum at $z = 1$, when $h_- = h_+$ and $\hat{\tau}_- = 0.5$, and for the same z gives the Curzon-Alborn efficiency, at which many existing heat engines operate [20, 33].

Table 11.2 Methods of conversion of heat to electricity using phase transformations in multiferroics

Phase 1	Phase 2	Physics	Notes
1. Ferromagnetic	Nonmagnetic	Faraday's law	Biasing by a permanent magnet; external coil
2. Ferroelectric	Nonferroelectric	Ohm's law	Biasing by a capacitor; polarization-induced current
3. Ferromagnetic; high anisotropy	Ferromagnetic; low anisotropy	Faraday's law	Biasing by a permanent magnet, external coil intermediate magnetic field;
4. Ferroelectric; high permittivity	Ferroelectric; low permittivity	Ohm's law	Biasing by a capacitor, intermediate electric field; polarization-induced current
5. Ferromagnetic; large M_s near T_c	Nonmagnetic	Faraday's law	Second-order transition; biasing by a permanent magnet [34]
6. Ferroelectric; large P_s near T_C	Nonpolar	Ohm's law	Second-order transition; biasing by a capacitor [35]
7. Metal	Insulator or semiconductor	Ohm's law	Biasing by a capacitor; polarization-induced current
8. Nonpolar; nonmagnetic	Nonpolar; nonmagnetic	Stress-induced transformation; Faraday's law	Shape-memory engine driving generator; biased by stress [36]

In all cases the transformation is induced by direct heating and cooling. M_s is the saturation magnetization. P_s the saturation polarization, and T_c the Curie temperature (magnetic or ferroelectric, resp.) (Reproduced from [1] with permission © John Wiley & Sons, Inc.)

11.6 Other Similar Energy Conversion Methods

The method discussed in this chapter mainly concerns the transformation from a ferromagnetic phase to a nonmagnetic one. In fact, there are many other similar kinds of phase transformation can be utilized, which are summarized in Table 11.2. Some of them have been prototyped, while many are waiting to be explored. Most of these multiferroic properties have been found in Heusler alloys.

Acknowledgments The work of YS and RJ was supported by the MURI project FA9550-12-1-0458 (managed by AFOSR), ONR N00014-14-1-0714 and NSF-PIRE OISE-0967140. The contribution of CL was supported by the DOE under award DE-FG02-06ER46275

References

1. V. Srivastava, Y. Song, K. Bhatti, R.D. James, The direct conversion of heat to electricity using multiferroic alloys. Adv. Energy Mater. **1**(1), 97–104 (2011)
2. P.J. Webster, K.R.A. Ziebeck, S.L. Town, M.S. Peak, Magnetic order and phase transformation in Ni_2MnGa. Philos. Mag. B **49**(3), 295–310 (1984)
3. K. Ullakko, J.K. Huang, C. Kantner, R.C. O'Handley, V.V. Kokorin, Large magnetic-field-induced strains in NiMnGa single crystals. Appl. Phys. Lett. **69**, 1966 (1996)
4. R. Kainuma, Y. Imano, W. Ito, H. Morito, Y. Sutou, K. Oikawa, A. Fujita, K. Ishida, S. Okamoto, O. Kitakami, T. Kanomata, Metamagnetic shape memory effect in a Heusler-type $Ni_{43}Co_7Mn_{39}Sn_{11}$ polycrystalline alloy. Appl. Phys. Lett. **88**(19), 192513–3 (2006a)
5. V.V. Khovaylo, T. Kanomata, T. Tanaka, Y. Amako, R. Kainuma, R. Y. Umetsu, H. Morito, H. Miki, Magnetic properties of Ni_{50} probed by Mössbauer spectroscopy. Phys. Rev. B, **80**: 144409 (2009)
6. V. Srivastava, X. Chen, R.D. James, Hysteresis and unusual magnetic properties in the singular Heusler alloy NiCoMnSn. Appl. Phys. Lett. **97**(1), 014101 (2010)
7. R. Kainuma, Y. Imano, W. Ito, Y. Sutou, H. Morito, S. Okamoto, O. Kitakami, K. Oikawa, A. Fujita, T. Kanomata, K. Ishida, Magnetic-field-induced shape recovery by reverse phase transformation. Nature **439**(7079), 957–960 (2006b)
8. V.K. Pecharsky, K.A. Jr. Gschneidner, Giant magnetocaloric effect in Gd_5). Phys. Rev. Lett. **78**: 4497–4497 (1997)
9. V.K. Pecharsky, K.A. Gschneidner Jr, Magnetocaloric effect and magnetic refrigeration. J. Magn. Magn. Mater. **200**, 44–56 (1999)
10. M. Gueltig, H. Ossmer, M. Ohtsuka, H. Miki, K. Tsuchiya, T. Takagi, M. Kohl, High frequency thermal energy harvesting using magnetic shape memory films. Adv. Energy Mater (2014)
11. J.M. Ball, R.D. James, Fine phase mixtures as minimizers of energy. Archive Ration. Mech. Anal. **100**, 13 (1987)
12. R.D. James, Z. Zhang, A way to search for multiferroic materials with "unlikely" combinations of physical properties, in *Magnetism and Structure in Functional Materials*. Materials Science, ed. by A. Planes, L. Mañosa, A. Saxena, vol. 9, Chap. 9 (Springer, Berlin, 2005), pp. 159–175
13. R. Zarnetta et al., Identification of quaternary shape memory alloys with near-zero thermal hysteresis and unprecedented functional stability. Adv. Funct. Mater. **20**(12), 1917–1923 (2010)
14. Z. Zhang, R.D. James, S. Müller, Energy barriers and hysteresis in martensitic phase transformations. Acta Mater. **57**(15), 4332–4352 (2009)
15. K. Gall, H.J. Maier, Cyclic deformation mechanisms in precipitated niti shape memory alloys. Acta Mater. **50**(18), 4643–4657 (2002)
16. Z. Moumni, A. Van Herpen, P. Riberty, Fatigue analysis of shape memory alloys: energy approach. Smart Mater. Struct. **14**(5), S287 (2005)
17. Y. Song, K. Bhatti, C. Srivastava, V. Leighton, R.D. James, Thermodynamics of energy conversion via first order phase transformation in low hysteresis magnetic materials. Energy Environ. Sci. **6**(4), 1315–1327 (2013). doi:10.1039/C3EE24021E
18. J. Glanz, Making a bigger chill with magnets. Science **78**, 2045 (1998)
19. T. Krenke, E. Duman, M. Acet, E.F. Wassermann, X. Moya, L. Manosa, Inverse magnetocaloric effect in ferromagnetic Ni-Mn-Sn alloys. Nat. Mater. **4**, 450–454 (2005)
20. F.L. Curzon, B. Ahlborn, Efficiency of a carnot engine at maximum power output: endoreversible thermodynamics. Am. J. Phys. **43**(1), 22 (1975)
21. M. Rubin, Optimal configuration of a class of irreversible heat engines. I. Phys. Rev. A **19**(3), 1272–1276 (1979)
22. K.H. Hoffmann, J.M. Burzler, S. Schubert, Review article: endoreversible thermodynamics. J. Non-Equilib. Thermodyn. **22**(4), 311–355 (1997)
23. C. Van den Broeck, Thermodynamic efficiency at maximum power. Phys. Rev. Lett. **95**(19), 2–4 (2005)
24. M. Esposito, K. Lindenberg, C. van den Broeck, Universality of efficiency at maximum power. Phys. Rev. Lett. **102**(13), 1–4 (2009)

25. Y. Song, Performance analysis of energy conversion via caloric effects in first-order ferroic phase transformations. Phys. Chem. Chem. Phys. **16**(25), 12750–12763 (2014)
26. Y. Sutou, Y. Imano, N. Koeda, T. Omori, R. Kainuma, K. Ishida, K. Oikawa, Magnetic and martensitic transformations of NiMnX (X = In, Sn, Sb) ferromagnetic shape memory alloys. Appl. Phys. Lett. **85**(19), 4358 (2004)
27. K. Koyama, K. Watanabe, T. Kanomata, K. Oikawa, K. Ishida, Observation of field-induced reverse transformation in ferromagnetic shape memory alloy $Ni_{50}Mn_{36}Sn_{14}$. Appl. Phys. Lett. **88**, 132505 (2006)
28. T. Krenke, M. Acet, E.F. Wassermann, X. Moya, Lluís Mañosa, A. Planes, Martensitic transitions and the nature of ferromagnetism in the austenitic and martensitic states of Ni-Mn-Sn alloys. Phys. Rev. B **72**: 014412 (2005)
29. K. Bhatti, S. Khatib-El, V. Srivastava, R.D. James, C. Leighton, Small-angle neutron scattering study of magnetic ordering and inhomogeneity across the martensitic phase transformation in $Ni_{50-x}Co_xMn_{40}Sn_{10}$ alloys. Phys. Rev. B **85**, 134450 (2012)
30. M. Ujihara, G.P. Carman, D.G. Lee, Thermal energy harvesting device using ferromagnetic materials. *Appl. Phys. Lett.***91**(9) (2007). doi:http://dx.doi.org/10.1063/1.2775096
31. S.R. Hunter, N.V. Lavrik, T. Bannuru, S. Mostafa, S. Rajic, P.G. Datskos, Development of mems based pyroelectric thermal energy harvesters, in *SPIE Defense, Security, and Sensing* (International Society for Optics and Photonics, 2011), pp. 80350V–80350V
32. A. Post, C. Knight, E. Kisi, Thermomagnetic energy harvesting with first order phase change materials. J. Appl. Phys. **114**(3) (2013). doi:http://dx.doi.org/10.1063/1.4815933
33. A. Shakouri, Recent developments in semiconductor thermoelectric physics and materials. Ann. Rev. Mater. Res. **41**(1), 399–431 (2011). doi:10.1146/annurev-matsci-062910-100445
34. N. Tesla, Pyromagneto-electric generator. US Patent 428,057. 13 May 1890
35. W.H. Clingman, R.G. Moore Jr, Application of ferroelectricity to energy conversion processes. J. Appl. Phys. **32**(4), 675 (1961). doi:10.1063/1.1736069
36. W.S. Ginell, J.L. McNichols Jr, J.S. Cory, Nitinol heat engines for low-grade thermal energy conversion. Mech. Eng. **101**(5), 28–33 (1979)

Part IV
Heusler Alloy Films: Film Growth

Chapter 12
Spin Polarization in Heusler Alloy Films

Yukiko K. Takahashi and Kazuhiro Hono

Abstract Spin polarization measurements are essential for searching new alloys with high spin polarization. While tunneling magnetoresistance (TMR) measurements have been used to estimate the spin polarization of ferromagnetic (FM) electrodes of magnetic tunneling junctions (MTJs), it requires thin film stack and tedious microfabrication processes to prepare TMR pillars. On the other hand, point contact Andreev reflection (PCAR) can estimate current spin polarizations of bulk alloys relatively easily. Low temperature spin polarization of a series of quaternary substitutional alloys, $Co_2Y(Z_{1-x}Z'_x)$, designed based on the Fermi level tuning, have been investigated using PCAR. Promising alloys were screened and they were evaluated as ferromagnetic (FM) layers of current-perpendicular-to-plane giant magnetoresistance (CPP-GMR) pseudo spin valves. Some quaternary Heusler alloys, such as $Co_2Fe(Al_{0.5}Si_{0.5})$, $Co_2Mn(Ge_{0.75}Ga_{0.25})$ and $Co_2Fe(Ge_{0.5}Ga_{0.5})$ were found to be excellent FM layer for CPP-GMR. Recent theoretical and experimental investigations on the correlation between anisotropy magnetoresistance (AMR) and MR outputs suggest that AMR measurements will be more effective way for screening alloys for high spin polarization, as it can estimate spin polarization at finite temperatures.

12.1 Introduction

Highly spin-polarized ferromagnetic (FM) materials are strongly desired as spin-polarized current sources of spintronics devices such as magnetic tunnel junctions (MTJs), [1] current-perpendicular-to-plane giant-magnetoresistance (CPP-GMR), [2–6] lateral spin valves (LSV), [7, 8] spin injectors to semiconductors [9] and spin torque oscillators (STO) [10]. Among various potential half-metals, Co-based Heusler alloys are particularly promising as the ferromagnetic (FM) electrodes of

Y.K. Takahashi (✉) · K. Hono
Magnetic Materials Unit, National Institute for Materials Science, 1-2-1 Sengen, Tsukuba 305-0047, Japan
e-mail: takahashi.yukiko@nims.go.jp

K. Hono
e-mail: kazuhiro.hono@nims.go.jp

C. Felser and A. Hirohata (eds.), *Heusler Alloys*,
Springer Series in Materials Science 222, DOI 10.1007/978-3-319-21449-8_12

magnetoresistive devices since they have high Curie temperatures (T_c) [11]. Sakuraba et al. demonstrated large TMR ratio of 570 % at 2 K in the MTJ that used Co_2MnSi as FM electrodes with an alumina tunnel barrier; however, the TMR value showed a large temperature degradation, i.e., only 67 % at room temperature (RT). Such a large temperature degradation of MR was also reported in CPP-GMR using Heusler alloys, and this has been attributed to the degradation of bulk spin polarization of Co-based Heusler alloys and magnon excitations at FM/NM interfaces [3]. As seen from these examples, half-metallicity at RT has not been demonstrated yet in any magnetoresistive devices using Heusler alloys.

One of the possible reasons for the low spin polarization at RT is the lack of perfect $L2_1$ order in real materials. The half-metallic band structure was predicted for perfectly $L2_1$-ordered structure, while most of the materials used as thin FM electrodes contain some disorders. Because of the small difference in the Gibbs free energy for B2 (disordering between Y and Z in X_2YZ structure), A2 (X, Y and Z disordering), DO_3 (X and Y disordering) and $L2_1$ phases, B2, DO_3, and A2 disorders are common in Co-based Heusler alloys even if the $L2_1$ phase is thermodynamically stable at room temperature. In addition, thin films that are commonly grown with a sputtering method have the A2 structure in the as-deposited state and either in-situ or post-deposition annealing is required for the A2 phase to transform to the B2 or $L2_1$ ordered phase. Thus, certain degree of disorder in Co-based Heusler alloys is unavoidable. The disadvantage in the disordered structure is the reduction of the spin polarization at finite temperature because of the appearance of DOS in the minority band [3]. The discrepancy between the theoretical calculation and the experimental results in the magneto-transport properties in MR devices is considered to be due to the presence of such disorders in multi-layered thin films. Thus, from the practical point of view, search for Heusler alloys with high robustness of spin polarization against disordering and temperature is necessary.

12.2 Spin Polarization Measurements

The definition of the spin polarization (P) is

$$P = \frac{N\uparrow - N\downarrow}{N\uparrow + N\downarrow}, \tag{12.1}$$

where $N\uparrow$ and $N\downarrow$ are the spin-up and spin-down density of states (DOS) at the Fermi level (E_F). Although the definition of P is simple, experimental measurements of P are not so easy. There are several methods for measuring P, such as superconducting tunneling spectroscopy (STS), [12] Julliere's model for MTJ, [13] and point contact Andreev reflection (PCAR) [14]. Spin resolved x-ray photoelectron spectroscopy has also been used, but this method is limited to the measurements of spin states at surfaces. The first STS measurement of P was carried out by Meservey and Tedrow [12]. By applying high magnetic field to microfabricated

FM/oxide/superconductor (SC) junctions, DOS of the SC splits by Zeeman energy, causing asymmetric conduction curves. They obtained P of 44, 34 and 11 % for Fe, Co and Ni, respectively. 34 % for Co and 11 % for Ni are now considered to be underestimation.

The spin polarizations of electrode materials of MTJs were often estimated using Julliere model, [13]

$$TMR = \frac{2P_1P_2}{1-P_1P_2}, \tag{12.2}$$

where P_1 and P_2 are the tunneling spin polarization through FM/barrier/FM junctions. However, note that P_1 and P_2 are not the bulk spin polarization of materials, but are the polarization values of tunneling electrons; hence, it is susceptible to FM/barrier interface structure and the spin filtering effect through a tunnel barrier, e.g. MgO. For example, $P > 0.8$ for Fe/MgO/Fe interface, in spite of the low spin polarization of Fe of ~ 0.4. This is because the MgO barrier has a spin filtering effect for Δ_1 band electrons, [15] so only the electrons in the Δ_1 band in Fe, which is half-metallic, are tunneled through the MgO barrier. In addition, the method requires a time-consuming microfabrication process to prepare TMR pillars. To estimate the spin polarization of bulk and thin film materials promptly for screening promising materials, point contact Andreev reflection (PCAR) is a relatively easy method for which neither microfabrication nor thin film processing is necessary.

12.3 Point Contact Andreev Reflection (PCAR)

The spin polarization measurement by PCAR utilizes the Andreev reflection that occurs at a point contact between a normal conductor and a superconductor. Thus, this method requires cryogenic cooling for measurements. Figure 12.1a shows the schematic illustration of the density of states (DOS) of a non-magnetic metal (NM) and a superconductor (SC). Electrons can flow from NM to SC by forming Cooper pairs of spin-up and spin-down electrons when the energy of the incident electrons is smaller than the superconducting band gap. In the normal conductor with no polarization of spins ($P = 0$), the same number of spin-up and spin-down electrons are present at E_F, so they can form Cooper pairs and thus the conductance is enhanced to double within the superconducting gap as shown in Fig. 12.1c. To conserve the momentum in the system, holes are reflected to the normal conductor, which is called Andreev reflection. On the other hand, in the ferromagnetic material (FM), the Andreev reflection is somewhat suppressed due to the different DOS of spin-up and spin-down. When the FM is perfectly spin-polarized, there is no electron to form a Cooper pair as shown in Fig. 12.1b, so the conductance in the superconducting gap is completely suppressed. Therefore, the conductance curves vary depending on the spin polarization of the material. The spin polarization is deduced by fitting the conductance curve to the modified BTK model [16] under the assumption that the electron scattering occurs due to the surface oxidation.

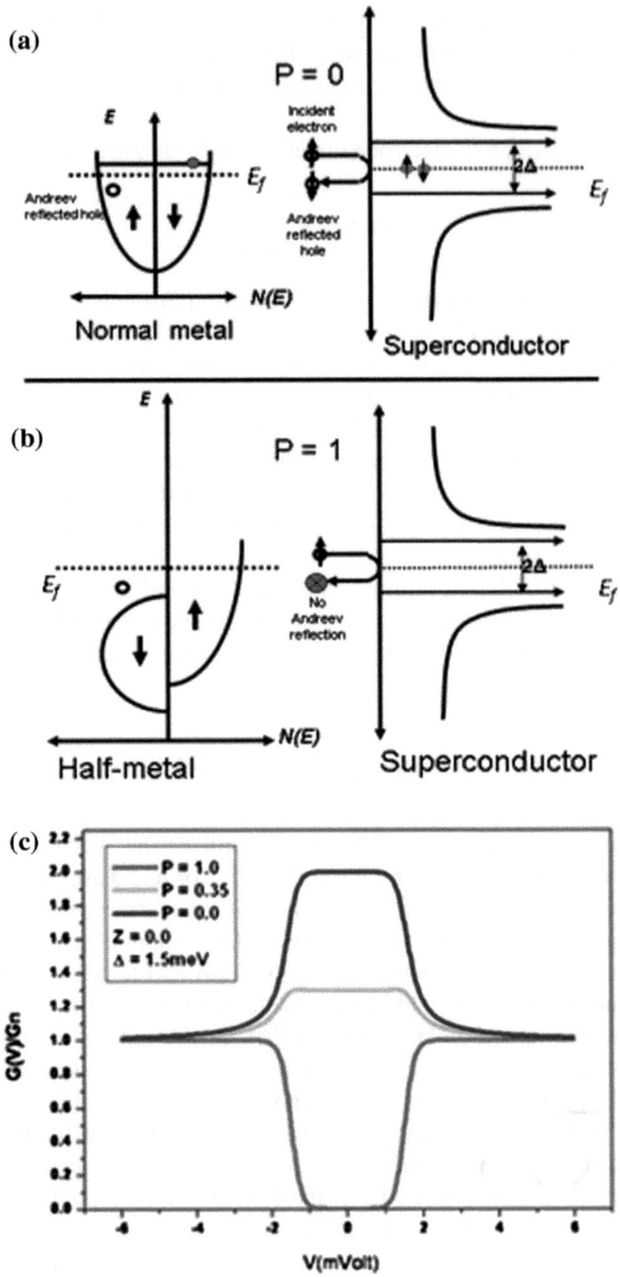

Fig. 12.1 Schematic images of Andreev reflections with point contact of (**a**) normal conductor and superconductor and (**b**) half-metal and superconductor. **c** Calculated conductance curves with different spin polarization. Courtesy of A. Rajanikanth

Based on the BTK model, [17] the current at the interface between the normal conductor and superconductor is expressed by

$$I = 2eSv_F N(E_F) \int_{-\infty}^{\infty} [1 + A(E, Z) - B(E, Z)][f(E - eV, T) - f(E, T)]dE, \tag{12.3}$$

where S, $N(E_F)$, $f(E, T)$ and Z represent an area of the interface between the normal conductor and the superconductor, DOS at the E_F in the normal conductor, Fermi distribution function and a scattering factor, respectively. $A(E, Z)$ and $B(E, Z)$ are the probability of Andreev reflection and the probability of the normal reflection. The derivative conductance $G = dI/dV$ is expressed by

$$G \propto \int_{-\infty}^{\infty} \frac{d}{dE} f(E - eV, T)\big[1 + A(E, Z) - B(E, Z)\big]dE. \tag{12.4}$$

At 0 K, (12.2) can be simplified to

$$G \propto 1 + A(E, Z) - B(E, Z). \tag{12.5}$$

Strigkers et al. [16] modified the BTK model to FM material. In the modified BTK model, the current is divided into non-spin-polarized and perfectly spin-polarized current (I_N and I_H).

$$I_{total} = (1 - P)I_N + PI_H, \tag{12.6}$$

where P is the spin polarization of the material. Therefore, the conductance at 0 K is expressed by

$$G_{total} \propto (1 - P)\big[1 + A_N - B_N\big] + P\big[1 + A_H - B_H\big], \tag{12.7}$$

where

$$1 + A_N - B_N = \begin{cases} \frac{2(1+\beta^2)}{\beta^2 + \left(1+2Z^2\right)^2}, & |eV| < \Delta, \\ \frac{2\beta}{1+\beta+2Z^2}, & |eV| < \Delta, \end{cases} \tag{12.8}$$

$$1 + A_H - B_H = \begin{cases} 0, & |eV| < \Delta, \\ \frac{1+\beta(1+2Z^2)}{(1+\beta)(1+2Z^2)+2Z^4}, & |eV| < \Delta, \end{cases} \tag{12.9}$$

and $\beta = \sqrt{|\Delta^2 - E^2|}$ [16]. The spin polarization is estimated by fitting the experimentally obtained conductance curve to equations (12.6)–(12.9).

To observe the Andreev reflection, a point contact of several 10 nm, which should be shorter than the mean free path of electrons, is required because the transport should be ballistic. Typically, electro-polished sharp Nb tips are used to create point contacts so the measurement temperature must be lower than the critical temperature of Nb, 9.2 K. However, a typical contact resistance was ~20 Ω, corresponding to a diameter of about 28 nm, which is larger than the mean free path of electrons. So the conductance is considered to occur through multiple diffusive contacts in actual experiments. In such a diffusive conduction, the spin polarization measured by PCAR is formulated as

$$P_{PCAR} = \frac{<Nv_F^2>\uparrow - <Nv_F^2>\downarrow}{<Nv_F^2>\uparrow + <Nv_F^2>\downarrow}, \tag{12.10}$$

where N is the DOS and v_F^2 is the Fermi velocity. Note that this is different from P of DOS defined in (12.1), which is defined as the difference in DOS. Since v_F^2 is the conductance of electron, P_{PCAR} is nearly equivalent to the bulk spin asymmetry $\beta = (\sigma\uparrow - \sigma\downarrow)/(\sigma\uparrow + \sigma\downarrow)$, which is deduced from transport measurements in CPP-GMR. The experimental electrical circuit is shown in Fig. 12.2. Since the contact resistance of SC/sample interface is small (1 ~ 100 Ω), the current was measured by the lock-in technique.

Table 12.1 shows the spin polarization of Fe, Co and FeCo alloy measured by PCAR and STS. The spin polarizations of these materials show good agreement with literature values. Since the d-electrons in the Heusler alloy is localized around E_F and

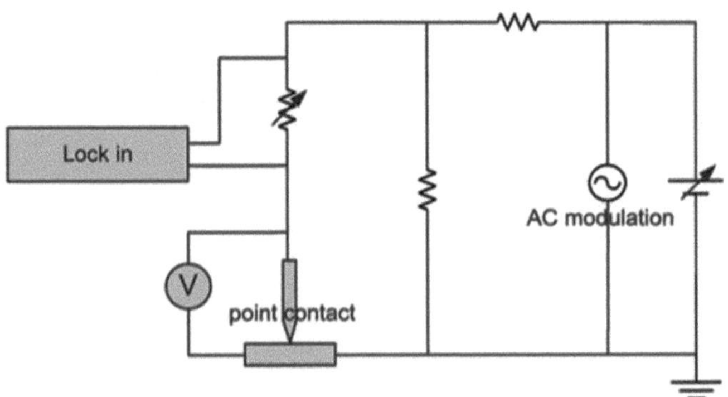

Fig. 12.2 Schematic illustration of the electric circuit in PCAR measurement

Table 12.1 Spin polarization of Fe and Co

	PCAR	STS	Reference PCAR
Fe	0.48	0.42 ± 0.02 [18–20]	0.43 ± 0.03 [9]
Co	0.48	0.40 ± 0.02 [18–20]	0.45 ± 0.02 [9]

does not contribute to the conductance in PCAR, the spin polarization estimated by PCAR does not match with that of DOS calculation. In addition, the spin polarization estimated by MTJ using Julliere's model does not match to that estimated by PCAR. The dominant electrons for transport in MTJ are tunneling electrons which received much effect from the barrier and that of PCAR are of metallic conduction electrons. Therefore, the material selection by PCAR should be effective for the applications in which bulk spin polarization plays a dominant role in device performance, such as CPP-GMR, STO and LSV.

12.4 Search for New Heusler Alloys with High Spin Polarization

12.4.1 Fermi Level Tuning by Quaternary Substitution

The right column in Table 12.2 shows the spin polarizations of ternary Heusler alloys measured by PCAR [21]. The spin polarization of the ternary alloys are around 60 %. Although Co_2MnSi was demonstrated to be half-metal at low temperature in the

Table 12.2 Spin polarization deduced by PCAR for various ternary and quaternary Heusler alloys and their Curie temperatures (Reprinted with permission from [21]. © 2012 Elsevier B.V.)

Quaternary alloys	P	Tc (K)	Ternary alloys	P	Tc (K)
$Co_2Mn(Ge_{0.75}Ga_{0.25})$	0.74	895	Co_2CrAl	0.62	600
$Co_2Mn(Ga_{0.5}Sn_{0.5})$	0.72	770	Co_2CrGa	0.61	
$Co_2Fe(Si_{0.75}Ge_{0.25})$	0.70	990	Co_2MnSn	0.60	800
$Co_2Fe(Ga_{0.5}Ge_{0.5})$	0.69	1080	Co_2MnAl	0.60	
$Co_2(Cr_{0.02}Fe_{0.98})Ga$	0.67		Co_2MnGa	0.60	700
$Co_2Mn(Ge_{0.25}Sn_{0.75})$	0.67		Co_2FeSi	0.60	1100
$Co_2(Mn_{0.95}Fe_{0.05})Sn$	0.65		Co_2FeAl	0.59	
$(Co_{1.93}Fe_{0.062})MnGe$	0.68		Co_2MnGe	0.58	900
$Co_2(Mn_{0.5}Fe_{0.5})Ga$	0.70	990	Co_2FeGe	0.58	1000
$Co_2(Cr_{0.02}Fe_{0.98})Si$	0.65	990	Co_2FeGa	0.58	1100
$Co_2Mn(Ti_{0.25}Sn_{0.75})$	0.64	480	Co_2TiSn	0.57	364
$Co_2Mn(Al_{0.5}Sn_{0.5})$	0.63		Co_2MnSi	0.56	900
$Co_2Mn(Ga_{0.25}Si_{0.75})$	0.63		Fe_2VAl	0.56	
$Co_2Mn(Si_{0.25}Ge_{0.75})$	0.63		Co_2VAl	0.48	
$Co_2(Mn_{0.5}Fe_{0.5})Si$	0.61				
$Co_2Mn(Al_{0.5}Si_{0.5})$	0.60				
$Co_2Fe(Ga_{0.5}Si_{0.5})$	0.60				
$Co_2Fe(Al_{0.5}Si_{0.5})$	0.60				

TMR measurement by Sakuraba et al. [1], the current spin polarization of Co_2MnSi was deduced to be only 0.56 by PCAR. One reason for the discrepancy is the spin polarization measured by TMR reflects DOS at E_F and the one deduced by PACR is of electron current. Nevertheless the spin polarization of these ternary alloys appears to be too low. This might be due to either low degree of $L2_1$ order in the ternary alloy or the location of the E_F with respect to the minority spin band gap. The spin polarization will be affected by the presence of disorder, defects and lattice parameter change when E_F is close to the edge of the band-gap [22]. Garanakis proposed that high spin polarization could be maintained by tuning the E_F to the center of the band-gap [18]. Co_2FeSi ($x = 0$) was predicted to be half-metal by the *ab initio* calculations taking account of an effective Coulomb exchange interaction [23]. In addition, T_c of Co_2FeSi is the highest among all known Heusler alloys (1100 K) [24]. However, E_F of Co_2FeSi is located close to the conduction band of the minority band. Since band edges are known to smear out with the presence of structural disorder, spin polarization was expected to degrade in this case. By substituting Si with Al, the E_F shifts to the middle position of the band gap as shown in Fig. 12.3a because the number of the valence electron of Al is one less than that of Si. Figure 12.3b shows experimentally obtained P of the $Co_2Fe(Al_xSi_{1-x})$ alloys shows a maximum with $x = 0.5$ due to the improved robustness of the spin polarization against disordering [25]. Based on the idea of the E_F tuning, some new Heusler alloys with high spin polarization were explored as shown in the left column in the Table 12.1. Among these, $Co_2Mn(Ge_{1-x}Ga_x)$ [19] and $Co_2Fe(Ge_{1-x}Ga_x)$ [5, 21] systems looked promising, so how these compositions were optimized is described below in more details.

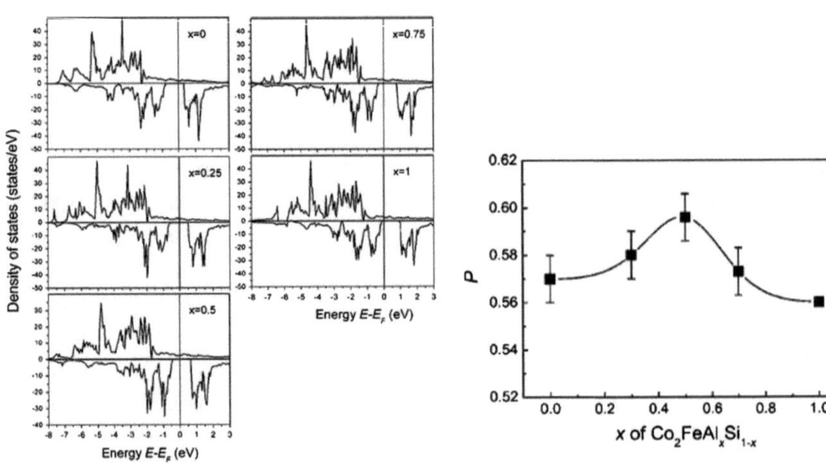

Fig. 12.3 The total density of states (DOS) of Co2FeAlxSi1-x obtained from the GGA+U theorem and the dependence of spin polarization on the Al concentration measured by PCAR. (Reprinted with permission from [25] © 2013 American Institute of Physics)

12.4.2 $Co_2Mn(Ge_{1-x}Ga_x)$

Co_2MnGe was theoretically predicted to be half-metal with a high T_c (905 K) [20, 26–28]. However, there was no experimental evidence for the half-metallic character of the Co_2MnGe alloy. PCAR measurements of $L2_1$ ordered Co_2MnGe deduced the spin polarization of 59 % [29]. The low spin polarization was considered to be due to the presence of DO_3 disorder. According to *ab initio* calculations, replacing the Mn sites by Co atoms induces a spin-down in-gap state [30]. Since the formation energy of the Co antisite is relatively low, DO_3 disordering easily occurs, causing the deterioration of the half-metallicity. In order to tune E_F, Ga with a valence electron number of 3 was substituted with Ge with the valence electron umber of 4. Klaer et al. reported the half-metallicity in the whole range in $Co_2Mn(Ge_{1-x}Ga_x)$ by the spin-resolved partial DOS calculated from the x-ray absorption spectroscopy (XAS) and x-ray magnetic circular dichroism (XMCD) data [31]. Spin resolved DOS of calculations of $Co_2Mn(Ge_{1-x}Ga_x)$ predicted that Co_2MnGe is half-metal, but the spin polarizations based on total DOS and the DOS of s-electron decrease as increasing x [19]. In spite of such predictions based on the DOS calculations for a perfect $L2_1$ order, experimental results was totally different [19].

Figure 12.4a shows the normalized conductance curves of $Co_2Mn(Ge_{1-x}Ga_x)$. Since these curves were fitted with the scattering parameter (Z) of zero, the spin polarization can be directly compared based on the minimum values of the conductance curve. Compared to the suppressions of the conductance in Co_2MnGe and Co_2MnGa, the drop of the conductance curve at zero bias voltage in $Co_2Mn(Ge_{0.75}Ga_{0.25})$ is substantially higher, suggesting that the spin polarization of $Co_2Mn(Ge_{0.75}Ga_{0.25})$ is much higher than those of the terminal alloys. By fitting the conductance curves to the modified BTK model, [16] the spin polarization values were deduced for different alloy compositions. Figure 12.4b shows the spin polarization of $Co_2Mn(Ge_{1-x}Ga_x)$ as a function of Ga composition x. It shows very high spin polarization of 74 % in $Co_2Mn(Ge_{0.75}Ga_{0.25})$, which is the highest value reported for Heusler alloys using PCAR.

This result is somewhat contradictory from the DOS calculation result [19]. The Rietveld fitting showed that some DO_3 disorder was present in the Co_2MnGa (Co-Mn site disorder), and it decreased with the increase of x in the $Co_2Mn(Ge_{1-x}Ga_x)$. The substitution of Ge with Ga appears to increase the degree of $L2_1$ order in the series of alloys. Figure 12.5 shows differential thermal analysis (DTA) curves of the $Co_2Mn(Ge_{1-x}Ga_x)$ alloys. Co_2MnGe shows two endothermic peaks at around 900 and 1300 K, which correspond to the T_c and the melting temperature, respectively. Since no order/disorder transformations were observed up to the melting temperature, the Co_2MnGe alloy is considered to be an intermetallic compound. However, substantial amount of DO_3 disorder is present in the lattice. With increasing Ga composition, T_c decreases from 900 K for Co_2MnGe to 700 K for Co_2MnGa. At the Ga concentration higher than 75 %, the order/disorder transformation reaction was observed below the melting temperature. Since the driving force for the $L2_1$ order becomes smaller with decreasing order-disorder temperature, the alloy with larger x is expected to

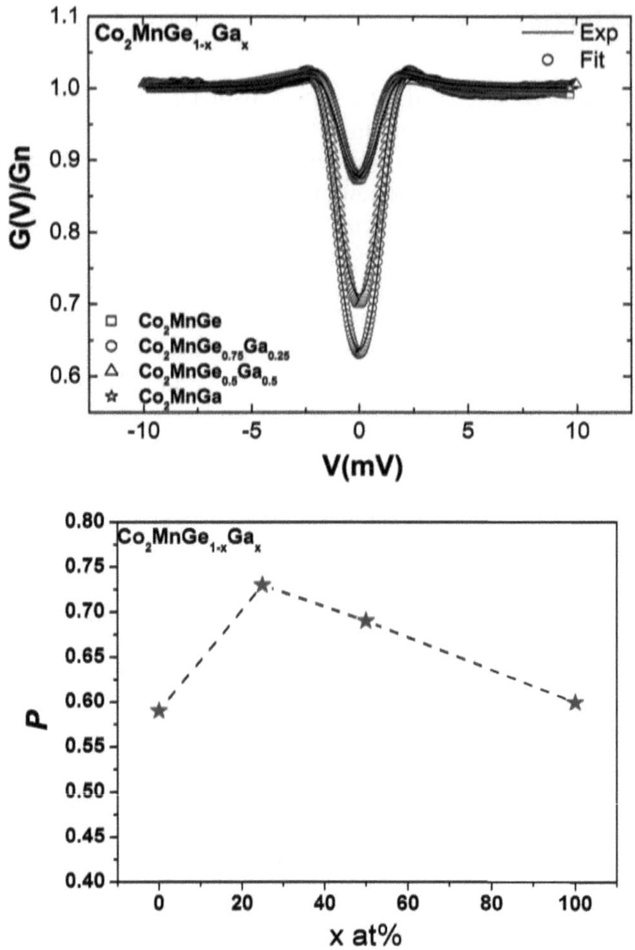

Fig. 12.4 Normalized conductance curves of $Co_2Mn(Ge_{1-x}Ga_x)$ ($x = 0, 0.25, 0.5$ and 1) alloys and the reduced spin polarization values for different Ga concentration, x [19]

show smaller kinetics of $L2_1$ ordering. The alloy with $x = 0.25$ decreases DO_3 disorder with sufficient kinetic of order to the $L2_1$ structure. Such structural factors appear to have affected the spin polarization of the $Co_2Mn(Ge_{1-x}Ga_x)$ alloys.

12.4.3 *$Co_2Fe(Ge_{1-x}Ga_x)$*

Unlike the Co_2Mn-based Heusler alloys, Co_2Fe-based Heusler alloys have a lower $L2_1$ order-disordering tendency, hence the driving force for $L2_1$ ordering appears to be low. However, it shows good tolerance against interdiffusion in CPP-GMR devices

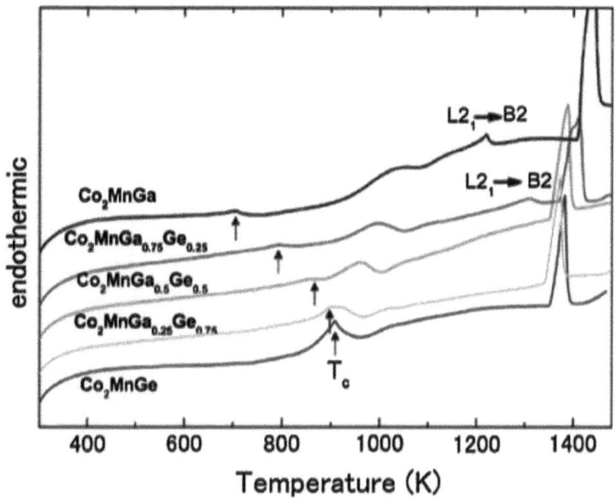

Fig. 12.5 Differential thermal analysis (DTA) curves of various Ga composition in $Co_2Mn(Ge_{1-x}Ga_x)$. Courtesy of B. Varaprasad

with Ag spacer. Since Ag has a low solubility with Co and Fe, the interdiffusion of FM elements into the Ag spacer does not occur even at the elevated temperatures required for the $L2_1$ ordering (>720 K). On the other hand, Ag has a large solubility with Mn, so Mn in Co_2MnSi diffuse into a Ag spacer, causing the interlayer exchange coupling between two Co_2MnSi layers and a spin-glass behavior at low temperature [32]. Therefore, Co_2Fe-based Heusler alloys with high spin polarization may be better suited as FM layer for CPP-GMR devices. According to *ab initio* calculations, Co_2FeGe is supposed to be half-metal [21]. However, a $L2_1$ single phase was not obtained experimentally. On the other hand, Co_2FeGa forms a $L2_1$ single phase, but it is not a half-metal [33]. By substituting Ge with Ga, the driving force for the $L2_1$ ordering is expected to increase. At the same time, a $L2_1$ single phase becomes stable in $Co_2Fe(Ge_{1-x}Ga_x)$ with E_F in the minority electron band gap for the minority electron. Figure 12.6 shows the calculated spin resolved DOS of Co_2FeGe, $Co_2Fe(Ge_{0.5}Ga_{0.5})$ and Co_2FeGa alloys with $L2_1$ and B2 structures [21]. In the Co_2FeGa, both $L2_1$ and B2 ordered structures show half-metallicity. The E_F of B2 ordered Co_2FeGe is near the band edge, so the spin polarization will be degraded at finite temperature. Co_2FeGa shows DOS in spin-down band around the E_F. Therefore, Co_2FeGa is not expected to show high spin polarization. In fact, the spin polarization of Co_2FeGa measured by PCAR was only 59 % [33]. In $Co_2Fe(Ge_{0.5}Ga_{0.5})$, both $L2_1$ and B2 structures show half-metallicity, even though the half-metallic gap reduces in the B2 structure. E_F of B2-ordered $Co_2Fe(Ge_{0.5}Ga_{0.5})$ is in the center of the minority band gap, so high tolerance of spin polarization against disordering and temperature is expected.

XRD patterns (data not shown here) showed all the alloys had the $L2_1$ structure with a secondary phase in the Co_2FeGe alloy [21]. The degree of order estimated

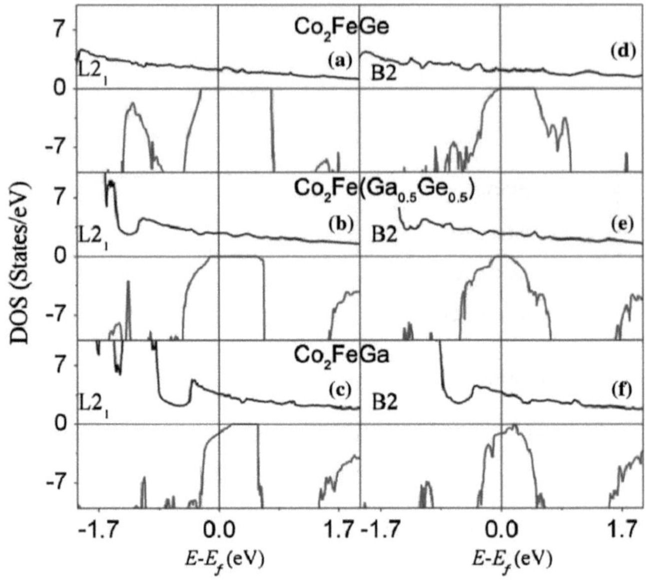

Fig. 12.6 Spin resolved DOS of Co_2FeGe, $Co_2Fe(Ge_{0.5}Ga_{0.5})$ and Co_2FeGa alloys with $L2_1$ and B2 structures near E_F. (Reprinted with permission from [21]. © 2012 Elsevier B.V.)

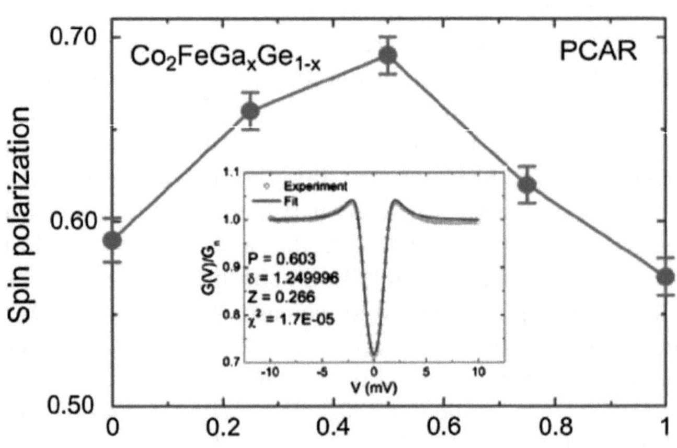

Fig. 12.7 Change in spin polarization of $Co_2Fe(Ge_{1-x}Ga_x)$ alloy as a function of x (Ga) content. Inset shows typical conductance and fitted curves of $Co_2Fe(Ge_{0.5}Ga_{0.5})$ alloy. (Reprinted with permission from [5] © 2011 American Institute of Physics)

by the integrated x-ray intensity ratios showed similar values for all the alloys, 0.68 for B2 order and 0.32 for $L2_1$ order. Figure 12.7 shows the spin polarization change as a function of x for $Co_2Fe(Ge_{1-x}Ga_x)$ [5]. The spin polarization increases with increasing x, showing a maximum of 0.69 at $x = 0.5$, which is the second highest spin polarization in Co_2Fe-based Heusler alloys measured by PCAR. Then, the spin polarization decreases with x. The inset shows the typical conductance curves for $Co_2Fe(Ge_{0.5}Ga_{0.5})$ alloy.

As shown above, since $Co_2Mn(Ge_{0.75}Ga_{0.25})$ and $Co_2Fe(Ge_{0.5}Ga_{0.5})$ showed high spin polarization and T_c, their magneto-transport properties were further investigated using tri-layer CPP-GMR pseudo spin valves.

12.5 Magneto-Transport Measurements Using CPP-GMR Pseudo Spinvalves

Bulk spin polarization of conduction electron can be estimated by measuring CPP-GMR. Valet and Fert derived the general equation for MR output of CPP-GMR by rigorously solving the diffusion equation under the periodic boundary condition of FM/NM multilayers [34]. Assuming that there is no spin relaxation through the FM/NM multilayer and the spin diffusion length is much thicker than the thicknesses of FM and NM layers, the equation for FM/NM/FM tri-layer CPP-GMR can be simplified to the following formula,

$$\Delta RA = \frac{4(\beta \rho_F^* t_F + 2\gamma AR_{F/N}^*)^2}{2\rho_F^* t_F + \rho_N t_N + 4AR_{F/N}^*}. \tag{12.11}$$

This equation can be further simplified to

$$\Delta RA \approx 2\overset{2}{\beta} \rho_F^* t_F + 2\gamma^2 AR_{F/N}^* = 2\frac{\beta^2}{1-\beta^2}\rho_F t_F + 2\frac{\gamma^2}{1-\gamma^2}AR_{F/N}, \tag{12.12}$$

where β and γ are the bulk and interface spin asymmetries, ρ_F and ρ_N are the resistivities of FM and spacer materials ($\rho_F^* = \rho_F/(1-\beta^2)$), t_N and t_F are the thicknesses of the spacer and FM layers, $AR_{F/N}$and is the interfacial resistivity between FM and spacer layers ($AR_{F/N}^* = AR_{F/N}/(1-\gamma^2)$). Therefore, high MR output is expected in CPP-GMR using FM material with high bulk spin asymmetry of $\beta \sim 1$ and a spacer layer that cause high interfacial spin asymmetry of $\gamma \sim 1$. Here, β is the bulk spin polarization of current, i.e., $\beta = (\sigma\uparrow - \sigma\downarrow)/(\sigma\uparrow + \sigma\downarrow)$, whose physical origin is the same as the P deduced from PCAR measurements in the diffusive mode.

Figure 12.8 shows *ΔRA* of CPP-GMR using $Co_2Mn(Ge_{0.75}Ga_{0.25})$ (hereafter, CMGG) Heusler alloy as FM layers and Ag as a spacer layer as a function of the annealing temperature. Since the as-deposited CMGG has the B2 ordered structure, the annealing is necessary for improving the degree of $L2_1$ order. Note that CMGG orders to the $L2_1$ structure by annealing above 350 °C, which is relatively low annealing temperature compared to that required for Co_2Fe-based Heusler alloys. With the increase of the annealing temperature, *ΔRA* increases from 3 mΩμm^2 at 250 °C to 6 mΩμm^2 at 400 °C and then decreases. The reason for the decrease of the *ΔRA*

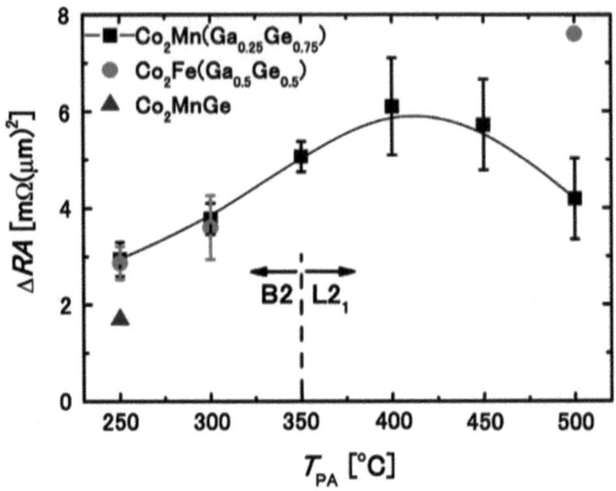

Fig. 12.8 Dependence of ΔRA of $Co_2Mn(Ga_{0.25}Ge_{0.75})$ (*squares*), several established Heusler alloys $Co_2Fe(Ga_{0.5}Ge_{0.5})$ (*circles*), and polycrystalline Co_2MnGe (*triangles*) on annealing temperature. (Reprinted with permission from [35] © 2013 American Institute of Physics)

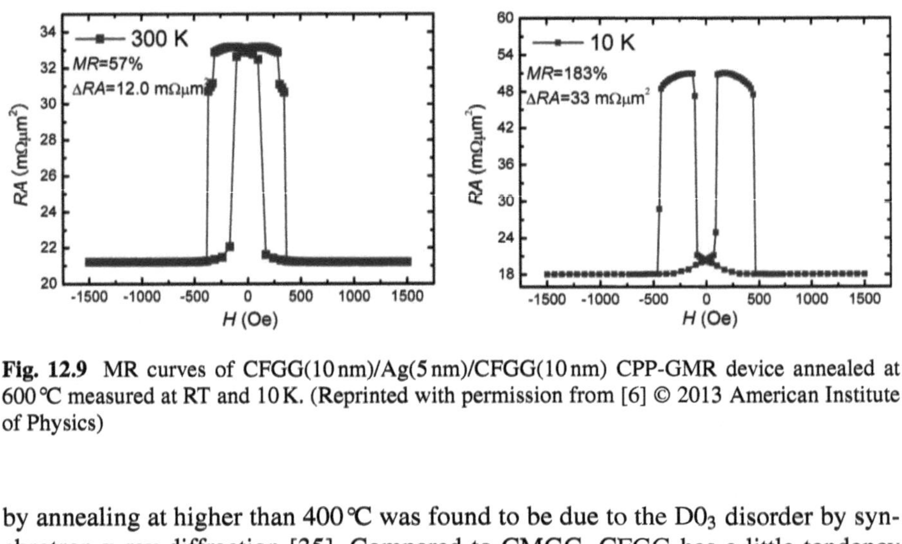

Fig. 12.9 MR curves of CFGG(10 nm)/Ag(5 nm)/CFGG(10 nm) CPP-GMR device annealed at 600 °C measured at RT and 10 K. (Reprinted with permission from [6] © 2013 American Institute of Physics)

by annealing at higher than 400 °C was found to be due to the $D0_3$ disorder by synchrotron x-ray diffraction [35]. Compared to CMGG, CFGG has a little tendency for the $D0_3$ disorder at elevated temperature as indicated by the filled circle. Thus, higher ΔRA can be achieved using CFGG, if high temperature annealing is allowed. In practical applications as magnetic sensors, the annealing temperature is limited to 350 °C due to the heat tolerance of the magnetic shield. Below 350 °C, ΔRA in this system is one of the highest values.

Figure 12.9 shows MR curves of a CPP-GMR device using $Co_2Fe(Ge_{0.5}Ga_{0.5})$ (hereafter, CFGG) as FM layers and Ag as a spacer layer that was annealed at 600 °C measured at RT and 10 K [6]. A large ΔRA value of 12 m$\Omega\mu m^2$ was obtained

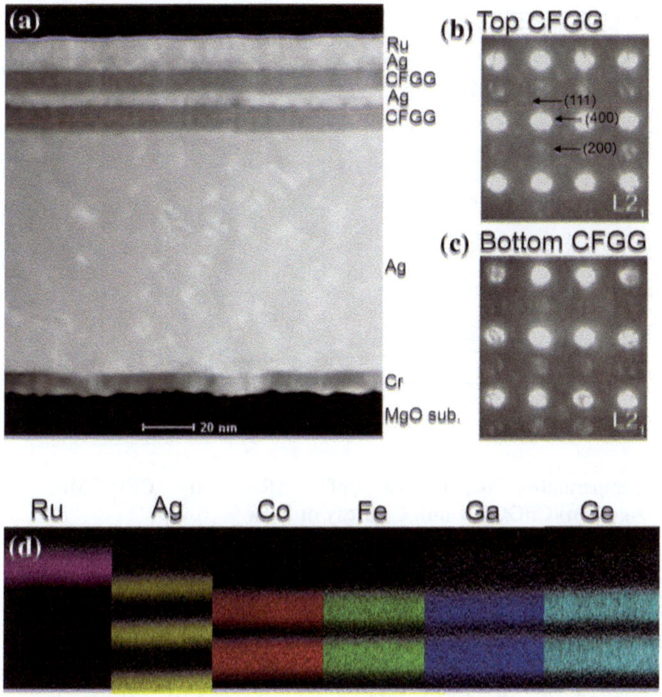

Fig. 12.10 **a** cross-sectional HAADF image of in-situ annealed at 600C in MgO(001) sub. /Cr(10 nm)/Ag(100 nm)/CFGG(10 nm)/Ag(5 nm)/CFGG(10 nm)/ Ag(5 nm)/ Ru(8 nm). Nanobeam diffraction of **b** *top* and **c** *bottom* CFGG layers. **d** EDS mapping results for elements of Ru, Ag, Co, Fe, Ga and Ge. (Reprinted with permission from [6] © 2013 American Institute of Physics)

by annealing the CFGG(10 nm)/Ag(5 nm)/CFGG(10 nm) multi-layer film at 600 ℃. Annealing at higher than 600 ℃ led to the reduction in ΔRA due to the structural degradation of the multi-layers; however, the tolerance of MR against the high temperature annealing is notable compared to that of CMGG shown in Fig. 12.8. The MR output and the corresponding MR ratio were further enhanced at low measurement temperature of 10 K. An extremely large MR output of $\Delta RA = 33\ \mathrm{m}\Omega\mu\mathrm{m}^2$ and MR ratio of 183 % was obtained from the CPP-GMR device at 10 K. Figure 12.10 shows high angle annular dark field image of the device, and nanobeam diffraction patterns obtained from the top and bottom CFGG layers and energy dispersive x-ray map of the CPP-GMR device that showed the highest ΔRA output. The nanobeam diffraction patterns show that both the CFGG layers were $L2_1$ ordered and sharp interfaces were kept even after the annealing at 600 ℃.

As shown above, the MR output of the CPP-GMR devices using CFGG and other Heusler alloys show large temperature dependence. Figure 12.11 shows ΔRA of the CFGG(10 nm)/Ag(5 nm)/CFGG(10 nm) CPP-GMR device annealed at various temperatures [6]. In the as-deposited state, the CFGG layer contain large A2 and B2 disorder with respect to the $L2_1$ order, so the FM layer is not a half-metal, but it

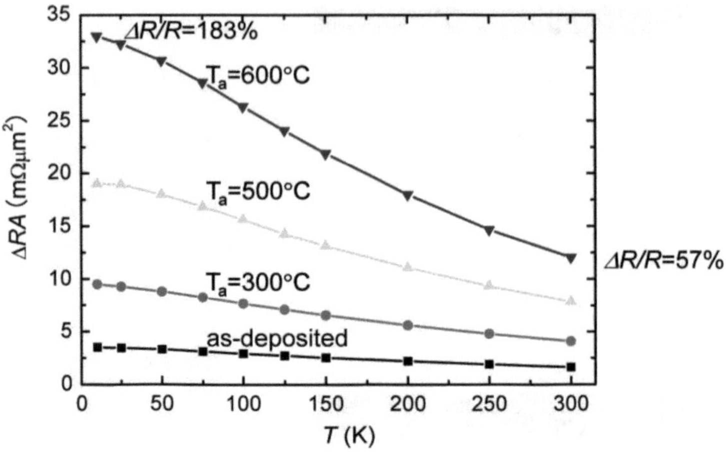

Fig. 12.11 Temperature dependence of ΔRA in CPP-GMR device with CFGG(10 nm)/Ag(5 nm)/CFGG(10 nm). Courtesy of S. Li

is just a normal FM layer. In this case, MR output does not show any temperature dependence, as the spin-polarization is due to the normal ferromagnetism. As the annealing temperature increases, the degree of B2 and $L2_1$ order increases, and the CFGG shows a half-metallic feature. This causes large MR output at low temperature. When the degree of $L2_1$ order becomes close to 1, the CFGG layer becomes half-metal at low temperature. Hence, very large MR output and MR ratio was obtained at low temperature. However, because of magnon excitations in bulk and CFGG/Ag interfaces, both β and γ are considered to decrease at finite temperature.

In order to estimate β and γ at RT and 10 K, ΔRA was plotted as a function of CFGG thickness as shown in Fig. 12.12. At both temperatures, ΔRA increases with CFGG thickness and saturate at around 7 nm, which is due to the short spin diffusion length of CFGG. By fitting these data to Valet-Fert model, β and γ were determined independently following the procedure described by Nakatani et al. [3] and Taniguchi et al. [36] As a result, $\beta = 0.83 \pm 0.02$ and $\gamma = 0.6 \pm 0.22$ at RT and $\beta = 0.93 \pm 0.004$ and $\gamma = 0.84 \pm 0.14$ at 10 K were deduced. Note that the bulk spin polarization β of CFGG is close to 1 at 10 K, meaning that it is half-metal at 10 K. However, the β degraded to 0.83 at RT. Accordingly, the γ values also decreases to 0.6. Figure 12.13 shows the changes of *ΔRA* and saturation magnetization (M_s) of CFGG as a function of temperature. In the temperature range between 10 K to RT, M_s changes little due to high T_c. Nevertheless, a large decrease in ΔRA occurs in this temperature range. In the figure, β_{RT}/β_{10K} and $(\beta_{RT}/\beta_{10\,K})$ [2] are also shown. If the decrease of *ΔRA* is only because of the degradation of bulk spin polarization, such a large drop in *ΔRA* is not expected. This suggests that the large drop of *ΔRA* at RT is also due to the magnenon excitation at FM/NM interface, i.e., the temperature dependence of γ. To solve this problem, we need to find the origin of the large temperature dependence of β and γ.

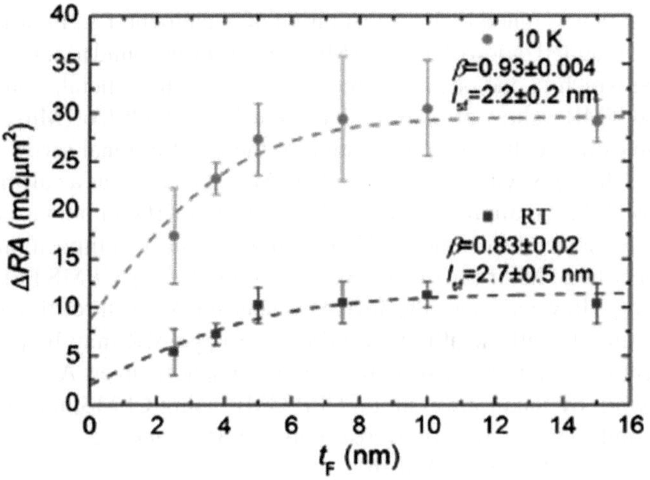

Fig. 12.12 CFGG layer thickness tF dependence of ΔRA fitted to the Valet-Fert model, measured at RT (*squares*) and 10 K (*circle*). (Reprinted with permission from [6] © 2013 American Institute of Physics)

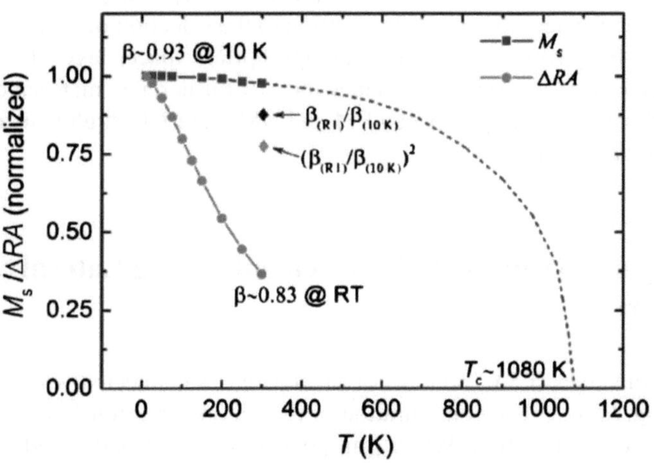

Fig. 12.13 Temperature dependence of ΔRA in CPP-GMR device with CFGG(10 nm)/Ag(5 nm)/CFGG(10 nm) and Ms of CFGG. Courtesy of S. Li

12.6 Other Quaternary Heusler Alloys

There is one more promising quaternary Heusler alloys proposed by Balke et al., $Co_2(Mn_{1-x}Fe_x)Si$ [37]. Co_2MnSi and Co_2FeSi are well known Heusler alloys because both of them are predicted to be half-metal theoretically and T_c are substantially higher than RT, i.e., 985 and 1100 K. In this quaternary alloy, the Fermi level or valence electron was tuned by the substitution of Y element, i.e., Fe was substituted with Mn. The magnetization of the $Co_2(Mn_{1-x}Fe_x)Si$ was found to follows the Slater-Pauling rule, which is one of the indirect evidence for half-metal. Later, Kubota et al. measured TMR values of alumina barriered MTJs with $Co_2(Mn_{1-x}Fe_x)Si$ FM electrodes to estimate the spin polarization of the substitutional series of the alloys and identified the alloy with $x = 0.4$, i.e., $Co_2(Mn_{0.6}Fe_{0.4})Si$ has a half-metallic band gap with low Gilbert damping constant. Since low Gilbert damping is considered to be another character of half-metals [38]. Using this alloy composition, Sato et al. reported a very large MR ratio of 74.8 % in CPP-GMR using CFMS FM layers and a Ag spacer [39]. However, since the device resistance RA was three times higher than the typical value for all metallic CPP-GMR, the large MR may be partly because of the current confined path effect from oxidized spacer layer. According to more systematic work by Sakuraba et al. on CPP-GMR using CFMS, the ΔRA value of 13 m$\Omega\mu m^2$ and 58 % of MR ratio were reported, [40] which are comparable to those reported for CFGG/Ag CPP-GMR [6].

Table 12.3 shows ΔRA, β, γ and spin diffusion length, λ_{SF}, of CPP-GMR with various quaternary Heusler alloys and spacers. So far, the spacer material is very limited, only Ag and Cu. The β values deduced from CPP-GMR show good correlations with those deduced from PCAR. γ changes depending on the combination of FM layer and spacer materials. In addition to the selection of high spin polarized materials for FM layer, a selection of an appropriate spacer layer that give rise to large γ value is also necessary to enhance the MR further, i.e., material combination with a good band matching between FM and NM materials, high resistive FM with high spin polarization.

12.7 Magneto Transport Measurements Using Lateral Spin Valves

A lateral spin valves (LSV) consists of two FM wires connected with a NM wire as shown in Fig. 12.14a. The configuration of the LSV is an extension of a CPP-GMR device to a lateral direction. When spin polarized current is injected from FM1 to NM, the pure spin current without charge diffuses to the NM wire. If the distance between the two FM wires is within the spin diffusion length of the NM material, the spin accumulation occurs at the interface between FM2 and NM. While the local configuration of the LSV is fundamentally analogous to a CPP-GMR device, the non-local configuration enables the generation and detection of a pure spin current without any charge current. The FM material of interest such as Heusler alloys can be annealed before making a contact with a NM layer, thus the interfacial reaction that commonly occurs in multi-layer stacks for CPP-GMR by high temperature annealing can be avoided. To describe the spin accumulation at the FM2/NM interface, Takahashi and Maekawa proposed the one spin diffusion model [46]. By solving the equations of the one dimensional spin diffusion taking into account the continuity of spin and

Table 12.3 ΔRA, RA, spin polarizations of FM (β) and interface (γ) and spin diffusion length (l_{SF}) estimated by CPP-GMR devices with highly spin polarized Heusler alloys annealed at various temperatures

FM /Spacer	T_a (℃)	T_{meas} (K)	ΔRA (mΩ μm^2)	RA (mΩ μm^2)	β	γ	l_{SF} (nm)	Reference
Co_2MnSi /Ag	550	RT	11.5		–	–	–	[41]
	500	RT	9		0.50	0.87	–	
		100	7.5		0.72	0.87	–	
	350	RT	8.5	51.4	0.46	0.82	–	
		100	14		0.65	0.86	–	
$Co_2Fe(Ga_{0.5}Ge_{0.5})$ /Ag	500	RT	8.7 ± 1.6	21	0.73 ± 0.02	0.60 ± 0.05	2.1 ± 0.4	[5, 6, 42]
		10	22.9 ± 2.2	16.5	0.90 ± 0.01	0.84 ± 0.01	1.8 ± 0.4	
	600	RT	11.3 ± 1.3	21	0.83 ± 0.02	0.63 ± 0.07	2.7 ± 0.5	
		10	30.5 ± 4.9	18	0.93 ± 0.004	0.88 ± 0.05	2.2 ± 0.2	
$Co_2Fe(Ga_{0.5}Ge_{0.5})$ /Cu	300	RT	3.5 ± 0.9	18	0.41 ± 0.03	0.61 ± 0.04	1.6 ± 0.3	[43]
		10	8.4 ± 2.3	16.5	0.54 ± 0.02	0.74 ± 0.02	2.2 ± 0.3	
Co_2FeSi/Ag	550	RT	5.3 ± 0.8	27	0.57 ± 0.04	0.84	1.7 ± 0.4	[44]
		25	22.7		0.81 ± 0.03	0.94	1.7 ± 0.3	
$Co_2Fe(Al_{0.5}Si_{0.5})$ /Ag	500	RT	8	23	0.70 ± 0.08	0.77	2.2 ± 1.0	[3]
		14	16		0.77 ± 0.05	0.93	3.0 ± 0.8	
$Co_2(Fe_{0.4}Mn_{0.6})Si$/Ag	500	RT	12.5 ± 2.5	–	–	–	–	[45]
Co_2 $(Cr_{0.1}Fe_{0.9})Si$ /Ag	500	RT	5.6 ± 0.6	27	–	–	–	[44]
		25	17.4 ± 2.9	24	–	–	–	
$Co_2MnGa_{0.5}Sn_{0.5}$ /Ag	400	RT	3.2	31	0.36	–	2.7	[54]
		14	6.1	26.8	0.47	–	3.1	
$Co_2MnGa_{0.25}Ge_{0.75}$ /Ag	400	RT	6.1	15	0.52	0.74	3.0 ± 0.3	[35]
	250	RT	3	22	0.33	0.63	2.8 ± 0.1	

Courtesy of T. Furubayashi

charge at the interfaces, the spin accumulation at the interface of the FM2 and NM wires can be expressed as follows;

$$\Delta R_S = 4R_{NM}\frac{\left(\frac{P_J}{1-P_J{}^2}\frac{R_1}{R_{NM}}+P_{FM}\frac{R_{FM}}{R_{NM}}\right)\left(\frac{P_J}{1-P_J{}^2}\frac{R_2}{R_{NM}}+P_{FM}\frac{R_{FM}}{R_{NM}}\right)e^{-\frac{d}{\lambda_{NM}}}}{\left(1+\frac{2}{1-P_J{}^2}\frac{R_1}{R_{NM}}+2\frac{R_{FM}}{R_{NM}}\right)\left(1+\frac{2}{1-P_J{}^2}\frac{R_1}{R_{NM}}+2\frac{R_{FM}}{R_{NM}}\right)-e^{-\frac{2d}{\lambda_{NM}}}} \tag{12.12}$$

where $R_{FM} = \rho_{FM}\lambda_{FM}/A_J$ and $R_{NM} = \rho_{NM}\lambda_{NM}/A_{NM}$ are spin resistances of ferromagnets and channel, respectively, with ρ, λ, and A correspond to resistivity, spin diffusion length, and effective cross section for spin current, respectively. The P_{FM} and P_J are the bulk ferromagnet and FM/NM interface spin polarizations respectively, and $R_1$and R_2 are the contact resistances of injector and detector respectively. When the interface is transparent, this formula can be simplified as follows;

$$\Delta R_S = \frac{4P_F2}{(1-P_F2)}R_{NM}\left(\frac{R_{FM}}{R_{NM}}\right)^2\frac{e^{\frac{d}{\lambda_{NM}}}}{1-e^{\frac{2d}{\lambda_{NM}}}} \tag{12.13}$$

It can be easily understood that the large output is expected by using highly spin polarized material. Figure 12.14b shows the spin signal change as a function of the distance between two ferromagnetic wires in a LSV device with CFGG and Cu [8]. A large spin signal of 17.3 mΩ was obtained for the distance of 350 nm. By fitting these data to the one dimensional spin diffusion model, P_j of CFGG was estimated to be 0.59. Note that the fitting has to be made using the general expression of the one spin diffusion model described by the equation (12.12) without any simplification as widely performed in other reports, [47–49] because the interface contribution is not negligible when high-spin polarization FM materials are used. The P_j of CFGG/Cu is almost the same as the interfacial spin asymmetry of the CFGG/Cu CPP-GMR devices at RT, ~0.61 [42].

12.8 Spin Polarization and AMR

While some successful materials searches have been conducted using PCAR, it can measure spin polarizations only at low temperature (~4.2 K). All Heusler alloys so far investigated showed large degradation of spin polarization at RT. In order to search for room temperature half-metals, the spin polarization of materials must be readily estimated at various temperatures. Recently Kokado et al. developed a new theoretical model to analyze anisotropic magnetoresistance (AMR) effect in various

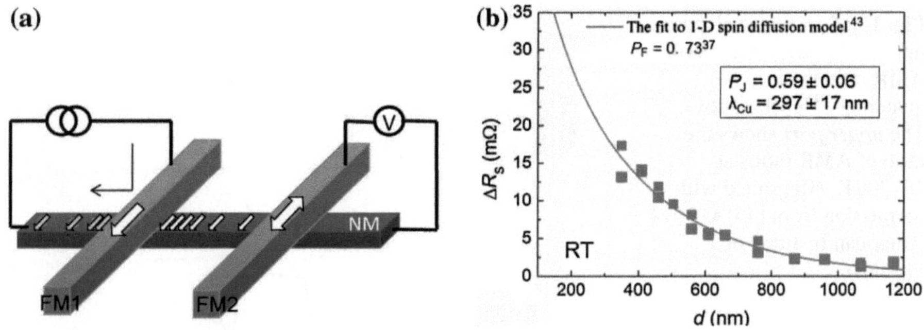

Fig. 12.14 Schematic view of a lateral spin valve device (a) and spin signal change as a function of two CFGG wires at RT (b). The solid line represents the fitting result by one dimensional spin diffusion model [7]

kinds of FM materials [50, 51]. In their theoretical framework, half-metallic materials always exhibit a negative sign of AMR ratios and its absolute valueis related to spin asymmetry as

$$AMR = \frac{\rho_{//} - \rho_{\perp}}{\rho_{//}} \propto \gamma \left(D_{\uparrow}^{(d)} - D_{\downarrow}^{(d)}\right)\left(\frac{1}{\rho\downarrow} - \frac{1}{\rho\uparrow}\right), \tag{12.14}$$

where $\gamma = (3/4)(l/H_{ex})^2$, l is the spin-orbit coupling constant, H_{ex} is the exchange field, and $D^{(d)}$ is the DOS of d-electron. Using this relationship, AMR that can be measured without any tedious micro-fabrications can be a fingerprint of the half-metallic nature of FM materials. Based on their prediction, Yang et al. found a sign change of AMR from negative to positive when x for $Co_2Fe_xMn_{1-x}Si$ (CFMS) films became larger than 0.8, suggesting a disappearance of half-metallic nature for $x>$ 0.8 [52]. This result was consistent with the half-metallicity of CFMS investigated in the previous studies on TMR and CPP-GMR using CFMS [37, 53]. Furthermore, Sakuraba et al. studied how the sign and magnitude of AMR change with the number of valence electron (N_V)in Co_2MnZ and Co_2FeZ alloys systematically and found an excellent correlation between the AMR ratios of Heusler alloy films and the ΔRA values of the CPP-GMR devices using them [45].

Figure 12.15 shows AMR in the Co_2MnZ and Co_2FeZ films for the valence electron number N_V [45]. The ratio of AMR at 10–300 K is plotted in the upper column. It clearly shows that the sign change of AMR occurs at around $N_V = 28.3$ and 30.3. $Co_2MnAl_{0.8}Si_{0.2}$ and Co_2FeSi films show sign changes from negative at 10 K to positive at 300 K. These sign changes are observed in the alloy whose N_v is close to that of the half-metallic gap, i.e., $Co_2FeGe_{0.25}Si_{0.75}$ alloy ($N_V = 30$). The magnitude of negative AMR reflects a spin-polarization at E_F, therefore this large temperature dependence can be explained by the depolarization due to the thermal fluctuation of electron occupation based on Fermi-Dirac distribution. Note that large negative AMR ratios were observed at around $N_V = 28.8 - 29.4$ in the well $L2_1$-ordered compounds such as CMS and CMGG, and CFGG. In this N_v region, the tempera-

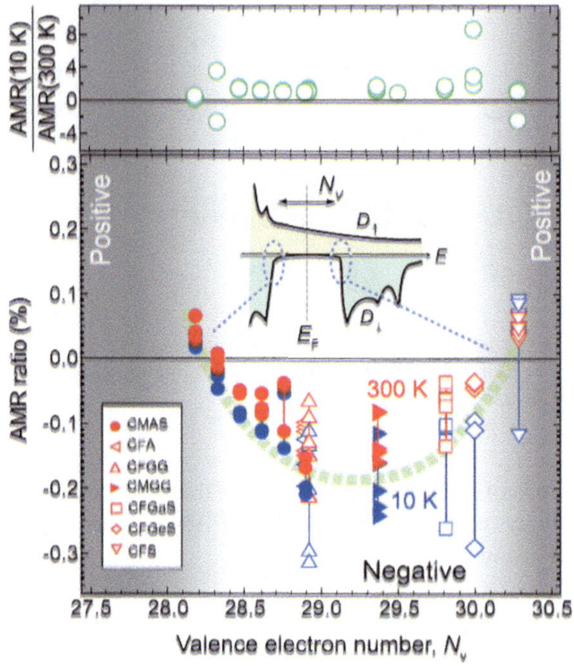

Fig. 12.15 Valence electron number NV dependence of AMR ratio in various quaternary Heusler alloys. The *upper part* shows the ratio of AMR ratios at 10–300 K. (Reprinted with permission from [45] © 2014 American Institute of Physics)

ture dependence of AMR ratio is not so large as that in the region near gap edges because there is less depolarization effect against temperature at around the center of half-metallic gap. This work has convincingly shown that AMR can be a new tool for searching new Heusler alloys with high spin polarization. The unique feature of this method is that it can measure the temperature degradation of spin polarization of Heusler alloys.

12.9 Summary

Exploration of highly spin polarized FM Heusler alloys based on the band tuning concept using substitutional series of quaternary Heusler alloy, $X_2Y(Z_{1-x}Z'_x)$ and $X_2(Y_{1-x}Y'_x)Z$ has been reviewed. Since PCAR does not require thin film processing and microfabriations for measurements, screening of high-spin polarization materials can be readily done using bulk alloys. Examples of quaternary alloys with relatively high spin polarization, $Co_2Fe(Al,Si)$, $Co_2Mn(Ge,Ga)$, and $Co_2Fe(Ge,Ga)$ were reviewed. The spin polarization measurements of these Heusler alloys by magneto-transport measurements of CPP-GMR pseudo spin valves and lateral spin valves were also shown. One critical issue of FM Heusler alloys revealed from these transport measurements is the degradation of MR outputs at room temperature. So a facile method to estimate the temperature dependence of spin polarization is strongly desired. Recent theoretical and experimental investigations have shown good correlations between MR output in CPP-GMR and anisotropy magnetoresistance (AMR). Although PCAR can measure only the spin polarization at around 4 K, AMR can estimate the temperature change of spin polarization. Thus, AMR measurements will become more useful tool for searching new alloys with high spin polarization at room temperature.

References

1. Y. Sakuraba, M. Hattori, M. Oogane, Y. Ando, H. Kato, A. Sakuma, T. Miyazaki, H. Kubota, Appl. Phys. Lett. **88**, 192508 (2006)
2. T. Furubayashi, K. Kodama, H. Sukegawa, Y.K. Takahashi, K. Inomata, K. Hono, Appl. Phys. Lett. **93**, 122507 (2008)
3. T.M. Nakatani, T. Furubayashi, S. Kasai, H. Sukegawa, Y.K. Takahashi, S. Mitani, K. Hono, Appl. Phys. Lett. **96**, 212501 (2010)
4. T. Iwase, Y. Sakuraba, S. Bosu, K. Saito, S. Mitani, K. Takanashi, Appl. Phys. Express **2**, 063003 (2009)
5. Y.K. Takahashi, A. Srinivasan, B. Varaprasad, A. Rajanikanth, N. Hase, T.M. Nakatani, S. Kasai, T. Furubayashi, K. Hono, Appl. Phys. Lett. **98**, 152501 (2011)
6. S. Li, Y.K. Takahashi, T. Furubaashi, K. Hono, Appl. Phys. Lett. **103**, 042405 (2013)
7. Y.K. Takahashi, S. Kasai, S. Hirayama, S. Mitani, K. Hono, Appl. Phys. Lett. **100**, 052405 (2012)
8. Iktiar, S. Kasai, A. Itoh, Y.K. Takahashi, T. Ohkubo, S. Mitani, K. Hono. J. Appl. Phys. **115**, 173912 (2014)
9. K. Kasahara, Y. Fujita, S. Yamada, K. Sawano, M. Miyao, K. Hamaya, Appl. Phys. Express **7**, 033002 (2014)
10. A. Takeo et al. Intermag (2014)
11. J. Kubler, A.R. Williams, C.B. Sommers, Phys. Rev. B **28**, 1745 (1983)
12. R. Meservey, P. Tedrow, D. Fried, Phys. Rev. Lett. **25**, 1270 (1970)
13. M. Julliere, Phys. Lett. A **54**, 225 (1975)
14. R.J. Soulen Jr, J.M. Byers, M.S. Osofsky, B. Nadgorny, T. Ambrose, S.F. Cheng, P. Broussard, C. Tanaka, J. Nowak, J. Moodera, A. Barry, J.M. Coey, Science **282**, 85 (1998)
15. W.H. Butler, X.-G. Zhang, T.C. Schulthess, J.M. MacLaren, Phys. Rev. B **63**, 054416 (2001)
16. G.J. Strijkers, Y. Ji, F.Y. Yang, C.L. Chien, J.M. Byers, Phys. Rev. B **63**, 104510 (2001)
17. G.E. Blonder, M. Tinkham, T.M. Klapwijk, Phys. Rev. B **25**, 4515 (1982)
18. I. Galanakis, J. Phys.: Condens. Matter **16**, 3089 (2004)
19. B.S.D.C.S. Varaprasad, A. Rajanikanth, Y.K. Takahashi, K. Hono, Appl. Phys. Express **3**, 23002 (2010)
20. S. Ishida, S. Fujii, S. Kashiwagi, S. Asano, J. Phys. Soc. Jpn. **64**, 2152 (1995)
21. B.S.D.C.S. Varaprasad, A. Srinivasan, Y.K. Takahashi, M. Hayashi, A. Rajanikanth, K. Hono, Acta Mater. **60**, 6257 (2012)
22. G.H. Fecher, B. Balke, S. Ouardi, C. Felser, G. Schönhense, E. Ikenaga, J.-J. Kim, S. Ueda, K. Kobayashi, J. Phys. D:Appl. Phys. **40**, 1576 (2007)
23. H.C. Kandpal, G.H. Fecher, C. Felser, G. Schonhense, Phys. Rev. B **73**, 094422 (2006)
24. S. Wurmehl, G.H. Fecher, H.C. Landpal, V. Ksenofontiv, C. Felser, H.J. Lin, J. Morais, Phys. Rev. B **72**, 184434 (2005)
25. T.M. Nakatani, A. Rajanikanth, Z. Gercsi, Y.K. Takahashi, K. Inomata, K. Hono, J. Appl. Phys. **102**, 033916 (2007)
26. S. Picozzi, A. Continenza, A.J. Freeman, Phys. Rev. B **66**, 094421 (2002)
27. I. Galanakis, P.H. Dederichs, N. Papanikolaou, Phys. Rev. B **66**, 174429 (2002)

28. P.J. Webster, J. Phys. Chem. Solids **32**, 1221 (1971)
29. A. Rajanikanth, Y.K. Takahashi, K. Hono, J. Appl. Phys. **101**, 023901 (2007)
30. S. Picozzi, A. Continenza, A.J. Freeman, Phys. Rev. B **69**, 94423 (2004)
31. P. Klaer, M. Kallmayer, C.G.F. Blum, T. Graf, J. Barth, B. Balke, G.H. Fecher, C. Felser, H.J. Elmersi, Phys. Rev. B **80**, 144405 (2009)
32. Y. Sakuraba, K. Izumi, S. Bosu, K. Saito, K. Takanashi, J. Phys. D: Appl. Phys. **44**, 064009 (2011)
33. M. Zhang, E. Bruck, F.R. de Boer, Z. Li, G. Wu, J. Phys. D: Appl. Phys. **37**, 2049 (2004)
34. T. Valet, A. Fert, Phys. Rev. B **48**, 7099 (1993)
35. Y.K. Takahashi, N. Hase, M. Kodzuka, A. Itoh, T. Koganezawa, T. Furubasyashi, S. Li, BSDChS Varaprasad, T. Ohkubo, K. Hono, J. Appl. Phys **113**, 223901 (2013)
36. T. Taniguchi, H. Imamura, T.M. Nakatani, K. Hono, Appl. Phys. Lett. **98**, 042503 (2011)
37. B. Balke, G.H. Fecher, H.C. Kandpal, C. Felser, Phys. Rev. B **74**, 104405 (2006)
38. T. Kubota, S. Tsunegi, M. Oogane, S. Mizukami, T. Miyazaki, H. Naganuma, Y. Ando, Appl. Phys. Lett. **94**, 122504 (2009)
39. J. Sato, M. Oogane, H. Naganuma, Y. Ando, Appl. Phys. Express **4**, 113005 (2011)
40. Y. Sakuraba, M. Ueda, Y. Miura, K. Sato, S. Bosu, K. Saito, M. Shirai, T.J. Konno, K. Takanashi, Appl. Phys. Lett. **101**, 252408 (2012)
41. H.S. Goripati, T. Furubayashi, S.V. Karthik, T.M. Nakatani, Y.K. Takahashi, K. Hono, J. Appl. Phys. **109**, 043901 (2011)
42. H.S. Goripati, T. Furubaashi, Y.K. Takahashi, K. Hono, J. Appl. Phys **113**, 043901 (2013)
43. H.S. Goripati, *Thesis* (University of Tsukuba, Tsukuba, 2013)
44. N. Hase, *Thesis* (University of Tsukuba, Tsukuba, 2012)
45. Y. Sakuraba, S. Kokado, Y. Hirayama, T. Furubayashi, H. Sukegawa, S. Li, Y.K. Takahashi, K. Hono, Appl. Phys. Lett. **104**, 172407 (2014)
46. S. Takahashi, S. Maekawa, Phys. Rev. B **67**, 052409 (2003)
47. G. Bridoux, M.V. Costache, J. Van de Vondel, I. Neumann, S.O. Velenzuela, Appl. Phys. Lett. **99**, 102107 (2011)
48. T. Kimura, H. Hashimoto, S. Yamada, M. Miyao, K. Hamaya, NPG Asia Mater. **4**, e9 (2012)
49. S. Oki, S. Yamada, N. Hashimoto, M. Miyao, T. Kimura, K. Hamaya, Appl. Phys. Express **5**, 063004 (2012)
50. S. Kokado, M. Tsunoda, K. Harigaya, A. Sakuma, J. Phys. Soc. Jpn. **81**, 024705 (2012)
51. S. Kokado, M. Tsunoda, Adv. Mater. Res. **750–752**, 978 (2013)
52. F.J. Yang, Y. Sakuraba, S. Kokado, Y. Kota, A. Sakuma, K. Takanashi, Phys. Rev. B **86**, 020409(R) (2012)
53. Y. Sakuraba, M. Ueda, Y. Miura, K. Sato, S. Bosu, K. Saito, M. Shirai, T.J. Konno, K. Takanashi, Appl. Phys. Lett. **101**, 252408 (2012)
54. N. Hase, PhD Thesis, University of Tsukuba, 2012

Part V
Heusler Alloy Films: Magnetic Properties

Chapter 13
$Co_2Mn_{0.6}Fe_{0.4}Si$: A Heusler Compound Opening New Perspectives in Magnon Spintronics

Thomas Sebastian and Burkard Hillebrands

Abstract This chapter addresses magnon propagation in the Heusler compound $Co_2Mn_{0.6}Fe_{0.4}Si$ and the corresponding perspectives for the emerging field of magnon spintronics. The concept of magnon spintronics requires the utilization of advanced materials providing, in particular, a low magnetic Gilbert damping and compatibility with industrial standards concerning the fabrication of micro- and nano-structures. We present how this challenge can be addressed by the use of low-damping $Co_2Mn_{0.6}Fe_{0.4}Si$ films on the basis of recent studies using micro-focus Brillouin light scattering spectroscopy. The low damping in this Heusler compound not only allows for the realization of increased propagation distances. The pronounced occurrence of nonlinear phenomena might even lead the way towards novel concepts and functionalities in magnonic devices.

13.1 Introduction

In this chapter we present recent results on the spin-wave propagation in magnetic microstructures based on the full Heusler compound $Co_2Mn_{0.6}Fe_{0.4}Si$ (CMFS). As we will see, these results are not only interesting from the scientific point of view but also open new perspectives for the realization of a spin-wave based logic—the goal of *magnon spintronics*.

In the last years, the field of magnon spintronics gained increasing interest caused by the demonstration of advanced concepts for the excitation and detection of magnons and recent developments in micro- and nano-patterning techniques.

T. Sebastian · B. Hillebrands (✉)
Fachbereich Physik and Forschungszentrum OPTIMAS,
Technische Universität Kaiserslautern, 67663 Kaiserslautern, Germany
e-mail: hilleb@uni-kl.de

T. Sebastian
e-mail: tomseb@physik.uni-kl.de

T. Sebastian
Helmholtz-Zentrum Dresden-Rossendorf, Institute of Ion Beam
Physics and Materials Research, Bautzner Landstraße 400, 01328 Dresden, Germany

C. Felser and A. Hirohata (eds.), *Heusler Alloys*,
Springer Series in Materials Science 222, DOI 10.1007/978-3-319-21449-8_13

In contrast to the charge based spin transport in conventional spintronics [1], the idea behind magnon spintronics is the utilization of pure spin currents, e.g. spin waves, to transfer and process information [2–8].

The potential introduction of new devices to the market does not only depend on the quality of these devices itself but also on their compatibility with existing technology. Since spin dynamics typically show frequencies in the Gigahertz range, one of the advantages of magnon spintronics is its compatibility with existing electronic circuitry in present devices, e.g. mobile phones. By a proper choice of materials, magnon spintronics is also compatible with the entire field of conventional complementary metal oxide semiconductor (CMOS) electronics. Therefore it can be regarded as an extension of existing concepts that can be introduced step by step, rather than a substitute.

Another important advantage of magnon spintronics over conventional spintronics and electronics is the different character of the information transport. Electric currents are accompanied by unavoidable electron scattering processes which cause Joule heating and, therefore, the loss of energy. In contrast to this, magnons are eigenstates of the magnonic system and losses are caused only by the coupling to other systems, e.g. the phononic system, or nonlinear processes in the magnonic system itself. These loss channels can be reduced by the utilization of advanced materials and the realization of adequate experimental conditions. Therefore, an energy efficient information transport and processing based on magnons can be realized.

Among the most prominent phenomena related to magnon spintronics is the conversion of electric currents to pure spin currents [9], which is caused by spin Hall effect and spin-transfer torque phenomena [10–14], and vice versa by a combination of spin pumping and the inverse spin Hall effect [15–20]. These effects offer new interesting schemes for the excitation and manipulation as well as the detection of magnons.

Great potential is also given by the interactions between heat currents and spin currents (like the spin Seebeck effect) [21–25]. In particular, the excitation and amplification of spin waves caused by a temperature gradient might lead to the recycling of otherwise lost waste heat in devices—a perspective leading to an even more energy efficient magnonic logic.

While some of these effects are mainly studied on the macro scale, recent developments in micro- and nano-patterning techniques allow for the realization of new concepts on the micro scale as, for example, the excitation of spin waves by spin-torque nano-oscillators [26–31]. In addition, new concepts for the layout and fabrication of sophisticated waveguide structures like magnonic crystals, or for two-dimensional spin-wave propagation show the potential regarding complete building blocks of a magnon based logic [32–34].

The materials commonly used in related studies are yttrium iron garnet (YIG) [35] and $Ni_{81}Fe_{19}$. The ferrimagnetic YIG material has been used for decades in the field of microwave techniques, because it shows a very low Gilbert damping that allows for the propagation of spin waves on macroscopic length scales. Unfortunately, the material also shows severe disadvantages concerning possible technical applications. The complicated crystallographic structure of YIG hinders the fabrication of high quality samples on the nano- and even the micro-scale. YIG films are usually prepared by liquid-phase epitaxy or pulsed-laser deposition—growth processes that are rather expensive and incompatible with industrial demands. In particular, the material is incompatible with CMOS-electronics. In addition, yttrium is a rare earth metal, a class of materials in the higher price tag region. Therefore, YIG serves as an excellent demonstrator or model system but its use regarding technical applications is currently still rather limited.

On the other hand, the magnetic losses in $Ni_{81}Fe_{19}$ and other metallic ferromagnets, which can be combined with CMOS-electronics, only allow for spin-wave propagation in the range of a few micrometers. Therefore, at present, the effects and technical abilities mentioned above can be illustrated in these materials, but not brought to a practicable application. However, regarding perspective technical applications, it is crucial to optimize these effects and to transfer them to the nano scale. This situation shows the necessity of the identification and incorporation of advanced materials as an important step towards the realization of perspective devices based on pure spin currents. As illustrated by the following results, the full Heusler compound CMFS can solve the challenges of magnon spintronics.

Among the possible candidates of magnetic materials that are compatible with the demands of industrial growth- and micro-/nano-structuring processes, Heusler compounds are very promising alternatives. In the field of magneto resistive devices for sensing and data storage, Heusler compounds have already been successfully incorporated in the last years [36, 37]. In addition, experiments on the conversion of pure spin currents to electrical signals and on the interaction with heat currents underline the potential of this class of materials [15, 23].

In particular, most of the cobalt-based full Heusler materials show sufficiently high Curie temperature as well as a higher spin polarization and a lower magnetic Gilbert damping than most conventional metallic 3d-ferromagnets [38]. While the Curie temperature is crucial for all technical applications, an increased spin polarization is advantageous for magneto resistive devices. Of special importance for the transport of information via magnons is the decreased magnetic Gilbert damping that gives rise to an increased decay length, decreased thresholds for nonlinear phenomena, and decreased threshold for direct-current based excitation schemes. A thorough investigation and understanding of spin-wave propagation in Heusler materials can open new perspectives for the field of magnon spintronics.

In the following sections, we will illustrate the opportunities and advantages offered by the utilization of Heusler compounds in magnon spintronics. In Sect. 13.2 we give a brief introduction to spin dynamics. Section 13.3 is devoted to the description of the sample layout and an introduction to the experimental technique of Brillouin light scattering (BLS) spectroscopy, which was used to obtain the following results. Section 13.4 addresses the propagation of spin waves in a CMFS microstructure in the linear regime. Section 13.5 addresses nonlinear spin-wave propagation in a CMFS waveguide. Nonlinear phenomena offer outstanding possibilities for the processing of microwave signals, such as frequency conversion. In Sect. 13.5 we will discuss the nonlinear emission of spin-wave caustics at twice and three times the excitation frequency. This interesting interplay of different phenomena in the spin system illustrates the possibilities offered by the utilization of nonlinearities in magnon spintronics. The chapter will be closed by a short summary and outlook in Sect. 13.6.

13.2 Brief Introduction to Spin Dynamics

Spin dynamics are governed by the Landau-Lifshitz and Gilbert (LLG) equation

$$\frac{d\mathbf{M}}{dt} = -|\gamma|\mu_0 (\mathbf{M} \times \mathbf{H}_{\text{eff}}) + \frac{\alpha}{M_S}\left(\mathbf{M} \times \frac{\mathbf{M}}{dt}\right), \tag{13.1}$$

which describes the precessional motion of a magnetic moment **M** about an effective field $\mathbf{H}_{\text{eff}}$. In this equation γ is the gyromagnetic ratio and μ_0 is the vacuum permeability. While the first term of the LLG equation is responsible for the motion of **M** around $\mathbf{H}_{\text{eff}}$, the second one incorporates the magnetic Gilbert damping α of this motion. This term eventually leads to an alignment of the magnetic moment **M** parallel to the effective field direction. For real systems of magnetic moments, which are coupled by exchange and dipolar interaction, additional effects like demagnetizing fields and the proper choice of boundary conditions have to be included in the model. While there is no analytical solution for this nonlinear problem, there are several approaches to the definite solution of this equation in the linear regime.

In the following we will briefly introduce an analytical equation for the linear spin-wave dispersion in magnetic thin films derived by B. Kalinikos and A. Slavin in 1986 [39], which is used for the description of our experimental findings in Sects. 13.4 and 13.5.

$$f(k,\phi) = \frac{|\gamma|}{2\pi}\sqrt{\left[\mu_0 H_{eff} + DK^2\right]\left[\mu_0 H_{eff} + Dk^2 + \mu_0 M_s F_{nn}(k,\phi,\mathrm{t})\right]}, \tag{13.2}$$

where k is the wave vector, D is the exchange stiffness, and t the thickness of the thin film. The function F_{nn} describes the anisotropy of this dispersion relation given by the relative orientation of the wave vector and the magnetization direction $\phi = \angle(\mathbf{M}, \mathbf{k})$. Figure 13.1 illustrates the application of this approach to a CMFS thin film with a thickness of 30 nm in an external magnetic bias field of $\mu_0 H_{\text{ext}} = 48\,\text{mT}$. As we will see in Sect. 13.5, this anisotropy of the dispersion gives rise to interesting phenomena.

Even though this model was developed for the case of homogeneous thin films, its application to systems with finite lateral dimensions shows a good accordance to experimental findings. In particular, for the case of spin-wave waveguides, the introduction of one finite lateral dimension and the resulting quantization of the wave vector over the width of the waveguide turned out to be a very successful approach.

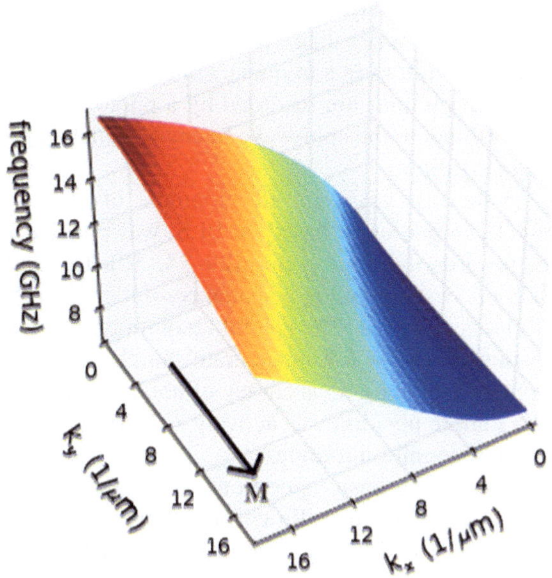

Fig. 13.1 Spin-wave dispersion surface in a CMFS film for an external bias magnetic field of $\mu_0 H_{ext} = 48$ mT calculated according to the analytical model by Kalinikos and Slavin [39]

Without going into detail, we want to emphasize that the above dispersion relation is a linear approximation. While this approximation given by (13.2) is only valid in the linear regime, approaches incorporating higher orders yield a huge variety of nonlinear effects, some of them even more pronounced than in other nonlinear systems [40–45]. In particular, we want to indicate the possibility for the excitation of higher harmonics as presented in Sect. 13.5.

Since a thorough discussion of spin dynamics would by far exceed the scope of this chapter, this introduction should only serve to provide the most basic information for the understanding of the following results. For a more complete treatment of the field of spin dynamics the reader is referred to [45–47].

13.3 Sample Layout and Instrumentation

All results presented in this chapter were obtained by Brillouin light scattering (BLS) spectroscopy [48]. In the following, we will give a brief introduction to this experimental technique. BLS is the inelastic scattering of photons from the fundamental excitations of the magnetic system, the magnons. Magnons are the quasi-particles or quanta of spin waves. The conservation of energy and momentum in this scattering process as well as the phase sensitivity of the process are the basis for the investigation of spin dynamics.

The frequency analysis of the inelastically scattered light and, thus, the investigated spin-wave modes, is performed with a (3 + 3) tandem-Fabry-Perot interferometer, designed by John R. Sandercock [49]. The advanced interferometry technique results in a contrast better than 10^{10} for the separation of the elastically and inelastically scattered photons in a frequency range from 500 MHz up to 1 THz. The inelastically scattered photons are recorded by a photo detector. The resulting BLS intensity is proportional to the spin-wave intensity which is given by the square of the dynamic magnetization.

In the last years, the development of microscopes based on magneto-optical effects, in particular based on BLS, was pushing the research field of spin dynamics towards the micro scale [29–34, 50]. In addition to the frequency resolution, BLS microscopy offers a spatial resolution corresponding to the fundamental limit of classical optics. The spatial resolution in the experiments presented below was 250 nm. To allow for an accurate sample positioning, the sample was mounted on a piezo stage with an accuracy of about 10 nm and monitored with a CCD camera. The sample position was actively stabilized during the measurement by automated image recognition routines.

Besides the basic operations of frequency- and space-resolved measurements, BLS microscopy also allows for the investigation of magnonic transport phenomena with time- and phase-resolution [50–54]. The latter will be briefly discussed together with experimental results in Sect. 13.4.

At this point, it is worth mentioning that BLS spectroscopy also offers the possibility to evaluate all material parameters that affect the LLG equation (13.1), namely the saturation magnetization M_S, the exchange stiffness D, and the gyromagnetic ratio γ. In fact, first BLS measurements on different Co-based Heusler thin films have been used for the evaluation of materials parameters [38, 55]. An exact knowledge of these material parameters is not only essential in the field of magnon spintronics but can also be crucial for the design of any other device based on dynamic magnetization processes like, for example, the switching of a magnetic random access memory cell.

The CMFS thin films used in our studies were sputter deposited and post-annealed at a temperature of 450°C to improve the crystal structure. This procedure yields a dominating crystallographic $L2_1$ order. Further details about the growth of the film and the properties of the material can be found in [56–59]. Measurements (not shown here) using the magneto-optical Kerr effect and vector network analyzer ferromagnetic resonance (FMR) techniques reveal a saturation magnetization of $M_s = 1003$ kA/m and a negligible anisotropy field of $H_{\text{ani}} = 1$ kA/m of the unstructured films.

The microstructured spin-wave waveguides were patterned using electron beam lithography and Ar ion milling. The widths of these waveguides were 4 and 5 μm in the case of the experiments in the linear and nonlinear regime, respectively. In order to excite spin waves in the magnetic microstructures, the shorted end of a coplanar waveguide (CPW) made of copper served as an antenna structure. Microwave currents flowing in this antenna result in dynamic Oersted fields that are suitable to excite spin dynamics in the Gigahertz range. An external magnetic bias field was applied transversely to the waveguide in all experiments. This results in a magnetization direction **M** perpendicular to the wave vector **k** of the propagating spin waves,

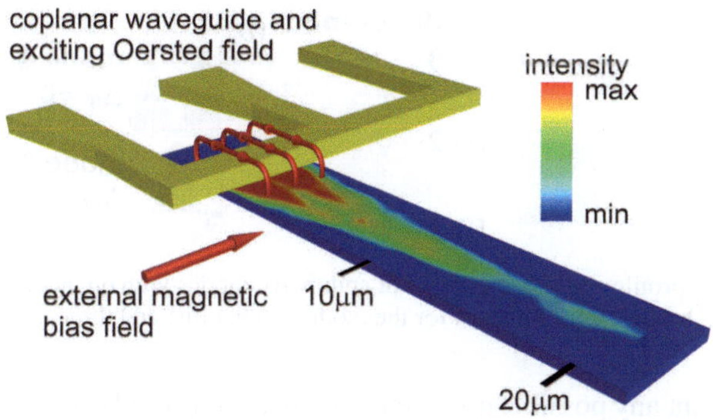

Fig. 13.2 Scheme of the sample layout. The CMFS waveguides are magnetized transversely by the external magnetic bias field. The shorted end of a coplanar waveguide serves to excite spin dynamics in the GHz range. The probing laser light of the BLS microscope is used to detect the spin-wave intensity with a spatial resolution of 300 nm. The intensity distribution shown on a 4 μm wide waveguide was observed for an external bias field of $\mu_0 H_{\text{ext}} = 40\,\text{mT}$ and $f = 6.0\,\text{GHz}$

$\phi = \angle(\mathbf{M}, \mathbf{k}) = 90°$. This geometry is usually referred to as Damon-Eshbach geometry [60]. The choice of this specific experimental scenario can be supported by two facts. Spin-wave modes in the Damon-Eshbach geometry exhibit a comparatively large group velocity and, thus, also a large decay length which is favorable in all experiments. In addition, the dynamic Oersted fields **h** below the exciting antenna, where the field amplitudes are large, are oriented perpendicular to the magnetization direction, which results in a large torque $\mathbf{h} \times \mathbf{M}$ and, therefore, an efficient excitation of spin waves. A sketch of the sample layout is shown in Fig. 13.2.

13.4 Spin-Wave Propagation in the Linear Regime

Since the propagation distances in most metallic ferromagnets and compounds are limited to a few micrometers, a first step towards the realization of perspective magnon-spintronic devices is the realization of an increased decay length in the linear regime of wave propagation [58].

The low Gilbert damping of CMFS makes the compound a promising candidate for the realization of this goal. CMFS has a Gilbert damping of ($\alpha(CMFS) = 3 \times 10^{-3}$) that is almost three times smaller than the commonly used $Ni_{81}Fe_{19}$ with $\alpha(Ni_{81}Fe_{19}) = 8 \times 10^{-3}$.

We set up a structure comprising a 4 μm wide spin-wave waveguide, patterned from a 30 nm thick film of CMFS, and a copper antenna placed across the waveguide. A sketch of the layout is presented in Fig. 13.2. In addition to this schematic, Fig. 13.2 includes the measured two-dimensional BLS intensity map for an external bias field of $\mu_0 H_{\text{ext}} = 40\,\text{mT}$, an excitation frequency $f = 6.0\,\text{GHz}$, and a microwave power of $P = 0.1\,\text{mW}$. The latter ensures spin dynamics with precession angles that are small

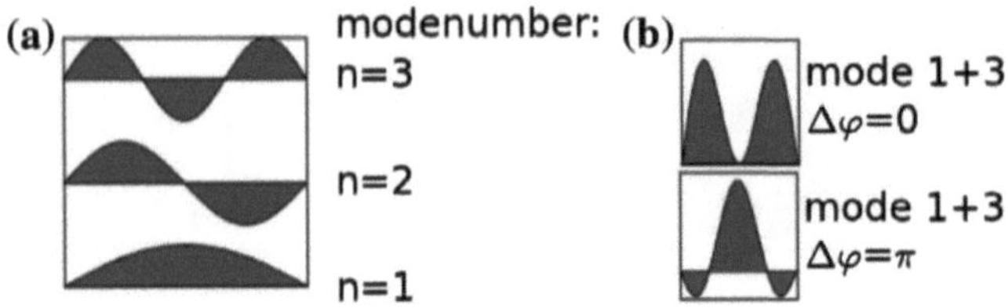

Fig. 13.3 **a** Mode profile and mode number of spin-wave modes with quantization over the width of the waveguide. **b** Interference pattern for the mode number one and three

enough to prevent any possible nonlinear effects which could open up additional loss channels for the spin-wave modes.

The intensity map was recorded by accumulating the BLS intensity in a small frequency range of $\Delta f = 300\,\text{MHz}$ around the excitation frequency $f = 6.0\,\text{GHz}$ as a function of the probing position on the waveguide. The most notable feature of this intensity map is the very long propagation range of more than $20\,\mu\text{m}$ where spin waves can be detected. Before the decay length of the propagating spin waves will be evaluated quantitatively, we address the observed intensity pattern.

This pattern shows variations of the intensity across the width of the waveguide. Similar patterns have been observed in other experiments on spin-wave propagation in waveguide structures and are caused by multi-mode propagation and the resulting interference [60]. This effect can be understood as follows: The finite width of the waveguide results in a quantization of the transverse wave vectors usually denoted by the number of antinodes across the waveguide as shown in Fig. 13.3a. Since spin-wave modes with different mode numbers n have different wave vectors along the propagation direction, the resulting interference pattern changes along the waveguide. Exemplary interference patterns for the mode number 1 and 3 are shown in Fig. 13.3b for different phase shifts of $\Delta\varphi = 0$ and $\Delta\varphi = \pi$. The specific pattern shown in Fig. 13.2 is mainly caused by the interference of exactly these transverse modes of the waveguide with the mode numbers one and three. It should be mentioned that an antenna structure as used in this experiment and shown in Fig. 13.2 excites odd mode number predominantly.

For a quantitative analysis of the decay length of the propagating spin waves in the CMFS waveguide, the BLS intensity was evaluated as a function of the distance from the exciting antenna. To minimize any influence of a possible multi-mode propagation, the BLS intensity in the recorded maps–as shown in Fig. 13.2—was integrated over the entire width of the waveguide for different distances. Exemplary results are presented in Fig. 13.4 for an external field $\mu_0 H_{\text{ext}} = 50\,\text{mT}$ and an excitation frequency $f = 6.9\,\text{GHz}$.

The graph shows the integrated BLS intensity as a function of the distance from the antenna in the range from 2 to $22\,\mu\text{m}$. The observed decay of the BLS intensity can be described using an exponential equation

$$I(x) = I_0 \exp\left(-\frac{2x}{\delta}\right) + b \tag{13.3}$$

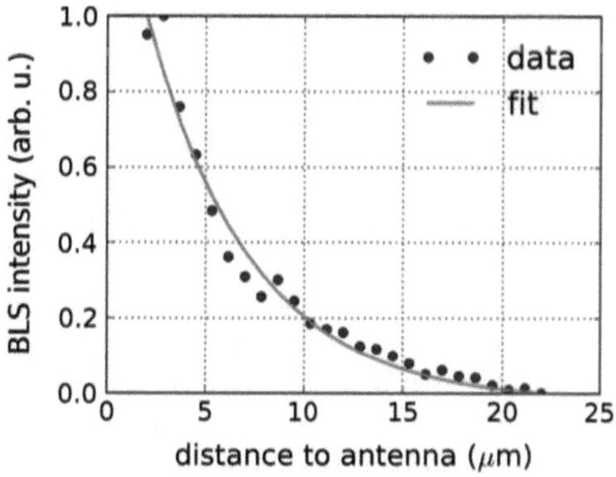

Fig. 13.4 BLS intensity integrated over the width of a 4 μm wide CMFS waveguide in dependence on the distance from the exciting antenna and fit according to (13.3). In this exemplary measurement the external field was $\mu_0 H_{ext} = 50\,\text{mT}$ and the excitation frequency was $f = 6.9\,\text{GHz}$

where x is the distance from the antenna, δ is the decay length of the spin-wave amplitude, and b is an offset caused by the noise-level of the measurement. Since the decay length of the spin wave refers to the amplitude—and not to the intensity—(13.3) includes the factor 2 in the exponent. Thus, we take into account that the experimentally observed BLS intensity is proportional to the square of the spin-wave amplitude. A fitting procedure according to (13.3) yields the decay length δ as a parameter. As indicated by the fit also shown in Fig. 13.4, the data points follow the expected exponential behavior.

The result of this exemplary measurement and for different experimental parameters are presented in Table 13.1. All measurements performed in the CMFS waveguide yield a decay length which is rather large compared with data found for comparable waveguide geometries made of the commonly used $Ni_{81}Fe_{19}$ (for comparison see [31, 61, 62] where $\delta(Ni_{81}Fe_{19}) < 6\,\mu m$).

The varying values for the decay length can be explained by the different experimental parameters used in the measurements. Since values like the external field $\mu_0 H_{ext}$ and the frequency f—among others—control the character of the excited spin-wave mode, different group velocities of the investigated magnons can be observed. The higher the group velocity, the larger is the decay length. Analytical calculations according to [39] for a magnetization direction perpendicular to the waveguide ($\phi = \angle(\mathbf{M}, \mathbf{k}) = 90°$) predict a behavior as illustrated by our data. The

Table 13.1 Summary of the results on the decay length in the CMFS Heusler compound and the results for $Ni_{81}Fe_{19}$ waveguides with comparable geometries from [31, 61, 62]

Frequency (GHz)	External field (mT)	Decay length (μm)
6.0	40	8.7
6.2	40	11.9
7.5	40	9.1
6.9	50	10.6
7.2	50	16.7
$Ni_{81}Fe_{19}$	References	<6 μm

group velocity tends to zero close to FMR resulting in a small decay length. Towards higher frequencies the group velocity (and therefore the decay length) shows a sharp maximum and a subsequent decay.

In our measurements, a remarkable maximum value of $\delta = 16.7\,\mu\text{m}$ was found at $\mu_0 H_{\text{ext}} = 50\,\text{mT}$ and $f = 7.2\,\text{GHz}$. Even for spin-wave modes close to FMR ($f = 6.0\,\text{GHz}$ at $\mu_0 H_{\text{ext}} = 40\,\text{mT}$ and $f = 6.9\,\text{GHz}$ at $\mu_0 H_{\text{ext}} = 50\,\text{mT}$) the decay length reaches values larger than in $Ni_{81}Fe_{19}$. The low Gilbert damping found in the Heusler compound CMFS and the related increase of the decay length are an important step towards the realization of possible future technical applications. We believe that this increased decay length is a direct consequence of the reduced damping in the investigated compound. Since, up to now, theoretical predictions of the damping constants [63] in similar Heusler materials are still one order of magnitude smaller than the experimental findings, research efforts in this direction seem to be very promising.

While an increased decay length of propagating spin waves is key for magnon based information transport, many concepts for spin-wave logic and the processing of information are based on the wave character of magnons. Thus, another important property of spin waves regarding their possible use in future applications is their coherence length [63]. Phase-resolved BLS is a powerful tool for the investigation of phase related properties [50–53]. In the following we describe the estimation of the coherence length of externally excited magnons in the CMFS waveguide via phase-resolved BLS microscopy. In this approach the phase information of the scattered photons is extracted via the interference of the scattered light with a reference beam. The frequency of this reference beam is shifted to match the frequency of the inelastically scattered light by using an electrooptical modulator (EOM). This ensures temporal coherence of reference and probing laser beam. Therefore, relative phase shifts $\Delta\beta$ of these two beams can only be caused by a spatially varying phase of the detected spin-wave mode at different probing positions.

The interference pattern for light scattered at the spin-wave mode at $\mu_0 H_{\text{ext}} = 40\,\text{mT}$ and $f = 8.5\,\text{GHz}$ with the reference beam is shown in Fig. 13.5a as a function of the position on the waveguide. The two different data sets (red and blue lines) have been recorded with an artificially introduced phase shift of the reference beam of $\Delta\beta = \pi/2$. The results show an almost sinusoidal behavior as expected for single mode propagation in the waveguide. The slight distortions of this trend are likely caused by a weak combination of higher order transverse modes in the waveguide—as already discussed above and shown in Fig. 13.2. In addition, the phase shift of $\pi/2$ results in the expected shift of the entire interference pattern.

Further evaluation of the data reveals a linear dependence of the phase on the propagation distance (see Fig. 13.5b) and, thus, coherence for the entire range experimentally accessible. Similar results (not shown here) were observed for other applied field strengths and frequencies and guarantee a reliable signal processing of interfering spin-wave signals over large distances.

This long ranged and coherent magnonic transport illustrates the advantages of the utilization of CMFS in magnon-spintronic devices in the linear regime. A maximum decay length of $16.7\,\mu\text{m}$ has been observed which is an increase of almost a factor of

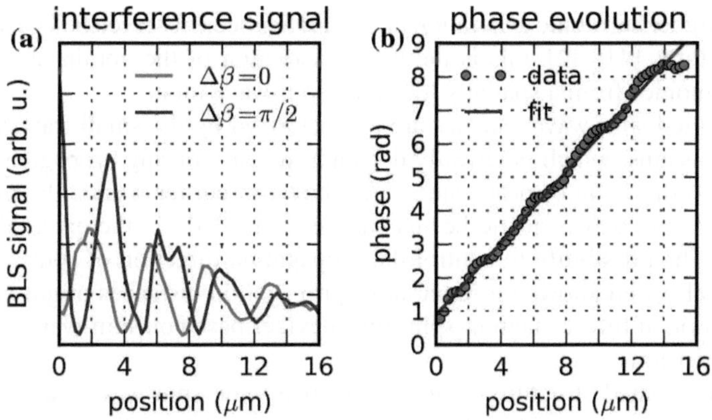

Fig. 13.5 Phase-resolved BLS measurement. The external field was $\mu_0 H_{\text{ext}} = 40\,\text{mT}$ and the excitation frequency was $f = 8.5\,\text{GHz}$. (**a**) Shows the interference of the reference beam and the scattered light. The phase of the reference beam was changed by $\pi/2$ between the two different measurements. (**b**) Shows the phase evolution as a function of the propagation distance

three compared to the decay length observed in typical experiments using $Ni_{81}Fe_{19}$ waveguides of comparable geometries. In addition, a coherence length on the same order of magnitude as the decay length was detected. This allows for the realization of concepts based on the interference of spin waves. These results qualify CMFS as a very promising candidate for future magnon spintronic devices in information transport and signal processing.

13.5 Spin-Wave Propagation in the Nonlinear Regime

Magnonic systems—governed by the intrinsically nonlinear LLG (13.1)—are known to exhibit a variety of nonlinear effects, some of them even more pronounced as in other fields of wave physics. In the last years, nonlinear spin dynamics in magnetic microstructures made of metallic ferromagnetic thin films or layer stacks have gained large interest [40–45]. The decreased magnetic losses in CMFS, already described in the preceding Section, give rise not only to increased propagation distances but also to the pronounced occurrence of nonlinear processes.

The phenomenon presented below is not only interesting from the scientific point of view but also regarding new schemes and perspective applications in magnon spintronics. Therefore, this example is well suited to illustrate the possibilities for information transport and processing offered by nonlinearities.

In the following, we report the nonlinear generation of higher harmonics from a localized spin-wave edge mode in a micro-structured Heusler waveguide causing the emission of caustic spin-wave beams [59].

While higher harmonic generation in general has been observed before in magnetic microstructures [41, 42], this is the first observation of the nonlinear emission of higher harmonics from a spatially localized spin-wave mode.

The emitted spin-wave caustics are characterized by the small transversal aperture of the beams, which practically does not increase during propagation, and the well-defined direction of propagation. As shown in earlier works, the propagation direction of spin-wave caustic beams can be controlled via the externally applied field [67]. The possibility to control the propagation direction of spin-wave beams in a material which allows for remarkable propagation distances might be a key for the realization of future magnon-spintronic devices based on spin-wave propagation in two dimensions.

The investigated structure comprises a 5 μm wide spin-wave waveguide and a copper antenna with a width of $\Delta x = 1$ μm placed across the waveguide. In the following description, the waveguide is positioned in the x–y-plane with the propagation direction in x-direction. The origin of the coordinate system is given by the position of the antenna between $x = -1$ μm and $x = 0$ μm. An external magnetic field of $\mu_0 H_{\text{ext}} = 48$ mT was applied transversely to the waveguide in y-direction resulting in Damon-Eshbach geometry [58] for spin waves propagating along the waveguide. In all measurements presented below, the microwave frequency for the excitation of the spin dynamics was fixed at $f_e = 3.5$ GHz and the microwave power was $P = 20$ mW.

In a first measurement, a spectrum was recorded in the center of the waveguide at a distance of 4.5 μm from the antenna. As can be seen in Fig. 13.6, the resulting spectrum exhibits not only the directly excited spin-wave mode at $f = 3.5$ GHz but also higher harmonics at twice and three times the excitation frequency. Even more interesting than the mere occurrence of the higher harmonics, the most striking feature of the results depicted in Fig. 13.6 is the intensity ratio of the different spin-wave modes. At this probing position the intensity of the directly excited spin wave at 3.5 GHz is lower than for the higher harmonics. As we will see, this is a consequence of the different propagation characteristics of the observed spin-wave modes and can be described in terms of the anisotropic spin-wave dispersion and caustic effects.

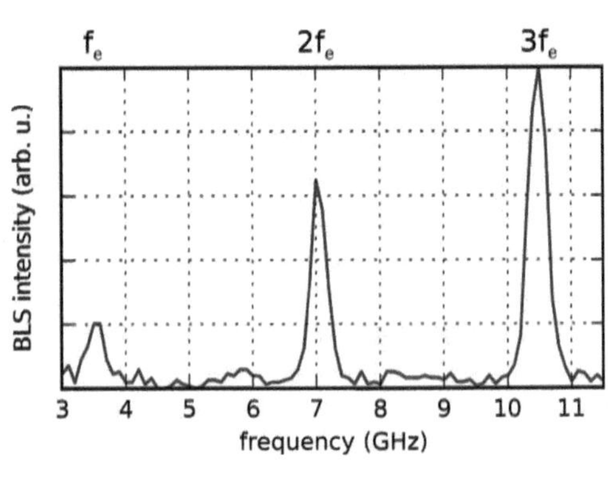

Fig. 13.6 BLS spectrum observed in the center of a 5 μm wide CMFS waveguide for an excitation frequency $f_e = 3.5$ GHz, a microwave power $P = 20$ mW, and an external field $\mu_0 H_{\text{ext}} = 48$ mT. The probing position is 4.5 μm from the antenna in the center of the waveguide

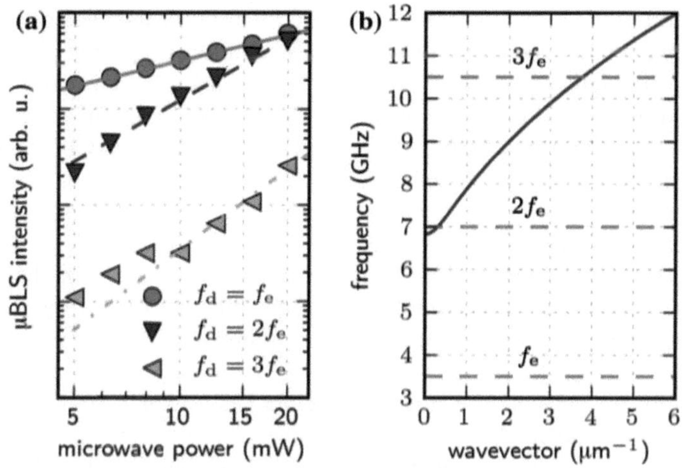

Fig. 13.7 **a** Power dependencies of the BLS intensity for the detection frequencies $f_d = 3.5\,\text{GHz}$, 7.0 GHz, and 10.5 GHz at a fixed excitation frequency $f_e = 3.5\,\text{GHz}$. Please note the log-log presentation of the data. The lines in the graph correspond to fits according to (13.4). These least square fits yield $s_{1f} = 0.9 \pm 0.1$, $s_{2f} = 2.1 \pm 0.1$, and $s_{3f} = 2.8 \pm 0.3$. **b** Calculated dispersion relation for the center of the CMFS waveguide according to [39] as well as the excitation frequency f_e and the detection frequencies f_d for the higher harmonics (*dashed lines*)

Before we will turn to the description of these propagation characteristics in detail, we will analyze the higher harmonic generation quantitatively. Therefore, we investigated the dependency of the BLS intensity on the applied microwave power. To do so, the BLS intensities have been recorded in a range of $\Delta f = 300\,\text{MHz}$ around the center frequencies of the individual modes. Figure 13.7a shows the results for the directly excited mode and the higher harmonics. The measurements have been performed close to the edge of the waveguide and near the antenna at $x = 0.7\,\mu\text{m}$ and $y = 0.8\,\mu\text{m}$. The data is presented on a log-log scale with fits according to

$$I_m(p) = A_m P^{s_m} + b, \tag{13.4}$$

where I_m is the BLS intensity, A_m a proportionality factor, P the applied microwave power, and b the noise-level in our measurement. In accordance with previous reports [41, 42], the observed nonlinear processes do not show a threshold power level, but reliable detection on the background of the noise is not possible for powers below $P = 5\,\text{mW}$ for the detection frequency $f_d = 3f_e = 10.5\,\text{GHz}$.

The slopes of the curves for the different spin-wave modes m are caused by the different power-laws specified by the exponent s_m. A least square fit of the data yields $s_{1f} = 0.9 \pm 0.1$, $s_{2f} = 2.1 \pm 0.1$, and $s_{3f} = 2.8 \pm 0.3$, respectively. These experimental findings close to the integer values 1, 2, and 3 are in accordance with previously reported experimental data and theoretical predictions for the nonlinear generation of higher harmonics [41, 42].

In addition, Fig. 13.7b shows the dispersion relation for the center of the CMFS waveguide calculated according to [39]. The excitation frequency f_e and the frequencies of the higher harmonics $2f_e$ and $3f_e$ are marked by dashed lines. As can be seen, and in contrast to the frequencies of the higher harmonics, the excitation frequency $f_e = 3.5\,\text{GHz}$ is far below the lower cut-off frequency of the spin-wave dispersion in the waveguide. Together with the intensity ratio of the three observed spin-wave modes shown in Fig. 13.6, this fact raises the question of how these modes propagate in the waveguide.

To understand the nature of the different spin-wave modes and their intensity ratio shown in Fig. 13.6, we recorded two-dimensional BLS intensity maps similar to the one shown in Fig. 13.2 for the detection frequencies $f_d = f_e$, $f_d = 2f_e$, and $f_d = 3f_e$.

The results are presented in Fig. 13.8. Figure 13.8a, which shows the intensity map of the directly excited mode, reveals a strong localization at the edges of the waveguide. This localization can be expected taking into account the dispersion calculation shown in Fig. 13.7b. The existence of a localized mode close to the edges of the waveguide can be understood as follows. In the center of the waveguide, the assumption of a homogeneous magnetization oriented parallel to the external field is a very good approximation and the calculation shown in Fig. 13.7b is valid. However, the demagnetizing fields created by the magnetization pointing transversely to the waveguide are responsible for a strongly inhomogeneous and decreased effective field near the edges. This field configuration allows for the existence of the localized edge modes energetically far below the spin-wave dispersion for propagating modes

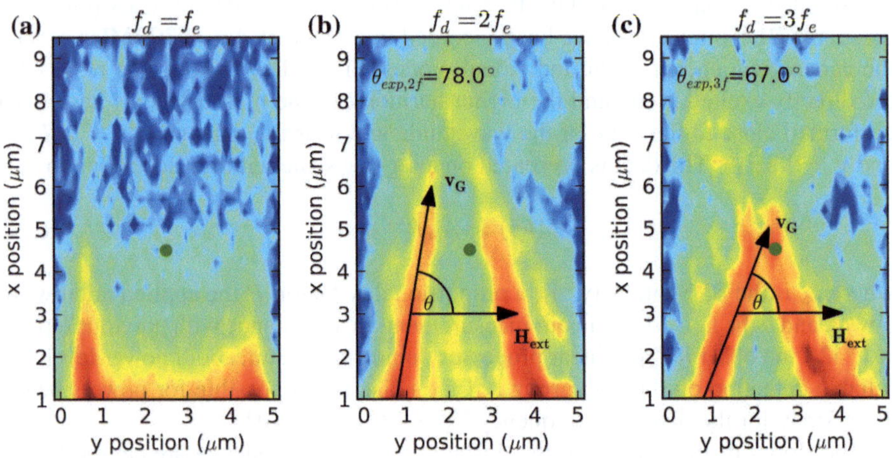

Fig. 13.8 BLS intensity maps for **a** $f_d = f_e = 3.5\,\text{GHz}$, **b** $f_d = 2f_e = 7.0\,\text{GHz}$, and **c** $f_d = 3f_e = 10.5\,\text{GHz}$. While we observed a localized edge mode for $f_d = f_e$, the intensity maps of the higher harmonics show strongly directed spin-wave beams along the angles $\theta = \angle(\mathbf{H}_{ext}, \mathbf{v}_G)$. The lines in these maps are guides to the eye to identify the propagation angles $\theta_{exp,2f}$ and $\theta_{exp,3f}$. The *green dots* in the graphs indicate the position of the probing laser beam in the measurement presented in Fig. 13.6 (Color figure online)

in the center of the waveguide [65, 66].Therefore, we conclude that the highest intensities close to the edges result from the resonant excitation of these edge modes by the microwave field. In contrast to this, the intensity close to the antenna in the center of the waveguide is attributed to forced excitation.

Figures 13.8b, c show intensity maps for the higher harmonics at the detection frequencies $f_d = 2f_e = 7.0\,\text{GHz}$ and $f_d = 3f_e = 10.5\,\text{GHz}$, respectively. In both cases the waves radiated from the position of the edge mode are strongly directed, have a small transversal aperture, and show nondiffractive behavior. The propagation direction of the higher harmonics can be defined by using the angle $\theta = \angle(\mathbf{H}_{\mathbf{ext}}, \mathbf{v}_\mathbf{G})$ formed by the group velocity of the spin waves $\mathbf{v}_\mathbf{G}$ and the external field $\mathbf{H}_{\mathbf{ext}}$. These propagation directions are responsible for the intensity distribution shown in the spectrum in Fig. 13.6. The position of the probing point in this measurement is indicated in the corresponding intensity maps with a green dot. Since the spin wave at $f_d = 3.5\,\text{GHz}$ is localized at the edges of the waveguide, its intensity is comparably weak in the center. In contrast, the higher harmonics have frequencies above the cut-off frequency of the spin-wave dispersion and can propagate in the center of the waveguide supporting the intensity ratio recorded in our measurement.

At this point of our analysis, the understanding of the propagation characteristics for the spin-wave modes at $2f_e$ and $3f_e$ is the remaining issue. Since the nonlinear frequency multiplication results in well-defined frequencies, we have to examine the possible spin-wave modes at $f = 2f_e = 7.0\,\text{GHz}$ and $f = 3f_e = 10.5\,\text{GHz}$. These modes can be found by an analytical calculation of the iso-frequency curves $f(k_x, k_y) = \text{const}$ based on the dispersion relation specified in (13.2). Calculations for $f = 2f_e = 7.0\,\text{GHz}$ and $f = 3f_e = 10.5$ GHz are illustrated in Fig. 13.9a where the iso-frequency curves are shown in the k_x-, k_y-plane.

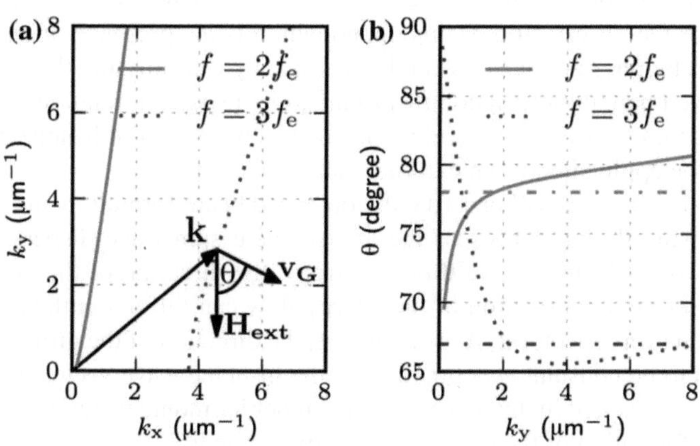

Fig. 13.9 Analytical calculations according to [38]. **a** Iso-frequency curves $f_{\text{const}} = f(k_x, k_y)$ for $f = 2f_e = 7.0\,\text{GHz}$ and $f = 3f_e = 10.5\,\text{GHz}$. Based on these calculations exemplary directions for $\mathbf{k}$, $\mathbf{v}_\mathbf{G}$, $\mathbf{H}_{\mathbf{ext}}$ and the propagation angle $\theta = \angle(\mathbf{H}_{\mathbf{ext}}, \mathbf{v}_\mathbf{G})$ are shown in the graph. **b** Propagation direction θ calculated from the iso-frequency curves shown in **a**. *Dash-dotted lines* correspond to the angles $\theta_{\text{exp},2f}$ and $\theta_{\text{exp},3f}$ observed in the experiment

As a first result of these calculations we can state that there is no initial wave vector $\mathbf{k}_i$ that could fulfill wave vector conservation for the simultaneous generation of the second and third harmonic, i.e. $\mathbf{k}(2f) = 2\mathbf{k}_i$ and $\mathbf{k}(3f_e) = 3\,\mathbf{k}_i$. In addition to the strong localization of the edge mode which leads to an uncertainty of its wave vector $\mathbf{k}_i$, this supports an approach to regard the edge mode as a source emitting partial waves with different wave vectors rather than a source generating higher harmonics with one well-defined wave vector. This approach follows a common description of nonlinear higher harmonic generation in terms of the dynamic stray fields created by the elliptical precession in the spin system [41, 42, 45]. Since these stray fields can, in a first approximation, be regarded as isotropic, the reason for the strongly directed spin-wave beams has to be found in the magnonic system itself.

The range of different wave vectors that form the observed beams of higher harmonics depends on the source of the higher harmonic generation. As we can see in Figs. 13.8a, b, the origin of these spin-wave beams is given by the position of the localized edge mode, which can be regarded as this source. The intensity of this edge mode and, thus, the nonlinear higher harmonic generation is strongest directly below the exciting antenna. The width of the antenna is $\Delta x = 1\,\mu\text{m}$ and the spread of the edge mode in y-direction can be estimated from the BLS intensity map in Fig. 13.8a to be approximately $\Delta y = 1\,\mu\text{m}$. As a first approximation for the maximum wave vector that can be excited by a source of $1 \times 1\,\mu\text{m}^2$, we calculate the Fourier transform of this geometry and obtain $k_{\text{max}} = 6.28\,\mu\text{m}^{-1}$. All calculations presented below take into account a wave vector range that includes this maximum value.

The nondiffractive propagation of strongly directed beams with small transversal aperture is known from various fields of wave physics, e.g. optics, and corresponding beams are called caustics. Caustic beams are formed by partial waves with different wave vectors but coinciding directions of their group velocities. A crucial precondition for the observation of this phenomenon is the presence of an anisotropic dispersion relation. In optics this can be realized by transmission through or reflection on curved boundaries. In contrast to this, the dispersion relation of spin waves in magnetic thin films is intrinsically anisotropic. Moreover, this anisotropy can be steered by the application of an external magnetic field.

In the following, we will describe the observed beam formation and propagation directions using the properties of the approximate spin-wave dispersion given in (13.2) [39, 67, 68]. Because of this anisotropy, the direction of the group velocity $\mathbf{v}_G$ and, therefore, the direction of the energy flow can differ significantly from the direction of its wave vector $\mathbf{k}$. As we will see, the direction of the group velocity can be assumed to be constant as a function of the in-plane wave vector for most wave vectors that are excited in the process of the higher harmonic generation. This finally leads to the formation of the caustics in our experiment.

Using the data obtained by our calculations of the iso-frequency curves shown in Fig. 13.9a, we can calculate the propagation direction θ the spin-wave beams by $\theta = \angle(\mathbf{H}_{\text{ext}},\mathbf{v}_G) = \arctan(v_y/v_x)$. Figure 13.9b shows the calculated propagation angle θ in the CMFS waveguide as a function of k_y. The most important feature in the trend of θ is the small variation of $\Delta\,\theta \leq 2°$ for both frequencies $f = 2f_e$ and

$f = 3f_e$. This almost constant propagation direction of the different partial waves is the reason for the strongly directed beams.

Our calculations yield $\theta_{calc,2f} = 79°$ and $\theta_{calc,3f} = 66°$ as mean values in the wave vector range $k_y = 2-7\mu m^{-1}$, respectively. The dash-dotted lines in Fig. 13.9b represent the propagation angles $\theta_{exp,2f} = 78°$ and $\theta_{exp,3f} = 67°$ of the spin-wave beams observed experimentally in the BLS intensity maps shown in Figs. 13.8b, c. The comparison of our experimental findings and the analytical calculations shows an agreement well within the expected accuracy of our experimental setup and is, therefore, supporting our conclusion of the formation of spin-wave caustic beams.

As can be seen in Fig. 13.9b, higher harmonics with $k_y \leq 2\mu m^{-1}$ are emitted with strongly varying directions from the edge mode. We therefore conclude that these spin-wave modes do not contribute to the formation of the caustic beams and can be regarded as a negligible background in our measurement.

In summary, we reported the nonlinear higher harmonic generation from a localized spin-wave mode in a micro-structured CMFS waveguide leading to the emission of strongly directed spin-wave beams or caustics. All experimental results are in agreement not only with theoretical predictions and related experimental findings but also with our quantitative analysis using the dispersion relation for spin waves in magnetic thin films.

Regarding possible technical applications, these results might show a promising way to realize spin-wave frequency converters. Moreover the strongly directed emission of these frequency converters can be steered by the direction of the externally applied field.

From a theoretical point of view, these observations are possible in in-plane magnetized thin films of arbitrary materials. However, this experiment was the first report on this complex interplay of phenomena involving nonlinearities in magnetic microstructures. Like the increased decay length in the linear regime, this fact can be attributed to the low Gilbert damping in CMFS. We therefore believe that further thorough investigations of spin dynamics in CMFS will allow for the observation of various nonlinear phenomena.

13.6 Summary and Outlook

This chapter illustrates recent results on spin-wave propagation in microstructures based on the full Heusler compound CMFS. Motivated by the extremely suitable material parameters offered by CMFS, these studies realized the incorporation of the material in microstructures for magnonic transport. A key aspect of these studies is the compatibility of Heusler-based devices with existing CMOS-electronics. The expected advantages of the utilization of the low-damping material have been demonstrated in the linear and nonlinear regime. In the linear regime a decay length increased by almost a factor of three was observed. In the nonlinear regime, the emission of higher harmonics from a localized mode was observed. These first results reveal the potential for novel phenomena related to the decreased Gilbert damping.

In addition, recent novel findings of nonlinear spindynamics in CMFS underline the potential of the material system [70]. By using four-magnon instabilities as a probing tool, it has been demonstrated that a non-Gilbert-damping mechanism, which can be attributed to the interaction of the considered magnon modes with the thermal magnon bath, plays an important role for the spindynamics in CMFS. The observation of this additional intrinsic damping mechanism gives further insight not only into the spin-wave system of CMFS but also into the fundamental mechanisms of nonlinear spindynamics and damping mechanisms in general.

While the further optimization of low-damping Heusler materials and the thorough investigation of spin-wave propagation in CMFS are interesting tasks, the class of Heusler materials still offers other promising possibilities for future magnon spintronics.

The low Gilbert damping in combination with the high spin polarization of the Co-based compounds makes them interesting candidates for the utilization in spin-torque oscillators as well. In addition, materials with huge perpendicular anisotropy and comparatively small Gilbert damping like manganese-based compounds [69] could allow for devices based on spin-wave propagation in an alternative geometry. This geometry has hardly been addressed experimentally in microstructures due to a lack of suitable materials.

In summary, we believe that the studies presented here can open the field of magnon spintronics towards the utilization of advanced materials and might lead the way for the development and optimization of sophisticated structures for magnonic transport.

References

1. S.A. Wolf, D. Awschalom, R. Buhrman, J. Daughton, S. Von Molnar, M. Roukes, A.Y. Chtchelkanova, D. Treger, Science **294**, 1488 (2001)
2. M.P. Kostylev, A.A. Serga, T. Schneider, B. Leven, B. Hillebrands, Appl. Phys. Lett. **87**, 153501 (2005)
3. A. Khitun, M. Bao, K.L. Wang, I.E.E.E. Trans., Magn. **44**, 9 (2008)
4. T. Schneider, A.A. Serga, B. Leven, R.L. Stamps, M.P. Kostylev, B. Hillebrands, Appl. Phys. Lett. **92**, 022505 (2008)
5. A. Khitun, M. Bao, K.L. Wang, J. Phys. D **43**, 264005 (2010)
6. V.V. Kruglyak, S.O. Demokritov, D. Grundler, J. Phys. D **43**, 264001 (2010)
7. A. Khitun, K.L. Wang, J. Appl. Phys. **110**, 034306 (2011)
8. A. Khitun, J. Appl. Phys. **111**, 054307 (2012)
9. Y. Kajiwara, K. Harii, S. Takahashi, J. Ohe, K. Uchida, M. Mizuguchi, H. Umezawa, H. Kawai, K. Ando, K. Takanashi, S. Maekawa, E. Saitoh, Nature **464**, 262–6 (2010)
10. L. Berger, Phys. Rev. B **54**, 9353–9358 (1996)
11. J.C. Slonczewski, J. Magn. Magn. Mater. **159**, L1–L7 (1996)
12. J. Hirsch, Phys. Rev. Lett. **83**, 1834–1837 (1999)
13. K. Ando, S. Takahashi, K. Harii, K. Sasage, J. Ieda, S. Maekawa, E. Saitoh. Phys. Rev. Lett. **101**, 036601 (2008)
14. V.E. Demidov, S. Urazhdin, H. Ulrichs, V. Tiberkevich, A. Slavin, D. Baither, G. Schmitz, S.O. Demokritov, Nat. Mater. **11**, 1 (2012)

15. Y. Tserkonvnyak, A. Brataas, G.E. Bauer, Phys. Rev. Lett. **88**, 117601 (2002)
16. Y. Tserkonvnyak, A. Brataas, G.E. Bauer, Phys. Rev. B **66**, 224403 (2002)
17. E. Saitoh, M. Ueda, H. Miyajima, G. Tatara, Appl. Phys. Lett. **88**, 182509 (2006)
18. K. Ando, T. Yoshino, E. Saitoh, Appl. Phys. Lett. **94**, 152509 (2009)
19. M.B. Jungfleisch, A.V. Chumak, V.I. Vasyuchka, A.A. Serga, B. Obry, H. Schultheiss, P.A. Beck, A.D. Karenowska, E. Saitoh, B. Hillebrands, Appl. Phys. Lett. **99**, 182512 (2011)
20. H. Chudo, K. Ando, K. Saito, S. Okayasu, R. Haruki, Y. Sakuraba, H. Yasuoka, K. Takanashi, E. Saitoh, J. Appl. Phys. **109**, 073915 (2011)
21. K. Uchida, S. Takahashi, K. Harii, J. Ieda, W. Koshibae, K. Ando, S. Maekawa, E. Saitoh, Nature **455**, 778–81 (2008)
22. K. Uchida, T. Ota, K. Harii, S. Takahashi, S. Maekawa, Y. Fujikawa, E. Saitoh, Solid State Commun. **150**, 524–528 (2010)
23. S. Bosu, Y. Sakuraba, K. Uchida, K. Saito, T. Ota, E. Saitoh, K. Takanashi, Phys. Rev. B **83**, 1–6 (2011)
24. B. Obry, V.I. Vasyuchka, A.V. Chumak, A.A. Serga, B. Hillebrands, Appl. Phys. Lett. **101**, 192406 (2012)
25. M.B. Jungfleisch, T. An, K. Ando, Y. Kajiwara, K. Uchida, V.I. Vasyuchka, A.V. Chumak, A.A. Serga, E. Saitoh, B. Hillebrands, Appl. Phys. Lett. **102**, 062417 (2013)
26. J. Katine, F. Albert, R. Buhrman, E. Myers, D.C. Ralph, Phys. Rev. Lett. **84**, 3149 (2000)
27. M. Tsoi, A. Jansen, J. Bass, W.-C. Chiang, V. Tsoi, P. Wyder, Nature **406**, 46 (2000)
28. S.I. Kiselev, J.C. Sankey, I.N. Krivorotov, N.C. Emley, R.J. Schoelkopf, R.A. Buhrman, D.C. Ralph, Nature **425**, 380–383 (2003)
29. H. Schultheiss, X. Janssens, M. van Kampen, F. Ciubotaru, S.J. Hermsdoerfer, B. Obry, A. Laraoui, A.A. Serga, L. Lagae, A.N. Slavin, B. Leven, B. Hillebrands, Phys. Rev. Lett. **103**, 157202 (2009)
30. V.E. Demidov, S. Urazhdin, S.O. Demokritov, Nat. Mater. **9**, 984 (2010)
31. M. Madami, S. Bonetti, G. Consolo, S. Tacchi, G. Carlotti, G. Gubbiotti, F.B. Mancoff, M.A. Yar, J. Åkerman, Nat. Nanotechnol. **1**, 635 (2011)
32. A.V. Chumak, P. Pirro, A.A. Serga, M.P. Kostylev, R.L. Stamps, H. Schultheiss, K. Vogt, S.J. Hermsdoerfer, B. Laegel, P.A. Beck, B. Hillebrands, Appl. Phys. Lett. **95**, 262508 (2009)
33. P. Clausen, K. Vogt, H. Schultheiss, S. Schäfer, B. Obry, G. Wolf, P. Pirro, B. Leven, B. Hillebrands, Appl. Phys. Lett. **99**, 162505 (2011)
34. K. Vogt, H. Schultheiss, S. Jain, J. Pearson, A. Hoffmann, S. Bader, B. Hillebrands, Appl. Phys. Lett. **101**, 042410–042410 (2012)
35. A.A. Serga, A.V. Chumak, B. Hillebrands, J. Phys. D: Appl. Phys. **43**, 264002 (2010)
36. S. Tsunegi, Y. Sakuraba, M. Oogane, K. Takanashi, Y. Ando, Appl. Phys. Lett. **93**, 112506 (2008)
37. N. Tezuka, N. Ikeda, F. Mitsuhashi, S. Sugimoto, Appl. Phys. Lett. **94**, 162504 (2009)
38. S. Trudel, O. Gaier, J. Hamrle, B. Hillebrands, J. Phys. D **43**, 193001 (2010)
39. B. Kalinikos, A. Slavin, J. Phys. C: Solid State **19**, 7013 (1986)
40. S.O. Demokritov, A.A. Serga, V.E. Demidov, M.P. Kostylev, B.A. Kalinikos, B. Hillebrands, Nature **426**, 159 (2003)
41. V.E. Demidov, M.P. Kostylev, K. Rott, P. Krzysteczko, G. Reiss, S.O. Demokritov, Phys. Rev. B **83**, 054408 (2011)
42. V.E. Demidov, H. Ulrichs, S. Urazhdin, S.O. Demokritov, V. Bessonov, R. Gieniusz, A. Maziewski, Appl. Phys. Lett. **99**, 012505 (2011)
43. H.T. Nembach, K.L. Livesey, M.P. Kostylev, P. Martin-Pimentel, S.J. Hermsdoerfer, B. Leven, J. Fassbender, B. Hillebrands, Phys. Rev. B **84**, 184413 (2011)
44. H. Schultheiss, K. Vogt, B. Hillebrands, Phys. Rev. B **86**, 054414 (2012)
45. A.G. Gurevich, G.A. Melkov, *Magnetization Oscillations and Waves* (CRC, Boca Raton, 1996)
46. J. Jorzick, S.O. Demokritov, C. Mathieu, B. Hillebrands, F. Rousseaux, A.N. Slavin, Phys. Rev. B **60**, 15194 (1999)
47. B. Hillebrands, Brillouin light scattering from layered magnetic structures, in: *Light Scattering in Solids VII*. Topics in Applied Physics, ed. by M. Cardona, G. Güntherodt, vol. 75 (Springer, Heidelberg, 1999)

48. S.O. Demokritov, B. Hillebrands, A.N. Slavin, Phys. Rep. **348**, 441–489 (2001)
49. B. Hillebrands, Rev. Sci. Instrum. **70**, 1589 (1999)
50. V.E. Demidov, S.O. Demokritov, M. Laufenberg, P.P. Freitas, B. Hillebrands, Appl. Phys. Lett. **85**, 2866 (2004)
51. A.A. Serga, T. Schneider, B. Hillebrands, S.O. Demokritov, M.P. Kostylev, Appl. Phys. Lett. **89**, 063506 (2006)
52. F. Fohr, A.A. Serga, T. Schneider, J. Hamrle, B. Hillebrands, Rev. Sci. Instrum. **80**, 043903 (2009)
53. K. Vogt, H. Schultheiss, S.J. Hermsdoerfer, P. Pirro, A.A. Serga, B. Hillebrands. Appl. Phys. Lett. **95**, 182508 (2009)
54. H. Schultheiss, C.W. Sandweg, B. Obry, S.J. Hermsdoerfer, S. Schäfer, B. Leven, B. Hillebrands, J. Phys. D **41**, 164017 (2008)
55. J. Hamrle, O. Gaier, S.-G. Min, Y. Sakuraba, Y. Ando, B. Hillebrands, J. Phys. D: Appl. Phys. **42**, 084005 (2009)
56. T. Kubota, S. Tsunegi, M. Oogane, S. Mizukami, T. Miyazaki, H. Naganuma, Y. Ando, Appl. Phys. Lett. **94**, 122504 (2009)
57. M. Oogane, T. Kubota, Y. Kota, S. Mizukami, H. Naganuma, A. Sakuma, Y. Ando, Appl. Phys. Lett. **96**, 252501 (2010)
58. T. Sebastian, Y. Ohdaira, T. Kubota, P. Pirro, T. Brächer, K. Vogt, A.A. Serga, H. Naganuma, M. Oogane, Y. Ando, B. Hillebrands, Appl. Phys. Lett. **100**, 112402 (2012)
59. T. Sebastian, T. Brächer, P. Pirro, T. Kubota, A.A. Serga, H. Naganuma, M. Oogane, Y. Ando, B. Hillebrands, Phys. Rev. Lett. **110**, 067201 (2013)
60. R.W. Damon, J.R. Eshbach, J. Phys. Chem. Solids **19**, 308 (1961)
61. V.E. Demidov, S.O. Demokritov, K. Rott, P. Krzysteczko, G. Reiss, Phys. Rev. B **77**, 064406 (2008)
62. P. Pirro, T. Brächer, K. Vogt, B. Obry, H. Schultheiss, B. Leven, B. Hillebrands, Phys. Status Solidi B **10**, 2404 (2011)
63. C. Liu, C.K.A. Mewes, M. Chshiev, T. Mewes, W.H. Butler, Appl. Phys. Lett. **95**, 022509 (2009)
64. H. Schultheiss, S. Schäfer, P. Candeloro, B. Leven, A.N. Slavin, B. Hillebrands, Phys. Rev. Lett. **100**, 2–5 (2008)
65. G. Gubbiotti, M. Conti, G. Carlotti, P. Candeloro, E.D. Fabrizio, K.Y. Guslienko, A. André, C. Bayer, A.N. Slavin, J. Phys. Cond. Matt. **16**, 7709 (2004)
66. C. Bayer, J.P. Park, H. Wang, M. Yan, C.E. Campbell, P.A. Crowell, Phys. Rev. B **69**, 134401 (2004)
67. T. Schneider, A.A. Serga, A.V. Chumak, C.W. Sandweg, S. Trudel, S. Wolff, M.P. Kostylev, V.S. Tiberkevich, A.N. Slavin, B. Hillebrands, Phys. Rev. Lett. **104**, 197203 (2010)
68. V.E. Demidov, S.O. Demokritov, D. Birt, B. O'Gorman, M. Tsoi, X. Li, Phys. Rev. B **80**, 014429 (2009)
69. S. Mizukami, F. Wu, A. Sakuma, J. Walowski, D. Watanabe, T. Kubota, X. Zhang, H. Naganuma, M. Oogane, Y. Ando, T. Miyazaki, Phys. Rev. Lett. **106**, 117201 (2011)
70. P. Pirro, T. Sebastian, T. Brächer, A.A. Serga, T. Kubota, H. Naganuma, M. Oogane, Y. Ando, B. Hillebrands, Phys. Rev. Lett. **113**, 227601 (2014)

Chapter 14
Mössbauer Analysis

Ko Mibu, Masaaki Tanaka and Kohei Hamaya

Abstract As a powerful experimental tool to examine local crystallographic and magnetic environments of Heusler alloy films, Mössbauer spectroscopy is introduced briefly in this section. This method, which is applicable easily when the alloy contains Fe or Sn as a constituent element, can give us unique information on the local degree of structural order, the magnetic stability at the interfaces, and so on.

Mössbauer spectroscopy is an experimental method where local electronic states in solids can be studied through absorption spectra of γ-rays by the constituent nuclei in the solids [1–3]. The energy levels of nuclei are influenced by the surrounding electrons through the hyperfine interactions so that the absorption spectra contain information on local electronic states, including magnetism. The applicability of this experimental method is highly dependent of the sort of nuclei or isotopes and the most easily applicable nuclei in the level of university laboratories are ^{57}Fe (the natural abundance 2.2 %) and ^{119}Sn (the natural abundance 8.6 %). Since some Heusler alloys contain Fe or Sn as an ingredient, Mössbauer spectroscopy can be a powerful tool to elucidate local electronic states of Heusler alloys. Some pioneering works on bulk Heusler alloys are introduced already in a review book on Mössbauer spectroscopy published in 1971 [1] and many papers have been published so far. The number of papers including Mössbauer spectroscopic study on Heusler alloys started increasing again in 2004, boosted by the progress in spintronics, and some of these recent papers have been highly sited up to date [4, 5]. In this section, we focus exclusively on Mössbauer spectroscopic studies about epitaxial Heusler alloy films,

K. Mibu (✉) · M. Tanaka
Graduate School of Engineering, Nagoya Institute of Technology,
Nagoya 466-8555, Japan
e-mail: k_mibu@nitech.ac.jp

M. Tanaka
e-mail: mtanaka@nitech.ac.jp

K. Hamaya
Graduate School of Engineering Science, Osaka University,
Toyonaka 560-8531, Japan
e-mail: hamaya@ee.es.osaka-u.ac.jp

C. Felser and A. Hirohata (eds.), *Heusler Alloys*,
Springer Series in Materials Science 222, DOI 10.1007/978-3-319-21449-8_14

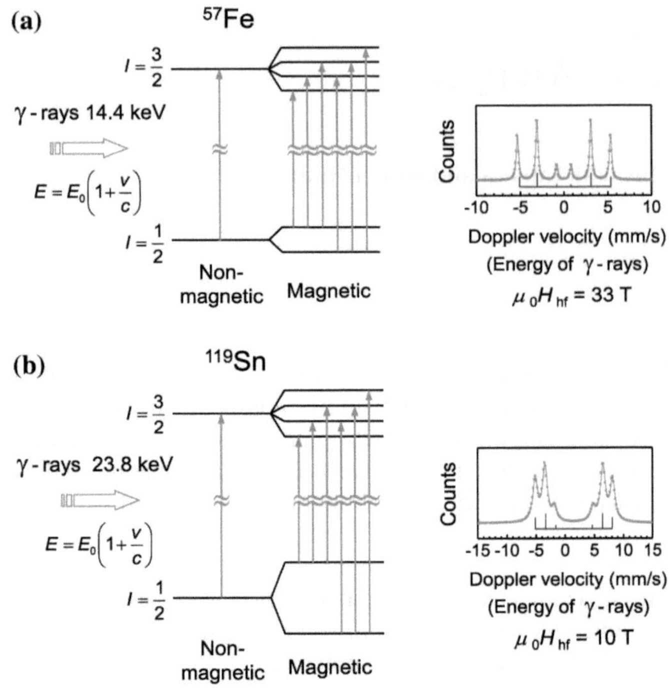

Fig. 14.1 Mössbauer-related energy levers for (**a**) ^{57}Fe and (**b**) ^{119}Sn nuclei in nonmagnetic and magnetic environments. Typical Mössbauer spectra with a magnetically-split six-line pattern are also shown on the *right*

which started appearing in literature around 2007, mainly on the studies performed by our research group.

Figure 14.1 shows typical Mössbauer-related energy levers of ^{57}Fe and ^{119}Sn nuclei and corresponding ^{57}Fe and ^{119}Sn Mössbauer spectra in magnetic environments. Both nuclei have the ground state with the nuclear spin $I = 1/2$ and the first exited state with $I = 3/2$. When the nuclei feel an effective magnetic field, the energy revels are split into 2 and 4 sublevels, respectively. On the occasion of γ-ray absorption, 6 transitions are usually allowed between the sublevels, so that a six-line pattern is observed in the absorption spectra. The effective magnetic field in magnetic materials is called "magnetic hyperfine field" because it originates from the hyperfine interaction between the nuclei and electrons. For usual experiments, a radioisotope is used as the γ-ray source and the energy of the highly monochromatic γ-rays is modulated through the Doppler effect by giving the velocity to the γ-ray source. The abscissa of the spectra is therefore expressed by the Doppler velocity, which is linearly correlated with the energy of the incident γ-rays, instead of being expressed by the units of energy. The center of the gravity of the spectra reflects the charge density of electrons at the nuclear sites and the deviation from the standard position (usually the deviation relative to the center of the spectrum of α-Fe or $CaSnO_3$) is called "isomer shift". The energy levels of the nuclei, hence the absorption spectra, are also affected by the electric field gradient at the nuclear sites, but this effect is negligibly small in the case of Heusler alloys, where the crystallographic structures basically have cubic symmetry. For epitaxial thin films grown on single crystal substrates, absorption spectra are usually recorded by detecting internal conversion electrons and other secondary electrons, which are resonantly emitted from the sample right after the absorption of γ-rays, in a scattering geometry. This method is called "conversion electron Mössbauer spectroscopy (CEMS)" and the spectra appear as upward peaks as in Fig. 14.1. For magnetic thin films with the magnetic moment in the film plane, as is usual in ferromagnetic Heusler alloy films, the direction of the magnetic hyperfine field is also along the film plane. In such cases the intensity ratio of the six lines becomes 3:4:1:1:4:3, as shown in Fig. 14.1.

In stoichiometric X_2YZ Heusler alloys with an ideal $L2_1$-ordered structure, the atoms of each constituent element occupy a specific crystallographic site, i.e., the 8c site of the $Fm\overline{3}m$ space group for X, the 4b site for Y, the 4a site for Z, so that each element has a single sort of local environments. The Mössbauer spectra for one of the constituent element therefore consist of one six-line pattern when the alloy is magnetically ordered or one single-line pattern when the alloy is nonmagnetic. In off-stoichiometric cases or disordered cases, there appears distribution in the local environments of each constituent element so that the spectra become multiple six-line or single-line patterns or, instead, show line broadening. For such cases, each spectrum component (subspectrum) is usually attributed to a nuclear site which is classified by the sort and number of the atoms at the nearest-neighboring sites. For example, in ferromagnetic Co_2FeZ (Z = Si, Ge, Sn) alloys, each Fe atom is surrounded by 8 nearest-neighboring Co atoms when the Fe is at the proper site (the 4b site) in an ideal $L2_1$ structure, whereas it is surrounded by 4 nearest-neighboring Fe atoms and 4 nearest-neighboring Z atoms when it is at the Co site (the 8c site) due to off-stoichiometry or structural disorder. The Fe nuclei at the latter site are expected to feel smaller magnetic hyperfine field than those at the former site. The effects from the 6 next-nearest-neighboring atoms are usually small in comparison with those from the 8 nearest-neighboring atoms. In this way, Mössbauer analyses can be used to study the local or short-range environments of the constituent atoms in Heusler alloys, as NMR analyses can be [6]. This situation has been well-demonstrated for bulk $L2_1$ Heusler alloys, for example, in [7] and [8]. The first part of this section is devoted to these kinds of analyses on structural order of epitaxial Heusler alloy films. Another effective way to apply Mössbauer spectroscopy to Heusler alloy films is to check the local electronic states, especially magnetism, at the interfaces, which can be highly correlated with the magnetoresistance features of Heusler-alloy-based multilayers or magnetic tunnel junctions. This point is discussed in the latter part of this section.

The validity of the Mössbauer analysis for the study of site occupation is well-demonstrated for "pseudo-Heusler" or "binary Heusler" Fe_3Si alloy films with $D0_3$ structure [9–12]. In this structure, the 8c site and the 4b site in the $Fm\overline{3}m$ space group are occupied by the atoms of the same element, in the case of Fe_3Si, by Fe. Here, each Fe atom at the 8c site has 4 Fe atoms and 4 Si atoms at the nearest-neighboring sites, and that at the 4b site has 8 Fe atoms at the nearest-neighboring sites. It is

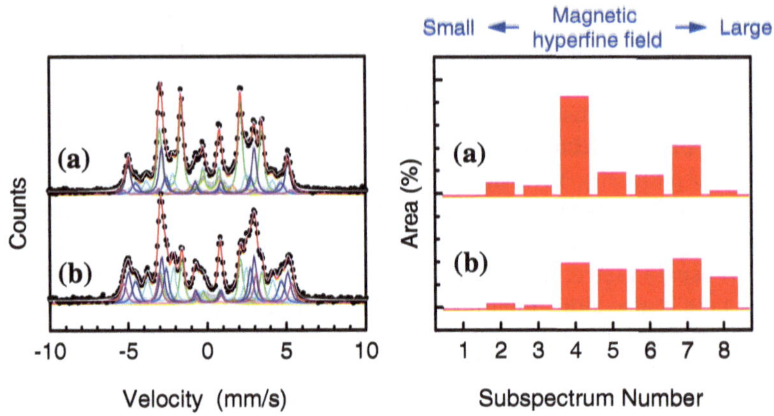

Fig. 14.2 ^{57}Fe Mössbauer spectra of epitaxial (**a**) $Fe_3Si(111)$ (25 nm) and (**b**) $Fe_{3.2}Si_{0.8}(111)$ (30 nm) films on Ge(111) substrates prepared by low-temperature molecular beam epitaxy [10]. Each spectrum is fitted with 7 magnetically-split six-line patterns (7 subspectra), each of which is attributed to the Fe site with a different number of nearest-neighboring Fe and Si atoms. The relative area of each subspectrum is shown on the *right*

known from an experimental study for bulk Fe_3Si [13] that the magnetic hyperfine field for the Fe nuclei at the 8c site (∼20 T) is about 2/3 of that for the Fe atoms at the 4b site (∼31 T). The electron charge density at both sites is also different. Eventually Mössbauer spectrum appears as superposition of two magnetically-split sex-line patterns with different magnetic hyperfine fields and different isomer shifts, with the integrated peak intensity ratio of 2:1. For thin films, where the samples are prepared through non-equilibrium process, it is usually not easy to exclude off-stoichiometry and structural disorder. Figure 14.2 shows the ^{57}Fe Mössbauer spectra of epitaxial $Fe_3Si(111)$ (25 nm) and $Fe_{3.2}Si_{0.8}(111)$ (30 nm) films on Ge(111) substrates prepared by low-temperature molecular beam epitaxy [10]. For the film growth, the Fe source was enriched with the ^{57}Fe isotope up to 20 %, to obtain better S/N ratios in shorter measurement time. The spectra can be fitted with superposition of 7 six-line patters, i.e., 7 magnetic subspectra, as shown in the figure. The intensity ratios of the subspectra are shown on the right, with each subspectrum numbered from 2 to 8 in the order of the magnitude of magnetic hyperfine fields (5.0, 13.9, 20.0, 24.7, 28.5, 31.3, and 32.8 T in the case of $Fe_3Si(111)$). Each subspectrum can be attributed to the Fe site with a different number of nearest-neighboring Fe and Si atoms, according to the analyses on the spectra of bulk samples [13]. The Fe site which has more Fe atoms at the nearest-neighboring sites has a larger magnetic hyperfine field and a smaller (more negative) isomer shift. Subspectra 2, 3, 4, 5, 6, and 7 correspond to the Fe sites in the $D0_3$ phase with the number of nearest-neighboring Fe atoms 0 ∼ 2, 3, 4, 5, 6, 7 ∼ 8, respectively. Subspectrum 8 is from the Fe site in the A2 or bcc phase. A nonmagnetic single-line subspectrum, which is later assigned as Subspectrum 1 for other Heusler alloy systems, is not clearly separable in these spectra. Obviously, for the stoichiometric $Fe_3Si(111)$ film, the Fe site with 4 Fe atoms (Subspectrum 4)

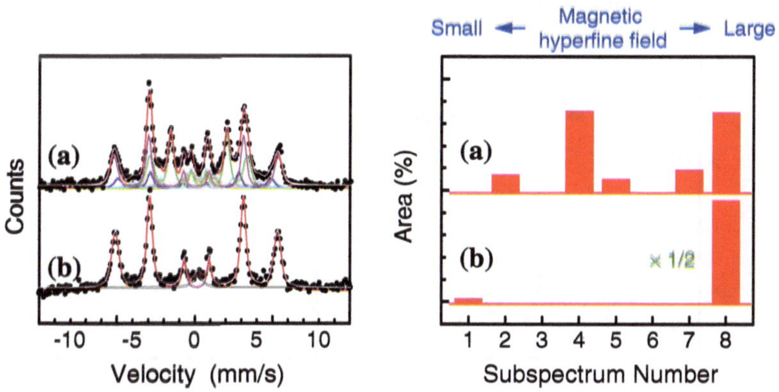

Fig. 14.3 ^{57}Fe Mössbauer spectra of epitaxial (**a**) $Fe_2CoSi(111)$ (25 nm), and (**b**) $Co_2FeSi(111)$ (50 nm) films prepared on Ge(111) substrates by low-temperature molecular beam epitaxy [14, 15]. Each spectrum is fitted with 5 or 2 subspectra. The relative area of each subspectrum is shown on the *right*

and that with 8 Fe atoms (Subspectrum 7) at the nearest neighbors are dominant, suggesting that the epitaxial films basically have a highly ordered $D0_3$ structure. The existence of the component with other Fe coordination numbers indicates that a certain degree of disorder exists in these films, and the degree of disorder is found larger for the off-stoichiometric $Fe_{3.2}Si_{0.8}(111)$ film. With a simple assumption, the local degree of structural order is estimated to be 67 % for the stoichiometric $Fe_3Si(111)$ film grown at 130 °C [10]. This value is sufficiently high for the film grown at such a low temperature. The degree of order tends to increase as the thickness of the films increases, which suggests that the structural disorder gets smaller as the position becomes away from the interface [12]. In this way, it is possible to examine which growth conditions are optimal to grow crystallographically well-ordered Heusler alloy films from Mössbauer spectroscopic measurements. Similar analyses are reported also for epitaxial $Fe_3Si(001)$ films grown on MgO(001) or GaAs(001) substrates [9, 11].

When the Fe atoms in Fe_3Si are replaced by Co atoms to be Fe_2CoSi and Co_2FeSi, the ^{57}Fe Mössbauer spectra change as in Fig. 14.3 [14, 15]. The spectrum of the Fe_2CoSi (111) (25 nm) film on Ge(111) can be fitted with 5 magnetic subspectra as in Fig. 14.3a [14]. Each subspectrum corresponds to the Fe site with a different number of nearest-neighboring Fe atoms. The subspectrum with the largest magnetic hyperfine field of 32.8 T is named as Subspectrum 8, as in the case of Fe_3Si. This subspectrum is from the Fe site with 8 magnetic atoms (Fe or Co) at the nearest-neighboring sites, the crystallographically equivalent component to Subspectrum 7 in Fe_3Si. The magnetic hyperfine field at this Fe site of Fe_2CoSi is enhanced by the existence of Co. Subspectrum 4 with the magnetic hyperfine field of 19.4 T, which can be attributed to the Fe site with 4 magnetic atoms and 4 Si atoms at the nearest neighbors, is also dominant. If the 8c site is occupied by Fe and Co atoms ideally with the ratio of 1:1, the ratio between Subspectrum 4 and 8 should be

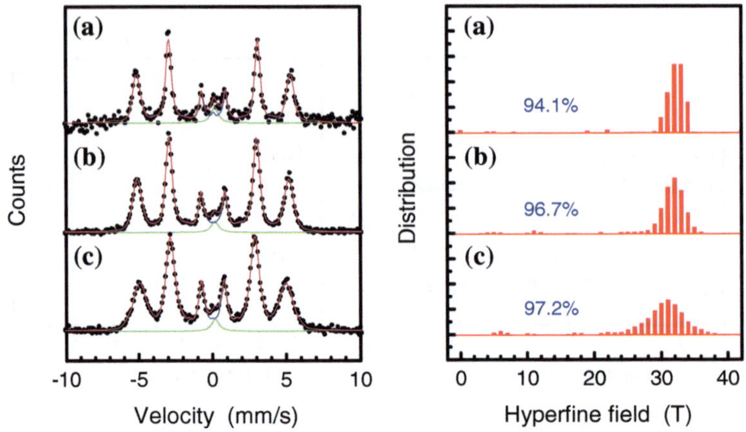

Fig. 14.4 ^{57}Fe Mössbauer spectra of epitaxial (**a**) $CoFe_2Si(111)$ (50 nm) [17], (**b**) $Co_2FeSi_{0.5}Al_{0.5}(111)$ (25 nm) and (**c**) $Co_2FeAl(111)$ (25 nm) films prepared on Si(111) substrates by low-temperature molecular beam epitaxy. Each spectrum is fitted with a distribution of magnetic hyperfine fields with a small amount of nonmagnetic singlet. The distribution is shown on the *right*, with the total area of the magnetic component indicated in percentage

1:1. The observed spectrum and its analysis in Fig. 14.3a are close to this situation, although there also exist other Fe environments. The spectrum of the Co_2FeSi (111) (50 nm) film (made of non-enriched natural Fe) on Ge(111), on the other hand, can be fitted with one magnetic subspectrum with 32.3 T (Subspectrum 8) and one nonmagnetic subspectrum with 0 T (Subspectrum 1), as shown in Fig. 14.3b [15]. The disappearance of Subspectrum 4 shows that the Fe atoms in the 8c site are perfectly repelled by the Co atoms. Therefore, it can be concluded that this film has a highly $L2_1$ or B2 ordered structure without having an A2-type disorder. Note that ^{57}Fe Mössbauer spectra are not so sensitive to the difference between the $L2_1$ and B2 structures because the Fe site has 8 Co atoms at the nearest-neighboring sites in both structures. The effect from the next-nearest-neighboring atoms possibly causes slight peak broadening, which is discussed in the next paragraph. The appearance of Subspectrum 1 means that there is a small amount of nonmagnetic component, probably at the interface region [15]. Note that the effect of interfacial diffusion by annealing is examined for epitaxial $Co_2Cr_{0.6}Fe_{0.4}Al$ films grown on MgO(001) substrates with and without an Fe buffer layer [16].

Figure 14.4 shows the ^{57}Fe Mössbauer spectra of $Co_2FeSi(111)$ (50 nm) (made of non-enriched natural Fe) [17], $Co_2FeSi_{0.5}Al_{0.5}(111)$ (25 nm), and $Co_2FeAl(111)$ (25 nm) films on Si(111) substrates prepared by low-temperature molecular beam epitaxy. The spectra appear as magnetically-split six-line patterns with different line widths. Although each spectrum may be composed of some subspectra, the separation among the subspectra is not so clear, unlike the spectra in Figs. 14.2 and 14.3. Each spectrum is therefore fitted with a distribution of magnetic hyperfine fields. Coexistence of a small amount of nonmagnetic singlet, as observed in the spectrum in Fig. 14.3b, is taken into consideration. The obtained distribution of magnetic hyperfine fields is also shown in the figure. The component around 20 T (equivalent position to Subspectrum 4) is negligible, which indicates that the amount of Fe atoms at the 8c site is negligibly small. The distribution becomes wider, hence the parks in the spectrum become broader, as Si atoms are replaced by Al atoms. If the degree of B2 disorder (the disorder between Fe and Z in the 4b and 4a sites) is the same for all the three samples, the local environments of Fe, when considered up to the next-nearest-neighboring atoms, should be the most diverse in the $Co_2FeSi_{0.5}Al_{0.5}$ film. The observed result, however, shows that the Co_2FeAl film has larger peak broadening. This fact implies that the degree of the B2 disorder is larger in the Co_2FeAl film than in the Co_2FeSi film, and that the Co_2FeSi film has a highly $L2_1$ ordered structure. Actually, electron diffraction patterns support the existence of the $L2_1$ordered phase in the Co_2FeSi film [17].

For the application of magnetic Heusler alloy films to the multilayered giant-magnetoresistance structures or magnetic tunnel junctions, it is important to know whether the magnetic order is maintained at the interfaces on nonmagnetic layers. Mössbauer spectroscopy is a powerful tool to study local crystallographic structure and magnetism at buried interfaces in layered structures (Chap. 5 in [3]). As already shown in this section, the coexistence of a small amount of nonmagnetic phase among the major magnetic phases can be detected as a single line in Mössbauer spectra. In magnetic tunnel junctions with (001)-oriented X_2YZ Heusler alloy layers and a MgO(001) barrier, the tunnel magnetoresistance ratio is yet not satisfactorily large as expected from the half metallicity of Heusler alloys and the ratio decreases largely when the measurement temperature increases from low temperatures to room temperature. To find a key to solve this problem, it is quite informative if the sort of the terminating atoms at the interface (whether X or YZ) and the stability of magnetization at the interface at room temperature are clarified experimentally. Mössbauer analysis can give some information on these points. In our group, X_2YZ Heusler alloy films with the (001) orientation have been prepared on MgO(001) substrates by depositing an atomic layer of X, a half an atomic layer of Z and Y alternately in a controlled manner [18, 19]. By this preparation method, the degree of order may be improved and also it might be possible to control the interfacial atomic species.

Mössbauer spectroscopic experiments were performed for epitaxial Co_2MnSn (001) films, where ^{119}Sn is the Mössbauer isotope, prepared on MgO(001) substrates by the atomically controlled alternate deposition. For the film growth, the Sn source was occasionally enriched with the ^{119}Sn isotope, to 97 % at the maximum. Figure 14.5 shows the ^{119}Sn Mössbauer spectrum for a Co_2MnSn(40.2 nm) film, together with that of a reference bulk Co_2MnSn sample prepared by arc melting. Since the peaks from individual subspectra are not well separated unlike the ^{57}Fe Mössbauer spectra in Figs. 14.2 and 14.3, each spectrum is fitted with a distribution of magnetic hyperfine fields as shown in the figure. (Note that the natural line width is larger for ^{119}Sn Mössbauer spectra than that for ^{57}Fe Mössbauer spectra, as shown in Fig. 14.1.) Although Sn is basically a nonmagnetic element, large magnetic hyperfine fields are induced at the Sn nuclear sites of the ferromagnetic alloys through the Fermi contact interaction between the spin polarized conduction electrons

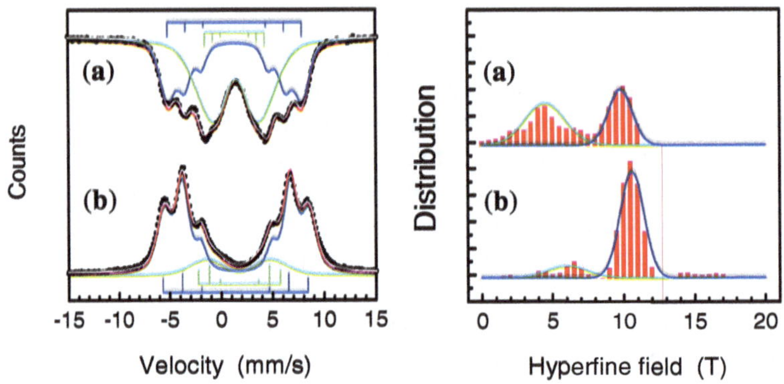

Fig. 14.5 ^{119}Sn Mössbauer spectra and distribution of magnetic hyperfine fields for (**a**) bulk Co_2MnSn Heusler alloy prepared by arc melting (measured with a transmission geometry), and (**b**) an epitaxial $Co_2MnSn(001)$ (40.2 nm) film prepared on a MgO(001) substrate by atomically controlled alternate deposition at 500 °C [19]

and the nuclei. In the arc-melted sample, there are two peaks in the hyperfine-field distribution (at ~4.5 and 10 T), suggesting that there are two Sn environments with different numbers of nearest-neighboring magnetic atoms. Similar spectra are historically reported by other groups [20–22]. Although the origin of the component with the smaller hyperfine field is not clearly assigned, there seems to be Sn atoms at the 8c site. In the film sample made by the atomically controlled alternate deposition, the peak with the smaller magnetic hyperfine field in the distribution histogram becomes much smaller. This means that the Sn atoms in this film sample are with more uniform environments than those in the arc-melted sample. This uniformness in the Sn environments makes it possible to examine the interface magnetism of the Co_2MnSn layers through Mössbauer analyses. Figures 14.6 and 14.7 show the ^{119}Sn Mössbauer spectra and corresponding distribution of magnetic hyperfine fields for $Co_2MnSn(001)$ (2.2 nm)/Ag(001) (3 nm) and $Co_2MnSn(001)$ (2.2 nm)/Cr(001) (3 nm) multilayers. Since the Heusler alloy layers are thinner than those in the sample in Figs. 14.2, 14.3, 14.4, the Mössbauer spectra become more interface-sensitive in these multilayered samples. The deposition order of the Heusler alloy layer is different among the 4 samples, i.e., (a) Co, Mn, Sn, … , Co, (b) Co, Sn, Mn, … , Co, (c) Mn, Sn, Co, … , Mn, Sn, and (d) Sn, Mn, Co, … Sn, Mn. The interfacial atomic layer is "nominally" designed to be Co for the former two samples, and to be MnSn for the latter two samples. The spectra show a couple of experimental facts. First, the spectra look the same independent of the order of atomic deposition and there is no firm evidence that the interface atomic species are controlled as designed. Second, the spectra and the corresponding distribution of magnetic hyperfine fields for the Co_2MnSn/Ag multilayer are similar to those of the thicker $Co_2MnSn(001)$ film in Fig. 14.5b, whereas those for the Co_2MnSn/Cr multilayers are quite different and the reduction of magnetic hyperfine fields is more dominant probably due to the stronger interface effect. The fact that the Co_2MnSn/Ag multilayers have no

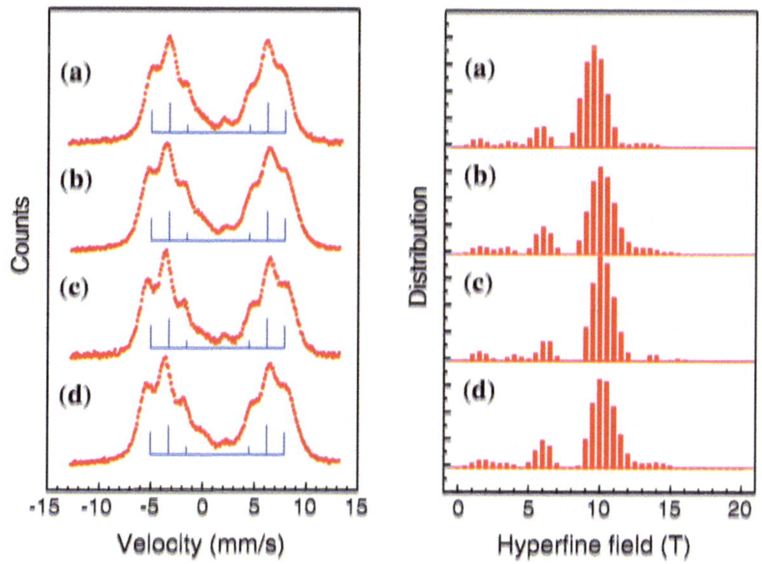

Fig. 14.6 ^{119}Sn Mössbauer spectra and distribution of magnetic hyperfine fields for epitaxial $Co_2MnSn(001)$ (2.2 nm)/Ag(001) (3 nm) multilayers on MgO(001) substrates. **a**, **b**, **c** and **d** are for the samples with the Co_2MnSn layers prepared with different deposition sequences (see the text)

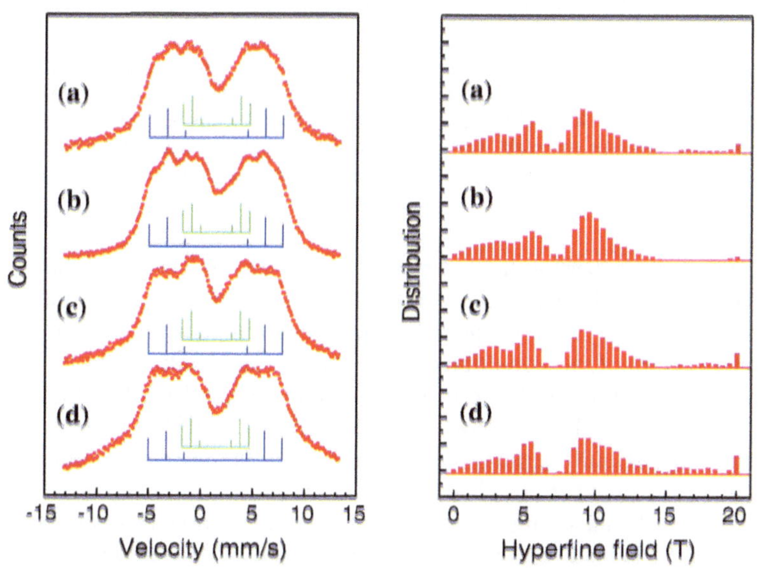

Fig. 14.7 ^{119}Sn Mössbauer spectra and distribution of magnetic hyperfine fields for epitaxial $Co_2MnSn(001)$ (2.2 nm)/Cr(001) (3 nm) multilayers on MgO(001) substrates. **a**, **b**, **c** and **d** are for the samples with the Co_2MnSn layers prepared with different deposition sequences (see the text)

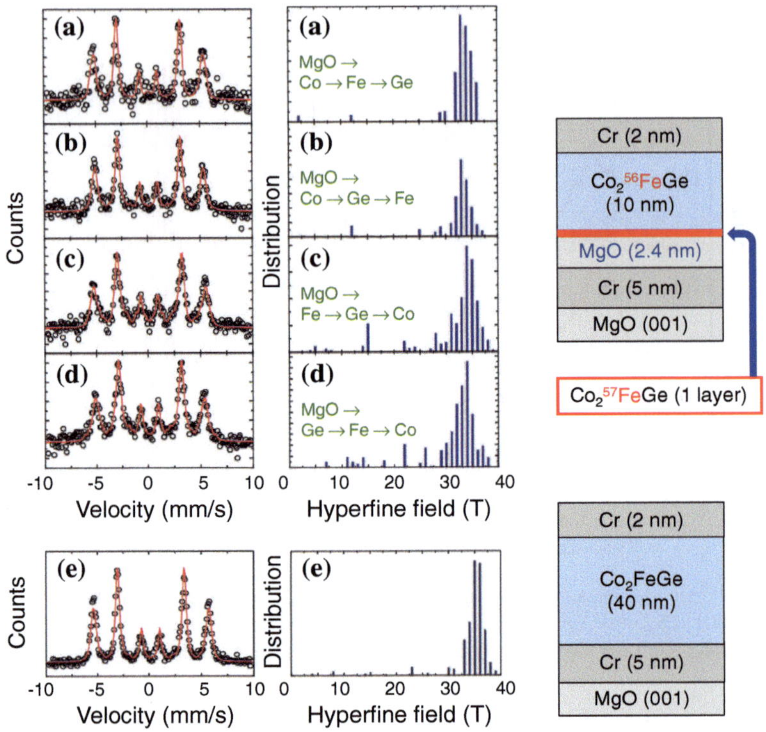

Fig. 14.8 **(a)–(d)** ^{57}Fe Mössbauer spectra and distribution of magnetic hyperfine fields for epitaxial Co_2FeGe(001) (10 nm) films on MgO(001) with ^{57}Fe enriched interfaces prepared by atomically controlled alternate deposition [24]. **a**, **b**, **c** and **d** are for the samples prepared with different deposition sequences (see the text). **(e)** ^{57}Fe Mössbauer spectrum and distribution of magnetic hyperfine fields for epitaxial Co_2FeGe(001) (40 nm) film prepared by the same method

reduced magnetic hyperfine field at room temperature suggests that the interfacial magnetization of the Heusler layers, at least in this sample, has no strong reduction at room temperature. Actually there was no strong temperature dependence in the Mössbauer spectra of these multilayers below room temperature.

Similar experiments can be performed for epitaxial Co_2FeZ(001) films [23, 24], where ^{57}Fe is the Mössbauer isotope. Figure 14.8 shows the ^{57}Fe Mössbauer spectra of epitaxial Co_2FeGe(001) films prepared on MgO(001) substrates by alternate deposition of the constituent atoms [24]. Samples with an ^{57}Fe enriched interface, where the Fe at the interfacial atomic layer is composed of the Mössbauer isotope ^{57}Fe, and the rest of the Fe is made of a non-Mössbauer isotope ^{56}Fe, were prepared for more interface-sensitive experiments. The results for the interface-sensitive Co_2FeGe(001) (10 nm) layers are shown in Figs. 14.8a–d. The deposition order in the Heusler alloy layer is different among the 4 samples, i.e., (a) Co, Fe, Ge, … , Co, (b) Co, Ge, Fe, … , Co, (c) Fe, Ge, Co, … , Fe, Ge, and (d) Ge, Fe, Co, … Ge, Fe. Again, there is no clear deposition-order dependence or designed-termination-atom dependence in the spectra. The spectra and distribution of magnetic hyperfine fields for the films with ^{57}Fe-enriched interfaces are sharp enough, and the value of the magnetic hyperfine field at the peak in distribution is comparable to that of the thicker $Co_2FeGe(001)$ (40 nm) film (non-interface-sensitive sample) in Fig. 14.8e. These results indicate that there is no strong reduction in the magnetization at the Heusler alloy interface at room temperature. On the other hand, the tunneling magnetoresistance effect of the Co_2FeGe/MgO/Fe tunnel junctions, which were prepared with the same procedure, shows strong temperature dependence [24], as reported usually for Heusler-alloy-based magnetic tunnel junctions. There are some suggestions that Heusler alloy films have reduced magnetization at the interface at room temperature and that can be the reason for the reduction of magnetoresistance at room temperature in Heusler-alloy-based magnetoresistance structures [25, 26]. The Mössbauer results imply that the reduction of magnetoresistance at room temperature is not simply due to the temperature-dependent decrease in the interface magnetization [24].

In summary of this section, Mössbauer spectroscopy can be a powerful experimental tool to examine the local environments of Fe or Sn in Heusler alloy films. It can be a tool to judge the local degree of structural order, the magnetic stability at the interfaces, and so on. Recent development of synchrotron-radiation-based Mössbauer spectroscopy would open up new possibility for the researches on electronic states of Heusler alloy films in certain external conditions [27], and also lead to easier accessibility to other Mössbauer elements than Fe and Sn.

References

1. N.N. Greenwood, T.C. Gibb, *Mössbauer Spectroscopy* (Chapman and Hall Ltd., London, 1971)
2. P. Gütlich, E. Bill, A.X. Trautwein, *Mössbauer Spectroscopy and Transition Metal Chemistry* (Springer, Berlin, 2011)
3. Y. Yoshida, G. Langouche (eds.), *Mössbauer Spectroscopy, Tutorial Book* (Springer, Berlin, 2013)
4. S. Wurmehl, G.H. Fecher, H.C. Kandpal, V. Ksenofontov, C. Felser, H.-J. Lin, J. Morais, Phys. Rev. B **72**, 184434 (2005)
5. B. Balke, G.H. Fecher, H.C. Kandpal, C. Felser, K. Kobayashi, E. Ikenaga, J.-J. Kim, S. Ueda, Phys. Rev. B **74**, 104405 (2006)
6. K. Inomata, M. Wojcik, E. Jedryka, N. Ikeda, N. Tezuka, Phys. Rev. B **77**, 214425 (2008)
7. V. Jung, B. Balke, G.H. Fecher, C. Felser, Appl. Phys. Lett. **93**, 042507 (2008)
8. V. Ksenofontov, M. Wójcik, S. Wurmehl, H. Schneider, B. Balke, G. Jakob, C. Felser, J. Appl. Phys. **107**, 09B106 (2010)
9. B. Krumme, C. Weis, H.C. Herper, F. Stromberg, C. Antoniak, A. Warland, E. Schuster, P. Srivastava, M. Walterfang, K. Fauth, J. Minár, H. Ebert, P. Entel, W. Keune, H. Wende, Phys. Rev. B **80**, 144403 (2009)
10. K. Hamaya, T. Murakami, S. Yamada, K. Mibu, M. Miyao, Phys. Rev. B **83**, 144411 (2011)
11. S.I. Makarov, B. Krumme, F. Stromberg, C. Weis, W. Keune, H. Wende, Appl. Phys. Lett. **99**, 141910 (2011)
12. S. Yamada, J. Sagar, S. Honda, L. Lari, G. Takemoto, H. Itoh, A. Hirohata, K. Mibu, M. Miyao, K. Hamaya, Phys. Rev. B **86**, 174406 (2012)
13. M.B. Stearns, Phys. Rev. **129**, 1136 (1963)

14. K. Tanikawa, S. Oki, S. Yamada, K. Mibu, M. Miyao, K. Hamaya, Phys. Rev. B **88**, 014402 (2013)
15. K. Kasahara, K. Yamamoto, S. Yamada, T. Murakami, K. Hamaya, K. Mibu, M. Miyao, J. Appl. Phys. **107**, 09B105 (2010)
16. V. Ksenofontov, C. Herbort, M. Jourdan, C. Felser, Appl. Phys. Lett. **92**, 262501 (2008)
17. S. Yamada, K. Hamaya, K. Yamamoto, T. Murakami, K. Mibu, M. Miyao, Appl. Phys. Lett. **96**, 082511 (2010)
18. W. Zhang, N. Jiko, T. Okuno, K. Mibu, K. Yoshimura, J. Magn. Magn. Mater. **309**, 132 (2007)
19. K. Mibu, D. Gondo, T. Hori, Y. Ishikawa, M.A. Tanaka, J. Phys.: Conf. Ser. **217**, 012094 (2010)
20. J.M. Williams, J. Phys. C **1**, 473 (1968)
21. R.A. Dunlap, R.H. March, G. Stroink, Can. J. Phys. **59**, 1577 (1981)
22. A.G. Gavriliuk, G.N. Stepanov, S.M. Irkaev, J. Appl. Phys. **77**, 2648 (1995)
23. M.A. Tanaka, Y. Ishikawa, Y. Wada, S. Hori, A. Murata, S. Horii, Y. Yamanishi, K. Mibu, K. Kondou, T. Ono, S. Kasai, J. Appl. Phys. **111**, 053902 (2012)
24. M.A. Tanaka, D. Maezaki, T. Ishii, A. Okubo, R. Hiramatsu, T. Ono, K. Mibu, J. Appl. Phys. **116**, 163902 (2014)
25. T. Saito, T. Katayama, T. Ishikawa, M. Yamamoto, D. Asakura, T. Koide, Y. Miura, M. Shirai, Phys. Rev. B **81**, 144417 (2010)
26. Y. Miura, K. Abe, M. Shirai, Phys. Rev. B **83**, 214411 (2011)
27. K. Mibu, M. Seto, T. Mitsui, Y. Yoda, R. Masuda, S. Kitao, Y. Kobayashi, E. Suharyadi, M.A. Tanaka, M. Tsunoda, H. Yanagihara, E. Kita, Hyp. Int. **217**, 127 (2013)

Chapter 15
Element-Specific Magnetic and Electronic Properties of Epitaxial Heusler Films

Hans-Joachim Elmers

Abstract X-ray magnetic circular dichroism (XMCD) provides an experimental access to element-specific electronic properties of thin epitaxial films of Heusler alloys. The combination of bulk-sensitive transmission and surface-sensitive electron yield methods reveals quantitative information on bulk and interface properties. Extreme cases of magnetically dead and life layers at interfaces illustrate the high potential of XMCD. The dependence of XMCD on disorder, structure and composition is discussed as well as dynamic properties investigated by a combination of ferromagnetic resonance and XMCD. Examples are shown where spectroscopic information provided by XMCD is exploited to discuss subtle changes of the partial density of states (PDOS) as a function of composition, magnetization direction and temperature.

15.1 Introduction

Magnetic films and multilayers with half-metallic ferromagnetic (HMF) Heusler alloys enjoy a large scientific interest because of their high potential for applications in spin valves and other spintronics devices [1–7]. Recent experimental progress in the fabrication of tunneling magnetoresistance (TMR) devices showing extremely large TMR values provides evidence for a large spin polarization in Heusler alloys [3, 5, 7–10].

Magnetic shape memory (MSM) materials represent the second research area where Heusler alloys have attracted substantial scientific interest. MSM materials offer an additional control parameter, the external magnetic field. In particular, the Ni_2MnGa system is in the focus of research interest [11]. The unique combination of large reversible strain, high switching frequency and high work output makes the compound a promising actuator material. Since the MSM effect results from an intrinsic mechanism, MSM devices based on epitaxial films possess great potential for implementation in microsystems [12, 13].

H.-J. Elmers (✉)
Johannes Gutenberg-Universität Mainz, 55128 Mainz, Germany
e-mail: elmers@uni-mainz.de

C. Felser and A. Hirohata (eds.), *Heusler Alloys*,
Springer Series in Materials Science 222, DOI 10.1007/978-3-319-21449-8_15

Despite the increased knowledge in the field of Heusler alloys many open questions have to be solved for further improving material properties for existing and emerging new applications [14]. Concerning the impact of electron correlation effects the development of new theoretical tools requires a thorough testing against experimental results [15]. The precise influence of disorder and reduced lattice symmetry on magnetic properties has not been clarified yet. New emerging applications like topological insulators [16] and thermomagnetic applications [17] require basic research for new materials.

X-ray magnetic circular dichroism (XMCD) in photo absorption spectroscopy (XAS) is a powerful tool for studying the element-specific magnetism even at buried interfaces [18, 19]. Many of the above mentioned problems can be investigated with this method. Due to the limited mean free path of 2.5 nm for photo emitted electrons XAS provides surface sensitive information through thin capping layers. The comparison with simultaneously measured data in transmission mode allows an independent determination of interface and bulk properties of thin films. Element-specific information on spin and orbital magnetic moments derived from the spectra allow a detailed test of theoretical predictions and can be used for optimizing sample preparation. As an example, the antiferromagnetic exchange coupling between nearest-neighbour Mn/Cr atoms at antisite defects causes an antiparallel alignment of moments [20]. Since in this case the Mn/Cr moment counts twice, this type of defect will drastically reduce exclusively the Mn/Cr moment and can only be detected by XMCD. Finally, large magnetic fields can be applied during the spectroscopic measurement, thus allowing to rotate the magnetization in any desired direction.

Spectroscopic information can be extracted from the XAS/XMCD measurements. In principle, the L-edge absorption spectra for left and right circularly polarized X-ray light reflects the spin-resolved partial density of states (PDOS) at the $3d$ transition metal atoms [21, 22]. For strongly localized states, e.g. in an oxide, the strong interaction between the core-hole and the conduction band in the final state leads to an additional splitting of the spectra, often denoted as multiplet effects [23]. These multiplet effects may effectively mask the band structure and in this case it is impossible to disentangle the PDOS and multiplet contributions. Telling et al. [24] pointed out that the existence of local moments at the Y site in Heusler compounds also give rise to a pronounced multiplet structure in the absorption spectra. However, previous investigations at the Co and Ni L-edge in intermetallic compounds clearly revealed PDOS related features in the absorption spectra of Heusler alloys [25, 26]. For Co-based Heusler compounds a proper consideration of the final state effects partly recovers the spin-resolved PDOS for unoccupied Co states from the XMCD spectra [27, 28]. This procedure enables a direct quantitative comparison of ab-initio calculations with experimental results in respect to the existence and position of the minority band gap relative to the Fermi level.

This chapter is organized as follows: Sect. 15.2 reports on the experimental approach for disentangling bulk and interface properties of thin films accompanied by two extreme examples of magnetically dead and life layers at the interface. The dependence of XMCD on disorder, structure and composition is discussed in Sect. 15.3. Section 15.4 refers to dynamic properties investigated by XMCD. Finally, Sect. 15.5

covers the spectroscopic information provided by XMCD, that can be exploited to discuss subtle changes of the PDOS as a function of composition, magnetization direction and temperature.

15.2 Interface and Bulk Magnetization

15.2.1 Experimental Approach

For many application devices the relevant properties are located at the interface while the bulk properties are less important. For tunneling magnetoresistance devices the topmost atomic layer at the interface exclusively determines the TMR value. Therefore, an independent determination of magnetic properties at the surface and in the bulk is important for the optimization of materials and processes.

In order to gain this information the XMCD signal is simultaneously measured in bulk-sensitive transmission (TM) and surface-sensitive total electron yield (TEY) mode [29, 30]. The transmission signal is measured via the X-ray luminescence within the substrate, thus allowing the investigation of epitaxially grown films, in contrast to previous approaches using parylene foils or SiN membranes. As TEY is surface sensitive while the transmission signal averages along the complete film normal one infers the depth dependence of the magnetization.

Additionally, this procedure solves an additional problem of XMCD. Quantitative information on element-specific spin and orbital magnetic moment obtained by XMCD comprises uncertainties because of the experimentally unaccessible number of d-holes N_h. Additional problems may arise for total electron yield measurements. For thin films with a common in-plane easy axis a favorable geometry is provided by the x-ray light shining at grazing incidence onto the sample thus requiring only small fields to reach saturation. In this case the sum rule analysis leads to large errors caused by the x-ray self absorption. Here, transmission and total yield signals can be evaluated with the identical sum-rule analysis revealing relative values of surface and bulk properties without uncertainties. Common errors include finite polarization of X-ray light, unknown number of unoccupied d-states, background subtraction for transition into s-states, and errors due to separation of L_3 and L_2 transitions can be avoided. An external magnetic field of 1.6 T saturates the sample magnetization perpendicular to the film surface and parallel to the incident x-ray beam.

For illustration Fig. 15.1a shows the incident-photon-flux-normalized transmission XAS spectra of Co_2FeSi/Al_2O_3 Heusler alloy films for thicknesses $d = 2.3$ nm to $d = 68$ nm averaged over the two magnetization directions. For the thickest film the transmission signal at the L_3 edge is less than half the intensity measured at the pre-edge. Assuming that the luminescence signal of the substrate $I^{\pm}_{lum}$ is proportional to the transmitted X-ray intensity, the X-ray absorption coefficient k can be calculated using the equation $k^{\pm}(h\nu) = -ln[I^{\pm}(h\nu)/I_{\mathrm{ref}}(h\nu)]$ (Fig. 15.1a). The reference spectra $I_{\mathrm{ref}}(h\nu)$ was measured by the bare substrate crystal and found to increase

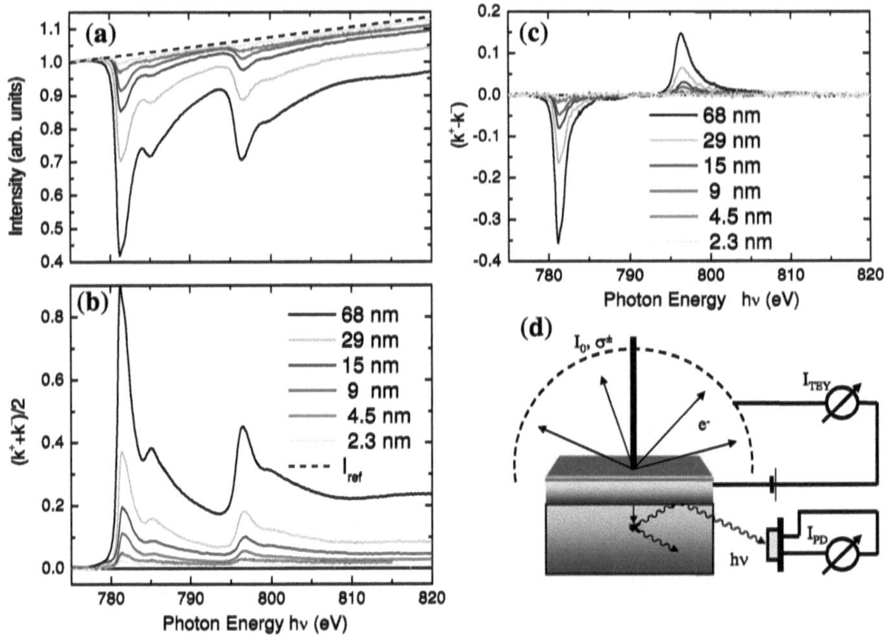

Fig. 15.1 **a** Intensity of the luminescence light emerging from the substrate measured at the Co $L_{2,3}$-edge of Co_2FeSi films on Al_2O_3. Thicknesses of the films are indicated in the figure. The intensity was normalized at the pre-edge. **b** Corresponding absorption coefficient $k = -ln(I/I_{ref})$ for various film thicknesses. **c** Circular magnetic dichroism signal $(k^+ - k^-)$ for thicknesses indicated in the Figure. **d** Cartoon of the experimental setup used for the x-ray absorption experiment in transmission and TEY. (from [30])

linearly with the photon energy. $I_{ref}(h\nu)$ was then normalized at the pre-edge region of the corresponding element. This normalization corresponds to an infinitely large penetration depth at the pre-edge.

Please note, that the interface properties discussed in this Section represent an integral information over a region determined the mean free path of the detected electrons, i.e. ca. 2.5 nm. A more detailed information on the magnetization profile can be gained from an analysis of resonant x-ray diffraction data [31, 32].

Experiments on $Co_2Cr_{0.6}Fe_{0.4}Al$ films that were deposited in a not fully optimized process show that the magnetic moments calculated from the surface sensitive TEY are reduced by 17 % averaged over the electron escape depth of about 25 Å with respect to the corresponding moments determined from the transmission signal (see Table 15.1) [29]. Since a homogeneous magnetization reduction at the surface is unlikely the reduction infers an even larger reduction for the topmost layer and may explain the observed low TMR values and spin polarization at the Fermi level of these films. Depositing in chambers with lower base pressure and adjusting the stoichiometry of the compounds carefully, these problems could partly be avoided [35, 40].

Table 15.1 Comparison of element-specific magnetic moments derived from the transmission spectra (TM) and from the surface sensitive total electron yield spectra (TEY) for the indicated $X_2Y_{1-x}Y'_xZ$ epitaxial Heusler films

		μ_{spin} X	μ_{orb} X	μ_{spin} Y	μ_{orb} Y	μ_{spin} Y'	μ_{orb} Y'	μ_{sum} XMCD	μ_{sum} SQUID	Reference
Co_2CrAl	TM	0.70	0.07	0.79	0.00	–	–	2.33	1.3	[35]
	TEY	0.68	0.07	0.46	0.00	–	–	1.96		
$Co_2Cr_{0.6}Fe_{0.4}Al$	TM	0.89	0.07	2.64	0.05	0.20	0.00	3.12	3.4	[29]
	TEY	0.74	0.08	1.90	0.10	0.10	0.00	2.50		
$Co_2Cr_{0.6}Fe_{0.4}Al$	TM	0.96	0.09	2.50	0.08	0.80	0.10	3.67	3.5	[35]
	TEY	0.99	0.06	2.51	0.10	0.96	0.11	3.79		
$/Al_2O_3$	TEY	0.99	0.06	2.42	0.10	0.10	0.00	3.17		
Co_2FeAl	TM	1.36	0.14	2.28	0.12	–	–	5.40	5.0	[36]
	TEY	1.27	0.12	2.52	0.04	–	–	5.34		
Co_2MnAl	TM	0.84	0.03	3.02	0.00	–	–	4.76	4.2	[37]
	TEY	0.73	0.03	2.81	0.00	–	–	4.33		
$Co_2FeAl_{0.3}Si_{0.7}$	TM	1.12	0.08	2.45	0.10	–	–	4.75	4.9	[38]
	TEY	1.26	0.12	2.21	0.10	–	–	4.92		
$Co_2FeAl_{0.5}Si_{0.5}$	TM	1.41	0.09	2.43	0.10	–	–	5.5*	5.5	[39]
	TEY	1.36	0.10	2.56	0.10	–	–	5.5*		
Co_2MnSi	TM	0.87	0.04	3.20	0.00	–	–	5.02	5.0	[27]
	TEY	0.80	0.05	3.04	0.11	–	–	4.85		
$Co_2Fe_{0.5}Mn_{0.5}Si$	TM	1.20	0.08	2.85	0.10	3.00	0.00	5.56	5.5	[27]
	TEY	1.20	0.08	2.80	0.12	2.80	0.06	5.46		
Co_2FeSi	TM	1.37	0.14	2.66	0.08	–	–	5.75	6.0	[27]
	TEY	1.35	0.07	2.80	0.12	–	–	5.54		
$Co_2FeGa_{0.5}Ge_{0.5}$	TM	1.25	0.07	2.50	0.06	–	–	5.2	5.3	[37]
	TEY	1.14	0.06	2.64	0.10	–	–	5.14		

In some cases (indicated by *) XMCD values measured at higher temperatures are scaled to the corresponding low-temperature SQUID value. The sum moment μ_{XMCD} results from a weighted sum of the atomic moments. The small induced moment (ca. $-0.1\ \mu_B$) on the main group element [34], oriented antiparallel to the main magnetization, is here neglected. Values are given in μ_B per atom for element-specific moments and μ_B per formula unit for the magnetization

A similar reduction of surface magnetization was detected for $Co_2FeSi(110)$ films grown by RF magnetron sputtering on Al_2O_3 $(11\bar{2}0)$ substrates [30]. From a comparison of the two experimental techniques TEY and TM one finds that the magnetization at the interface between Heusler film and Al capping is reduced by 8 % integrated over the electron penetration depth of 2.5 nm. Combining this result with the total interface reduction evaluated from the thickness dependence of the magnetic moment, one obtains a second interesting information. At the interface between the Heusler film and the substrate the magnetization is reduced, too, corresponding to 0.55 nm of magnetically dead layers.

Epitaxial Co_2FeAl [36] and $Co_2FeAl_{0.5}Si_{0.5}$ [39] films for which large TMR values have been measured and a large spin polarization was inferred from other methods show a negligible difference between bulk and interface magnetization. Interestingly, in these films the Co moment is slightly reduced and the Fe moment slightly increased at the interface (TEY) compared to their bulk (TM) values. Hence, the interface moments are even closer to the theoretical prediction for the fully $L2_1$—ordered compound.

Similar favorable trends were observed for laser ablated $Co_2Fe_xMn_{1-x}Si$ and $Co_2FeGa_{1-x}Ge_x$ films (see Table 15.1) [27, 37]. Transmission and TEY Co XAS signals (Fig. 15.2c–d) show the typical features, i.e. a satellite peak at 4 eV above the L_3 absorption edge, observed already for bulk samples, which is indicative for a high degree of local order [41, 42]. The extra peak was explained by a hybridization of Co d-band states with *sp*-states of the main group element. For all samples of this series the element-specific spectra do not show a prominent deviation between TEY and TM data indicating the high quality of the interface. The element specific moments (see Table 15.1) do not show large differences except for the orbital moments. The Fe and Mn surface orbital moments are larger compared to the corresponding bulk values. The Co orbital bulk moments are smaller in the case of Co_2MnSi and a larger in the case of Co_2FeSi compared to the surface orbital moment. This observation might indicate a deviation from the cubic crystal symmetry in the surface region invoked by the deposition process or by the capping layer.

15.2.2 Magnetic Dead Layers at the Interfaces

One of the most important parameter to be controlled during the fabrication process of Heusler films is the composition [43, 44]. Characterization of the composition by bulk sensitive methods is in general not sufficient as the composition at the interface may severely deviate from that in the bulk. An example is shown for the case of thin Ni_2MnGa films [45].

Epitaxial Ni_2MnGa films of 100 nm thickness were deposited by dc-sputtering onto $Al_2O_3(11\bar{2}0)$ substrates held at 773 K during deposition as optimized for a high degree of $L2_1$ ordering. The Heusler films were capped in-situ by 4 nm Al using sputter deposition at a lower temperature of 373 K in order to prevent interdiffusion [25, 26].

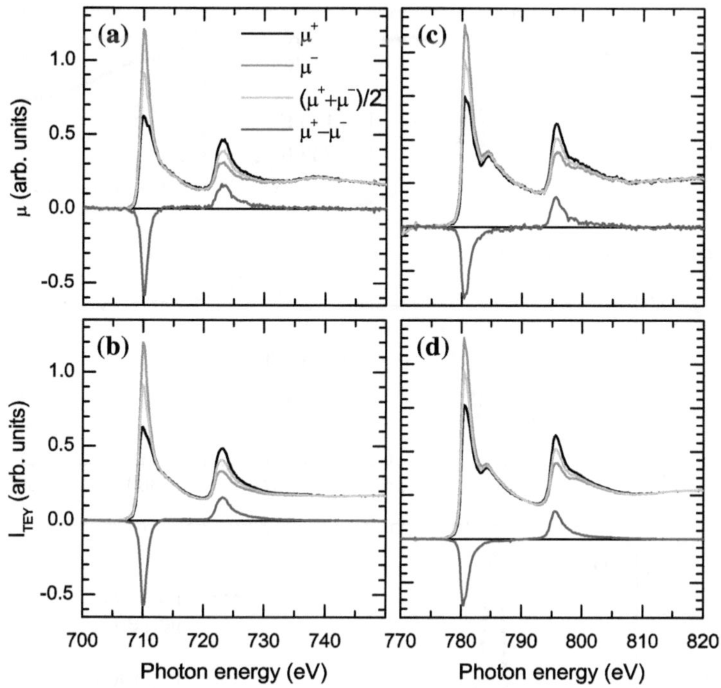

Fig. 15.2 **a** Absorption coefficient at the Fe $L_{2,3}$ edge for magnetization direction parallel (antiparallel) to the x-ray polarization, $\mu^{+(-)} \propto -\ln(I^{+(-)}/I_{ref})$, for a Co_2FeSi film on MgO. The XMCD spectrum, $\mu^+ - \mu^-$ is shown on the same scale. **b** Simultaneously measured total electron yield spectra $I^{+(-)}_{TEY}$ for the same sample. **c–d** Corresponding data measured at the Co $L_{2,3}$ edge. (from [33])

The Ni XAS measured in transmission (Fig. 15.3a) shows the pronounced increase of the satellite peak at 3.8 eV above the L_3 absorption maximum when the compound enters the austenitic high temperature phase, as discussed in Sect. 15.3 [25]. The presence of the satellite peak is indicative of local $L2_1$ order and cubic symmetry. For the surface sensitive TEY-XAS the corresponding satellite peak shows a smaller and broader shape and the difference between high and low temperature phase is smaller compared to the transmission signal. The Ni XMCD signal derived from the TEY-XAS shows a less pronounced peak structure and reduced values compared to the transmission data.

The Mn XAS (Fig. 15.3b) shows no pronounced difference between the high and low temperature phase. The step height of the background signal at 680 eV of the transmission XAS is smaller by a factor of two compared to the Ni XAS, as expected from the composition. The TEY-XAS, normalized with the same factor as the Ni TEY-XAS shown in Fig. 15.3a, is significantly smaller than the transmission signal. The observed decrease of the Mn TEY-XAS indicates a significant reduction of Mn concentration in the surface region. In addition the two broadened shoulders above

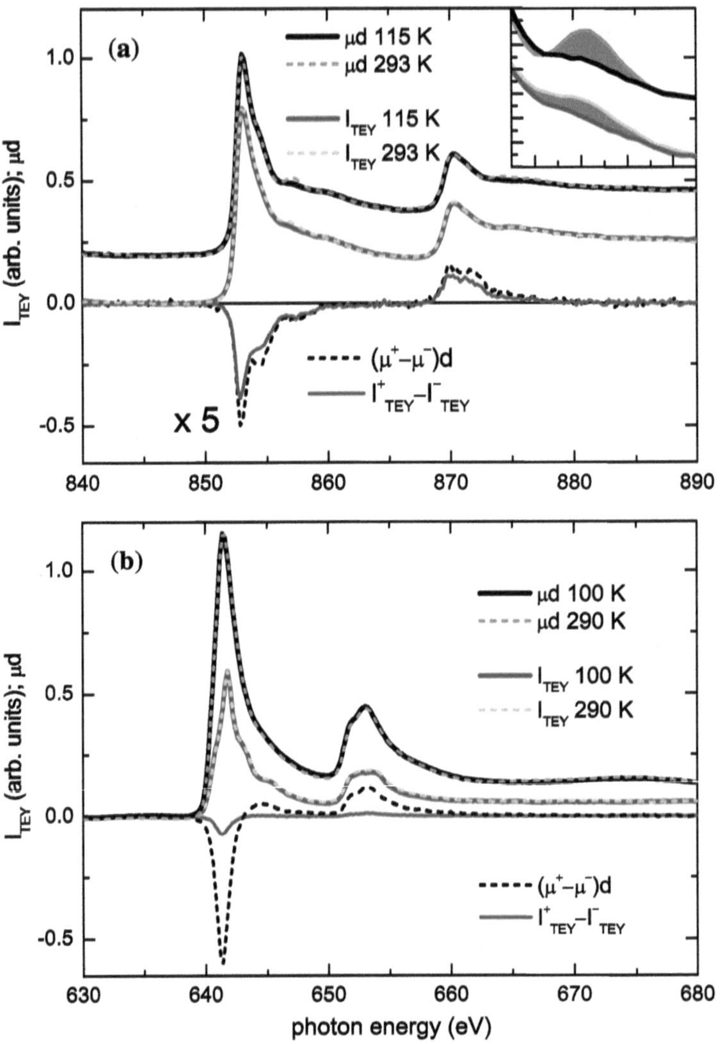

Fig. 15.3 **a** Comparison of Ni XAS and corresponding XMCD obtained at 100 and 300 K, i.e. below and above T_m, by TEY and transmission. The TEY XAS was normalized to the transmission signal at the step height of the background signal at 890 eV. The *inset* shows a magnified view of the satellite peak at 4 eV above the L_3 absorption edge, emphasizing the reversible change of the electronic structure at T_m. **b** Mn XAS and corresponding XMCD obtained at 100 and 300 K measured for the same sample in the same run as in (**a**). For direct comparison, the Mn TEY XAS was multiplied with the same factor as the Ni TEY XAS in (**a**). (from [45])

the L_3 edge indicate the presence of Mn in an oxidized state. The oxidation was attributed to residual gas during the fabrication process. From a comparison with a fully oxidized state one estimated that 10 % of the Mn atoms are oxidized. Thus the large reduction of the corresponding TEY-XMCD signal cannot be explained by the initial oxidation.

From the inset shown in Fig. 15.3a it is obvious that the increase of the satellite peak in the high-temperature phase phase, which is indicative for the, martensitic transition can hardly be seen near the surface. Temperature dependent measurements revealed a smooth transition from low to high temperature in contrast to the bulk behavior where a clear phase transition could be seen.

The reduced Mn concentration near the film surface must be attributed either to a selective evaporation of Mn after the sputter process or to a diffusion of Mn from the surface region into the film. It is well known that Mn has a large diffusion rate compared to other elements. However, a deviation from a constant concentration throughout the sample requires a different chemical potential for surface and bulk for which no obvious reason exists. The vapor pressure of Mn is the highest of all constituents and amounts to $1.3 \cdot 10^{-6}$ Pa at 773 K. This roughly corresponds to an evaporation rate of $2 \cdot 10^{12}$ Mn atoms/(cm^2·s). During sputter deposition the effect of evaporation is negligible as the deposition rate of 2 nm/s $\approx$ 4 (unit cells/s) provides at least an arrival rate of $5 \cdot 10^{15}$ Mn atoms/(cm^2 s). On the other hand, the observed reduction of Mn concentration at the surface corresponds to a loss of $1.5 \cdot 10^{15}$ Mn atoms/cm^2. Thus, a slow cooling rate explains the reduced Mn concentration at the interface.

As a consequence, the decreased Mn concentration suppresses the martensitic phase transition at the surface. The remaining temperature dependence of the satellite peak in the TEY XAS might be attributed to a structural change that is caused by a morphological change of the surface at T_m and to an inhomogeneous composition near the surface. A magnetization decrease with decreasing Mn concentration has been observed for bulk samples, too [46–48], and was attributed to the smaller number of Mn atoms. In addition, the small Mn moment hints to a destroyed local order and to an antiferromagnetic alignment of remaining Mn moments [47]. The stronger temperature dependence of the surface moment agrees well with the reduction of the Curie temperature observed for bulk samples with reduced Mn concentration [49]. The stronger decrease of the Mn moment compared to the Ni moment might be attributed to the increased disorder at the interface leading to antiferromagnetically coupled Mn atoms at anti-site positions.

15.2.3 Magnetic Life Layers at the Surface Above T_C

Instead of a suppression of interface magnetization one may also encounter a significant increase [35]. An extreme case is described for Co_2CrAl films, where one finds zero bulk magnetization at room temperature, i.e. below the Curie temperature of the bulk. In contrast, the magnetization at the interface shows a significant value.

The epitaxial Co_2CrAl thin films were prepared by rf-sputtering at room temperature on MgO(100) substrates [50]. After the deposition the samples where annealed in situ under UHV conditions at 700 °C and finally capped by 4 nm Al. The films are ordered in the B2 structure as determined by x-ray diffraction.

The bulk magnetic moment as derived by SQUID measurement amounts to a maximum value of $\mu(4K) = 1.3\ \mu_B$/ f.u.. This is much smaller than the theoretically predicted ground state value of $\mu = 3\ \mu_B$/ f.u.. but consistent with the experimentally observed values from other bulk [51] and thin film samples [52]. For high annealing temperature which was optimized for structural ordering determined from XRD, SQUID results in a Curie temperature of $T_C = 290$ K.

The interesting result obtained by XMCD on these samples is the increased magnetic ordering temperature in comparison to the bulk value. At 300 K a clear XMCD signal is observed at the Co edge (Fig. 15.4) when the x-ray absorption is measured in the total yield mode, although XMCD measured in transmission reveals the absence of bulk ferromagnetism. An interface magnetic moment larger than the bulk value is observed for all temperatures but the difference almost vanishes at very low temperature.

An explanation for this effect is an improved surface order on the Co sites induced by an energetically favorable layer termination as motivated by scanning tunneling microscopy investigations [50]. According to bulk sensitive XRD a higher annealing temperature results in more narrow diffraction peaks with higher integrated intensities and also the intensity ratio of the (200) and (400) peaks increases, which is usually

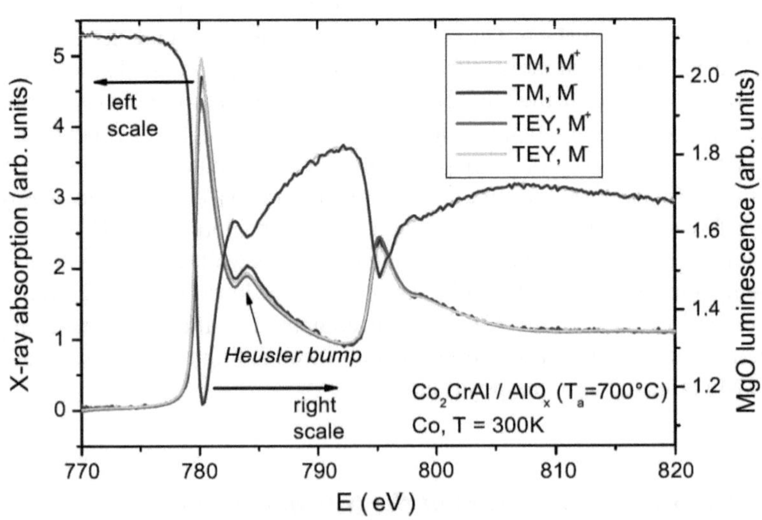

Fig. 15.4 *Left scale* X-ray energy dependent (Co-edges) XMCD-TEY absorption intensity (*red* and *green curves*) and XMCD-TM absorption intensity (*blue* and *cyan curves*) for up M^+ and down M magnetization of Co_2CrAl covered by Al. The shoulder feature labeled Heusler bump is characteristic for the Heusler structure. *Right scale* X-ray energy dependent XMCD-TM luminescence intensity. (from [35]) (Color figure online)

associated with an increased degree of B2 order. However, the x-ray scattering cross sections of Co and Cr are very similar. Thus Co/Cr disorder, which is theoretically expected to result in antiferromagnetically coupled Cr cannot be detected by standard x-ray diffraction. This type of disorder explains the absence of Cr magnetization and also the lower Curie temperature in the bulk of the film. A similar increase of the Curie temperature by ordering has also been observed for Cu_2MnAl [53, 54].

15.3 Structural Influence on Absorption Spectra

15.3.1 Dependence on Atomic Order

The influence of local atomic order on the XAS signal has first been discussed in detail in [41]. Here, we report results for $Co_2FeAl_{0.5}Si_{0.5}$ thin films. $Co_2FeAl_{0.5}Si_{0.5}$ thin films are very attractive since they can have a spin polarization as high as 0.91 at room temperature with a weak dependence on temperature [55]. Annealing of $Co_2FeAl_{0.5}Si_{0.5}$ films has been shown to be an efficient way to optimize several important properties for the production of high-quality tunneling junctions [56–59]. The improvement of local structure with increasing annealing temperature was shown using nuclear magnetic resonance [57]. Here, the relation between local atomic order and changes in the x-ray absorption spectra is discussed.

In order to relate XMCD spectra and structural properties a sample series has been grown on single-crystal Mg(001) substrates. The stacking structure of the sample is as follows: MgO^{sub}(001) / MgO^{buffer}(20 nm) / $Co_2FeAl_{0.5}Si_{0.5}$(30 nm) / $MgO^{barrier}$(0 or 3 nm) / Al^{cap}(3 nm). Samples were prepared by a magnetron sputtering as described in detail in [39]. The samples were annealed *in-situ* for 30 min after $Co_2FeAl_{0.5}Si_{0.5}$ deposition at temperatures of 480, 525, 550, 575, and 600 °C. The $Co_2FeAl_{0.5}Si_{0.5}$ films were capped either by an Al layer or by a 3 nm MgO layer the latter being more representative of magnetic tunneling junction devices [6, 60]. X-ray diffraction revealed a transition from B2 order to an improvement of the degree of $L2_1$ ordering with increasing annealing temperature.

Representative XAS Fe and Co $L_{2,3}$ edge spectra are presented in Fig. 15.5a, b for samples annealed at 550 °C, with and without an MgO barrier deposited on the $Co_2FeAl_{0.5}Si_{0.5}$ layer.

It has been demonstrated that the spectral weight of this satellite peak at 3.8 eV above the Co L_3 absorption edge directly correlates with the degree of atomic ordering in Co-based Heusler alloys, [41, 61] where higher intensities are observed for well-ordered ($L2_1$) samples. The satellite peak is more pronounced at the L_3 edge than at the L_2 edge because of the Coster-Kronig lifetime broadening, and was explained by a hybridization of Co d-band states with sp states of the main group (non-magnetic) element [42]. The local atomic order increases with increasing annealing temperature in agreement with the results from X-ray diffraction [57, 58, 60] and nuclear magnetic resonance characterization [57].

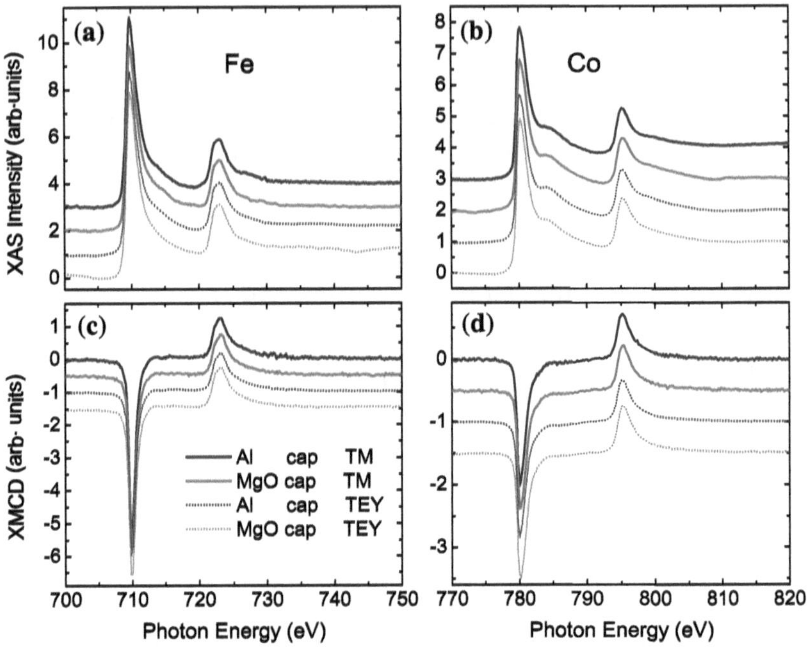

Fig. 15.5 X-ray absorption spectra of $Co_2FeAl_{0.5}Si_{0.5}$ annealed at 550 °C at the Fe (**a**) and Co (**b**) $L_{2,3}$ edge averaged from total electron yield (TEY) and transmission intensities (TM) measured at 300 K for magnetization direction parallel (antiparallel) to the x-ray polarization, $(I^+ + I^-)/2$. **c**, **d** Corresponding XMCD spectra, $I^+ - I^-$. The *inset* in (**b**) indicates the determination of the spectral weight of the satellite peak in the Co XAS spectra. *Solid lines* TM spectra; *dashed lines* TEY spectra. *Blue* is Al-capped, while *orange* is MgO-capped. (from [39]) (Color figure online)

The fact that the satellite peak is more pronounced for the transmission signal indicates that the bulk of the film shows a higher degree of local atomic order than the surface region. For the surface of the series of samples capped with an MgO barrier, one observes a particularly small satellite peak. This observation indicates that the local atomic order at the film surface is degraded by the interfacial MgO layer, thus suggesting a route towards the optimization of high-quality tunneling magneto-resistance device fabrication.

Interesting information can be gained from the comparison of surface and volume magnetic moments. Both film series comprise similar values for surface and bulk in the case of the magnetic spin moments located at the Fe atoms. For the Al-capped films the spin moments located at the Co atoms are equal, too. This is in contrast to previous observations for $Co_2Cr_{0.6}Fe_{0.4}Al$ films, [29] but consistent with improved deposition parameters [35] for the same composition and also for $Co_2FeSi_xAl_{1-x}$ films [27]. In contrast, the MgO-capped films show an increased Co spin moment for annealing temperatures below 550 °C. The enhanced Co moment near the surface is a direct consequence of the increased disorder in this region. While a Co moment of 1.3 μ_B is expected for an ideally ordered Heusler compound, the pure Co bulk value is considerably larger (1.6 μ_B). Therefore, disorder tends to increase the Co moment [38]. For an annealing temperatures of 575 °C the difference between surface and bulk moment nearly vanishes, thus indicating a better ordering also at the MgO—interface. Upon further annealing at 600 °C the Co interface moment rises again with respect to the bulk moment indicating that order on the Co sites is reduced again by transposition of Co and Fe. The existence of such a mechanism was demonstrated by Mössbauer-spectroscopy on the Cr containing related compound $Co_2Cr_0.6Fe_0.4Al$ [62].

The XMCD results demonstrate that bulk characterization methods are of limited significance for the optimization of Heusler thin films incorporated into tunneling junctions or spin valves. The measurements of the annealing temperature dependent surface magnetic moment of the $Co_2FeAl_{0.5}Si_{0.5}$ films explain the observation, that the maximum tunneling magnetoresistance of junctions with this electrode are obtained with annealing temperatures in the intermediate range [56].

15.3.2 Structural Phase Transitions

Martensitic phase transitions are displacive first order transitions from cubic high temperature phases (austenite) to tetragonally distorted low temperature phases (martensite). Due to their technological importance they have a long history of scientific interest. Ni_2MnGa is an especially interesting martensite structure, as it is a magnetic shape memory (MSM) material [11]. It can change its macroscopic shape under application of an external magnetic field. Due to an exceptionally large c/a ratio of the tetragonal distortion large length changes up to 10 % can be achieved in single crystals [63]. Because of the subtle change of the electronic structure at the martensitic phase transition only few experimental results confirm the variation of the DOS. X-ray absorption is particularly sensitive to this change because of its element-specific character.

Ni_2MnGa/Al_2O_3 films show a martensitic phase transition at $T_m = 275(3)$ K [25] as determined by X-ray diffraction. The structural change at T_m is accompanied by an unconventional temperature dependence of the Ni spin and orbital magnetic moments. The Ni orbital moment shows a pronounced peak-like maximum and the spin moment reveals a step-like increase close to T_m [25, 45].

XAS shows a significant change at the martensitic phase transition. A satellite peak observed at 3.8 eV above the absorption edge for the austenitic phase is nearly suppressed in the martensitic state (see bottom panel of Fig. 15.6). To first order the absorption intensity at the L_3 absorption edge is proportional to the density-of-states function of unoccupied states above the Fermi edge. The same holds for the L_2-edge which is, however, broadened compared to the spectra at the L_3 edge. Accordingly, the observed change of the spectra can be explained by a change of the electronic structure. The suppression of the satellite peak in the martensitic state is due to the lift of degeneracies in the Ni 3d related unoccupied electronic states as predicted by ab initio calculations [64] .

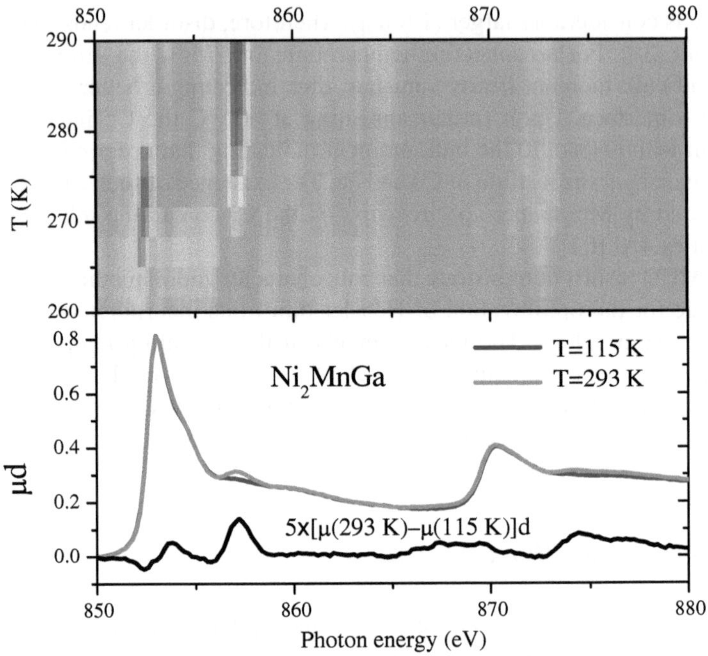

Fig. 15.6 Difference of the Ni XAS spectra measured at the indicated temperature and at low temperature (115 K) indicated by a color code (*blue* means negative values, *yellow* means positive values. Ni XAS spectra for 115 K (*blue line*) and 293 K (*red line*) and the corresponding difference spectrum (*black line*) is shown below. (from [65]) (Color figure online)

For a detailed study of the spectral change XAS is measured for a series of temperatures covering the phase transition region around T_m with data acquisition at 3 K steps [65]. Figure 15.6 emphasizes the temperature dependence of the electronic structure in the region of T_m. In order to emphasize the temperature dependence the difference is plotted as a color map with minima indicated in blue and maxima indicated in red. The Fermi edge is located near the initial increase at the L_3 edge. The prominent feature is the rise of the 3.8 eV satellite close to $T_m = 275$ K. In addition to this prominent feature the difference reveals a minimum close to E_F and an additional maximum above the L_3 absorption maximum appearing near the phase transition temperature. The initial minimum of the difference indicates a decrease of the local DOS at E_F when the system enters the high temperature cubic phase. This is also in agreement with the theoretical prediction [64]. In the same temperature region the minimum at E_F reveals a subtle shift of 250 meV to higher energy. The shift of the minimum indicates a shift of electronic states across the Fermi edge which might be the origin of the martensitic phase transition.

15.3.3 Compositional Dependence

In this section we explain, how XMCD can be used to characterize the compositional dependence of element-selected magnetic moments in Heusler alloy films. As an example we refer to the case of Ni_2MnGa derived alloys. This is an important case, because the magnetic and mechanical properties of these films can be effectively improved by adding further elements to the alloy. The martensitic transformation temperature (T_m) of Ni_2MnGa–based compounds, being below room temperature for the stoichiometric Ni_2MnGa compound (202 K), is very sensitive to the composition (see i.e. [66]) and increases up to 410 K for slightly off-stoichiometric compounds $Ni_{1.96}Mn_{1.22}Ga_{0.82}$[67, 68]. The addition of further constituents, i.e. Fe and Co [69] can remarkably change T_m and further magnetic properties, too. Theoretical investigations [49] revealed the importance of the electron-to-atom ratio e/a for the martensitic transformation. XMCD is a valuable tool for a testing these theoretical predictions.

In the following results are shown for the composition $Ni_{1.96}Mn_{1.22}Ga_{0.82}$ motivated by the increase of T_m observed before in bulk samples of this composition, and $(Ni_2MnGa)_{1-x}(Co_2FeSi)_x$ ($x = 0.025$ and $x = 0.2$) motivated by the expected increase of magnetization while keeping the e/a ratio constant. Details of the sample preparation are described in [65].

Due to the increased Mn content the martensitic phase transition in $Ni_{1.96}Mn_{1.22}Ga_{0.82}$ films occurs at $T_m = 410$ K instead of 275 K for Ni_2MnGa. We do not observe any temperature dependence of the XAS in the case of $Ni_{1.96}Mn_{1.22}Ga_{0.82}$. Figure 15.7 compares the XAS/XMCD results for $Ni_{1.96}Mn_{1.22}Ga_{0.82}$ with those for the stoichiometric compound both measured at 115 K, i.e. in the martensitic phase. The step height of the Mn absorption edge for $Ni_{1.96}Mn_{1.22}Ga_{0.82}$ is larger than for Ni_2MnGa as expected from the increased Mn content. Differences in the spectral shape of the two samples are marginal.

In contrast, the 3.8 eV satellite feature in the Ni XAS is completely suppressed by the increased Mn content. This is in agreement with the fact that the martensite structure is supported by the non-stoichiometric composition leading to a more pronounced deviation from the cubic symmetry.

The XMCD signal of both Ni and Mn is smaller for the non-stoichiometric compound. The sum rule analysis results in a total reduction of the magnetic moment of 40 % at 300 K and still 10 % at 115 K, indicating a stronger temperature dependence for $Ni_{1.96}Mn_{1.22}Ga_{0.82}$. The ratio of the Ni and Mn spin moments of 0.12 remains the same for Ni_2MnGa and $Ni_{1.96}Mn_{1.22}Ga_{0.82}$. The observed reduction of magnetic moments for $Ni_{1.96}Mn_{1.22}Ga_{0.82}$ by XMCD is in agreement with the magnetization results obtained by magnetometry. A reduction of magnetic moments for non-stoichiometric Ni_2MnGa compounds was predicted by theory [47]. According to this model the magnetization should decrease linearly with the deviation from the average number of valence electrons per atom, e/a, starting from a maximum value of 4 μ_B per atom at $e/a = 7.5$ for Ni_2MnGa. For $Ni_{1.96}Mn_{1.22}Ga_{0.82}$ one accordingly expects a value of 3.5 μ_B per atom [47]. The relative decrease, 3.5/4,

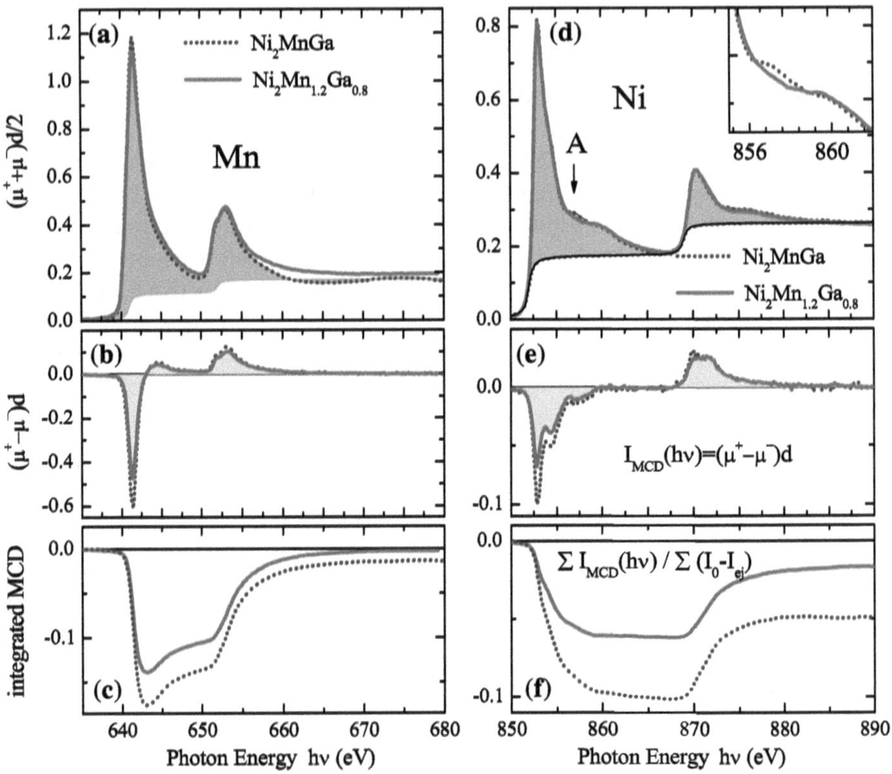

Fig. 15.7 **a** Absorption coefficient $\mu d = -\ln(I/I_{ref})$ of X-ray light transmitted through a 100 nm $Ni_{1.96}Mn_{1.22}Ga_{0.82}$ (110) film on $\alpha-Al_2O_3$ at the Mn $L_{2,3}$ edge (full *green line*) in comparison to the stoichiometric compound Ni_2MnGa (*dashed blue line*). The data was measured at $T = 115$ K ($T \ll T_m$). The step function (*black line*) used for the sum rule analysis is shown in addition. **b** Corresponding XMCD spectra $(\mu^+ - \mu^-)d$. **c** Integrated XMCD spectra from the data shown in (**b**). **d–f** Corresponding data measured at the Ni $L_{2,3}$ edge. (from [65]) (Color figure online)

is in rough agreement with our result for films, 3/3.3, however, absolute values are considerably smaller for both compounds. Smaller values for the magnetization of Ni_2MnGa films have been reported before [25] and were attributed to the remaining atomic disorder of the Heusler structure. The magnetization reduction in non-stoichiometric compounds with Mn excess was mainly attributed to an antiferromagnetic alignment of Mn moments that occupy antisite positions. In this model the Ni moment should remain constant independent on the composition. In contrast, our experimental result shows a similar reduction of both Mn and Ni moments in $Ni_{1.96}Mn_{1.22}Ga_{0.82}$. This observation indicates a more complicated reaction of the atomic moments on compositional changes.

As a second example we discuss the case of a constant number of valence electrons per atom, $e/a = 7.5$, but partly exchange of atoms on all sublattices of the Heusler structure. $Ni_{1.6}Co_{0.4}Mn_{0.8}Fe_{0.2}Ga_{0.8}Si_{0.2}$ films have been prepared as

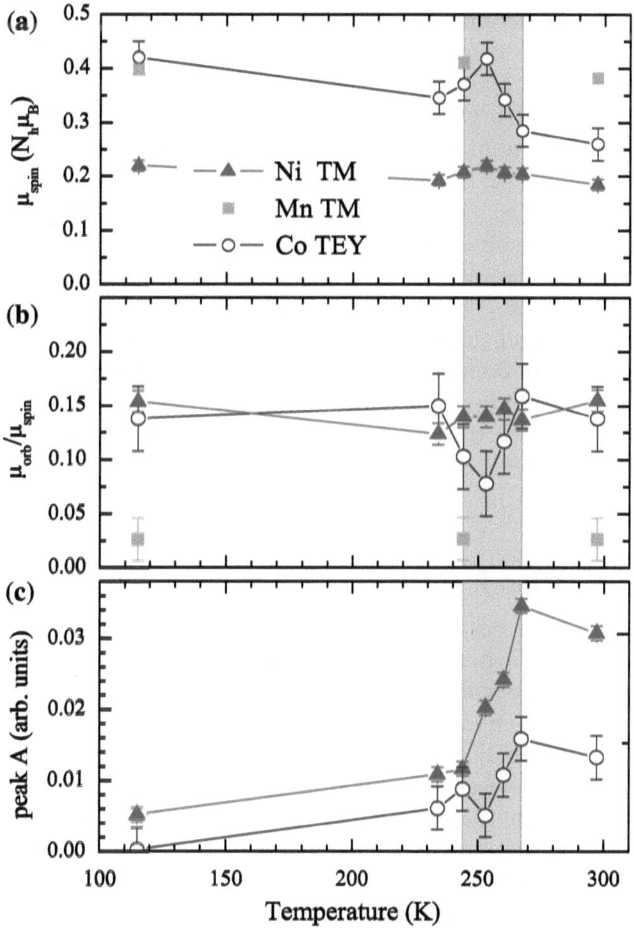

Fig. 15.8 **a** Effective spin moments per d-hole, N_h, for Mn (*red squares*), Co (*blue circles*) and Ni (*cyan triangles*), excluding a jj-mixing correction factor in the case of Mn; and (**b**) orbital to spin momentum ratio for Mn (*red squares*), Co (*blue circles*) and Ni (*cyan triangles*) for a 100 nm $(Ni_2MnGa)_{0.975}(Co_2FeSi)_{0.025}$ (110) film on $\alpha-Al_2O_3$. **c** The increase of the peak area (after linear background subtraction) of the satellite peak A observed 3.8 eV above the Ni L_3 (*cyan triangles*) and Co L_3 (*blue circles*) absorption edge indicates the martensitic phase transition. The temperature range of increasing peak area is marked in *gray*. (from [65]) (Color figure online)

multilayers of Co_2FeSi and Ni_2MnGa at high deposition temperature leading to a strong interdiffusion of the two compounds [65]. The martensitic transition temperature of T_m = 256 K is significantly smaller compared to the value of Ni_2MnGa films. On the other hand the magnetization increases with increasing Co content to 4.5 μ_B/ f.u..

The temperature dependence of Ni and Mn magnetic moments are depicted in Fig. 15.8. The Co moment shows an opposite behavior to the Ni and Mn moments. While the Co spin moment reveals a pronounced maximum and the orbital moment a minimum in the temperature region of the phase transition, the Ni moment shows a weak maximum. This behavior might be induced by the strong decrease of local symmetry at the phase transition.

The sum rule analysis results in a total magnetic moment of 4 μ_B per atom for Mn and 0.55 μ_B per atom for Ni. These values are considerably larger than the experimental value for Ni_2MnGa films and even larger than the theoretical ground state value for Ni_2MnGa. The Fe moment is equal to values observed for Co_2FeSi films and only the Co moment remains smaller.

The sum moment for increases from 3.3 μ_B/f.u. (formula unit) for $x = 0$ to 5.1 μ_B/f.u. for $x = 0.2$. The weighted average value expected from a multilayer without interdiffusion amounts to just 3.6 μ_B/f.u., assuming 6 μ_B/f.u. for Co_2FeSi. This comparison again demonstrates the strong influence of the Co and Fe atoms partly replacing Ni and Mn on the corresponding sublattices. The strong increase of the Mn moment for $x = 0.2$ is the result of the suppression of antiferromagnetic alignment of Mn moment on anti-site positions.

15.4 Magnetization Dynamics

The dynamical magnetic properties of half-metallic materials are particularly interesting because spin-dependent transport and relaxation phenomena may considerably deviate from conventional ferromagnetic materials [70–72]. Element-specific XMCD offers a unique disentanglement of contributions from different layers in multilayer spintronic devices and even from different sublattices.

Liu et al. [72] predicted a Gilbert damping parameter of 10^{-4} in half-metals, which is one order of magnitude lower than in traditional 3d-transition metal ferromagnets. Indeed, very low experimental values of the Gilbert damping were measured for half-metallic NiMnSb [73] and Co-based Heusler films [74–76]. In addition, theoretical predictions [77] suggest that the different exchange interaction on X and Y sublattice sites in the $L2_1$ or B2 structure may well induce a different dynamical sublattice response.

Time-integrated [78, 79] and time-resolved [80–84] set-ups have been used to combine XMCD and ferromagnetic resonance (XFMR). Time-resolved XFMR utilizes the pulsed nature of the synchrotron radiation and thereby offers additional information on the phase between microwave excitation and precessing magnetization [82].

Arena et al. [82] demonstrated the high sensitivity to small-angle excitations and a determination of the relative phase lag between oscillations of individual layers and elements in a $Ni_{81}Fe_{19}$/Cu/$Co_{93}Zr_7$ trilayer. Similar measurements revealed a small ferromagnetic coupling in a CoFe/Ru/NiFe trilayer [85].

A novel experimental approach for time-resolved XFMR has been reported in [86], allowing dynamic investigations of epitaxial films. In this case the luminescence yield of the MgO substrate was used as an absorption signal similar to the case of static measurements described above. The advantage is that epitaxial film structures buried under a Cu conducting layer can be investigated and that external magnetic fields do not disturb the transmission signal in contrast to the electron yield mode.

For the dynamic measurement the photon energy is tuned to the maximum XMCD signal at an absorption edge of the selected element. The microwave generator operating at multiples of the synchrotron ring clock frequency of 500 MHz is phase-locked with the X-ray pulses of 50 ps length using the bunch marker trigger. The microwave current in a coplanar waveguide then excites the magnetization that precesses around the direction of the external field, i.e. along the waveguide. The phase of the microwave signal is adjusted by an electronic delay generator. In order to measure the excited magnetization amplitude, the microwave phase is modulated by 180° at a frequency of $f_{ref} = 44$ Hz and lock-in technique is used for the signal detection. The amplitude of the intensity signal measured by the lock-in amplifier is proportional to the XMCD signal and thus a measure for the magnetization component m_ξ along the excitation axis (across the coplanar waveguide) at the time when the x-ray pulse arrives.

By adjusting the phase of the exciting microwaves to 90° with respect to the X-ray bunches, m_ξ represents the imaginary part of the dynamic magnetic susceptibility. By sweeping the bias field across the resonance condition of the $Co_2(Cr_{0.6}Fe_{0.4})Al$ film one obtains the resonance curves shown in Fig. 15.9a, c, e. The result reveals identical values for the resonance field H_r and linewidth ΔH for Co, Fe and Cr.

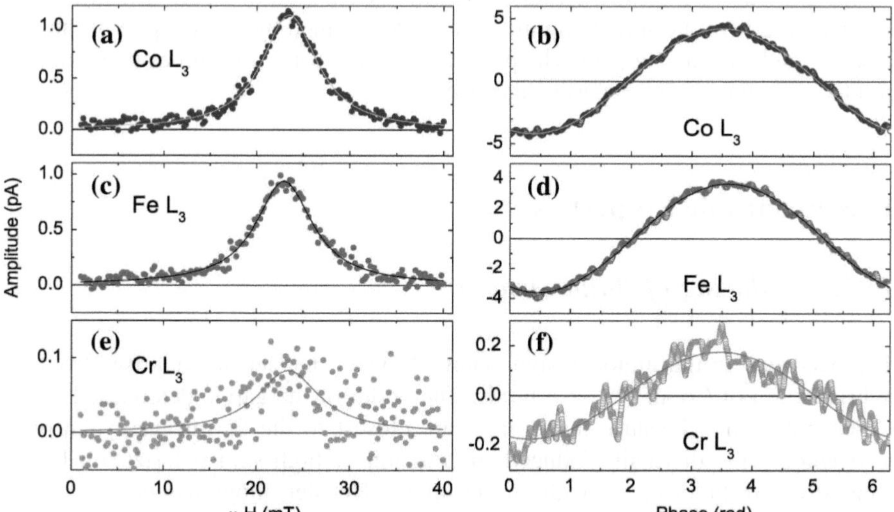

Fig. 15.9 **a**, **c** and **e** Amplitude of the oscillating magnetization component as a function of the applied external field measured for a $Co_2(Cr_{0.6}Fe_{0.4})Al$ film at fixed exciting frequency. Full *lines* indicate a fit with a Lorentzian function. **b**, **d** and **f** Amplitude of the oscillating magnetization component as a function of the phase delay between the exciting microwave and the X-ray pulse from the synchrotron. Full *lines* indicate a fit with a sine function The amplitude is measured with the photon energy tuned to the Cr (a,b), Fe (c,d) and Co (e,f) L_3 edge. (from [86])

By scanning the delay time, the time dependence of m_ξ is measured at constant resonance field, revealing a sinusoidal signal thus reflecting a regular precession. A sinusoidal fit results in a negligible phase difference between the Co and the Cr/Fe sublattice of $\Delta\phi = 2° \pm 2.7°$.

This method is used to investigate the influence of a weakly coupled second ferromagnetic layer on the damping properties of a thin $Co_2(Cr_{0.6}Fe_{0.4})Al$ film. Two identical samples were fabricated in one deposition run with the exception of an additional CoFe capping layer for the second sample. The comparison of the frequency-dependent linewidths of the two samples reveals that the weakly antiferromagnetically coupled CoFe layer causes an increase of the ferromagnetic resonance linewidth. The increase consists of a constant offset and a component that linearly increases with frequency. The increased offset is attributed to a lateral variation of the interlayer thickness leading to a variation of H_{iec} and thus to a frequency-independent contribution to ΔH. The linear increase corresponds to $\partial(\mu_0 \Delta H)/\partial f = 1.06(16)$ mT/GHz and is attributed to non-local damping due to spin pumping. Electron angular momentum is transfered between the layers and causes an additional relaxation torque [87]. This effect leads to an additional damping term proportional to the microwave frequency in agreement with our observation. The additional damping of $\alpha_s = \alpha(\mathrm{II}) - \alpha(\mathrm{I}) = 0.0072(30)$ is almost of the same strength as the intrinsic damping and when the magnitude is scaled according to thickness and magnetization one finds a six times larger effect than the non-local damping observed for two Fe films separated by an 8 nm thick Au interlayer [88]. The magnitude of the spin pumping effect is determined by the real part of the spin mixing conductance $\mathrm{Re}\{g^{\uparrow\downarrow}\}$. The present result indicates a six times larger mixing conductance than in the experiment of [88].

15.5 Electronic Properties

15.5.1 Tailoring of Band Structure

The band gap in the minority-spin states of HMF Heusler alloys arises from the hybridization of Co and Y 3*d* orbitals. The width of the gap is determined by the Co-Co interaction because these states are closest to the Fermi energy. In order to overcome the thermally induced suppression of high spin polarization further band structure tailoring through doping of the Heusler alloys has been proposed [4, 60, 89]. An especially interesting example is $Co_2FeAl_{1-x}Si_x$ [60] for which ab-initio calculations predict a Fermi energy E_F in the center of the minority gap for $x = 0.5$, in contrast to $x = 0$ and $x = 1$ with E_F positioned close to the upper or lower boundary of the gap [90, 91]. Therefore, a direct study of the band gap is of particular importance. Although spin-resolved photoemission [92–94] or scanning tunneling spectroscopy can directly probe the spin polarization at a half-metal surface, these methods have no access to the crucial buried interfaces in spintronic devices.

XMCD in principle provides the requested spectroscopic information. The L-edge absorption spectra for left and right circularly polarized X-ray light reflect the spin-resolved partial density of states (PDOS) at the $3d$ transition metal atoms [21, 22]. Previous investigations at the Co and Ni L-edge in intermetallic compounds clearly revealed PDOS related features in the absorption spectra of Heusler alloys [25, 26]. Extracting the PDOS from the Y sublattice is more difficult [99]. On the Y sublattice electronic states are typically more localized [24] and the strong interaction between the core-hole and the conduction band in the final state leads to an additional splitting of the spectra, often denoted as multiplet effects [23]. These multiplet effects may effectively mask the band structure and in this case it is impossible to disentangle the PDOS and multiplet contributions.

The experimental spin-resolved unoccupied Co PDOS is shown in Fig. 15.10 as derived from the Co L_3-edge absorption data. Shown is here the spin-resolved DOS of epitaxial (100) oriented films of $Co_2Fe_xMn_{1-x}Si$, $Co_2FeAl_{1-x}Si_x$ and $Co_2Cr_{0.6}Fe_{0.4}Al$ films grown on MgO(100) substrates using XMCD data [27]. The spin-resolved unoccupied PDOS function follows from the XAS spectra I^+ and I^- according to:

$$D^{\uparrow(\downarrow)}(1 - f_F) \propto I_{iso} - s + (-)\frac{1}{P_j}\frac{I^+ - I^-}{2}, \tag{15.1}$$

where I_{iso} denotes the isotropic absorption coefficient $(I^+ + I^-)/2$, s is the step function, f_F denotes the Fermi function and P_j is the spin polarization of the excited photoelectrons, i.e., $P_{L3} = 0.25$ and $P_{L2} = -0.5$. An electron correlation effect causes a shift of the majority PDOS of $\Delta E_c = 0.5$ eV with respect to the minority states as explained in [27, 28]. The position of the Fermi level was derived from the initial increase in the majority states considering the energy shift of ΔE_c. This procedure is confirmed by calculations considering finite state effects [100]. The advantage of the presented procedure of PDOS extraction is the readily available comparison to ground state calculations performed with various methods and approximations.

To account for the lifetime broadening of the XAS spectra one may deconvolve the experimental spectra with a Lorentzian function with a width of $\Gamma = 0.35$ eV. Note, that the Co minority PDOS is close to zero in a finite energy interval around E_F. The observation of a vanishing PDOS for $E < E_F$ is a consequence of a near coincidence of the Lorentzian function and the experimental data. Thus, the experimental data support the theoretical prediction of half-metal behavior particular for Co_2MnSi.

The PDOS determined for a series of samples reveals a variation of the position of E_F within the minority band gap with the substitution of the transition metal element or the main group element in agreement with earlier predictions. For the $Co_2(Fe_xMn_{1-x})Si$ films the spin-resolved Co PDOS reveals a minority maximum at $E_{v,max}$ approaching E_F with increasing x as depicted in Fig. 15.11. Ab-initio calculations have predicted this trend, although absolute values varied depending on the model assumptions and calculation schemes. LDA+U [101] fits better to the experiment, however, the value of U appears overestimated for Co_2MnSi. The best agreement with experiment is achieved for a LDA+DMFT calculation [98].

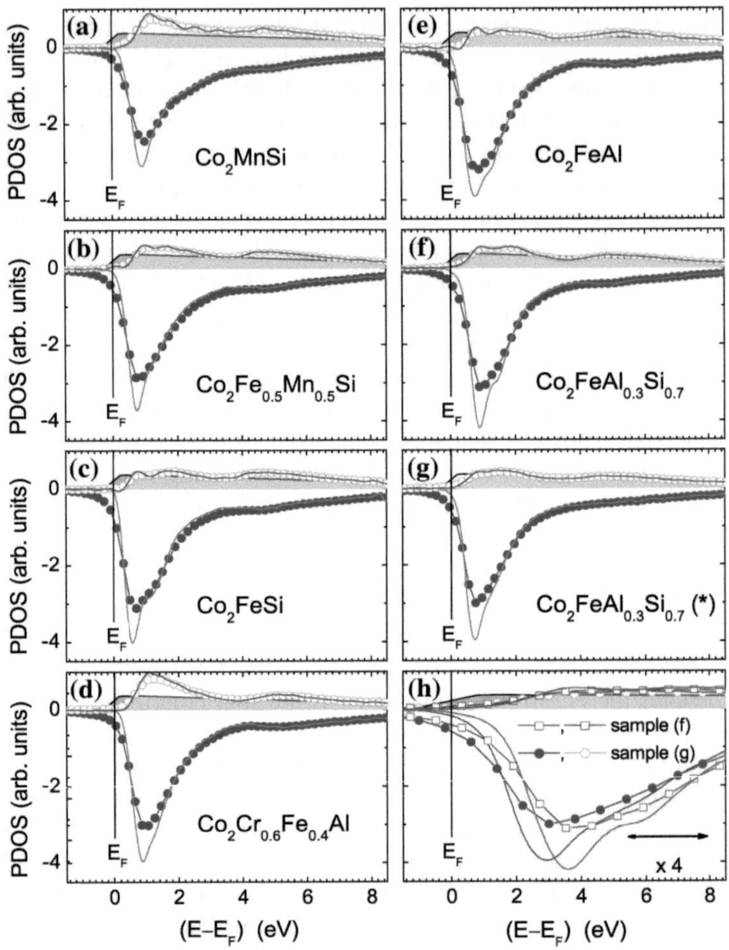

Fig. 15.10 Spin-resolved PDOS calculated from the XAS/XMCD data measured at the L_3-edge for thin film samples as indicated in the Figure. Majority PDOS (*green circles*) and minority PDOS (*red bullets*) are shown on a positive scale and negative scale, respectively. Full *lines* indicate deconvolved data using a Lorentzian function (0.4 eV width). *Thin black* and *gray* (*blue*) *lines* denote the majority PDOS stemming from the itinerant band with and without consideration of the core hole effect. *Shadings* denote the approximation of the itinerant band. **h** Comparison of the data shown in (**f**) and (**g**) on a magnified energy scale. (from [27]) (Color figure online)

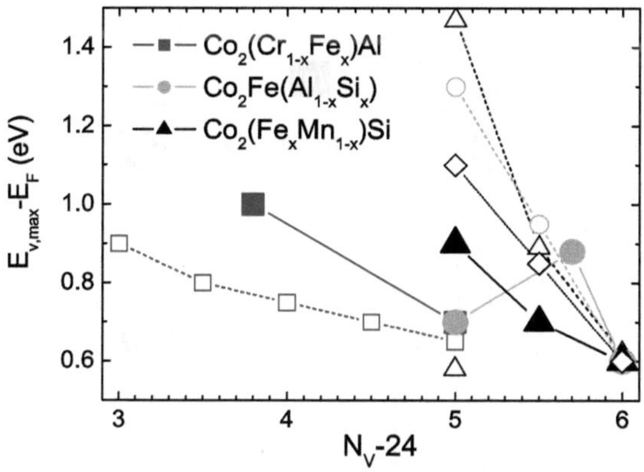

Fig. 15.11 Separation of the minority PDOS maximum and the Fermi energy $E_{v,max} - E_F$ for the indicated Heusler alloys (*full symbols*) compared with calculations (*open symbols*) using LDA (*squares* [95], *triangle* [96]), LDA+U (*triangles* [97], *circles* [90]) and LDA+DMFT (*diamonds* [98]). N_v indicates the number of valence electrons per formula unit. (from [27])

The comparison with theory allows an estimation of theoretical parameters, e.g. the electron-electron correlation potential. In addition one can observe a correlation of local order and electronic structure. For $Co_2FeAl_{1-x}Si_x$ films a broadening of the minority PDOS with increased local disorder was observed.

15.5.2 *Origin of Magnetic Anisotropies*

Magnetic anisotropies essentially determine the hysteresis behavior of ferromagnets. The understanding of magnetic anisotropies in thin films is of crucial importance for the development of magnetic devices based on new materials. While a huge amount of phenomenological data have been gathered, direct measurements of the origin of magnetic anisotropy, i.e. changes of the electronic states caused by the rotation of the magnetization vector with respect to the crystal lattice, have been reported only in a few cases [102]. Angular dependence of x-ray absorption spectra has been used to investigate its correlation with structural distortions in oxides [103]. Atomic multiplet calculations reveal the strong effect of core hole interactions in this case. In the following, it is shown that for metallic films with predominant itinerant electron states the spectroscopic information obtained by XMCD reveals information on the electronic states that are relevant for the anisotropy energy.

XMCD for Ni_2MnGa films with a typical thickness of 100 nm is measured with the photon beam aligned with a large external field [25]. In the martensitic state the averaged structure of the films is equivalent to a tetragonal distorted cubic lattice with the short axis along the surface normal. The photon flux transmitted through the film was detected via x-ray luminescence in the substrate [29, 30].

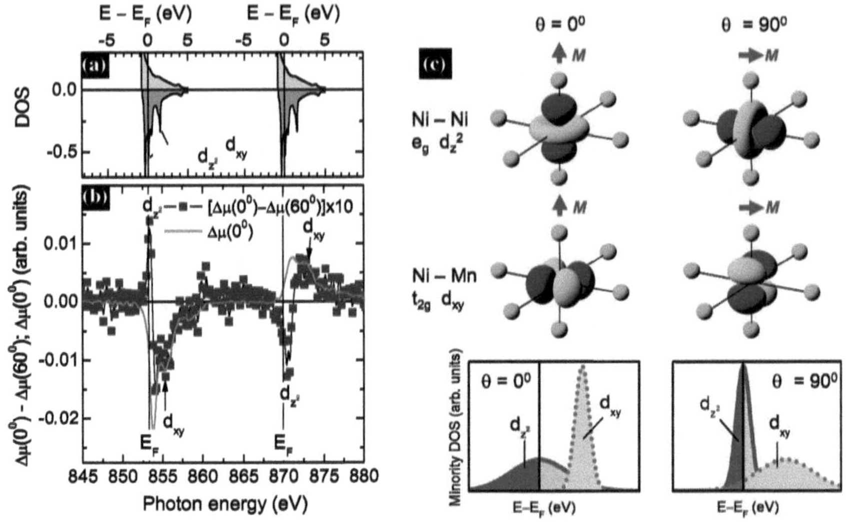

Fig. 15.12 **a** Site and spin-projected d-electron state densities for Ni (in arbitrary units) resulting from ab-initio calculations [64] shifted along the energy axis for direct comparison to the experimental data. Positive (negative) values indicate the majority (minority) density of states. **b** Difference of the XMCD signal measured at $\theta = 60°$ and normal incidence for $T < T_m$ (*dark* (*blue*) *squares*). Full (*red*) *line* shows for comparison the XMCD signal measured at normal incidence. **c** Sketch of the orbital symmetry of electron states with $d_{3z^2-r^2}$ (d_{z^2}) and d_{xy} symmetry indicating the dominating contribution to the DOS maxima for out-of-plane and in-plane magnetization direction, respectively. The two lower panels illustrate the opposite behavior of states with d_{z^2} and d_{xy} symmetry for perpendicular (*left*) and in-plane (*right*) magnetization. *Dark shaded areas* indicate occupied states. (from [104]) (Color figure online)

Spectroscopic changes in the Ni spectra with the magnetization angle are emphasized by plotting the difference of the XMCD signal at normal incidence and at 60° (Fig. 15.12b). The difference has opposite sign at the L_3 and L_2 edge, which is a direct hint that the change is caused by changes of the spin-resolved DOS rather than by orbital asymmetry. Interpreting the difference as changes in the minority density of states (DOS) at the Ni site and considering the opposite spin polarization of excited electrons at both absorption edges, a positive difference [$\Delta\mu(0°) > \Delta\mu(60°)$] at the L_3 edge (negative at the L_2 edge) thus implies a decreased minority DOS for perpendicular magnetization.

The comparison with a calculation by Ayuela et al. [64] (Fig. 15.12a) identifies the DOS maximum close to E_F with a $d_{3z^2-r^2}$ (d_{z^2}) state. The d_{z^2} state shows with its long axis parallel to the magnetization (see Fig. 15.12c). With normal magnetization the nearest neighbor atoms are close-by because of the tetragonal distortion. Hence, the hybridization is large, the bandwidth is also large and the peak is small. If the magnetization is rotated along the in-plane direction, the d_{z^2} state hybridizes less with the neighboring atoms positioned at a larger distance. Accordingly, the bandwidth is smaller and the DOS d_{z^2} peak is larger. For the minority d_{xy} state at 1.5 eV above E_F

the situation is opposite. For perpendicular magnetization the hybridization is small and the DOS peak is large. For in-plane magnetization the hybridization is larger and the DOS peak decreases.

The increased bandwidth of states with d_{z^2} symmetry in the case of $\theta = 0$ leads to a reallocation of occupied states (see Fig. 15.12) to lower kinetic energy because these states are nearly half-filled. This reallocation thus decreases the total energy of the system thus explaining the occurrence of magnetic anisotropy with an easy axis along the perpendicular (short) axis. A state with d_{z^2} symmetry does not contribute to the orbital moment because its magnetic quantum number is zero.

The decreased bandwidth of states with d_{xy} symmetry for $\theta = 0$ reflects a decrease of corresponding states below E_F. This decrease counteracts the behavior of the d_{z^2} states discussed above but this effect is much smaller as the maximum is positioned at a higher energy. Because of the non-zero magnetic quantum number of the d_{xy} states the reduction of occupied states causes a reduction of the orbital magnetic moment. As the spin-orbit interaction favors parallel orbital and spin moments the reduction means a reduction of minority orbital moment, i.e. in fact an increase of the total orbital moment in agreement with the experimental result.

The change of the electron density-of-states in Ni_2MnGa upon rotation of the magnetization vector with respect to the crystal lattice illustrates that the bulk magnetocrystalline anisotropy in Ni_2MnGa is caused by a reallocation of electron states with $3d_{z^2}$ symmetry located predominantly at the Ni atom. On the other hand the Ni orbital moment anisotropy is due to a varying occupation of electron states with $3d_{xy}$ symmetry.

XMCD provides a unique insight to the physics of magnetic anisotropy. Many spectroscopic methods use electrons as a probe (ARPES) or exciting particle (inverse PES). In this case it is not possible to apply the large magnetic field that is necessary to align the magnetization along the hard magnetic axis.

15.5.3 Electronic Correlation Effects

The temperature dependence of the electronic states in half-metals might contribute to the strong decrease of the spin polarization and its experimental investigation is highly interesting. In a single-particle model changes of the electronic structure are only expected in the vicinity of the Fermi level due to thermally excited occupation of previously unoccupied states which is in agreement with a high energy photoemission study for a half-metallic Co_2MnSi/MgO interface [105]. Instead, a model beyond the single-particle picture of the local density approximation considering finite temperature many-body effects [89, 106] results in a considerable temperature dependence of the DOS.

In the following it is shown how XMCD can be used to extract the temperature dependence of the spin-resolved PDOS of epitaxial Mn_2VAl films [107], that were predicted to be half-metallic comprising a band gap in the majority spin band (Mn_2VAl) [108].

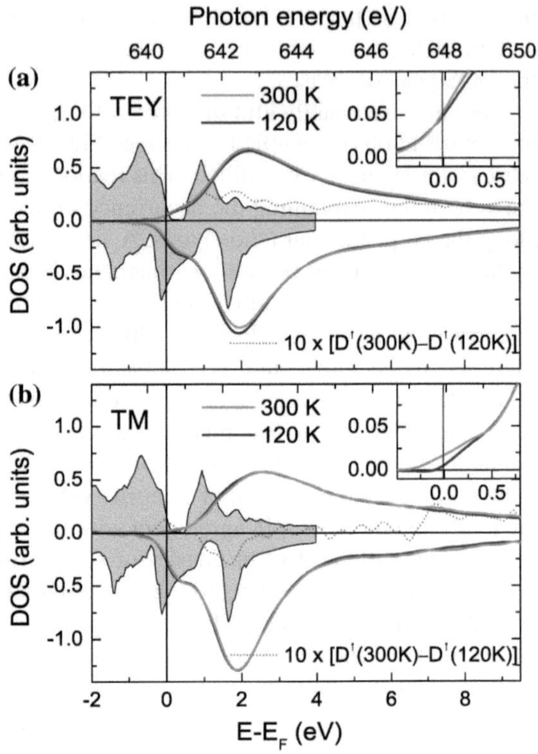

Fig. 15.13 **a** Spin-resolved unoccupied PDOS calculated from the XAS/XMCD TEY data measured at the Mn L_3-edge of the Mn_2VAl film. The majority and minority PDOS is shown on a positive and negative scale, respectively. The difference of the low and high temperature majority PDOS is ten-fold enhanced, emphasizing the temperature induced spectral changes near the majority band gap. The theoretical data from [106] are shown as *shaded areas*. The *inset* shows the magnified energy region around E_F for majority states. **b** Spin-resolved PDOS derived from the transmission data. (from [107])

The result for the Mn related PDOS in Mn_2VAl is shown in Fig. 15.13a, b and compared to theoretical data. The minority states clearly reproduce the ab-initio calculation considering the life-time broadening of the spectral features. The large maximum at $E - E_F$=1.9 eV corresponds to the peak in the theoretical data at 1.7 eV. The pronounced shoulder closer to E_F is the remainder of the maximum in the theoretical data just below E_F. The pronounced maximum of the majority states observed at 2.2 eV above E_F (see Fig. 15.13a, b) is at variance with theory. This might be explained by static electron correlation effects that are not properly considered by the evaluation procedure or by the theoretical calculation.

The majority state data are different for surface and bulk sensitive measurements. While the TEY data (Fig. 15.13a) reveals a pronounced non-zero PDOS at E_F, the transmission data (Fig. 15.13b) is almost zero at E_F for low temperatures and shows a small positive value at room temperature. The difference $D^{\uparrow}(300K) - D^{\uparrow}(120K)$ has clearly a maximum below E_F. This temperature dependence can not be explained by a thermal broadening of the Fermi function, which is much smaller at 300 K and would show up antisymetrically below and above E_F. A comparable temperature dependence is clearly absent for the case of the TEY related data. This fact coincides with the strong decrease of the Mn and V interface magnetic moments by a factor of two. One might speculate that the reason is related to the effect observed in Sect. 15.3

for Ni_2MnGa. Films deposited at room temperature without annealing showed a strongly reduced atomic and magnetic order in the bulk of the films. Therefore, the result suggests that the interface region is metallic instead of half-metallic.

The striking result of an increase of spectral weight just below the Fermi level for the case of the TM data of Mn_2VAl with a majority band gap is in qualitative agreement with the model of non-quasiparticle (NQP) states, calculated for real systems using local density approximation plus dynamical mean-field theory (LDA+DMFT) [106].

For epitaxial Co_2FeAl films comprising a minority spin band [95], instead, an unexpected change of the PDOS was observed, too [107]. In this case the minority PDOS increases close to E_F with increasing temperature. The difference for low and high temperature reveals an extra peak slightly above E_F (instead of below in the case of Mn_2VAl) again as predicted by the non-quasiparticle model [89].

The observed temperature induced changes of the spin-resolved DOS of two Heusler compounds can not be explained by thermal excitations within a single-particle model. The increase of the DOS within the band gap possibly contributes to the observed strong temperature dependence of the TMR effect.

15.6 Summary and Outlook

In the past decade XMCD has gained a wealth of information on epitaxial Heusler films. Thereby, XMCD represents a unique experimental method because it provides information on element-specific and quantitative magnetic properties that cannot be measured by any other means. This chapter emphasizes the most important experimental developments and evaluation methods explaining representative examples.

Using the sum rule analysis XMCD reveals element-specific spin and orbital magnetic moments. Bulk and interface properties of epitaxial films are determined independently evaluating both the transmitted x-ray intensity via the substrate luminescence and the total electron yield via the sample current. Combined with an analysis of a thickness series, interface properties at both interfaces can be derived separately. Beyond the magnetic moments derived by integration, a thorough analysis of the absorption spectra allows conclusions on the local atomic order, translational symmetries and composition. Subtle changes of the electronic structure caused by a solid state phase transition in the Ni-Mn-Ga family of shape memory materials lead to easily detectable features in the spectra.

The pulsed x-ray intensity provided from synchrotron light sources can be used to detect dynamical properties in the frequency range of several GHz limited only by the length of the x-ray pulses. Future developments of laser-based x-ray light sources promise an increase of the bandwidth in the future.

A calculation scheme is presented that allows for an extraction of the spin-resolved density of states from the measured spectra thus allowing for easy comparison to theoretical results. Using this scheme on results for composition series with partly replaced elements on the different lattice sites confirm the theoretical prediction of tailoring the band structure of Heusler alloys for specific applications. In particular for half-metallic compounds the position of the Fermi edge relative to the band gap can be tuned by the number of valence electrons. The experimental results unfold the role of electron correlation effects compelling many-particle models for theoretical description. Temperature dependent measurements confirm the existence of non-quasiparticle states that were predicted to occur within the half-metallic band gap. Spectral changes due to a rotation of the magnetization vector with respect to the crystal lattice reveal the microscopic origin of magnetic anisotropies in Heusler alloys.

Epitaxial Heusler alloy films have turned out to provide promising properties in the field of spintronics and shape memory alloys. Emerging fields of new application for Heusler alloys that have just started to be explored include topological insulators and magneto-thermal devices. The currently rapidly increasing number of beamlines worldwide will certainly enlarge the scientific community working in the field of X-ray techniques discovering new properties of Heusler alloys. This chapter has focused exclusively on XMCD used to explore and help to improve these properties. Besides XMCD, magnetic linear dichroism (XMLD) turns out to reveal interesting details of the electronic structure. In addition to x-ray absorption, diffraction and microscopic methods reveal further important aspects. Further technical developments exploit ultrashort x-ray pulses enabling understanding of ultrafast dynamical properties.

Acknowledgments The author would like to thank Stefan Cramm and Ingo Krug for many useful discussions and support at the beamline UE56/1-SGM at the German Light source BESSY II. Many results presented here have been developed in the PhD theses of Michael Kallmayer and Peter Klaer. The presented work was partly supported by the Deutsche Forschungsgemeinschaft (EL-172/12-1-3) and the BMBF (ES3XBA/5).

References

1. I. Galanakis, P.H. Dederichs, N. Papanikolaou, Slater-pauling behavior and origin of the half-metallicity of the full-heusler alloys. Phys. Rev. B **66**, 174429 (2002)
2. K. Inomata, S. Okamura, R. Goto, N. Tezuka, Large tunneling magnetoresistance at room temperature using a heusler alloy with the b2 structure. Jpn. J. Appl. Phys. **42**, L419 (2003)
3. S. Kämmerer, A. Thomas, A. Hütten, G. Reiss, Co2mnsi heusler alloy as magnetic electrodes in magnetic tunnel junctions. Appl. Phys. Lett. **85**, 79 (2004)
4. C. Felser, G.H. Fecher, B. Balke, Spintronics: a challenge for materials science and solid-state chemistry. Angew. Chem. Int. Ed. **46**, 668 (2007)
5. Y. Sakuraba, M. Hattori, M. Oogane, Y. Ando, H. Kato, A. Sakuma, T. Miyazaki, H. Kubota, Giant tunneling magnetoresistance in co_2mnsi/al-o/co_2mnsi magnetic tunnel junctions. Appl. Phys. Lett. **88**, 192508 (2006)
6. W. Wang, H. Sukegawa, R. Shan, K. Inomata, Fabrication of fully epitaxial magnetic tunnel junctions using $l2_1$-ordered $co_2feal_{0.5}si_{0.5}$ electrodes and their tunneling magnetoresistance characteristics. Appl. Phys. Lett. **93**, 122506 (2008)
7. S. Tsunegi, Y. Sakuraba, M. Oogane, K. Takanashi, Y. Ando, Large tunnel magnetoresistance in magnetic tunnel junctions using a co_2mnsi heusler alloy electrode and a mgo barrier. Appl. Phys. Lett. **93**, 112506 (2008)

8. N. Tezuka, N. Ikeda, F. Mitsuhashi, S. Sugimoto, Improved tunnel magnetoresistance of magnetic tunnel junctions with heusler $co_2fe al_{0.5} si_{0.5}$ electrodes fabricated by molecular beam epitaxy. Appl. Phys. Lett. **94**, 162504 (2009)
9. T. Taira, T. Ishikawa, N. Itabashi, K.-I. Matsuda, T. Uemura, M. Yamamoto, Spin-dependent tunnelling characteristics of fully epitaxial magnetic tunnel junctions with a heusler alloy co_2mnge thin film and a mgo barrier. J. Phys. D **42**, 084015 (2009)
10. N. Tezuka, N. Ikeda, S. Sugimoto, K. Inomata, Giant tunnel magnetoresistance at room temperature for junctions using full-heusler $co_2fe al_{0.5} si_{0.5}$ electrodes. Jpn. J. Appl. Phys. **46**, L454 (2007)
11. K. Ullakko, J.K. Huang, C. Kantner, R.C. OHandley, VV. Kokorin, Large magnetic-field-induced strains in Ni2MnGa single crystals. Appl. Phys. Lett. **69**(13), 1966–1968 (1996)
12. T. Eichhorn, R. Hausmanns, P. Klaer, M. Kallmayer, H.J. Elmers, G. Jakob, Structure and microscopic magnetism of epitaxial Ni-Mn-Ga films. Adv. Eng. Mater. **14**(8), 687–695 (2012)
13. M. Kohl, S.R. Yeduru, F. Khelfaoui, B. Krevet, A. Backen, S. Fahler, T. Eichhorn, G. Jakob, A. Mecklenburg, Recent progress in FSMA microactuator developments. Mater. Sci. Forum **635**, 145 (2010)
14. T. Graf, S.S.P. Parkin, C. Felser, Heusler Compounds-A material class with exceptional properties. IEEE Trans. Magn. **47**(2, Part 2), 367–373 (2011)
15. M. Meinert, J. Schmalhorst, M. Glas, G. Reiss, E. Arenholz, T. Boehnert, K. Nielsch, Insights into the electronic structure of Co2FeSi from x-ray magnetic linear dichroism. Phys. Rev. B **86**(5), 054420 (2012)
16. S. Chadov, X. Qi, J. Kuebler, G.H. Fecher, C. Felser, S.C. Zhang, Tunable multifunctional topological insulators in ternary Heusler compounds. Nat. Mat. **9**(7), 541–545 (2010)
17. J. Barth, G.H. Fecher, B. Balke, S. Ouardi, T. Graf, C. Felser, A. Shkabko, A. Weidenkaff, P. Klaer, H.J. Elmers, H. Yoshikawa, S. Ueda, K. Kobayashi, Itinerant half-metallic ferromagnets co_2tiz (z=si, ge, sn): Ab initio calculations and measurement of the electronic structure and transport properties. Phys. Rev. B **81**, 064404 (2010)
18. J. Stöhr, X-ray magnetic circular dichroism spectroscopy of transition metal thin films. J. Electron Spectrosc. Relat. Phenom. **75**, 253 (1995)
19. J. Stöhr, Exploring the microscopic origin of magnetic anisotropies with x-ray magnetic circular dichroism (XMCD) spectroscopy. J. Magn. Magn. Mater. **200**, 470 (1999)
20. Y. Miura, K. Nagao, M. Shirai, Atomic disorder effects on half-metallicity of the full-heusler alloys co2(cr1-xfex)al: a first-principles. Phys. Rev. B **69**, 144413 (2004)
21. H. Ebert, Magneto-optical effects in transition metal systems. Rep. Prog. Phys. **59**, 1665 (1996)
22. V.N. Antonov, A.P. Shapak, A.N. Yaresko, X-ray magnetic circular dichroism in d and f ferromagnetic materials: recent theoretical progress. Part i (review articlce). Low Temp. Phys. **34**, 1 (2008)
23. F.M.F. de Groot, J.C. Fuggle, B.T. Thole, G.A. Sawatzky, 2p x-ray absorption of 3d transition-metal compounds—an atomic multiplet description including the crystal field. Phys. Rev. B **42**, 5459 (1990)
24. N.D. Telling, P.S. Keatley, G. van der Laan, R.J. Hicken, E. Arenholz, Y. Sakuraba, M. Oogane, Y. Ando, K. Takanashi, A. Sakuma, T. Miyazaki, Evidence of local moment formation in co-based heusler alloys. Phys. Rev. B **78**, 184438 (2008)
25. G. Jakob, T. Eichhorn, M. Kallmayer, H.J. Elmers, Correlation of electronic structure and martensitic transition in epitaxial ni_2mnga films. Phys. Rev. B **76**, 174407 (2007)
26. M. Kallmayer, K. Hild, T. Eichhorn, H. Schneider, G. Jakob, A. Conca, M. Jourdan, H.J. Elmers, Solid state reaction at the interface between heusler alloys and Al cap accelerated by elevated temperature and rough surface. Appl. Phys. Lett. **91**, 192501 (2007)
27. M. Kallmayer, P. Klaer, H. Schneider, E. Arbelo-Jorge, C. Herbort, G. Jakob, M. Jourdan, H.J. Elmers, Spin-resolved unoccupied density of states in epitaxial heusler-alloy films. Phys. Rev. B **80**, 020406(R) (2009)
28. P. Klaer, M. Kallmayer, C.G.F. Blum, T. Graf, J. Barth, B. Balke, G.H. Fecher, C. Felser, H.J. Elmers, Tailoring the electronic structure of half-metallic heusler alloys. Phys. Rev. B **80**, 144405 (2009)

29. M. Kallmayer, H. Schneider, G. Jakob, H.J. Elmers, K. Kroth, H.C. Kandpal, U. Stumm, S. Cramm, Reduction of surface magnetization of co_2cr_0.6fe0.4al heusler alloy films. Appl. Phys. Lett. **88**, 072506 (2006)
30. M. Kallmayer, H. Schneider, G. Jakob, H.J. Elmers, B. Balke, S. Cramm, Interface magnetization of ultrathin epitaxial co_2fesi films. J. Phys. D **40**, 1552 (2007)
31. A. Bergmann, J. Grabis, A. Nefedov, K. Westerholt, H. Zabel, X-ray resonant magnetic scattering study of [Co2MnGe/Au](n) and [Co2MnGe/V](n) multilayers. J. Phys. D-Appl. Phys. **39**(5), 842–850 (2006)
32. J. Grabis, A. Bergmann, A. Nefedov, K. Westerholt, H. Zabel, Element-specific characterization of the interface magnetism in [Co2MnGe/Au](n) multilayers by x-ray resonant magnetic scattering. Phys. Rev. B **72**(2), 024438 (2005)
33. M. Kallmayer, Rntgenabsorptionsspektroskopie und Zirkulardichroismus von Heusler—Verbindungen. Ph.D. thesis, Johannes Gutenberg—Universitt Mainz (2010)
34. C. Antoniak, H.C. Herper, Y.N. Zhang, A. Warland, T. Kachel, F. Stromberg, B. Krumme, C. Weis, K. Fauth, W. Keune, P. Entel, R.Q. Wu, J. Lindner, H. Wende. Induced magnetism on silicon in Fe3Si quasi-Heusler compound. Phys. Rev. B **85**(21), 214432 (2012)
35. M. Jourdan, E. Arbelo Jorge, C. Herbort, M. Kallmayer, P. Klaer, H.J. Elmers, Interface and bulk magnetism of $co_2cr_{0.6}fe_{0.4}$al and co_2feal thin films. Appl. Phys. Lett. **95**, 172504 (2009)
36. J. Hamrle, O. Gaier, S. Trudel, B. Hillebrands, P. Klaer, H.J. Elmers, Y. Ando. Magnetic and electronic properties of co_2feal films. Unpublished
37. E. Vilanova, G. Jakob, P. Klaer, M. Kallmayer, and H. J. Elmers. Magnetic properties of epitaxial co_2mnal films. *unpublished*, -:–, -
38. E. Arbelo-Jorge, M. Jourdan, M. Kallmayer, P. Klaer, H.J. Elmers, Magnetic and structural properties of $co_2feal_{1-x}si_x$ thin films. J. Phys. Conf. Ser. **200**, 072006 (2010)
39. S. Trudel, G. Wolf, J. Hamrle, B. Hillebrands, P. Klaer, M. Kallmayer, H.J. Elmers, H. Sukegawa, W. Wang, K. Inomata. Effect of annealing on Co2FeAl0.5Si0.5 thin films: a magneto-optical and x-ray absorption study. Phys. Rev. B **83**(10), 104412 (2011)
40. M. Kolbe, S. Chadov, E. Arbelo Jorge, G. Schoenhense, C. Felser, H.J. Elmers, M. Klaeui, M. Jourdan, Test of band structure calculations for Heusler compounds by spin-resolved photoemission spectroscopy. Phys. Rev. B **86**(2), 024422 (2012)
41. N.D. Telling, P.S. Keatley, G. van der Laan, R.J. Hicken, E. Arenholz, Y. Sakuraba, M. Oogane, Y. Ando, T. Miyazaki, Interfacial structure and half-metallic ferromagnetism in co2mnsi-based magnetic tunnel junctions. Phys. Rev. B **74**, 224439 (2006)
42. H.J. Elmers, G.H. Fecher, D. Valdaitsev, S.A. Nepijko, A. Gloskovskii, G. Jakob, G. Schönhense, S. Wurmehl, T. Block, C. Felser, P.C. Hsu, W.L. Tsai, S. Cramm, Element-specific magnetic moments from core-absorption magnetic circular dichroism of the doped heusler alloy $co_2cr_{0.6}fe_{0.4}$al. Phys. Rev. B **67**, 104412 (2003)
43. V.R. Singh, V.K. Verma, K. Ishigami, G. Shibata, T. Kadono, A. Fujimori, D. Asakura, T. Koide, Y. Miura, M. Shirai, G.F. Li, T. Taira, M. Yamamoto, Effects of off-stoichiometry on the spin polarization at the Co2Mn beta Ge0.38/MgO interfaces: x-ray magnetic circular dichroism study. Phys. Rev. B **86**(14), 144412 (2012)
44. H.C. Herper, B. Krumme, D. Ebke, C. Antoniak, C. Weis, A. Warland, A. Huetten, H. Wende, P. Entel, Co2+xFe1+xSi/MgO(001) Heusler alloys: influence of off-stoichiometry and lattice distortion on the magnetic properties in bulk and on MgO(001). J. Appl. Phys. **109**(7), 07E128 (2011)
45. P. Poersch, M. Kallmayer, T. Eichhorn, G. Jakob, H. J. Elmers, C.A. Jenkins, C. Felser, R. Ramesh, M. Huth, Suppression of martensitic phase transition at the Ni(2)MnGa film surface. Appl. Phys. Lett. **93**(2), 022501 (2008)
46. V.V. Khovailo, V. Novosad, T. Takagi, D.A. Filippov, R.Z. Levitin, A.N. Vasil'ev, Magnetic properties and magnetostructural phase transitions in Ni2+xMn1-xGa shape memory alloys. Phys. Rev. B **70**(17), 174413 (2004)
47. J. Enkovaara, O. Heczko, A. Ayuela, R.M. Nieminen, Coexistence of ferromagnetic and antiferromagnetic order in Mn-doped Ni2MnGa. Phys. Rev. B **67**(21), 212405 (2003)

48. F. Albertini, L. Pareti, A. Paoluzi, L. Morellon, P.A. Algarabel, M.R. Ibarra, L. Righi, Composition and temperature dependence of the magnetocrystalline anisotropy in Ni2+xMn1+yGa1+z (x+y+z=0) Heusler alloys. Appl. Phys. Lett. **81**(21), 4032–4034 (2002)
49. P. Entel, V.D. Buchelnikov, V.V. Khovailo, A.T. Zayak, W.A. Adeagbo, M.E. Gruner, H.C. Herper, E.F. Wassermann, Modelling the phase diagram of magnetic shape memory Heusler alloys. J. Phys. D-Appl. Phys. **39**(5), 865–889 (2006)
50. C. Herbort, E. Arbelo, M. Jourdan, Preparation of CCFA films. J. Phys. D **42**, 084006 (2009)
51. S. Wurmehl, G.H. Fecher, K. Kroth, F. Kronast, H.A. Dürr, Y. Takeda, Y. Saitoh, K. Kobayashi, H.-J. Lin, G. Schönhense, C. Felser, Electronic structure and spectroscopy of the quaternary heusler alloy $co_2cr_{1-x}fe_xal$. J. Phys. D **39**, 803 (2006)
52. K. Inomata, S. Okamura, A. Miyazaki, M. Kikuchi, N. Tezuka, M. Wojcik, E. Jedryka, Structural and magnetic properties and tunnel magnetoresistance for Co-2(Cr,Fe)Al and Co2FeSi full-Heusler alloys. J. Phys. D-Appl. Phys. **39**(5), 816–823 (2006)
53. B. Krumme, H.C. Herper, D. Erb, C. Weis, C. Antoniak, A. Warland, K. Westerholt, P. Entel, H. Wende, Induced magnetic Cu moments and magnetic ordering in Cu2MnAl thin films on MgO(0 0 1) observed by XMCD. J. Phys. D-Appl. Phys. **44**(41), 415004 (2011)
54. D. Erb, G. Nowak, K. Westerholt, H. Zabel, Thin films of the Heusler alloys Cu2MnAl and Co2MnSi: recovery of ferromagnetism via solid-state crystallization from the x-ray amorphous state. J. Phys. D-Appl. Phys. **43**(28), 285001 (2010)
55. R. Shan, H. Sukegawa, W.H. Wang, M. Kodzuka, T. Furubayashi, T. Ohkubo, S. Mitani, K. Inomata, K. Hono, Demonstration of half-metallicity in fermi-level-tuned heusler alloy co[sub 2]feal[sub 0.5]si[sub 0.5] at room temperature. Phys. Rev. Lett. **102**, 246601 (2009)
56. N. Tezuka, N. Ikeda, S. Sugimoto, K. Inomata, 175% tunnel magnetoresistance at room temperature and high thermal stability using co[sub 2]feal[sub 0.5]si[sub 0.5] full-heusler alloy electrodes. Appl. Phys. Lett. **89**(25), 252508 (2006)
57. K. Inomata, M. Wojcik, E. Jedryka, N. Ikeda, N. Tezuka, Site disorder in $co_2 fe(al, si)$ heusler alloys and its influence on junction tunnel magnetoresistance. Phys. Rev. B **77**(21), 214425 (2008). Jun
58. N. Tezuka, N. Ikeda, A. Miyazaki, S. Sugimoto, M. Kikuchi, K. Inomata, Tunnel magnetoresistance for junctions with epitaxial full-heusler co[sub 2]feal[sub 0.5]si[sub 0.5] electrodes with b2 and l2[sub 1] structures. Appl. Phys. Lett. **89**, 112514 (2006)
59. H. Sukegawa, W. Wang, R. Shan, T. Nakatani, K. Inomata, K. Hono, Spin-polarized tunneling spectroscopy of fully epitaxial magnetic tunnel junctions using $co2feal0.5si0.5$ heusler alloy electrodes. Phys. Rev. B **79**(18), 184418 (2009)
60. W. Wang, H. Sukegawa, R. Shan, T. Furubayashi, K. Inomata, Preparation and characterization of highly l2(1)-ordered full-heusler alloy $co_2feal_{0.5}si_{0.5}$ thin films for spintronics device applications. Appl. Phys. Lett. **92**, 221912 (2008)
61. O. Gaier, J. Hamrle, B. Hillebrands, M. Kallmayer, P. Poersch, G. Schoenhense, H.J. Elmers, J. Fassbender, A. Gloskovskii, C.A. Jenkins, C. Felser, E. Ikenaga, Y. Sakuraba, S. Tsunegi, M. Oogane, Y. Ando, Improvement of structural, electronic, and magnetic properties of Co2MnSi thin films by He+ irradiation. Appl. Phys. Lett. **94**(15), 152508 (2009)
62. V. Ksenofontov, C. Herbort, M. Jourdan, C. Felser, Conversion electron Mossbauer spectroscopy of epitaxial Co2Cr0.6Fe0.4Al thin films. Appl. Phys. Lett. **92**, 262501 (2008)
63. A. Sozinov, A.A. Likhachev, N. Lanska, K. Ullakko, Giant magnetic-field-induced strain in NiMnGa seven-layered martensitic phase. Appl. Phys. Lett. **80**(10), 1746–1748 (2002)
64. A. Ayuela, J. Enkovaara, RM. Nieminen, Ab initio study of tetragonal variants in Ni2MnGa alloy. J. Phys.-Cond. Matt. **14**(21), 5325–5336 (2002)
65. M. Kallmayer, P. Poersch, T. Eichhorn, H. Schneider, C.A. Jenkins, G. Jakob, H.J. Elmers, Compositional dependence of element-specific magnetic moments in Ni2MnGa films. J. Phys. D-Appl. Phys. **42**(8), 084008 (2009)
66. B.R. Gautam, I. Dubenko, J.C. Mabon, S. Stadler, N. Ali, Effect of small changes in Mn concentration on phase transition temperatures and magnetic entropy variations in Ni2Mn0.75Cu0.25Ga Heusler alloys. J. Alloys Compd. **472**(1–2), 35–39 (2009)

67. V.A. Chernenko, E. Cesari, V.V. Kokorin, I.N. Vitenko, The development of new ferromagnetic shape-memory alloys in Ni-Mn-Ga system. Scripta Metall. Mater. **33**(8), 1239–1244 (1995)
68. C.B. Jiang, G. Feng, S.K. Gong, H.B. Xu, Effect of Ni excess on phase transformation temperatures of NiMnGa alloys. Mater. Sci. Eng. A **342**(1–2), 231–235 (2003)
69. V.V. Khovailo, T. Abe, V.V. Koledov, M. Matsumoto, H. Nakamura, R. Note, M. Ohtsuka, V.G. Shavrov, T. Takagi, Influence of Fe and Co on phase transitions in Ni-Mn-Ga alloys. Mater. Trans. **44**(12), 2509–2512 (2003)
70. D. Steil, S. Alebrand, T. Roth, M. Krau, T. Kubota, M. Oogane, Y. Ando, H.C. Schneider, M. Aeschlimann, M. Cinchetti, Band-structure-dependent demagnetization in the heusler alloy co2mn1-xfexsi. Phys. Rev. Lett. **105**, 217202 (2010)
71. G.M. Müller, J. Walowski, M. Djordjevic, G.X. Miao, A. Gupta, A.V. Ramos, K. Gehrke, V. Moshnyaga, K. Samwer, J. Schmalhorst, A. Thomas, A. Hütten, G. Reiss, J.S. Moodera, M. Münzenberg, Spin polarization in half-metals probed by femtosecond spin excitation. Nat. Mater. **8**, 56 (2009)
72. C. Liu, K.A. Claudia, K.A. Mewes, C. Mairbek, T. Mewes, W.H. Butler. Origin of low gilbert damping in half metals. Appl. Phys. Lett. **95**, 022509 (2009)
73. B. Heinrich, G. Woltersdorf, R. Urban, O. Mosendz, G. Schmidt, P. Bach, L. Molenkamp, E. Rozenberg, Magnetic properties of nimnsb(001) films grown on ingaas/inp(001). J. Appl. Phys. **95**, 7462 (2004)
74. S. Mizukami, D. Watanabe, M. Oogane, Y. Ando, Y. Miura, M. Shirai, T.J. Miyazaki, Low damping constant for co2feal heusler alloy films and its correlation with density of states. J. Appl. Phys. **105**, 07D306 (2009)
75. T. Kubota, S. Tsunegi, M. Oogane, S. Mizukami, T. Miyazaki, H. Naganuma, Y. Ando, Half-metallicity and gilbert damping constant in co2fexmn1-xsi heusler alloys depending on the film composition. Appl. Phys. Lett. **94**, 122504 (2009)
76. M. Oogane, T. Kubota, Y. Kota, S. Mizukami, H. Naganuma, A. Sakuma, Y. Ando, Gilbert damping constant of epitaxially grown co-based heusler alloy thin films. Appl. Phys. Lett. **96**, 252501 (2010)
77. M. Lezaic, Ph Mavropoulos, J. Enkovaara, G. Bihlmayer, S. Blügel, Thermal collapse of spin polarization in half-metallic ferromagnets. Phys. Rev. Lett. **97**, 026404 (2006)
78. G. Boero, S. Mouaziz, S. Rusponi, P. Bencok, F. Nolting, S. Stepanow, P. Gambardella, Element-resolved x-ray ferrimagnetic and ferromagnetic resonance spectroscopy. New J. Phys. **10**, 013011 (2008)
79. G. Boero, S. Rusponi, P. Bencok, R. Meckenstock, J.U. Thiele, F. Nolting, P. Gambardella, Double-resonant x-ray and microwave absorption: atomic spectroscopy of precessional orbital and spin dynamics. Phys. Rev. B **79**, 224425 (2009)
80. H. Stoll, A. Puzic, B.V. Waeyenberge, P. Fischer, J. Raabe, M. Buess, T. Haug, R. Höllinger, C.H. Back, D. Weiss, G. Denbeau, High resolution imaging of fast magnetization dynamics in magnetic nanostructures. Appl. Phys. Lett. **84**, 3328 (2004)
81. W.E. Bailey, L. Cheng, E. Vescovo, C.-C. Kao, D.A. Arena, Precessional dynamics of elemental moments in a ferromagnetic alloy. Phys. Rev. B **70**, 172403 (2004)
82. D.A. Arena, E. Vescovo, C.-C. Kao, Y. Guan, W.E. Bailey, Weakly coupled motion of individual layers in ferromagnetic resonance. Phys. Rev. B **74**, 064409 (2006)
83. D.A. Arena, E. Vescovo, C.-C. Kao, Y. Guan, W.E. Bailey, Combined time-resolved x-ray magnetic circular dichroism and ferromagnetic resonance studies of magnetic alloys and multilayers (invited). J. Appl. Phys. **101**, 09C109 (2007)
84. T. Martin, G. Woltersdorf, C. Stamm, H.A. Dürr, R. Mattheis, C.H. Back, G. Bayreuther, Layer resolved magnetization dynamics in interlayer exchange coupled ni81fe19/ru/co90fe10 by time resolved x-ray magnetic circular dichroism. J. Appl. Phys. **103**, 07B112 (2008)
85. T. Martin, G. Woltersdorf, C. Stamm, H.A. Dürr, R. Mattheis, C.H. Back, G. Bayreuther, Layer resolved magnetization dynamics in coupled magnetic films using time-resolved x-ray magnetic circular dichroism with continuous wave excitation. J. Appl. Phys. **105**, 07D310 (2009)

86. P. Klaer, F. Hoffmann, G. Woltersdorf, E. Arbelo Jorge, M. Jourdan, C.H. Back, H.J. Elmers, Element-specific ferromagnetic resonance in epitaxial Heusler spin valve systems. J. Phys. D-Appl. Phys. **44**(42), 425004 (2011)
87. R. Urban, G. Woltersdorf, B. Heinrich, Gilbert damping in single and multilayer ultrathin films: role of interfaces in nonlocal spin dynamics. Phys. Rev. Lett. **87**, 217204 (2001)
88. B. Heinrich, Y. Tserkovnyak, G. Woltersdorf, A. Brataas, R. Urban, G.E.W. Bauer, Dynamic exchange coupling in magnetic bilayers. Phys. Rev. Lett. **90**, 187601 (2003)
89. L. Chioncel, Y. Sakuraba, E. Arrigoni, M.I. Katsnelson, M. Oogane, Y. Ando, T. Miyazaki, E. Burzo, A.I. Lichtenstein, Nonquasiparticle states in co_2mnsi evidenced through magnetic tunnel junction spectroscopy measurements. Phys. Rev. Lett. **100**, 086402 (2008)
90. G.H. Fecher, C. Felser, Substituting the main group element in cobalt-iron based heusler alloys: $Co_2feal_{1-x}si_x$. J. Phys. D **40**, 1582 (2007)
91. Z. Gercsi, K. Hono, Ab initio predictions for the effect of disorder and quarternary alloying on the half-metallic properties of the selected co2fe-based heusler alloys. J. Phys. Condens. Matter **19**, 326216 (2007)
92. W.H. Wang, M. Przybylski, W. Kuch, L.I. Chelaru, J. Wang, Y.F. Lu, J.B.H.L. Meyerheim, J. Kirschner, Magnetic properties and spin polarization of co_2mnsi heusler alloy thin films epitaxially grown on gaas(001). Phys. Rev. B **71**, 144416 (2005)
93. H. Schneider, G. Jakob, M. Kallmayer, H.J. Elmers, M. Cinchetti, B. Balke, S. Wurmehl, C. Felser, M. Aeschlimann, H. Adrian, Epitaxial film growth and magnetic properties of co2fesi. Phys. Rev. B **74**, 174426 (2006)
94. M. Jourdan, J. Minar, J. Braun, A. Kronenberg, S. Chadov, B. Balke, A. Gloskovskii, M. Kolbe, H.J. Elmers, G. Schonhense, H. Ebert, C. Felser, M. Klaui, Direct observation of half-metallicity in the heusler compound co_2mnsi. Nat. Commun. **5**, 3974 (2014)
95. G.H. Fecher, H.C. Kandpal, S. Wurmehl, J. Morais, H.-J. Lin, H.J. Elmers, G. Schönhense, C. Felser, Design of magnetic materials: the electronic structure of the ordered, doped heusler compound $co_2cr_{1-x}fe_x$al. J. Phys.: Condens. Matter **17**, 7237 (2005)
96. K. Özdogan, B. Aktas, I. Galanakis, E. Sasioglu, Influence of mixing the low-valent transition metal atoms (y, y* = cr, mn, fe) on the properties of the quaternary $co_2[y_{1-x}y_x{}^*]z$ (z = al, ga, si, ge, or sn) heusler compounds. J. Appl. Phys. **101**, 073910 (2007)
97. B. Balke, G.H. Fecher, H.C. Kandpal, C. Felser, Properties of the quaternary half-metal-type heusler alloy $co_2mn_{1-x}fe_x$si. Phys. Rev. B **74**, 104405 (2006)
98. S. Chadov, G.H. Fecher, C. Felser, J. Minar, J. Braun, H. Ebert, Electron correlations in $co_2mn_{1-x}fe_x$si heusler compounds. J. Phys. D **42**, 084002 (2009)
99. P. Klaer, B. Balke, V. Alijani, J. Winterlik, G.H. Fecher, C. Felser, H.J. Elmers, Element-specific magnetic moments and spin-resolved density of states in CoFeMnZ (Z = Al, Ga; Si, Ge). Phys. Rev. B **84**(14), 144413 (2011)
100. M. Meinert, J. Schmalhorst, H. Wulfmeier, G. Reiss, E. Arenholz, T. Graf, C. Felser, Electronic structure of fully epitaxial Co2TiSn thin films. Phys. Rev. B **83**(6), 064412 (2011)
101. H.C. Kandpal, G.H. Fecher, C. Felser, Correlation in the transition-metal-based heusler compounds co_2mnsi and co_2fesi. Phys. Rev. B **73**, 094422 (2006)
102. D. Weller, J. Stohr, R. Nakajima, A. Carl, M.G. Samanat, C. Chappert, R. Megy, P. Beauvillain, P. Veillet, G.A. Held, Microscopic origin of magnetic anisotropy in Au/Co/Au probed with x-ray magnetic circular dichroism. Phys. Rev. Lett. **75**(20), 3752–3755 (1995)
103. G. van der Laan, R.V. Chopdekar, Y. Suzuki, E. Arenholz, Strain-induced changes in the electronic structure of MnCr2O4 thin films probed by x-ray magnetic circular dichroism. Phys. Rev. Lett. **105**(6), 067405 (2010)
104. P. Klaer, T. Eichhorn, G. Jakob, H.J. Elmers, Microscopic origin of magnetic anisotropy in martensitic Ni2MnGa. Phys. Rev. B **83**(21), 214419 (2011)
105. K. Miyamoto, A. Kimura, Y. Miura, M. Shirai, M. Ye, Y. Cui, K. Shimada, H. Namatame, M. Taniguchi, Y. Takeda, Y. Saitoh, E. Ikenaga, S. Ueda, K. Kobayashi, T. Kanomata, Absence of temperature dependence of valence-band spectrum of co_2mnsi. Phys. Rev. B **79**, 100405 (2009)

106. L. Chioncel, E. Arrigoni, M.I. Katsnelson, A.I. Lichtenstein, Majority-spin nonquasiparticle states in half-metallic ferrimagnet mn2val. Phys. Rev. B **79**, 125123 (2009)
107. P. Klaer, E. Arbelo, Jorge, M. Jourdan, W.H. Wang, H. Sukegawa, K. Inomata, H.J. Elmers, Temperature dependence of x-ray absorption spectra in the ferromagnetic heusler alloys mn_2feal. Phys. Rev. B **82**, 024418 (2010)
108. I. Galanakis, K. Ozdogan, E. Sasioglu, B. Aktas, Doping of mn_2val and mn_2vsi heusler alloys as a route to half-metallic antiferromagnetism. Phys. Rev. B **75**, 092407 (2007)

Part VI
Device Applications

Chapter 16
Giant Magnetoresistive Devices with Half-Metallic Heusler Compounds

Yuya Sakuraba and Koki Takanashi

Abstract Fundamental mechanism of giant magnetoresistive (GMR) effect is described. Recent experimental studies on the current-perpendicular-to-plane (CPP) GMR effect using half-metallic Heusler compounds are also reviewed. Additional important topics such as the studies for spin-torque oscillator using Heusler-based CPP-GMR and AMR effect in half-metallic Heusler films are also introduced in this chapter.

16.1 CIP and CPP-GMR

Giant magnetoresitive (GMR) effect is one of the most conventional phenomena in spintronics that was discovered by A. Fert and P. Grünberg in the late 1980s [1, 2]. There is no doubt that the discovery of GMR was the greatest achievement in spintronics that became a strong motivation for a lot of subsequent extensive researches through the historically-important application as magnetic read sensor for hard disk drives (HDD). GMR effect usually occurs in the film consisting of alternately stacking non-magnetic (NM) and ferromagnetic (FM) layers such as $[Co/Cu]_n$ and $[Fe/Cr]_n$ multilayers. There are two kinds of GMR effects depending on the flowing directions of electric current to the GMR film, i.e., current-in-plane (CIP) and current-perpendicular-to-plane (CPP) GMR. In this section, the fundamental mechanisms of CIP- and CPP-GMR are described.

In 1986, Grünberg et al. discovered that ferromagnetic and antiferromagnetic magnetic couplings periodically appear between Fe layers as a function of the thickness of Cr spacer in the Fe/Cr/Fe stacking films [1]. After that, Fert et al. found the

Y. Sakuraba
National Institute for Materials Science (NIMS), 1-2-1 Sengen, Tsukuba, Ibaraki 305-0047, Japan
e-mail: Sakuraba.Yuya@nims.go.jp

K. Takanashi (✉)
Institute for Materials Research, Tohoku university, 2-1-1 Katahira, Aoba-ku, Sendai 980-8577, Japan
e-mail: koki@imr.tohoku.ac.jp

C. Felser and A. Hirohata (eds.), *Heusler Alloys*,
Springer Series in Materials Science 222, DOI 10.1007/978-3-319-21449-8_16

giant magnetoresistance effect in the $[Fe/Cr]_n$ multilayers when the electric current flows to the in-plane direction of the film [2]. The observed MR ratio was about 20 % at room temperature (RT), which was much higher than that of anisotropic mangetoresistance (AMR) effect that was utilized for a read sensor for HDD at that moment. Therefore, it was named GMR effect at that time, and then it is usually called CIP-GMR in order to distinguish from CPP-GMR. The origin of CIP-GMR is the asymmetry of interfacial scattering between up and down spin electrons. Therefore, the period of metallic multilayer must be shorter than the mean free path of electrons to observe CIP-GMR. In other words, both the interfacial spin asymmetry and the mean free path are important parameters to determine the magnitude of MR ratio in CIP-GMR.

On the other hand, when the electric current flows to the perpendicular direction of the multilayers consisting of alternately stacked FM and NM layers, a resistance change depending on the relative direction of magnetizations also occurs, which is called CPP-GMR. In order to observe CPP-GMR, the multilayers must be patterned into a small pillar structure having the size below sub-micro meters with the top and bottom electrodes to apply an electric current to the perpendicular direction. In CPP-GMR, not only the electron scattering at the interface but also that in the bulk region of FM layers contribute to the MR effect. Therefore, a higher MR ratio can be expected in CPP-GMR than that in CIP-GMR in general, and the spin-diffusion length of a FM layer strongly affects the magnitude of MR ratio. By using the Boltzmann equation Valet and Fert established the model to describe the resistance change area product (ΔRA) in magnetic multilayers, in which the spin-dependent scatterings at the interface and in the bulk region are separately taken into account for both up and down spin electron channels [3]. For example, in the magnetic multilayers having the structure of NM/FM/NM/FM/NM as shown in Fig. 16.1, ΔRA can be written in

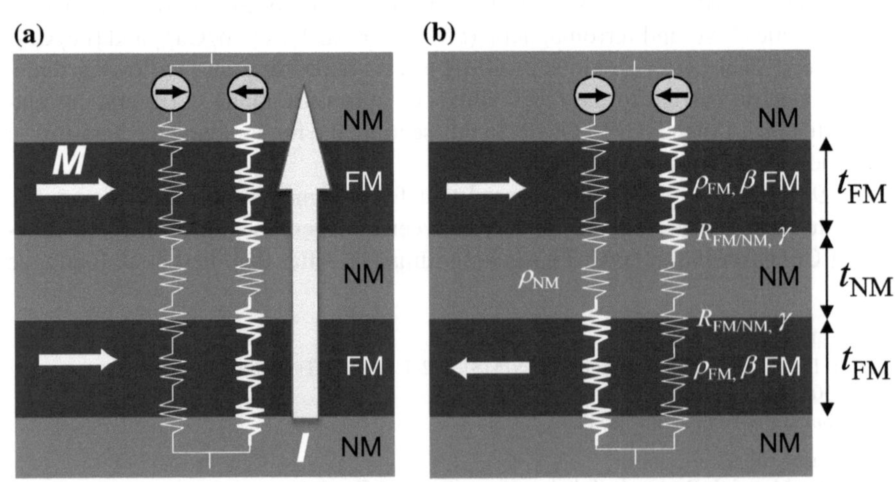

Fig. 16.1 Schematic views of a NM/FM/NM/FM/NM multilayer structure. The parallel two current circuit for up and down-spin channels considered in Valet-Fert model are also shown for both (**a**) parallel and (**b**) anti-parallel magnetization configurations

the following formula in the condition that the spin-diffusion lengths of FM and NM layers (l_{FM} and l_{NM}) are much larger than the thicknesses of FM and NM layers (t_{FM} and t_{NM})

$$\Delta RA = \frac{(2\beta\rho^*_{FM}t_{FM} + 2\gamma AR^*_{FM/NM})^2}{2\rho^*_{FM}t_{FM} + 2AR^*_{FM/NM} + \rho_{NM}t_{NM} + 2AR_{FM/NM}} \tag{16.1}$$

Here, γ and β are the coefficients representing the spin-asymmetry of electron scattering at the interface and in the bulk region, which are called interface and bulk spin-asymmetry coefficients, respectively. In other words, γ and β can be expressed using the interface resistance between FM and NM layers $R^{\uparrow(\downarrow)}_{FM/NM}$ and the bulk resistivity $\rho^{\uparrow(\downarrow)}_{FM}$ for up (down) spin electron as follows.

$$\gamma = \frac{R^{\downarrow}_{FM/NM} - R^{\uparrow}_{FM/NM}}{R^{\downarrow}_{FM/NM} + R^{\uparrow}_{FM/NM}}, \beta = \frac{\rho^{\downarrow}_{FM} - \rho^{\uparrow}_{FM}}{\rho^{\downarrow}_{FM} + \rho^{\uparrow}_{FM}} \tag{16.2}$$

$R^*_{FM/NM}$ and ρ^*_{FM} in (16.1) are written in the following expression: $R^*_{FM/NM} = R_{FM/NM}/(1-\gamma^2)$, $\rho^*_{FM} = \rho_{FM}/(1-\beta^2)$. The spin-diffusion lengths in various FM and NM materials l_{FM} and l_{NM} were well summarized by Bass and Pratt in their review [4]. It was repoted that l_{FM} in conventional 3*d* transition FM Fe, Co and Ni were about 8.5, 59 and 21 nm at low temepratute, respectively [5–7]. On the other hand, much shorter l_{FM} of about 3 nm was reported in Heusler compounds such as $Co_2FeAl_{0.5}Si_{0.5}$(CFAS) [8] and Co_2FeAl(CFA) [9], which is mainly due to the large resistivities ρ_{FM} of CFAS and CFA since l_{FM} is inversely proportional to ρ_{FM}. When the t_{FM} in the CPP-GMR stacking film becomes larger than l_{FM}, ΔRA is saturated because of spin relaxation in the FM layers but *RA* increases with t_{FM}, and thus the MR ratio decreases. Therefore, the optimization of t_{FM} with considering the l_{FM} is necessary to obtain the highest MR ratio in CPP-GMR devices.

16.2 CPP-GMR Devices Using Half-Metallic Heusler Compounds for Magnetic Read Sensor

Figure 16.2 shows the history of the advancement of the areal recording density in HDDs. In the beginning of 1990s, the areal density was just about 100 Mbit/in^2 but reached over 700 Gbit/in^2 in 2013 thanks to successive developments of high density recording media and highly sensitive magnetic readers using CIP-GMR and tunneling magnetoresistive (TMR) effects. However, it is predicted that present TMR head will soon face a limitation of its usage at around 1 – 2 Tbit/in^2 because it cannot satisfy a required low device resistance ($RA < 0.5\,\Omega \cdot \mu m^2$) with keeping a high MR ratio because the spin-filtering effect via MgO tunneling barrier is diminished by decreasing the thickness of the tunneling barrier. The CPP-GMR device can be an alternative for the next generation read head because it consists of only metallic

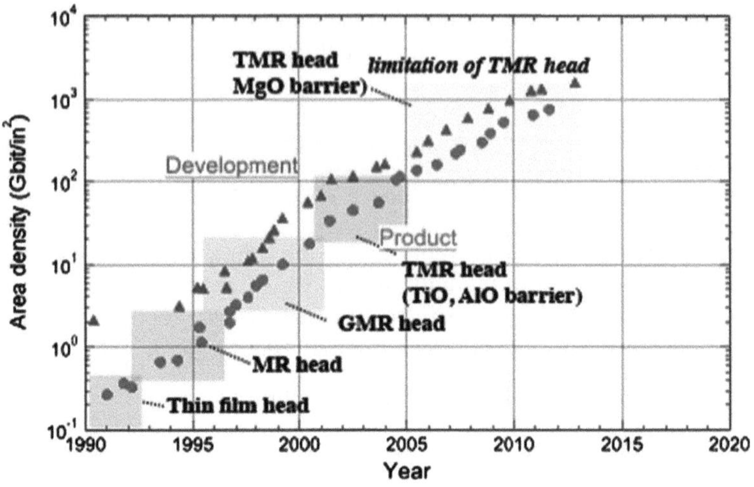

Fig. 16.2 History of the development of the areal recording density in HDDs

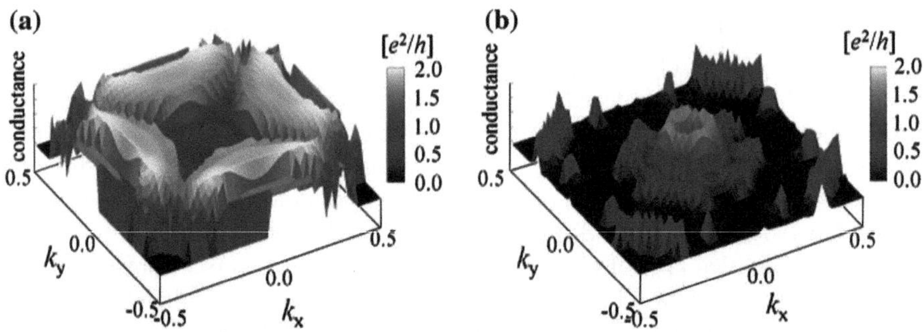

Fig. 16.3 Majority-spin conductances in the parallel magnetization configuration calculated for (**a**) (001)-CMS/Ag/CMS and (**b**) (001)-CMS/Cr/CMS as a function of in-plane wave vector $k_{//} = (k_x, k_y)$ [11]

layers showing a sufficiently small device resistance. One of the biggest issues in CPP-GMR device was a small MR ratio, usually less than 5 % at room temperature in the devices with conventional ferromagnets such as CoFe and NiFe, which is much smaller than that in TMR devices. It can be easily anticipated from (16.1) that, there are two strategies to improve the MR performance in CPP-GMR device, i.e., to obtain (i) large $R^*_{FM/NM}$ and ρ^*_{Fm} or (ii) large spin-asymmetries β and γ.

Therefore, to utilize the half-metallic Heusler compounds such as Co_2MnSi(CMS) is promising for enhancing MR properties remarkably because of their large resistivity and high spin-polarization of conduction electrons expected. In 2009, Iwase et al. reported the first achievement which clearly confirmed a benefit to use half-metallic Heusler electrodes for the CPP-GMR device; In their study a large MR ratio of 28 % at RT was observed in the CMS/Ag/CMS fully-epitaxial CPP-GMR device [10]. This enhancement of the MR ratio was explained by large inteface spin-asymmetry γ at the CMS/Ag interface as follows [11]. Figure 16.3 shows the calculated in-plane wave vector ($k_{//}$ -plane) dependence of the ballistic conductance of the majority spin-electron in the (001)-oriented CMS/Ag/CMS and CMS/Cr/CMS structures. It was clearly found that large total conductance in the $k_{//}$-plane is obtainable in the CMS/Ag/CMS structure compared with that in the CMS/Cr/CMS due to good electronic band matching between CMS and Ag to their (001)-direction. In other words, a small interface resistance for majority spin electron $R^{\uparrow}_{CMS/Ag}$ is expected at the CMS/Ag interfaces. Since the large inteface resistance for minority spin electrons $R^{\downarrow}_{CMS/Ag}$ is easily expected due to the half-metallic nature of CMS, the spin-asymmetry of interface resistance γ given in (16.2) must be high, and hence dominantly contributes to the observed large MR ratio. In addition to good band matching, the Ag spacer also has good lattice matching with the half-metallic Heusler compounds in the (001)-direction and no/low solubilities to general constituent atoms of half-metallic Heusler compounds such as Co, Mn, Fe, and Si, which are also important advantages against other NM spacer materials such as Cu. After Iwase et al. demonstrated the usefullness of Ag spacer, larger MR ratios have been observed by using other Heusler compounds with the Ag spacer as summarized in Table 16.1. Nakatani et al. reported a large MR ratio of 34 % in the CFAS/Ag/CFAS [21] and also

Table 16.1 Summary of the MR ratio and DRA in the CPP-GMR devices with Heusler compounds

Heusler	Stacking structure	MR ratio	ΔRA (mΩ • μm^2)	Reference
Co_2MnSi (CMS)	Epi-CMS/Cr/CMS	2.4 % (RT)	19 (RT)	[12] (2006)
	Epi-CMS/Cr/CMS	5.2 % (RT)	6.5 (RT)	[13] (2009)
	Poly-CMS/Cu/CMS	9.0 % (RT)	-	[14] (2008)
	Epi-CMS/Cu/CMS	8.6 % (RT), 30.7 % (6K)	14.2 (RT), 35.2@6K	[15] (2009)
	Epi-CMS/Ag/CMS	28.8 % (RT)	8.9 (RT)	[10] (2009)
	Epi-CMS/Ag/CMS	36.4 % (RT), 67.2 % (110K)	11.5 (RT), 18.4@110K	[11] (2010)
$Co_2Fe_{0.4}Mn_{0.6}Si$ (CFMS)	Epi-CFMS/Ag/CFMS	74.8 % (RT)		[16] (2011)
	Epi-CFMS/Ag/CFMS	58 % (RT), 184 % (30 K)	12 (RT)	[17] (2012)
Co_2MnGe (CMG)	Poly-CMG/Cu/CMG/ Cu/CMG	11 % (RT)	9.1 (RT)	[18] (2005)
	CMG/Rh2CuSn/CMG	6.7 (RT)	4.0 (RT)	[19] (2009)
$Co_2FeAl_{0.5}Si_{0.5}$ (CFAS)	Epi-CFAS/Ag/CFAS	6.9 % (RT), 14 % (6K)	7.4 (RT), 12.4 (6K)	[20] (2008)
	Epi-CFAS/Ag/CFAS	34 % (RT), 80 % (14K)	8 (290K), 17 (14K)	[21] (2010)
$Co_2MnGa_{0.5}Sn_{0.5}$ (CMGS)	Epi-CMGS/Ag/CMGS	8.8 % (RT), 17.2 % (12K)	4.0 (RT), 6.5 (12K)	[22] (2010)
$Co_2MnGa_{0.25}Ge_{0.75}$ (CMGG)	Epi-CMGG/Ag/CMGG	40.2 % (RT)	6.1 (RT), 15 (10K)	[23] (2013)
$Co_2FeGa_{0.5}Ge_{0.5}$ (CFGG)	Epi-CFGG/Ag/CFGG	57 % (RT), 183 % (10K)	12 (RT), 33 (10K)	[24] (2013)

oscillatory anti-ferromagnetic interlayer exchange coupling against the Ag spacer thikness [25], which is promising to apply to a trilayer magnetic read sensor showing a sessiors-type response to magnetic field. The highest MR ratios and ΔRAs beyond 50 % and 10 $m\Omega \cdot \mu m^2$ at RT, respectively, were reported in the devices with $Co_2Fe_{0.4}Mn_{0.6}Si$(CFMS) and $Co_2FeGa_{0.5}Ge_{0.5}$(CFGG) electrodes with an Ag spacer [16, 17, 24]. Recently, Kubota et al. and Du et al. used $L1_2$-Ag_3Mg alloy and B2-AgZn as the spacer materials, respectively, and achieved to realize dRA over 20 $m\Omega\mu m^2$ [26, 27]. It was observed in the devices with various Heusler compounds [11, 13, 17, 23, 24], that MR properties improve with increasing annealing temperature for Heusler electrodes until large inter-diffusion or $D0_3$-type disordering occurs, i.e., a qualitative positive realtionship between the MR ratio in the CPP-GMR device and the degree of $L2_1$-ordering was clearly confirmed in previous studies [11, 24]. Therefore, to realize high $L2_1$-ordering is important to obtain the highest MR performance in CPP-GMR devices in reality although the B2-ordering was predicted to be sufficient to obtain a half-metallic electronic structure theoretically [29, 30]. This result is in contrast to no remarkable dependence of the MR ratio in TMR devices on the degree of $L2_1$-ordering in [28]. This difference between CPP-GMR and TMR might originate from the strong sensitivity to the barrier interface in TMR, i.e., MR properties in TMR devices strongly depend on the interfacial chemical purity and/or the exsitence of dislocations and defects at the barrier interface rather than the degree of $L2_1$-ordering.

From the viewpoint of practical application for the magnetic read sensor using CPP-GMR, we must satisfy many requirements in addition to the MR output. For examples, it was suggested that the total thickness of the device must be less than 10.5 nm for the areal density of 5 Tbit/in^2 [31]. The highest annealing temeprature cannot exceed 350–400 °C to suppress the interdiffusion with the permalloy magnetic shield. Furthermore, the devices with high MR output should be grown on a thermally-oxdized Si substarate not on a single crystalline MgO substrate. Concerning the first requirement on the total thickness, Sakuraba et al. realized a large MR output (MR ratio of 43 % and ΔRA of 12 $m\Omega \cdot \mu m^2$) even in a very thin trilayer structure consisting of CFMS(4 nm)/Ag(3 nm)/CFMS(2 nm) as shown in Fig. 16.4, which is caused by the dominant contribution of interface scattering to the MR effect [17]. The limitation of annealing temperature for the second requirement is a serious problem since a high degree of $L2_1$-ordering is essential for obtaining a large MR output as mentioned above. It was repoted that the $L2_1$-ordering could be enhanced effectively by deposting CMS at elevated substrate temperatures of 250–300 °C, and hence a large MR ratio over 30 % was obtained without a general post-annealing process at high temperatures over 500 °C [32]. As for the last requirement about the substrate, improved MR properties have been recently realized using thermally-oxidized Si substrate [33–35]. Although further efforts to solve remaining issues are required, recent rapid progresses in half-metallic Heuser-based CPP-GMR made it more realistic to develop a next generation read head based on CPP-GMR device.

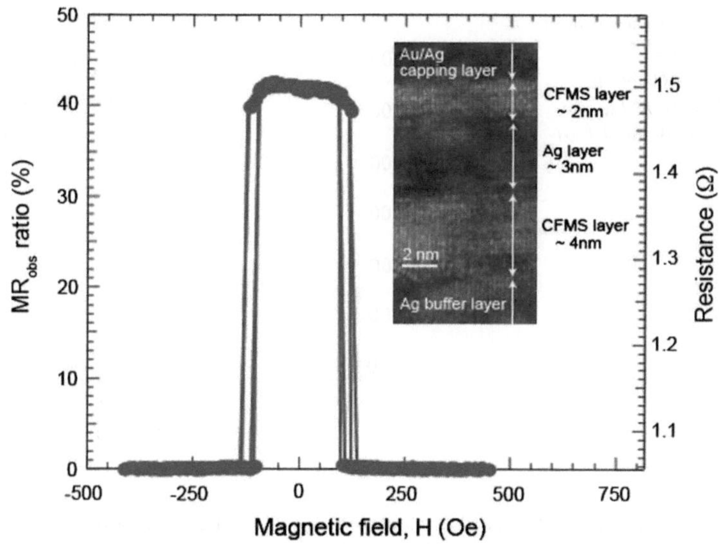

Fig. 16.4 MR curve for a sample with a CFMS(4 nm)/Ag(3 nm)/CFMS(2 nm) structure at RT. The inset shows a HR-TEM image of the sample [17]

16.3 CPP-GMR Devices Using Half-Metallic Heusler Compounds for Spin-Torque Oscillator

The transfer of angular momentum from conduction-electron spins to local spins, called spin-transfer torque (STT), creates a variety of magnetization dynamics in a nanometer scale magnetic device, such as a CPP-GMR and TMR devices. Automatic magnetization precession is a representative phenomenon using the magnetization dynamics induced by the balance between STT and the effective magnetic field, which is expected to be new spintronic devices such as a spin-torque oscillator (STO). A STO can be utilized for novel applications such as a nanoscale radio frequency (rf) oscillator for the on-chip communicator and microwave assisted magnetic recording (MAMR) technology. For the usage of the communicator, to realize a large rf output power and a narrow line-width leading to high quality (Q)-factor are critical issues. The output power of rf oscillation is proportional to the square of MR ratio in CPP-GMR/TMR devices. A line-width Δf depends on the homogeneity of the STT-induced spin dynamics in the oscillating FM layer, thereby small damping constant is preferable to realize a small Δf. Although a CPP-GMR device usually has the advantage to realize a small Δf compared with a TMR device, a tiny rf output power due to the small MR ratio is a serious problem. Therefore, large MR ratios observed in half-metallic Heusler-based CPP-GMR devices can be a solution to improve an rf output power in CPP-GMR devices. In addition, half-metallic Heusler compounds such as CMS usually show a small damping constant $\alpha < 0.005$ [36–38] because α is proportional to the total density of states at the Fermi level [39], which is also

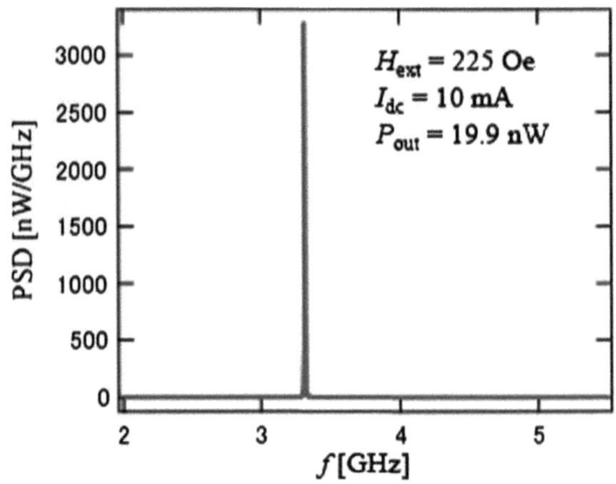

Fig. 16.5 The rf spectrum in the CFMS/Ag/CFMS CPP-GMR devices with applying dc current I_{dc} and external magnetic field H_{ext}

an important advantage to realize a narrow Δf. On the other hand, for the usage of a microwave source for MAMR technology, it is important to generate a large ac magnetic field at a small threshold current density, i.e., we should excite the magnetization oscillation in the FM layer having large magnetic volume (saturation magnetization and thickness product $M_S \times t_{FM}$ by applying a small dc current. Therefore, a large spin-polarization of conduction electrons generating from the half-metallic Heusler spin-injector is also useful for applying a microwave source for MAMR technology.

Figure 16.5 shows a typical rf spectrum excited by STT in a CFMS(20 nm)/Ag (5 nm)/CFMS(7 nm) CPP-GMR device that shows a large MR ratio of 32 % at RT. The observed rf output power here reaches about 20 nW, which is 2–3 orders of magnitude larger than that observed in the conventional CPP-GMR devices. Additionally, a narrow Δf of 3.2 MHz with a large Q-factor over 10^3 was also observed [40]. This result confirmed a clear advantage to realize both large rf output power and Q-factor by using Heusler-based CPP-GMR devices. Although similar promising results were reported using CFGG and CMS electrodes in [41, 42], the number of studies about STO in Heusler-based CPP-GMR devices are still limited so far in spite of their high potential. Therefore, extensive researches on this topic are strongly desired in future.

16.4 AMR Effect in Half-Metallic Heusler Compounds

The anisotropic magnetoresistance (AMR) effect is one of the most conventional MR effects in ferromagnets, in which the electric resistivity changes with a relative angle between the current and magnetization directions. The AMR ratio is generally defined by,

$$\frac{\Delta\rho}{\rho} = \frac{\rho_{//} - \rho_{\perp}}{\rho_{\perp}} \tag{16.3}$$

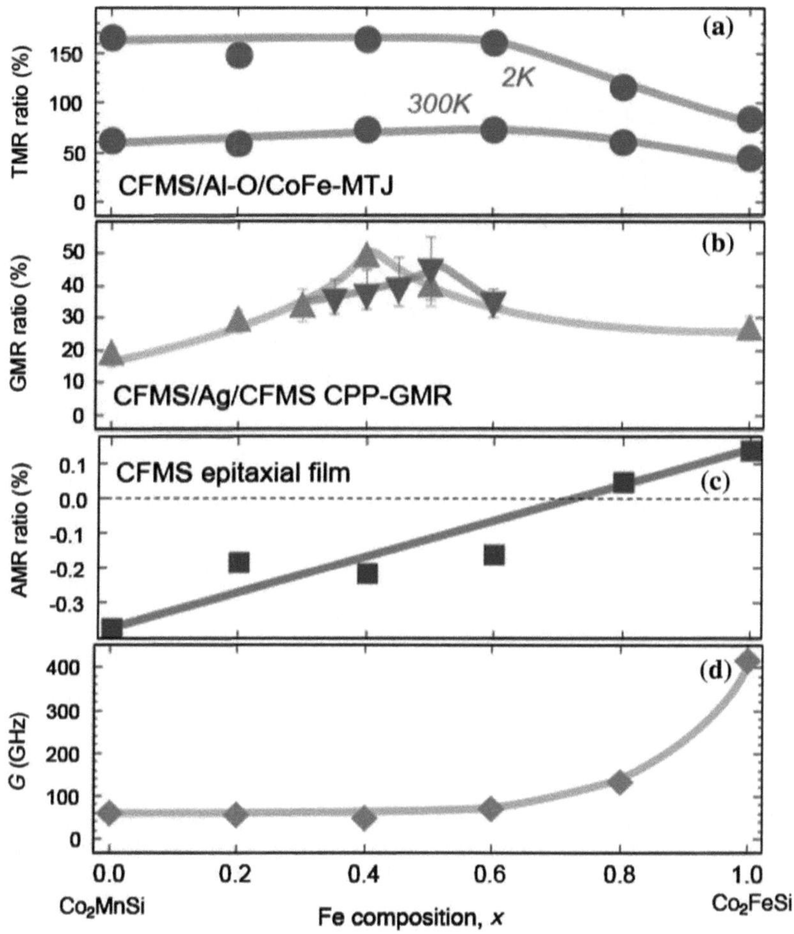

Fig. 16.6 Fe composition × dependence of (**a**) TMR ratio in the CFMS/Al-O/CoFe-MTJs [46], (**b**) GMR ratio in the CFMS/Ag/CFMS [17], (**c**) AMR ratio [43] and (**d**) Gilbert damping constant in the CFMS epitaxial films [38]

where $\rho_{//}(\rho_{\perp})$ represents the resistivity when the electric current flows parallel (perpendicular) to the magnetization. The AMR effect basically originates from *s*-*d* scattering from the conduction state (*s* state) to localized *d* states hybridized by spin-orbit interaction with the opposite spin state. Kokado et al. recently systematically investigated the relationship between the sign of the AMR ratio and the dominant *s*-*d* scattering process using an extended theory that specified the spin state of the *s* state in *s*-*d* resistivity [43]. As a result, they found that when the dominant *s*-*d* scattering process is $s \uparrow \rightarrow d \downarrow$ or $s \downarrow \rightarrow d \uparrow$, the sign of the AMR ratio tends to be positive ($\rho_{//} > \rho_{\perp}$) as confirmed from bcc Fe, fcc Co, and Ni[5]. In contrast, when the dominant scattering is $s \uparrow \rightarrow d \uparrow$ or $s \downarrow \rightarrow d \downarrow$, the sign tends to be negative ($\rho_{//} < \rho_{\perp}$). In half-metallic materials, one of the spin channels at the Fermi level is missing, hence

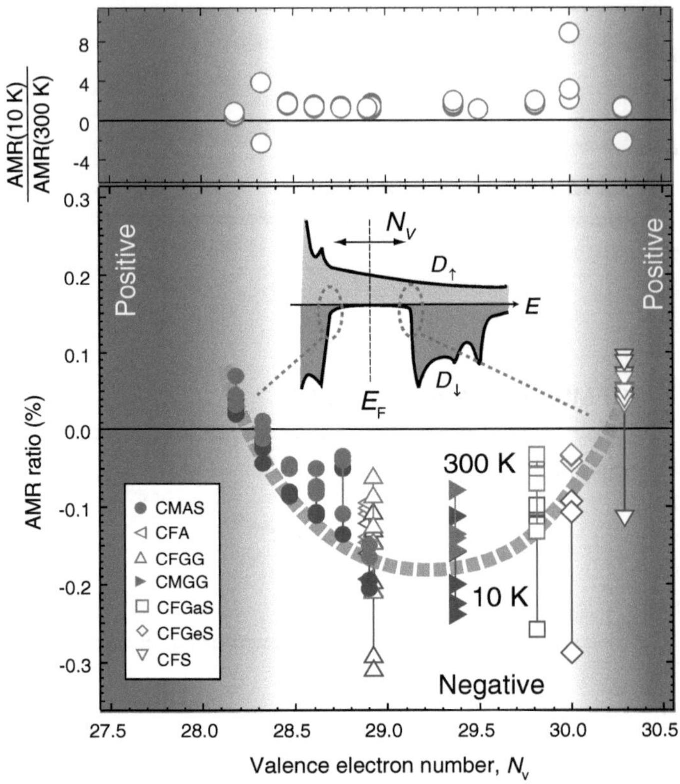

Fig. 16.7 Valence electron number N_V dependence of AMR ratio in all Co_2MnZ and Co_2FeZ films annealed at different temperature from 300–650 °C. Actual N_V estimated from the composition analysis was used for this plot. *Red* and *Blue* symbols represent AMR at 300 K and 10 K, respectively. The *upper* part shows the ratio of AMR ratios at 10–300 K [47] (Color figure online)

the sign of AMR ratio is always negative. In other words, the sign of AMR can be a fingerprint of half-metallic nature of FM materials, which is useful for developing new half-metallic materials without any time-consuming experiments. Yang et al. systematically investigated AMR ratios in $Co_2Fe_xMn_{1-x}Si$ with changing the Fe/Mn composition ratio x [44]. They found that the sign of AMR ratio is negative from x = 0.0 to 0.6 but suddenly changes to positive at x $\geqq$ 0.8. The x dependence of TMR, GMR, AMR ratio and the damping constant are summarized in Fig. 16.6 [45, 46]. Both TMR and GMR ratios drastically decreases at x > 0.6, indicating the reduction of spin-polarization in this regime. The damping constant measured by ferromagnetic resonance in the CFMS films also remarkably increases at x > 0.8, suggesting that the Fermi level across the localized d-states consisting of the conduction band of half-metallic gap in this regime. It was reasonably concluded from these results that, the negative sign of AMR ratio could be regarded as an indication of half-metallic nature in Heusler compounds. Recently Sakuraba et al. systematically investigated

AMR effect in various Co_2MnZ and Co_2FeZ epitaxial thin films [47]. Since the position of E_F shifts against the total valence electron number N_V in Co_2MnZ and Co_2FeZ with keeping the half-metallic electronic structure, they clearly confirmed that the sign of AMR ratio is negative when E_F locates in the half-metallic gap as shown in Fig. 16.7. They also found that there is a positive relationship between the magnitude of negative AMR ratio in CFGG and $Co_2MnGa_{0.25}Ge_{0.75}$ films and the observed ΔRA in the CPP-GMR device using those electrodes, which implies that the magnitude of negative AMR ratio is positively correlated with the spin-polarization of conduction electron. Therefore, large temperature dependence of AMR ratio at $N_V = 28.2$ and 30 in Fig. 16.7 (corresponding to the valence and conduction band edges of half-metallic gap, respectively) indicates the reduction of spin-polarization against temperature in these compounds because of thermal fluctuation of electron occupation. The knowledge of the relationship between AMR and spin-polarization will be important to develop new Heusler compounds having a higher spin-polarization and improve MR properties in a CPP-GMR device without performing any micro-fabrications.

References

1. P. Grünberg, R. Schreiber, Y. Pang, M.B. Brodsky, H. Sowers, Phys. Rev. Lett. **57**, 2442 (1986)
2. M.N. Baibich, J.M. Brot, A. Fert, F. Nguyen Van Dau, F. Petroff, P. Eitenne, G. Creuzet, A. Friederich, J. Chazelas, Phys. Rev. Lett. **61**, 2472 (1988)
3. T. Valet, A. Fert, Phys. Rev. B **48**, 7099 (1993)
4. J. Bass, W.P. Pratt, J. Phys. Cond. Matter **19**, 183201 (2007)
5. L. Piraux, S. Dubois, A. Fert, L. Belliard, Eur. Phys. J. B **4**, 413 (1998)
6. D. Bozed, Ph.D. Thesis Leeds University (2000)
7. C. Moreau, W.P.J. Pratt, N.O. Birge, Appl. Phys. Lett. **90**, 012101 (2007)
8. T.M. Nakatani, T. Furubayashi, S. Kasai, H. Sukegawa, Y.K. Takahashi, S. Mitani, K. Hono, Appl. Phys. Lett **96**, 212501 (2010)
9. S. Maat, M.J. Carey, J.R. Childress, J. Appl. Phys. **101**, 093905 (2007)
10. T. Iwase, Y. Sakuraba, S. Bosu, K. Saito, S. Mitani, K. Takanashi, Appl. Phys. Express **2**, 063003 (2009)
11. Y. Sakuraba, K. Izumi, Y. Miura, K. Futasukawa, T. Iwase, S. Bosu, K. Saito, K. Abe, M. Shirai, K. Takanashi, Phys. Rev. B **82**, 094444 (2010)
12. K. Yakushiji, K. Saito, S. Mitani, K. Takanashi, Y.K. Takahashi, K. Hono, Appl. Phys. Lett. **88**, 222504 (2006)
13. Y. Sakuraba, T. Iwase, K. Saito, S. Mitani, K. Takanashi, Appl. Phys. Lett. **94**, 012511 (2009)
14. T. Mizuno, Y. Tsuchiya, T. Machita, S. Hara, D. Miyauchi, K. Shimazawa, T. Chou, K. Noguchi, K. Tagami, I.E.E.E. Trans. Magn. **44**, 3584 (2008)
15. K. Kodama, T. Furubayashi, H. Sukegawa, T.M. Nakatani, K. Inomata, K. Hono, J. Appl. Phys. **105**, 07E905 (2009)
16. J. Sato, M. Oogane, H. Naganuma, Y. Ando, Appl. Phys. Express **4**, 113005 (2011)
17. Y. Sakuraba, M. Ueda, Y. Miura, K. Sato, S. Bosu, K. Saito, M. Shirai, T.J. Konno, K. Takanashi, Appl. Phys. Lett. **101**, 252408 (2012)
18. M. Saito, N. Hasegawa, Y. Ide, T. Yamashita, Y. Hayakawa, Y. Nishiyama, M. Ishizone, S. Yanagi, K. Nishimura, A. Takahashi, Digest of the Intermag conference, 2005 (unpublished), Paper No. FB-02

19. K. Nikolaev, P. Kolbo, T. Pokhil, X. Peng, Y. Chen, T. Ambrose, O. Mryasov, Appl. Phys. Lett. **94**, 222501 (2009)
20. T. Furubayashi, K. Kodama, H. Sukegawa, Y.K. Takahashi, K. Inomata, K. Hono, Appl. Phys. Lett. **93**, 122507 (2008)
21. T.M. Nakatani, T. Furubayashi, S. Kasai, H. Sukegawa, Y.K. Takahashi, S. Mitani, K. Hono, Appl. Phys. Lett. **96**, 212501 (2010)
22. N. Hase, BSDChS Varaprasad, T.M. Nakatani, H. Sukegawa, S. Kasai, Y.K. Takahashi, T. Furubayashi, K. Hono, J. Appl. Phys. **108**, 093916 (2010)
23. Y.K. Takahashi, N. Hase, M. Kodzuka, A. Itoh, T. Koganezawa, T. Furubayashi, S. Li, BSDChS Varaprasad, T. Ohkubo, K. Hono, J. Appl. Phys. **113**, 223901 (2013)
24. S. Li, Y.K. Takahashi, T. Furubayashi, K. Hono, Appl. Phys. Lett. **103**, 042405 (2013)
25. T.M. Nakatani, S. Mitani, T. Furubayashi, K. Hono, Appl. Phys. Lett. **99**, 182505 (2011)
26. H. Narisawa, T. Kubota, K. Takanashi, Appl. Phys. Express **8**, 063008 (2015)
27. Y. Du, Y. Furubayashi, T.T. Sasaki, Y. Sakuraba, Y.K. Takahashi, K. Hono, Appl. Phys. Lett. (in press, 2015)
28. Y. Sakuraba, M. Hattori, M. Oogane, H. Kubota, Y. Ando, A. Sakuma, N.D. Telling, P. Keatley, G. van der Laan, E. Arenholz, R.J. Hicken, T. Miyazaki, J. Magn. Soc. Jpn. **31**, 338 (2007)
29. S. PIcozzi, A.J. Freeman, A. Continenza, Phys. Rev. B **69**, 094423 (2004)
30. Y. Miura, K. Nagao, M. Shirai, Phys. Rev. B **69**, 144413 (2004)
31. M. Takagishi, K. Yamda, H. Iwasaki, H.N. Fuke, S. Hashimoto, I.E.E.E. Trans. Magn. **46**, 2086 (2010)
32. Y. Sakuraba, K. Izumi, T. Koganezawa, S. Bosu, R. Okura, M. Ueda, T. Kojima, K. Saito, K. Takanashi, IEEE Trans. Magn. **49**, 5464 (2013)
33. M.J. Carey, S. Maat, S. Chandrashekariaih, J.A. Katine, W. Chen, B. York, J.R. Childress, J. Appl. Phys. **109**, 0939 (2011)
34. Y. Du, BSDChS Varaprasad, Y.K. Takahashi, T. Furubayashi, K. Hono, Appl. Phys. Lett. **103**, 202401 (2013)
35. Z. Diao, M. Chapline, Y. Zheng, C. Kaiser, A.G. Roy, C.J. Chien, C. Shang, Y. Ding, C. Yang, D. Mauri, Q. Leng, M. Pakala, M. Oogane, Y. Ando, J. Magn. Magn. Mater. **356**, 73 (2014)
36. R. Yilgin, Y. Sakuraba, M. Oogane, S. Mizukami, Y. Ando, T. Miyazaki, Jpn. J. Appl. Phys. **46**, L205–L208 (2007)
37. S. Mizukami, D. Watanabe, M. Oogane, Y. Ando, Y. Miura, M. Shirai, T. Miyazaki, J. Appl. Phys. **105**, 07D306 (2009)
38. M. Oogane, T. Kubota, Y. Kota, S. Mizukami, H. Naganuma, A. Sakuma, Y. Ando, Appl. Phys. Lett. **96**, 252501 (2010)
39. V. Kambersky, Can. J. Phys. **48**, 2906 (1970)
40. T. Seki, Y. Sakuraba, H. Arai, M. Ueda, R. Okura, H. Imamura, K. Takanashi, Appl. Phys. Lett. **105**, 092406 (2014)
41. R. Okura, Y. Sakuraba, T. Seki, K. Izumi, M. Mizuguchi, K. Takanashi, Appl. Phys. Lett. **99**, 052510 (2011)
42. J. Sinha, M. Hayashi, Y.K. Takahashi, T. Taniguchi, M. Drapeko, S. Mitani, K. Hono, Appl. Phys. Lett. **99**, 162508 (2011)
43. S. Kokado, M. Tsunoda, K. Harigaya, A. Sakuma, J. Phys. Soc. Jpn. **81**, 024705 (2012)
44. F.J. Yang, Y. Sakuraba, S. Kokado, Y. Kota, A. Sakuma, K. Takanashi, Phys. Rev. B **86**, 020409(R) (2012)
45. Y. Sakuraba, M. Ueda, S. Bosu, K. Saito, K. Takanashi, J. Magn. Soc. Jpn. **38**, 45–49 (2014)
46. T. Kubota, S. Tsunegi, M. Oogane, S. Mizukami, T. Miyazaki, H. Naganuma, Y. Ando, Appl. Phys. Lett. **94**, 122504 (2009)
47. Y. Sakuraba, S. Kokado, Y. Hirayama, T. Furubayashi, H. Sukegawa, S. Li, Y.K. Takahashi, K. Hono, Appl. Phys. Lett. **104**, 172407 (2014)

Chapter 17
Magnetic Tunnel Junctions Using Heusler Alloys

Seiji Mitani

Abstract The remarkable feature of half-metallic Heusler alloys is high effective spin polarization of conduction electrons, which directly causes large tunnel magnetoresistance (over 300 % at room temperature and ~2,000 % at low temperature) as well as current-perpendicular-to-plane giant magnetoresistance. Studies on the magnetic tunnel junctions and their characteristics are overviewed, including a couple of specific topics such as spin-transfer-torque magnetization switching.

The basic structure of a magnetic tunnel junction (MTJ) consists of two ferromagnetic metal electrodes separated by a thin tunnel barrier layer, as shown in Fig. 17.1a. It is one of the most important devices in spintronics, and recent developments of hard disk drive (HDD) read heads and magnetoresistive random access memories (MRAMs) are owing to the increasing performance of MTJs [1, 2]. The tunnel magnetoresistance (TMR) is a phenomenon that tunnel resistance changes depending on magnetization configurations of the two ferromagnetic electrodes in MTJs. It originates from spin dependence of tunneling probability, and the TMR ratio is defined as $(R_{AP} - R_P)/R_P$, where R_P and R_{AP} are resistance of the MTJ in parallel and antiparallel configurations of magnetization, respectively. In the simplest model of TMR [3, 4], the total tunnel current is considered as the sum of currents flowing through up and down spin electron channels as shown in Fig. 17.1b, i.e., it is assumed that no spin mixing occurs during the tunneling process, and then TMR ratio is given as

$$TMR\ ratio = \frac{2P_1P_2}{1 - P_1P_2}, \tag{17.1}$$

where P_1 (P_2) is the spin polarization of conduction electrons in one (the other) electrode. The spin polarization P is derived from majority- (minority-) spin electron's density of states at Fermi level D_+ (D_-):

$$P = \frac{D_+ - D_-}{D_+ + D_-}. \tag{17.2}$$

S. Mitani (✉)
Magnetic Materials Unit, National Institute for Materials Science,
1-2-1 Sengen, Tsukuba 305-0047, Japan
e-mail: mitani.seiji@nims.go.jp

C. Felser and A. Hirohata (eds.), *Heusler Alloys*,
Springer Series in Materials Science 222, DOI 10.1007/978-3-319-21449-8_17

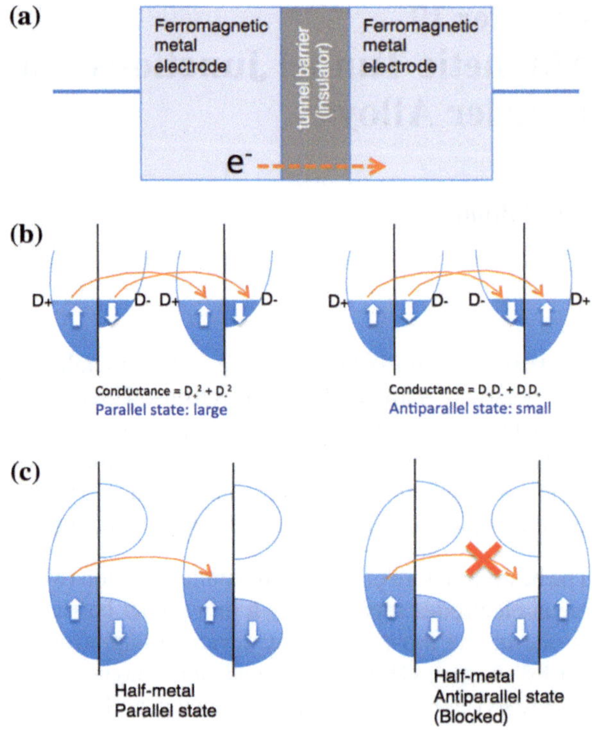

Fig. 17.1 Magnetic tunnel junction (MTJ) and tunnel magnetoresistance (TMR); **a** a schematic illustration of MTJ, **b** electron tunneling between two ferromagnetic electrodes through a tunnel barrier in parallel and antiparallel magnetization configurations, **c** electron tunneling in a MTJ using half-metallic electrodes

From (17.1) and (17.2), it can be understood that TMR ratios reflect electronic structures of the electrode materials. Here, it is noted that P is sometimes regarded as an effective spin polarization, since the TMR ratio and spin polarization P given by (17.1) and (17.2) may be too simplified in realistic cases. In relation to the TMR mentioned above, the most striking feature of some Co-based full Heusler alloys such as Co_2MnSi is a half-metallic band structure. The spin polarization of such half-metallic Heusler alloys is ideally 100 %, and therefore from (17.1) extremely large TMR is expected to occur in MTJs using half-metallic Heusler alloys. The point of its physical mechanism is the fact that electron tunneling is perfectly blocked in the antiferromagnetic magnetization configuration, as shown in Fig. 17.1c. A demonstration of large TMR in half-metallic materials was first made using LaSrMnO at low temperature [5].

A discovery of fairly large magnetoresistance in bulk Heusler alloy $Co_2Cr_{0.6}Fe_{0.4}Al$ triggered the study of TMR in Heusler alloy-based MTJs. Block et al. observed magnetoresistance exceeding 30 % in pressed powder compact samples of $L2_1$-ordered $Co_2Cr_{0.6}Fe_{0.4}Al$ at room temperature (RT) [6], which is considered TMR occurring across insulating grain boundary phases. The first clear demonstration of TMR in a Heusler alloy-based MTJ was performed by Inomata et al. using polycrystalline B2-$Co_2Cr_{0.6}Fe_{0.4}Al$ [7], and TMR ratios of 16 and 26 % at RT and 5 K, respectively, were reported for a $Co_2Cr_{0.6}Fe_{0.4}Al$/AlO/CoFe MTJ grown on

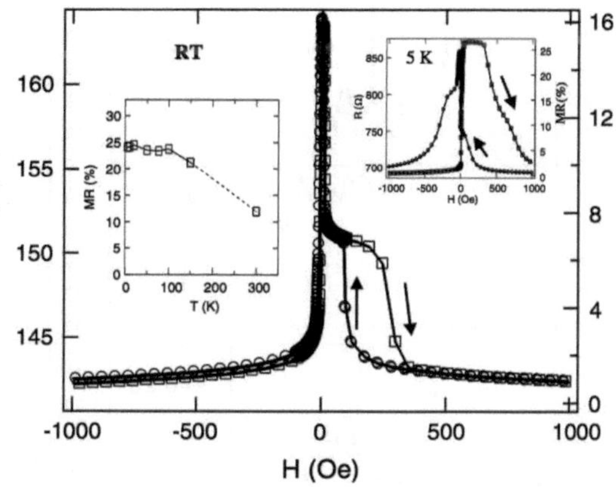

Fig. 17.2 TMR observed in a polycrystalline B2-ordered $Co_2Cr_{0.6}Fe_{0.4}Al$/AlO/CoFe MTJ. Reproduced from [7] (© 2003, The Japan Society of Applied Physics)

a thermally oxidized Si substrate (Fig. 17.2). This work was followed by several similar studies on Heusler alloy-based MTJs [8, 9]. A breakthrough was made by Sakuraba et al. [10], showing a giant TMR in Co_2MnSi/AlO/Co_2MnSi MTJs grown on a MgO(100) substrate (Fig. 17.3). The use of a MgO substrate made the film qualities such as $L2_1$-type chemical ordering and surface flatness be much improved. The TMR ratio observed at 2 K reached 570 %, proving the half-metallic nature of Co_2MnSi. Half-metallicity in MTJs was also examined in an engineered Heusler alloy $Co_2FeAl_{0.5}Si_{0.5}$, in which Fermi level was settled at the mid-point of the energy gap by changing the number of valence electrons, as shown in Fig. 17.4 [11, 12]. From comparison between $Co_2FeAl_{0.5}Si_{0.5}$/MgAlO/CoFe and CoFe/MgAlO/CoFe MTJs, it was shown that high spin polarization close to unity was obtained for $Co_2FeAl_{0.5}Si_{0.5}$ [13]. In this study, it is noted that the MgAlO barrier was not annealed at high temperature (remaining in low crystal quality) and therefore so-called coherent tunneling [2, 14] was not taken into account.

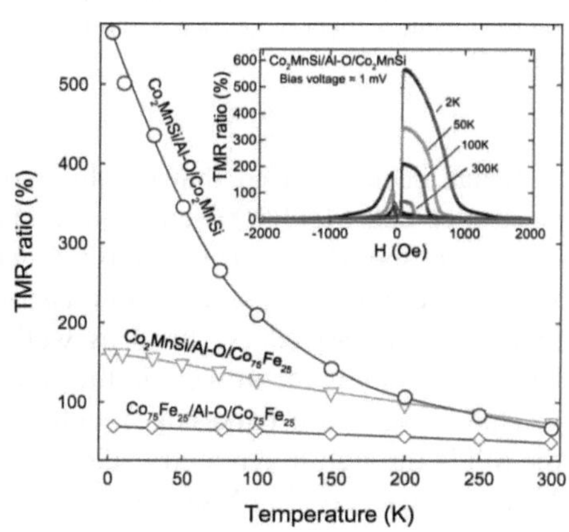

Fig. 17.3 Giant TMR and its temperature dependence in Co_2MnSi/AlO/Co_2MnSi MTJ. Reproduced from [10] (© 2006, American Institute of Physics)

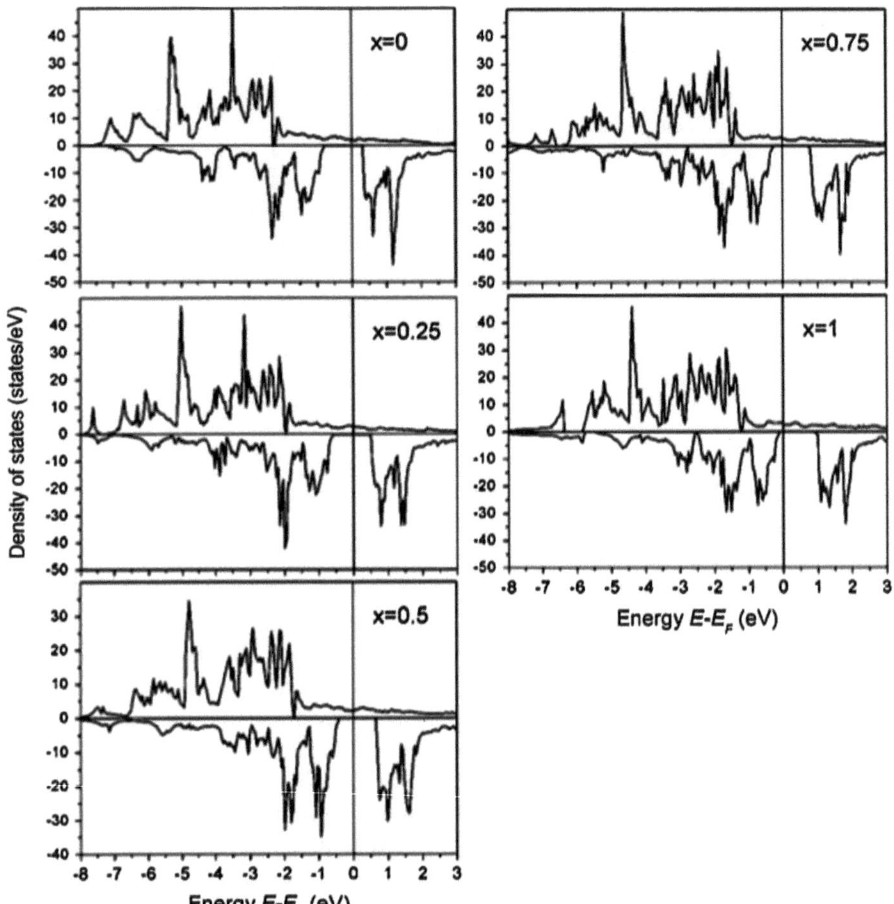

Fig. 17.4 Calculated spin-resolved density of states in $Co_2FeAl_xSi_{1-x}$. The position of Fermi level varies with the number of valence electrons. Reproduced from [12] (© 2007, American Institute of Physics)

It has been reported that combination of highly spin-polarized Heusler alloy electrodes and MgO(100) epitaxial tunnel barrier brings about larger TMR. This is because the MgO barriers cause coherent tunneling, i.e., selective tunneling of Δ_1 symmetry electrons, and simultaneously Co-based full Heusler alloys possess highly spin-polarized Δ_1 band as well as bcc Fe and CoFe. Even if total band half-metallicity is incomplete in the Heusler alloy (for example, due to Fermi level located close to the band edge), coherent tunneling effect may augment TMR in the MTJs through it symmetry selectivity. On the other hand, in highly conductive MTJs using a very thin MgO barrier (~4 atomic plane or less), the symmetry selectivity, i.e., coherent tunneling for large TMR, is degraded with decreasing the MgO barrier thickness. Even in such a case of thin MgO barrier, large TMR should be

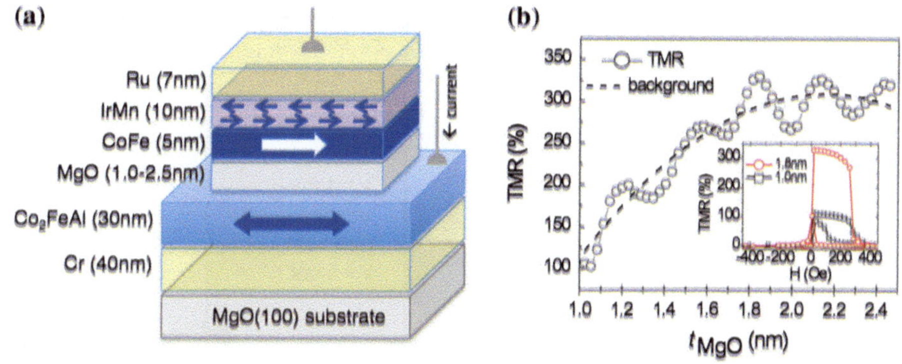

Fig. 17.5 **a** Cartoon of the stacking structure of MgO//Cr/Co_2FeAl/MgO/CoFe/IrMn/Ru MTJs used in the study of [17], **b** TMR and its oscillation as a function of MgO barrier thickness, reproduced from [17] (© 2010, American Physical Society)

maintained if a Heusler alloy with total band half-metallicity is employed for electrodes. MTJs using both a Heusler alloy electrode and a MgO barrier were investigated by Marukame et al. [15], and large TMR ratios of 90 and 240 % were obtained for a fully epitaxial $Co_2Cr_{0.6}Fe_{0.4}Al$/MgO/CoFe(100) MTJ at RT and 4.2 K, respectively. In Co_2FeAl/MgO/CoFe(100) MTJs with a spin-valve structure (Fig. 17.5a), Wang et al. [16] reported further large TMR ratios of 330 and 700 % at RT and 10 K, respectively, as well as TMR oscillation as a function of the MgO barrier thickness (Fig. 17.5b) [17]. In the case of Co_2FeAl, it is considered that coherent tunneling dominantly contributes to the large TMR. In a molecular beam epitaxy (MBE)-grown $Co_2FeAl_{0.5}Si_{0.5}$/MgO/CoFe(100) MTJs, Tezuka et al. [18] reported TMR ratios of 386 and 832 % at RT and 9 K, respectively, suggesting that better crystallinity is key to achieve high TMR ratios in Heusler/MgO MTJs. The largest TMR in Heusler alloy-based MTJs was demonstrated by Yamamoto's Group in Hokkaido University [19]. For example, their epitaxial Co_2MnSi/MgO/Co_2MnSi MTJ showed TMR ratios of 354 and 1995 % at RT and 4.2 K, respectively (Fig. 17.6). In addition to the record data, the effects of off-stoichiometry and misfit dislocations were also investigated [20], and they showed that the extremely large TMR was attributed to the Mn rich composition in the Co_2MnSi electrodes and to the low misfit dislocation density.

An interesting fact in the implementation of ultrathin bcc-CoFe insertion between an electrode and a barrier in MTJs is that TMR ratios obtained from Heusler alloy/bcc-CoFe/MgO structures can be larger than those from Heusler alloy/MgO or bcc-CoFe/MgO. This fact implies that TMR comes from not only the spin polarization of the interface ferromagnetic atoms but also the bulk spin polarization of the electrode. An example was reported for Co_2FeAl-based MTJs [21], in which TMR was enhanced by inserting a 0.5 nm thick CoFe layer between Co_2FeAl and MgO. The insertion of CoFeB reported in [22] may have a similar effect. Inserting a ultrathin Heusler alloy layer was also studied for controlling the interface electronic structure

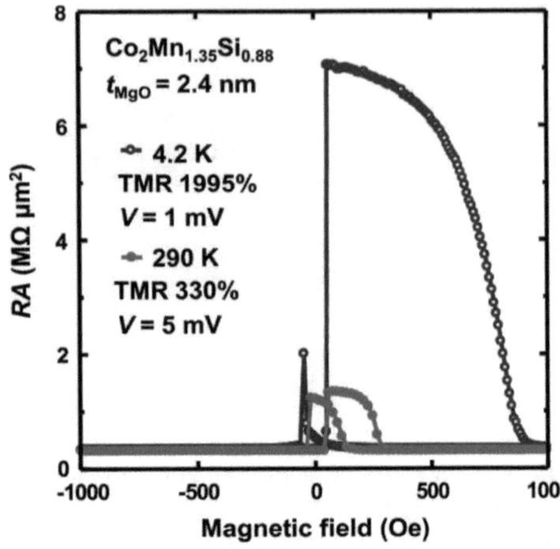

Fig. 17.6 A record TMR ratio in Heusler alloy-based MTJs ($Co_2MnSi/MgO/Co_2MnSi$). Reproduced from [19] (© 2012, American Institute of Physics)

in order to obtain larger TMR at RT [23]. In this study, Co_2MnAl that does not form any interface state is inserted between a highly spin polarized Co_2MnSi electrode and a MgO barrier. A lot of studies on TMR in Heusler alloy-based MTJs have been performed to date. Table 17.1 summarizes some selected observations of TMR previously reported in Co-based full Heusler alloys, in addition to those mentioned above [7–10, 13, 15–38]. MTJs using other Heusler alloys such as NiMnSb and related ordered alloys have also been studied from a view point of their half-metallicity or large magnetocrystalline anisotropy [39–41].

Strong temperature dependence of TMR in Heusler alloy-based MTJs is an important issue for better understanding of physical properties of Heusler alloys, and its mechanism is currently subject to debate. While the record TMR ratio in $Co_2MnSi/MgO/Co_2MnSi$ MTJ reaches approximately 2000 % at low temperature, it rapidly decreases down to around 300 % at RT [19]. As shown in Fig. 17.3, the large temperature dependence was also observed in the Co_2MnSi-based MTJs using an AlO barrier [10], and therefore it should be characteristic of Co_2MnSi Heusler alloy. From (17.1), one may consider that almost the half-metallic state ($P \sim 0.9$) at low temperatures is degraded to a high spin polarization state ($P \sim 0.7$ or lower) with increasing temperature up to RT. However, a systematic study on anisotropic magnetoresistance in Heusler alloys suggested that half-metallic nature sufficiently remains in Co_2MnSi even at RT [42]. Crucial contribution of nonquasiparticle states to the finite-temperature half-metallicity [43], interface state mediate tunneling at finite temperature [44], reduced exchange stiffness of interface manganese spin moments [45, 46] and suppressed interface manganese spin moments [47] etc. have been proposed to explain the observed temperature dependence of TMR. Here, it should be noted that the large temperature dependence of TMR is often observed in MTJs using Mn-containing Heusler alloy such as Co_2MnSi. This trend can be seen in Table 17.1. In MTJs using $Co_2FeAl_{0.5}Si_{0.5}$ or Co_2FeAl, on the other hand, temperature dependence of TMR is likely somewhat moderate, and it can be explained by a model in which temperature dependence of magnetization and spin polarization is

Table 17.1 TMR ratios observed in MTJs using Co-based full-Heusler alloys

Materials	TMR ratio	Year	Authors
Si/SiO_2//$Co_2Cr_{0.6}Fe_{0.4}Al$/AlO/CoFe	16 % (RT), 26 % (5 K)	2003	Inomata et al. [7]
SiO_2//V/Co_2MnSi/AlO/CoFe	33 % (RT), 61 % (10 K)	2004	Kammerer et al. [8]
Si/SiO_2//$Co_2CrFeAl$/AlO/CoFe	47 % (RT)	2005	Okamura et al. [9]
MgO//Cr/Co_2MnSi/AlO/Co_2MnSi	67 % (RT), 570 % (2 K)	2006	Sakuraba et al. [10]
MgO//Cr/$Co_2FeAl_{0.5}Si_{0.5}$/MgAlO/CoFe	102 % (RT), 162 % (26 K)	2009	Shan et al. [13]
MgO//$Co_2Cr_{0.6}Fe_{0.4}Al$/MgO/CoFe	90 % (RT), 240 % (4.2 K)	2006	Marukame et al. [15]
MgO//Cr/Co_2FeAl/MgO/CoFe	330 % (RT), 700 % (10 K)	2009	Wang et al. [16, 17]
MgO//$Co_2FeAl_{0.5}Si_{0.5}$/MgO/CoFe	386 % (RT), 832 % (9 K)	2009	Tezuka et al. [18]
MgO//Co_2MnSi/MgO/Co_2MnSi	354 % (RT), 1995 % (4.2 K)	2012	Liu et al. [19]
MgO//CoFe/MgO/Co_2MnSi	335 % (RT), 1049 % (4.2 K)	2012	Liu et al. [20]
MgO//MgO/Co_2MnSi/MgO/CoFe	173 % (RT), 448 % (4.2 K)	2012	Liu et al. [20]
MgO//CoFe/Co_2MnSi/MgO/CoFe	340 % (RT), 879 % (4.2 K)	2012	Liu et al. [20]
MgO//Cr/Co_2FeAl/MgO/CoFe/Co_2FeAl	360 % (RT), 785 % (10 K)	2010	Wang et al. [21]
MgO//Co_2MnSi/CoFeB/MgO/CoFe	222 % (RT), 510 % (10 K)	2009	Tsunegi et al. [22]
MgO//Co_2MnSi/Co_2MnAl/MgO/CoFe	180 % (RT), 600 % (10 K)	2010	Ozawa et al. [23]
MgO//Co_2MnSi/MgO/CoFe	217 % (RT), 753 % (2 K)	2008	Tsunegi et al. [24]
MgO//Co_2MnGe/MgO/CoFe	83 % (RT), 185 % (2 K)	2007	Hakamata et al. [25]
MgO//Co_2MnSi/MgO/CoFe	90 % (RT), 192 % (4.2 K)	2006	Ishikawa et al. [26]
MgO//Cr/Fe/MgO/Co_2FeSn	43 % (RT), 72 % (2 K)	2012	Tanaka et al. [27]
MgO//Co_2MnSi/MgO/Co_2MnSi	236 % (RT), 1135 % (4.2 K)	2009	Ishikawa et al. [28]
MgO//Co_2MnGe/MgO/CoFe	160 % (RT), 376 % (2 K)	2009	Taira et al. [29]
MgO//$Co_2Cr_{0.6}Fe_{0.4}Al$/AlO/CoFe	101 % (4 K)	2009	Herbort et al. [30]
MgO//Cr/Co_2FeSi/MgO/CoFeB	40 % (RT)	2008	Lim et al. [31]

(continued)

Table 17.1 (continued)

Materials	TMR ratio	Year	Authors
Si/SiO_2//CoFeB/MgO/Co_2FeSi	158 % (RT)	2008	Lim et al. [31]
Si/SiO_2//$Co_2Mn_{0.5}Fe_{0.5}$Si/AlO/CoFe	43 % (RT), 103 % (12 K)	2008	Schmalhorst et al. [32]
Sapphire//Co_2MnSi(110)/AlO/CoFe	40 % (RT), 120 % (2 K)	2008	Hattori et al. [33]
MgO//Co_2FeSi/AlO/CoFe	43 % (RT), 67 % (5 K)	2006	Gercsi et al. [34]
Si//Co_2FeSi/AlO/CoFe	28 % (RT), 44 % (30 K)	2014	Fujita et al. [35]
MgO//Cr/Co_2FeAl/MgO/Fe/CoFeB [PMA]	91 % (RT)	2012	Wen et al. [36]
MgO//Ru/Co_2FeAl/MgO/Fe/CoFeB [PMA]	132 % (RT)	2014	Wen et al. [37]
Si/SiO_2//MgO/Co_2FeAl/MgO/CoFe	166 % (RT), 252 % (48 K)	2011	Wen et al. [38]

[PMA] means that the MTJ is perpendicularly magnetized

described based on thermally excited spin waves [11, 21]. This model was originally proposed for understanding the temperature dependence of TMR in Co/AlO/Co and Co/AlO/NiFe MTJs [48].

Differential conductance (dI/dV) measurements are useful to investigate spin-dependent tunneling characteristics of MTJs although the physical mechanism of bias voltage dependence of TMR has not fully been understood. In a Co_2MnSi/AlO/CoFe MTJ, dI/dV was measured as a function of bias voltage V at low temperature, and the result obtained clearly showed the sign of half-metallic band gap in Co_2MnSi [49]. A Co_2FeAl/MgO/CoFe MTJ is also a representative example exhibiting a similar feature in dI/dV-V, which presumably reflects a half-metallic band gap, as shown in Fig. 17.7 [17]. However, the latter one corresponds to the absence of Δ_1 symmetry states around Fermi level in the minority-spin electron energy band, rather than a total band energy gap. It is also noted that for better understanding the specific feature deviated from a conventional parabola should be assigned to band structures. The details of dI/dV measurements in Co_2FeAl-based MTJs, as well as the temperature dependence of TMR, are described in [50].

In the end of this section, let us have a look at functions with which Heusler alloy-based MTJs have been equipped. Spin transfer torque (STT) magnetization switching in MTJs is an indispensable technique for developing high-density MRAMs. Sukegawa et al. first clearly demonstrated STT switching in MTJs using a Heusler alloy [51]. Co_2FeAl layers were used for a free or reference layer in the submicron-sized MTJs, and the critical currents needed for magnetization switching were determined. Although the intrinsic critical current density ($J_{c0} = 2.9 \times 10^7$ A/cm^2) in the MTJ with a Co_2FeAl free layer was not as small as that expected from the low magnetic damping constant of Co_2FeAl [52], typical STT switching behavior was observed. Damping measurements for the very thin Co_2FeAl free layer revealed that the large J_{c0} is due to the extrinsic effect to enhance the effective damping constant. Perpendicular magnetization is another function required for high density MRAMs. Interface perpendicular magnetic anisotropy (PMA) in Heusler alloy-based MTJs was also first demonstrated using very thin Co_2FeAl layers.

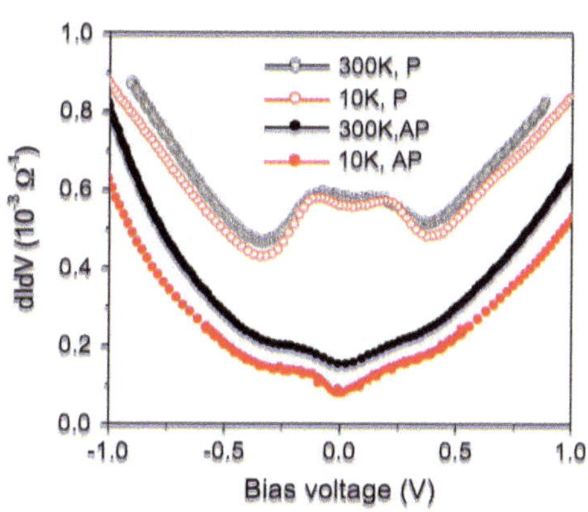

Fig. 17.7 Differential conductance curves (dI/dV-V) in a Co_2FeAl/MgO/CoFe MTJ. Reproduced from [17] (© 2010, American Physical Society)

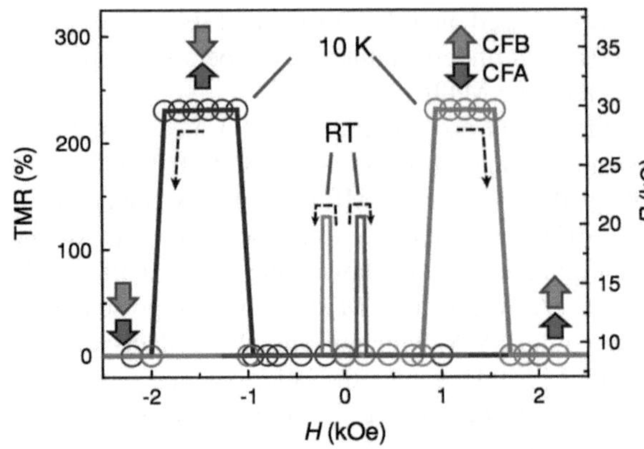

Fig. 17.8 TMR in perpendicularly magnetized a Ru-buffered $Co_2FeAl/MgO/Fe/CoFeB$ MTJ. Reproduced from [37] (© 2014, John Wiley and Sons, Inc.)

Wen et al. [36, 53] observed PMA energy density of 2–3 × 10^6 erg/cm^3 in a stack of MgO(100)/Cr/Co_2FeAl/MgO, and TMR ratios up to 91 % was obtained in the MTJs. Further large PMA and TMR were obtained by replacing the Cr buffer with a uniquely grown Ru buffer (Fig. 17.8) [37]. In addition to the above mentioned functions, recent progress in polycrystalline (highly (100)-oriented) Heusler alloy-based MTJs is worth referring to, from an application point of view. Even in MTJs grown on thermally oxidized Si substrates, appropriate buffer layers made TMR ratios comparable to those in the case of MgO(100) single crystal substrates [38]. Furthermore, STT switching was performed in low-resistive Co_2FeAl-based MTJs grown on a Si substrate [54], and integration with metal-oxide-semiconductor field effect transistors was demonstrated [55].

References

1. S.S.P. Parkin, K.P. Roche, M.G. Samant, P.M. Rice, R.B. Beyers, R.E. Scheuerlein, E.J. O'Sullivan, S.L. Brown, J. Bucchigano, D.W. Abraham, Lu Yu, M. Rooks, P.L. Trouilloud, R.A. Wanner, W.J. Gallagher, J. Appl. Phys. **85**, 5828 (1999)
2. S. Yuasa, D.D. Djayaprawira, J. Phys. D: Appl. Phys. **40**, R337 (2007)
3. M. Julliere, Phys. Lett. **54A**, 225 (1975)
4. S. Maekawa, U. Gafvert, IEEE Trans. Magn. **18**, 707 (1982)
5. M. Bowen, M. Bibes, A. Barthelemy, J.-P. Contour, A. Anane, Y. Lemaitre, A. Fert, Appl. Phys. Lett. **82**, 233 (2003)
6. T. Block, C. Felser, G. Jakob, J. Ensling, B. Muhling, P. Gutlich, R.J. Cava, J. Solid State Chem. **176**, 646 (2003)
7. K. Inomata, S. Okamura, R. Goto, N. Tezuka, Jpn. J. Appl. Phys. **42**, L419 (2003)
8. S. Kammerer, A. Thomas, A. Hutten, G. Reiss, Appl. Phys. Lett. **85**, 79 (2004)
9. S. Okamura, A. Miyazaki, S. Sugimoto, N. Tezuka, K. Inomata, Appl. Phys. Lett. **86**, 232503 (2005)
10. Y. Sakuraba, M. Hattori, M. Oogane, Y. Ando, H. Kato, A. Sakuma, T. Miyazaki, H. Kubota, Appl. Phys. Lett. **88**, 192508 (2006)
11. G.H. Fecher, C. Felser, J. Phys. D: Appl. Phys. **40**, 1582 (2007)

12. T.M. Nakatani, A. Rajanikanth, Z. Gercsi, Y.K. Takahashi, K. Inomata, K. Hono, J. Appl. Phys. **102**, 033916 (2007)
13. R. Shan, H. Sukegawa, W.H. Wang, M. Kodzuka, T. Furubayashi, T. Ohkubo, S. Mitani, K. Inomata, K. Hono, Phys. Rev. Lett. **102**, 246601 (2009)
14. H. Sukegawa, H. Xiu, T. Ohkubo, T. Furubayashi, T. Niizeki, W.H. Wang, S. Kasai, S. Mitani, K. Inomata, K. Hono, Appl. Phys. Lett. **96**, 212501 (2010)
15. T. Marukame, T. Ishikawa, K.-I. Matsuda, T. Uemura, M. Yamamoto, Appl. Phys. Lett. **88**, 262503 (2006)
16. W.H. Wang, H. Sukegawa, R. Shan, S. Mitani, K. Inomata, Appl. Phys. Lett. **95**, 182502 (2009)
17. W.H. Wang, E. Liu, M. Kodzuka, H. Sukegawa, M. Wojcik, E. Jedryka, G.H. Wu, K. Inomata, S. Mitani, K. Hono, Phys. Rev. B **81**, 140402(R) (2010)
18. N. Tezuka, N. Ikeda, F. Mitsuhashi, S. Sugimoto, Appl. Phys. Lett. **94**, 162504 (2009)
19. H.-X. Liu, Y. Honda, T. Taira, K.-I. Matsuda, M. Arita, T. Uemura, M. Yamamoto, Appl. Phys. Lett. **101**, 132418 (2012)
20. H.-X. Liu, Y. Honda, K.-I. Matsuda, M. Arita, T. Uemura, M. Yamamoto, Jpn. J. Appl. Phys. **51**, 093004 (2012)
21. W.H. Wang, H. Sukegawa, K. Inomata, Phys. Rev. B **82**, 092402 (2010)
22. S. Tsunegi, Y. Sakuraba, M. Oogane, H. Naganuma, K. Takanashi, Y. Ando, Appl. Phys. Lett. **94**, 252503 (2009)
23. E. Ozawa, S. Tsunegi, M. Oogane, H. Naganuma, Y. Ando, J. Phys.: Conf. Ser. **266**, 012104 (2011)
24. S. Tsunegi, Y. Sakuraba, M. Oogane, K. Takanashi, Y. Ando, Appl. Phys. Lett. **93**, 112506 (2008)
25. S. Hakamata, T. Ishikawa, T. Marukame, K.-I. Matsuda, T. Uemura, M. Arita, M. Yamamoto, J. Appl. Phys. **101**, 09J513 (2007)
26. T. Ishikawa, T. Marukame, H. Kijima, K.-I. Matsuda, T. Uemura, M. Arita, M. Yamamoto, Appl. Phys. Lett. **89**, 192505 (2006)
27. M.A. Tanaka, Y. Ishikawa, Y. Wada, S. Hori, A. Murata, S. Horii, Y. Yamanishi, K. MIbu, K. Kondou, T. Ono, S. Kasai, J. Appl. Phys. **111**, 053902 (2012)
28. T. Ishikawa, H.-X. Liu, T. Taira, K.-I. Matsuda, T. Uemura, M. Yamamoto, Appl. Phys. Lett. **95**, 232512 (2009)
29. T. Taira, T. Ishikawa, N. Itabashi, K.-I. Matsuda, T. Uemura, M. Yamamoto, J. Phys. D: Appl. Phys. **42**, 084015 (2009)
30. C. Herbort, E.A. Jorge, M. Jourdan, Appl. Phys. Lett. **94**, 142504 (2009)
31. W.C. Lim, G.A. Choi, T.D. Lee, S.A. Seo, IEEE Trans. Magn. **44**, 2595 (2008)
32. J. Schmalhorst, D. Ebke, A. Weddemann, A. Hutten, A. Thomas, G. Reiss, A. Turchanin, A. Golzhauser, B. Balke, C. Felser, J. Appl. Phys. **104**, 043918 (2008)
33. M. Hattori, Y. Sakuraba, M. Oogane, Y. Ando, T. Miyazaki, Appl. Phys. Express **1**, 021301 (2008)
34. Z. Gercsi, A. Rajanikanth, Y.K. Takahashi, K. Hono, M. Kikuchi, N. Tezuka, K. Inomata, Appl. Phys. Lett. **89**, 082512 (2006)
35. Y. Fujita, S. Yamada, Y. Maeda, M. Miyao, K. Hamaya, Thin Solid Films **557**, 386 (2014)
36. Z.C. Wen, H. Sukegawa, S. Kasai, M. Hayashi, S. Mitani, K. Inomata, Appl. Phys. Express **5**, 063003 (2012)
37. Z.C. Wen, H. Sukegawa, T. Furubayashi, J.W. Koo, K. Inomata, S. Mitani, J.P. Hadorn, T. Ohkubo, K. Hono, Adv. Mater. **26**, 6483 (2014)
38. Z.C. Wen, H. Sukegawa, S. Mitani, K. Inomata, Appl. Phys. Lett. **98**, 192505 (2011)
39. C.T. Tanaka, J. Nowak, J.S. Moodera, J. Appl. Phys. **81**, 5515 (1997)
40. C.E. ViolBarbosa, S. Ouardi, T. Kubota, S. Mizukami, G.H. Fecher, T. Miyazaki, X. Kozina, E. Ikenaga, C. Felser, J. Appl. Phys. **116**, 034508 (2014)
41. Q.L. Ma, A. Sugihara, K. Suzuki, X. Zhang, T. Miyazaki, S. Mizukami, SPIN **4**, 1440024 (2014)
42. Y. Sakuraba, S. Kokado, Y. Hirayama, T. Furubayashi, H. Sukegawa, S. Li, Y.K. Takahashi, K. Hono, Appl. Phys. Lett. **104**, 172407 (2014)

43. L. Chioncel, Y. Sakuraba, E. Arrigoni, M.I. Katsnelson, M. Oogane, Y. Ando, T. Miyazaki, E. Burzo, A.I. Lichtenstein, Phys. Rev. Lett. **100**, 086402 (2008)
44. T. Ishikawa, N. Itabashi, T. Taira, K.-I. Matsuda, T. Uemura, M. Yamamoto, Appl. Phys. Lett. **94**, 092503 (2009)
45. A. Sakuma, Y. Toga, H. Tsuchiura, J. Appl. Phys **105**, 07C910 (2009)
46. Y. Miura, K. Abe, M. Shirai, Phys. Rev. B **83**, 214411 (2011)
47. S. Tsunegi, Y. Sakuraba, K. Amemiya, M. Sakamaki, E. Ozawa, A. Sakuma, K. Takanashi, Y. Ando, Phys. Rev. B **85**, 180408(R) (2012)
48. C.H. Shang, J. Nowak, R. Jansen, J.S. Moodera, Phys. Rev. B **58**, R2917 (1998)
49. Y. Sakuraba, T. Miyakoshi, M. Oogane, Y. Ando, A. Sakuma, T. Miyazaki, H. Kubota, Appl. Phys. Lett. **89**, 052508 (2006)
50. K. Inomata, H. Sukegawa, in *Spintronics From Materials to Devices*, ed. by C. Felser, G.H. Fecher (Springer, New York, 2013)
51. H. Sukegawa, Z.C. Wen, K. Kondou, S. Kasai, S. Mitani, K. Inomata, Appl. Phys. Lett. **100**, 182403 (2012)
52. S. Mizukami, D. Watanabe, M. Oogane, Y. Ando, Y. Miura, M. Shirai, T. Miyazaki, J. Appl. Phys. **105**, 07D306 (2009)
53. Z.C. Wen, H. Sukegawa, S. Mitani, K. Inomata, Appl. Phys. Lett. **98**, 242507 (2011)
54. Z.C. Wen, H. Sukegawa, S. Kasai, K. Inomata, S. Mitani, Phys. Rev. Appl. **2**, 024009 (2014)
55. R. Nakane, Y. Shuto, H. Sukegawa, Z.C. Wen, S. Yamamoto, S. Mitani, M. Tanaka, K. Inomata, S. Sugahara, Solid State Electron. **102**, 52 (2014)

Chapter 18
Effect of Nonstoichiometry on the Half-Metallic Character of Co_2MnSi and Its Application to the Spin Sources of Spintronic Devices

Masafumi Yamamoto and Tetsuya Uemura

Abstract To take full advantage of the half-metallic character of Co-based Heusler alloys, the effect of defects associated with nonstoichiometry has to be understood. In this chapter, recent progress in understanding the effect of nonstoichiometry on the half-metallicy of Heusler alloys, in particular of Co_2MnSi (CMS), and its application to highly efficient spin sources for magnetic tunnel junctions (MTJs) and for spin injection into semiconductors is described. The effect of nonstoichiometry on the half-metallic character is experimentally investigated through the saturation magnetization per formula unit (μ_s) of $Co_2Mn_\alpha Si_\beta$ thin films and the tunneling magnetoresistance (TMR) ratio of fully epitaxial CMS/MgO/CMS MTJs (CMS MTJs) having $Co_2Mn_\alpha Si_\beta$ electrodes with various values of α. It was found that the μ_s value was in good agreement with the half-metallic Slater-Pauling value when α was increased to a Mn-rich composition. It was also shown that the TMR ratios at 4.2 and 290 K systematically increased with increasing α for the range from a Mn-deficient composition to a certain Mn-rich composition. A site-specific formula unit (SSFU) composition model, which assumes the formation of antisite defects, not vacancies, to accommodate nonstoichiometry, is described. The experimental α dependencies of μ_s and the TMR ratio have been consistently explained by first-principles calculations based on the SSFU composition model. These findings show that harmful defects in Co_2MnSi can be suppressed by appropriately controlling the film composition; i.e., Co_{Mn} antisites detrimental to the half-metallicity can be suppressed with a Mn-rich composition. By applying Mn-rich CMS electrodes for fully epitaxial MgO-based MTJs, giant TMR ratios of up to 1995 % at 4.2 K and up to 354 % at 290 K were demonstrated for CMS MTJs. Furthermore, an efficient spin

M. Yamamoto (✉) · T. Uemura
Division of Electronics for Informatics, Graduate School of Information Science and Technology, Hokkaido University, Sapporo 060-0814, Japan
e-mail: yamamoto@nano.ist.hokudai.ac.jp

T. Uemura
e-mail: uemura@ist.hokudai.ac.jp

C. Felser and A. Hirohata (eds.), *Heusler Alloys*,
Springer Series in Materials Science 222, DOI 10.1007/978-3-319-21449-8_18

injection into GaAs was demonstrated for lateral spin-transport devices by applying a half-metallic Mn-rich CMS electrode as a highly effective spin source. In addition, a nuclear field acting on the electron spins being produced by the dynamic nuclear polarization was electrically detected through the observation of transient oblique Hanle signals. Samples with a CMS spin source exhibited higher spin injection efficiency and a larger nuclear field compared to samples with a $Co_{50}Fe_{50}$ (CoFe) spin source, suggesting that the spin polarization of CMS is higher. These findings demonstrate that controlling defects through the film composition is critical to retain the half-metallicity of CMS. It is also clear that the half-metallic Co_2MnSi electrodes are promising as a highly efficient spin source for future spintronic devices.

18.1 Introduction

A highly efficient spin source is essential for constructing spintronic devices that manipulate the spin degree of freedom in addition to the charge of the electron. Half-metallic ferromagnets (HMFs) are characterized by the existence of an energy gap for one spin direction, providing complete spin polarization at the Fermi level (E_F) [1]. Because of this characteristic, HMFs are one of the most suitable spin source materials for spintronic devices. Co-based Heusler alloys (Co_2YZ, where Y is usually a transition metal and Z is a main group element) are among the most extensively applied to spintronic devices, including magnetic tunnel junctions (MTJs) [2–14] and giant magnetoresistance (GMR) devices [15–19], and for spin injection into semiconductors [20–26]. This is because of the theoretically predicted half-metallicity that many of them possess [27–29] and their high Curie temperatures, which are well above room temperature (RT) [30]. Furthermore, epitaxial Co_2YZ/MgO MTJs with Co_2YZ electrodes having Δ_1 states at E_F for the majority-spin band, such as Co_2MnSi [7], Co_2MnGe [5], $Co_2FeAl_{0.5}Si_{0.5}$ [13], and Co_2FeAl [14], benefit not only from the high spin polarization of the Co_2YZ electrodes arising from their half-metallic or nearly half-metallic nature, but also from the contribution of coherent tunneling of electrons in Δ_1 states to the enhancement of the tunneling magnetoresistance (TMR) ratio [31–35].

Co_2MnSi is one of the most extensively investigated ferromagnetic electrode materials among the Co-based Heusler alloys for MTJs [3, 6–12], GMR devices [15, 17, 18], and for spin injection into semiconductors [23, 26]. This is because of its theoretically predicted half-metallic nature [27, 28] with a large energy gap of 0.81 eV for its minority-spin band [28] and its high Curie temperature of 985 K [30]. Co_2MnSi-based epitaxial GMR devices with a Ag spacer also feature both the half-metallicity of Co_2MnSi and the matching of the Fermi surfaces of Co_2MnSi and Ag [17, 36].

In practice, to various degrees, nonstoichiometry in Co_2YZ thin films prepared by magnetron sputtering or molecular beam epitaxy is inevitable, leading to structural defects. Note that there is no minority-spin density of states at E_F for HMFs. If the minority-spin states at E_F are introduced by a structural defect in a potential HMF, the half-metallicity is destroyed. In this sense, the effect of such defects is critical for HMFs. Thus, it is highly important to clarify the effect of structural defects on the half-metallicity of potentially half-metallic Co_2YZ such as Co_2MnSi and Co_2MnGe. It is also required to find an approach that suppresses detrimental defects in order to maintain the half-metallicity.

The effect of defects in Co_2YZ on spin-dependent electronic structures has been investigated theoretically [37–40]. Picozzi et al. [37] theoretically predicted that the half-metallicity in Co_2MnSi and Co_2MnGe is lost for Co_{Mn} antisites, where a Mn site is replaced by a Co atom because of the appearance of minority-spin in-gap states just near E_F, while the half-metallicity is retained for Mn_{Co} antisites, where a Co site is replaced by a Mn atom. Galanakis et al. [39] theoretically predicted that the half-metallicity of Co_2MnSi is robust against disorder between Mn and Si in $Co_2Mn_{1+x}Si_{1-x}$ and $Co_2Mn_{1-x}Si_{1+x}$. Similarly, Miura et al. [38] theoretically predicted that disorder between Co sites and Cr sites in Co_2CrAl leads to a considerable reduction in the spin polarization at E_F, while disorder between Cr sites and Al sites does not significantly decrease the spin polarization.

There have been extensive experimental studies along with first-principles calculations recently on the effect of nonstoichiometry on the half-metallicy of Co-based Heusler alloys, in particular for Co_2MnSi (CMS) and Co_2MnGe (CMG). These studies include (1) the TMR characteristics of CMS/MgO/CMS MTJs (CMS MTJs) [9–12], $Co_{50}Fe_{50}$/MgO/CMS MTJs [41] and CMG/MgO/CMG MTJs [10], (2) the saturation magnetization per formula unit for CMG films [10] and CMS films [12], (3) the surface polarization for CMS films [42], (4) the Co and Mn spin magnetic moments investigated by x-ray absorption spectroscopy (XAS) and x-ray magnetic circular dichroism (XMCD) for CMG films [43, 44], and (5) the valence band investigated by hard x-ray photoelectron spectroscopy for CMG films [45] and CMS films [46].

In this chapter, we describe the recent progress in understanding the effect of nonstoichiometry on the half-metallicity of Heusler alloy thin films, in particular of Co_2MnSi, and its application to highly efficient spin sources for MTJs and for spin injection into semiconductors. As a basis for understanding the effect of nonstoichiometry in Co_2MnSi and Co_2MnGe, a model of the site-specific formula unit (SSFU) composition for nonstoichiometric $Co_2Mn_\alpha Si_\beta$ and $Co_2Mn_\alpha Ge_\beta$ was proposed [10, 12] that assumes the formation of antisite defects, not vacancies, to accommodate nonstoichiometry. This model was based on the theoretically predicted higher formation energies of vacancies at the nominal Co, Mn, and Si sites compared with various kinds of antisites for Co_2MnSi [37, 40]. The first-principles calculations based on the SSFU composition model were in good agreement with the various experimental results [12, 42, 44–46]. The origin of lower TMR ratios observed for MTJs with Mn-deficient CMS electrodes was ascribed to the formation of Co_{Mn} antisites [10, 12]. More specifically, a decrease in the TMR ratio observed for CMS MTJs having Mn-deficient electrodes was ascribed to small s- and p-orbital components of the local density of minority-spin in-gap states at the Fermi level that appeared for antisite Co_{Mn} atoms and Co atoms at the regular sites (Co_{Co}) [12]. In contrast, giant TMR ratios of up to 1995 % at 4.2 K and up to 354 % at 290 K were demonstrated for CMS MTJs having Mn-rich CMS electrodes [9–11]. These giant TMR ratios were explained by the suppression of residual Co_{Mn} antisites in CMS films by Mn-rich compositions [9–12]. Electrical spin injection from a half-metallic Mn-rich CMS electrode into a GaAs channel was also demonstrated through the observation of spin-valve signals and Hanle signals in a four-terminal nonlocal geometry [23, 26]. Furthermore, a nuclear field acting on electron spins being produced by the dynamic nuclear polarization was electrically detected through the observation of transient oblique Hanle signals [23]. Samples with a CMS spin source exhibited higher spin injection efficiency and a larger nuclear field compared to samples with a $Co_{50}Fe_{50}$ (CoFe) spin source, suggesting that the spin polarization of CMS is higher [23, 26]. This higher polarization is promising for realizing future spintronic devices and for advancing the understanding of spin interactions as well as spin-dependent transport properties in a semiconductor channel.

This chapter is organized as follows. Section 18.2 describes the effect of nonstoichiometry on the half-metallic character of Co_2MnSi. Section 18.2.1 describes the experimental methods. Section 18.2.2 describes the SSFU composition model based on the antisite formation to accommodate nonstoichiometry. Section 18.2.3 describes the structural properties of CMS films with various α values in $Co_2Mn_\alpha Si_\beta$. It is shown that the structural properties of these films from Mn-deficient to Mn-rich are in good agreement with the SSFU composition model. Section 18.2.4 describes how the saturation magnetization per formula unit, μ_s, varies with α. Section 18.2.5 describes the α dependence of the TMR ratios of CMS MTJs having $Co_2Mn_\alpha Si_\beta$ electrodes. In combination with the results of density-functional theory calculations with coherent potential approximation, it is shown that the experimental α dependence of μ_s and the TMR ratio can be consistently explained by the existence of harmful Co_{Mn} antisites for Mn-deficient CMS and the suppression of them for Mn-rich CMS.

Section 18.3 describes spin injection into GaAs from a half-metallic CMS electrode and the resultant dynamic nuclear polarization. Section 18.3.1 describes the experimental methods in spin injection from CMS into GaAs. Section 18.3.2 describes the experimental demonstration of all electrical injection and detection of spin polarized electrons through the observation of spin-valve signals and Hanle signals in a four-terminal nonlocal geometry. The bias-voltage dependence and temperature dependence of spin signals are then compared between CMS/CoFe/n-GaAs and CoFe/n-GaAs samples, and the spin injection efficiency of the CMS electrode is discussed. Section 18.3.3 describes the transient oblique Hanle signals, and discusses the model for the Overhauser field in a transient state. While previously the electrical detection of oblique Hanle signals has been done only in a steady state, here, oblique Hanle signals in a transient state, in which the magnetic field is swept faster than the relaxation time of the nuclear spins, are electrically detected. This gives us an important insight in terms of understanding nuclear spin dynamics, especially the transient response of nuclear spins to a change in the magnetic field and the characteristic time needed for the nuclear spins to reach a steady state through the dynamic nuclear polarization. Section 18.4 concludes the chapter.

18.2 Effect of Nonstoichiometry on the Half-Metallic Character of Co_2MnSi

18.2.1 Experimental Methods

In order to investigate the effect of nonstoichiometry on the structural, electronic and magnetic properties of CMS thin films, $Co_2Mn_\alpha Si_\beta$ films having various values of α and a constant value of β for each film series were prepared. The sample layer structure for these characterizations was (from the substrate side) MgO buffer (10 nm)/CMS (~30 nm)/MgO barrier (2 nm)/AlO_x (1 nm) cap layer, grown on a MgO(001) substrate. This layer structure corresponds to half of an MTJ. A MgO barrier was used to protect the CMS film surface because it has been confirmed that the CMS layer is not oxidized by MgO deposition by means of x-ray absorption spectroscopy and x-ray magnetic circular dichroism [47] and because of the demonstrated high quality of the CMS/MgO interface [9–11, 41].

The CMS thin film was deposited at room temperature (RT) using radio-frequency magnetron cosputtering from a nearly stoichiometric CMS target and a Mn target and was subsequently annealed in situ at 600 °C for 15 min. The Mn composition α in the $Co_2Mn_\alpha Si_\beta$ was systematically varied by adjusting the relative amount of Mn sputtered from a Mn target by cosputtering [9]. The film composition of the prepared CMS film was determined by inductively coupled plasma analysis with an accuracy of 2–3 % for each element except Si, for which the accuracy was 5 %.

Next, we will describe the preparation of MgO-buffered and $Co_{50}Fe_{50}$ (CoFe)-buffered CMS MTJs. The MTJ layer structure for MgO-buffered CMS MTJs was as follows: (from the substrate side) MgO buffer (10 nm)/CMS lower electrode (30 nm)/MgO barrier (2.0–3.0 nm)/CMS upper electrode (3–5 nm)/layers for exchange biasing/Ru cap (5 nm), grown on a MgO(001) single-crystal substrate [9]. The nominal thickness of the MgO tunnel barrier (t_{MgO}) was varied from 2.0 to 3.0 nm on each 20×20 mm^2 substrate by a linearly moving shutter. For MgO-buffered MTJs, a thick (30-nm) CMS lower electrode grown on a MgO-buffered MgO(001) substrate was used. The thick CMS layer was relaxed on the MgO-buffered MgO(001) substrate, causing a relatively large mismatch of -5.1 % between the fully relaxed CMS layer and the MgO barrier with a 45° in-plane rotation.

Given that the lattice mismatch between Co_2MnSi and CoFe is very small (-0.8 %) and that the lattice mismatch of -4.3 % between CoFe and MgO with a 45° in-plane rotation is smaller than -5.1 %, which is the lattice mismatch between a fully relaxed Co_2MnSi lower electrode and MgO, a thin (typically 3-nm) CMS lower electrode grown on a CoFe-buffered MgO(001) substrate was used for CoFe-buffered CMS MTJs [11]. This resulted in a much larger misfit dislocation spacing of 6.3 ± 0.1 nm at the lower and upper MgO barrier interface for the CoFe-buffered CMS MTJ [11] than the 4.3 ± 0.1 nm for the MgO-buffered CMS MTJ [48]. The MTJ layer structure for CoFe-buffered CMS MTJs was (from the substrate side) MgO buffer (10 nm)/CoFe buffer (30 nm)/CMS lower electrode (3 nm)/MgO barrier (1.4–3.2 nm)/CMS upper electrode (3 nm)/layers for exchange biasing/Ru cap (5 nm), grown on a MgO(001) single-crystal substrate. The layer structures for exchange biasing for these MTJ series were (above the Co_2MnSi upper electrode) Ru (0.8 nm)/$Co_{90}Fe_{10}$ (2 nm)/$Ir_{22}Mn_{78}$ (10 nm) or $Co_{50}Fe_{50}$ (1.1 nm)/$Ir_{22}Mn_{78}$ (10 nm). The slightly different layer structures for exchange biasing were not critical for the TMR characteristics.

Figure 18.1a shows a cross-sectional high-resolution lattice image of a CoFe-buffered CMS MTJ layer structure consisting of (from the lower side) CoFe buffer/lower CMS (3 nm)/MgO barrier (~2.5 nm)/upper CMS (3 nm)/Ru (0.8 nm)/ $Co_{90}Fe_{10}$ (2 nm)/$Ir_{22}Mn_{78}$ (10 nm)/Ru cap with Mn-rich $Co_2Mn_{1.29}Si_{1.0}$ electrodes [11]; it was taken along the [1–10] direction of the CMS layers. This image clearly shows that all layers were grown epitaxially and were single-crystalline. It was also shown that atomically flat and abrupt interfaces were formed. Figure 18.1b, c show nano-beam electron diffraction patterns (beam diameter: 2 nm) for the CMS lower and upper electrodes, where 111 spots specific to the $L2_1$ structure were observed in addition to 002 spots specific to the B2 and $L2_1$ structures, indicating that both the Mn-rich $Co_2Mn_{1.29}Si_{1.0}$ lower and upper electrodes had the $L2_1$ structure.

We defined the TMR ratio as $(R_{AP} - R_P)/R_P$, where R_{AP} and R_P are the tunneling resistances at a small dc voltage for the antiparallel and parallel magnetization configurations between the upper and lower electrodes.

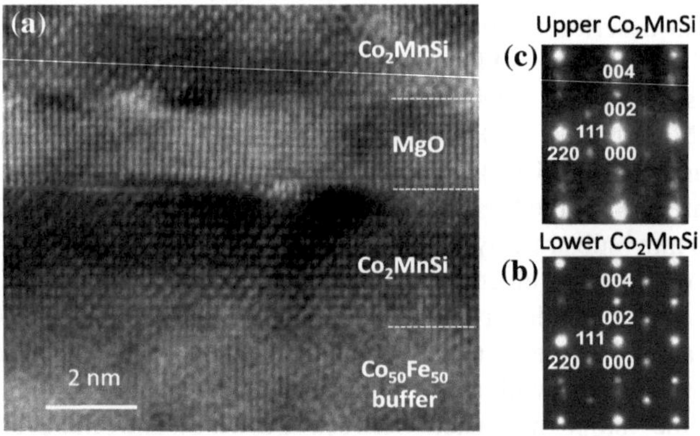

Fig. 18.1 **a** Cross-sectional high-resolution transmission electron microscopy lattice image of a CoFe-buffered Co_2MnSi (CMS)/MgO/CMS MTJ layer structure consisting of (from the lower side) $Co_{50}Fe_{50}$ (CoFe) buffer/lower CMS (3 nm)/MgO barrier (~2.5 nm)/upper CMS (3 nm)/Ru (0.8 nm)/$Co_{90}Fe_{10}$ (2 nm)/$Ir_{22}Mn_{78}$ (10 nm)/Ru cap (5 nm) with Mn-rich $Co_2Mn_{1.29}Si_{1.0}$ electrodes. Image was taken along the [1–10] direction of the Co_2MnSi. **b** and **c** show nano-beam electron diffraction patterns for the lower and upper Co_2MnSi electrodes, respectively. Beam diameter was 2 nm. Reprinted with permission from [11]. Copyright 2012, American Institute of Physics

18.2.2 Formula Unit Composition Model

As a basis for understanding the effect of nonstoichiometry in CMS, a site-specific formula unit (SSFU) composition model was developed [10, 12]. This model assumes the formation of only antisite defects, not vacancies, to accommodate nonstoichiometry. It also emphasizes the occupancy at each site in the $L2_1$ lattice. This model is based on the theoretically predicted higher formation energies of vacancies at the nominal Co, Mn, and Si sites compared with various kinds of antisites for Co_2MnSi [37, 40]. In this model, each site in the $L2_1$ structure is occupied, either by the nominal atom (Co, Mn, or Si) for that site or by an antisite defect atom, so the total number of atoms included in the formula unit is always four, even though the film is nonstoichiometric in terms of the Co:Mn:Si atomic ratio. Thus, this model preserves a generalized 2:1:1 stoichiometry in terms of the site occupancy and is called the antisite-based SSFU composition model.

Given this model, three types of SSFU compositions (Table 18.1) were proposed. A schematic diagram showing the regions of SSFU composition types I to III in the $\alpha - \beta$ plane for $Co_2Mn_\alpha Si_\beta$ with the restrictions of $0 < \alpha < 2$ and $0 < \beta < 2$ is plotted in Fig. 18.2. Let us first consider $Co_2Mn_\alpha Si_{1.0}$ having a fixed β of 1.0 and various α ranging from a Mn-deficient composition ($\alpha < 1$) to a Mn-rich one ($\alpha > 1$). For $\alpha < 1$, some of the Mn sites are filled by surplus Co and Si atoms, creating Co_{Mn} and Si_{Mn} antisites, both of which have lower calculated formation energy than

Table 18.1 General expressions for site-specific formula unit compositions of type I to type III for $Co_2Mn_\alpha Si_\beta$ with restrictions $0 < \alpha < 2$ and $0 < \beta < 2$

Type no.	General expression for site-specific formula unit composition	x	y
I	$Co_2[Mn_{1-x-y}Co_xZ_y]Z$	$\frac{2(2-(\alpha+\beta))}{2+\alpha+\beta}$	$\frac{3\beta-(2+\alpha)}{2+\alpha+\beta}$
II	$Co_2[Mn_{1-x}Co_x][Z_{1-y}Mn_y]$	$\frac{2(2-(\alpha+\beta))}{2+\alpha+\beta}$	$\frac{2+\alpha-3\beta}{2+\alpha+\beta}$
III	$[Co_{2-x}Mn_x]Mn[Z_{1-y}Mn_y]$	$\frac{2(\alpha+\beta-2)}{2+\alpha+\beta}$	$\frac{2+\alpha-3\beta}{2+\alpha+\beta}$

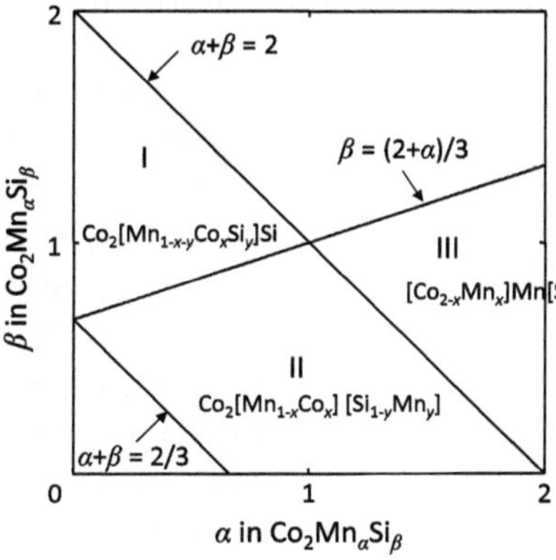

Fig. 18.2 Schematic diagram showing regions of site-specific formula unit composition types I to III in the $\alpha - \beta$ plane for $Co_2Mn_\alpha Si_\beta$ with restrictions of $0 < \alpha < 2$ and $0 < \beta < 2$. Parameters x and y for each type are given as functions of α and β in Table 18.1. Reprinted with permission from [12]. Copyright 2014, American Physical Society

Mn vacancies [37, 40]. The resulting SSFU composition is $Co_2[Mn_{1-x-y}Co_xSi_y]Si$, which is called a type I SSFU composition. $[Mn_{1-x-y}Co_xSi_y]$ is the composition on the nominal Mn site. For $\alpha > 1$, the excess Mn atoms occupy nominal Co and Si sites, creating Mn_{Co} and Mn_{Si} antisites, both of which have lower calculated formation energy than either a Co or Si site vacancy [37, 40]. The resulting SSFU composition is $[Co_{2-x}Mn_x]Mn[Si_{1-y}Mn_y]$, which we call type III. $[Co_{2-x}Mn_x]$ and $[Si_{1-y}Mn_y]$ are the compositions of the nominal Co and Si sites, respectively.

Next, let us consider Si-deficient $Co_2Mn_\alpha Si_\beta$ characterized by $\beta < (2+\alpha)/3$. There are two cases, set by the Co atomic ratio in the formula unit, r_{Co}. The ratio r_{Co} is given by $2/(2+\alpha+\beta)$, where the number of Co atoms included in one unit of $Co_2Mn_\alpha Si_\beta$, 2, is divided by the total number of atoms included in one unit of $Co_2Mn_\alpha Si_\beta$, $2+\alpha+\beta$. If $r_{Co} > 1/2$ (i.e., $\alpha+\beta < 2$), some of the Si sites are filled by Mn atoms, creating Mn_{Si} antisites, which have lower calculated formation energy than Co_{Si} antisites [40], and some of the Mn sites are filled by Co atoms, creating Co_{Mn} antisites, which have lower formation energy than Mn vacancies [40]. This results in a type II SSFU composition of $Co_2[Mn_{1-x}Co_x][Si_{1-y}Mn_y]$. If $r_{Co} < 1/2$ (i.e., $\alpha+\beta > 2$), which is the case for Mn-rich $Co_2Mn_\alpha Si_\beta$ with $\alpha > 2-\beta$, the SSFU composition is of type III.

The Co:Mn:Si atomic ratio in the resulting formula unit composition must be equal to the film composition expressed as $Co_2Mn_\alpha Si_\beta$ with given α and β. The parameters x and y in the SSFU composition are determined as functions of α and β from this requirement as shown in Table 18.1 Furthermore, the boundaries of the respective regions of type I to type III SSFU compositions in the α-β plane are determined from the requirements that the parameters x and y should be in the range of $0 < x < 1$, and $0 < y < 1$ (Fig. 18.2). The boundary given by $\alpha+\beta = 2$ corresponds to the SSFU composition where r_{Co} is 1/2. Similarly, the boundary given by $\beta = (2+\alpha)/3$ corresponds to the SSFU composition where the Si atomic ratio in the formula unit, r_{Si}, given by $r_{Si} = \beta/(2+\alpha+\beta)$, is 1/4. Thus, the Co atomic ratio in the formula unit for the regions of type I and type II in the α-β plane is larger than 1/2 and that for the region of type III is smaller than 1/2. The calculated SSFU compositions of $Co_2Mn_\alpha Si_{0.88}$ samples are listed in Table 18.2.

Table 18.2 SSFU compositions and formula unit type for $Co_2Mn_\alpha Si_{0.88}$ with various Mn compositions α

α in $Co_2Mn_\alpha Si_{0.88}$	SSFU composition	Type no.
0.72	$Co_2[Mn_{0.78}Co_{0.22}][Si_{0.98}Mn_{0.02}]$	II
1.00	$Co_2[Mn_{0.94}Co_{0.06}][Si_{0.91}Mn_{0.09}]$	II
1.15	$[Co_{1.99}Mn_{0.01}]Mn[Si_{0.88}Mn_{0.12}]$	III
1.32	$[Co_{1.90}Mn_{0.10}]Mn[Si_{0.84}Mn_{0.16}]$	III
1.37	$[Co_{1.88}Mn_{0.12}]Mn[Si_{0.83}Mn_{0.17}]$	III
1.53	$[Co_{1.81}Mn_{0.19}]Mn[Si_{0.80}Mn_{0.20}]$	III
1.68	$[Co_{1.75}Mn_{0.25}]Mn[Si_{0.77}Mn_{0.23}]$	III
1.84	$[Co_{1.69}Mn_{0.31}]Mn[Si_{0.75}Mn_{0.25}]$	III

18.2.3 Structural Properties

In this section, we describe the structural properties of $Co_2Mn_\alpha Si_\beta$ thin films having various values of α and show that they can be consistently explained by the SSFU composition model based on the antisites. To investigate the site occupancies of the nonstoichiometric $Co_2Mn_\alpha Si_\beta$ films, the ratio of the 002 peak intensity with respect to 004 peak intensity, I(002)/I(004), and the ratio of 111 peak intensity with respect to 022 peak intensity, I(111)/I(022), in x-ray diffraction (XRD) have been measured for various α for $Co_2Mn_\alpha Si_{0.88}$ film series [12]. The 002 peak is characteristic of the B2 and $L2_1$ structures, the 111 peak is characteristic of the $L2_1$ structure, and the 004 and 022 peaks are characteristic to the fundamental A2 structure. For these measurements, a two-dimensional position-sensitive proportional counter (two-dimensional detector) in the four-axis x-ray diffractometer was used. Note that the intensity ratio I(002)/I(004) is related to the order between the nominal Co plane and the nominal MnSi plane while I(111)/I(022) is related to the order between the nominal Mn site and the nominal Si site in the MnSi plane.

Figure 18.3 shows the measured intensity ratios I(002)/I(004) for four samples of $Co_2Mn_\alpha Si_{0.88}$ films with $\alpha = 0.72$, 1.0, 1.37 and 1.84 [12]. The respective SSFU compositions for these four samples are shown in Table 18.2. I(002)/I(004) decreased monotonically with increasing α. The atomic scattering factor of a Mn atom (atomic number 25) and a Co atom (27) are close, but that of a Si atom (14) is about half those of Mn and Co, so the 002 peak intensity for Co_2MnSi in the x-ray diffraction mainly depends on the occupation ratio of Si in the MnSi plane (γ_1). According to the SSFU compositions for $Co_2Mn_\alpha Si_{0.88}$, γ_1 decreases with increasing α from $\gamma_1 = 0.98/2$ for $\alpha = 0.72$ to $\gamma_1 = 0.75/2$ for $\alpha = 1.84$. Thus, the observed decrease

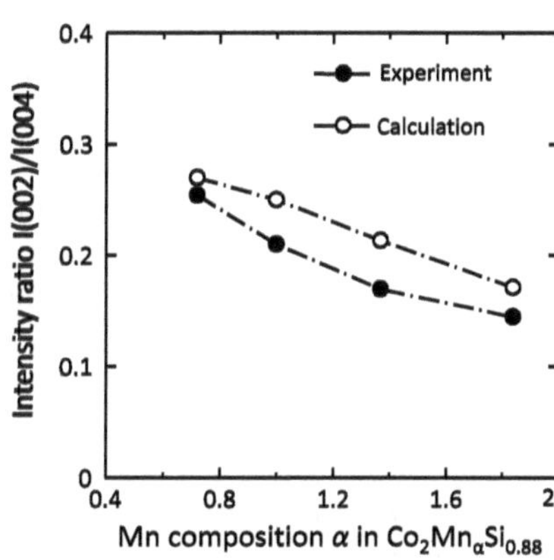

Fig. 18.3 Mn composition (α) dependence in $Co_2Mn_\alpha Si_{0.88}$ films of the experimental (*solid circles*) and calculated (*open circles*) ratios of x-ray diffraction intensity of 002 peak with respect to that of 004 peak, I(002)/I(004). Calculated ratios were obtained using the kinematical x-ray diffraction theory and formula unit compositions given in Table 18.2. Reprinted with permission from [12]. Copyright 2014, American Physical Society

in I(002)/I(004) can be ascribed to the decrease in the occupation of Si in the MnSi plane with increasing α, as shown by the SSFU compositions. The intensity ratios I(002)/I(004) for $Co_2Mn_\alpha Si_{0.88}$ films were also calculated as a function of α using kinematical XRD theory and the SSFU compositions. Although the main factor in the calculation is the structure factor that depends on the site occupations, the following factors were also taken into account; (1) the Lorentz factor proportional to $1/\sin(2\theta)$ for a single-crystalline film, (2) the polarization factor proportional to $1+\cos^2 2\theta$, and (3) the absorption factor proportional to $(1\text{-}\exp(-2\mu t/\sin\theta))$, where 2θ is the detection angle (or diffraction angle), t is a film thickness smaller than the effective x-ray penetration depth, and μ is the line absorption coefficient of the film. The calculated I(002)/I(004), shown in Fig. 18.3, decreased with increasing α. Thus, a fairly good agreement between the experimental and calculated dependencies of I(002)/I(004) was found, indicating the validity of the proposed SSFU compositions, in particular, regarding the order between the nominal Co plane and the nominal MnSi plane.

Furthermore, the α dependence of the measured intensity ratio I(111)/I(022) for the $Co_2Mn_\alpha Si_{0.88}$ films showed lower intensity ratios for $\alpha = 1.0$, 1.37 and 1.84 than that for $\alpha = 0.72$ [12]. Because the atomic scattering factor of the Si atom is much smaller than that of Mn and Co as described above, the 111 peak intensity for CMS mainly depends on the occupation of Si at the nominal Si site (γ_2), and the γ_2 values for $\alpha = 1.0$, 1.37, and 1.84 ranging from $\gamma_2 = 0.91$ for $\alpha = 1.0$ to $\gamma_2 = 0.75$ for $\alpha = 1.84$ are definitely lower than the $\gamma_2 = 0.98$ for $\alpha = 0.72$. Thus, the observed lower I(111)/I(022) for $\alpha = 1.0$, 1.37 and 1.84 than that for $\alpha = 0.72$ was consistent with the lower occupation of Si at the nominal Si site in the MnSi plane for $\alpha = 1.0$, 1.37 and 1.84 than that for $\alpha = 0.72$, as shown by the SSFU compositions. The calculated intensity ratios based on the SSFU compositions reproduced the measured lower intensity ratios I(111)/I(022) for $\alpha = 1.0$, 1.37, and 1.84 compared to $\alpha = 0.72$, indicating that the proposed SSFU compositions are basically valid regarding the order between the nominal Mn site and the nominal Si site.

To further characterize the structural properties, lattice images of Mn-deficient and Mn-rich Co_2MnSi films at atomic resolution using aberration-corrected Z-contrast scanning transmission electron microscopy (STEM) were observed. Figure 18.4a, b show typical Z-contrast STEM images for a Mn-deficient $Co_2Mn_{0.72}Si_{0.88}$ film (SSFU composition $Co_2[Mn_{0.78}Co_{0.22}][Si_{0.98}Mn_{0.02}]$) and a Mn-rich $Co_2Mn_{1.37}Si_{0.88}$ film ($[Co_{1.88}Mn_{0.12}]Mn[Si_{0.83}Mn_{0.17}]$) [12]. The atomic lattice with all three sites is clearly visible. For the Mn-deficient film with $\alpha = 0.72$, almost all the observed area of $\sim 16 \times 14\,\text{nm}^2$ showed the $L2_1$ structure. Figure 18.4c shows an enlarged image of a region of Fig. 18.4a, demonstrating the $L2_1$ structure. In contrast, for Mn-rich film with $\alpha = 1.37$, a typical image with the same area consists of mostly the B2 structure with small $L2_1$ regions. Figure 18.4d, e show enlarged images of two different regions of Fig. 18.4b, demonstrating the B2 and $L2_1$ structures, respectively. STEM images for a Mn-rich $Co_2Mn_{1.32}Si_{0.88}$ film ($[Co_{1.90}Mn_{0.10}]Mn[Si_{0.84}Mn_{0.16}]$) showed a similar dominant B2 structure with small $L2_1$ regions as the Mn-rich $Co_2Mn_{1.37}Si_{0.88}$ film.

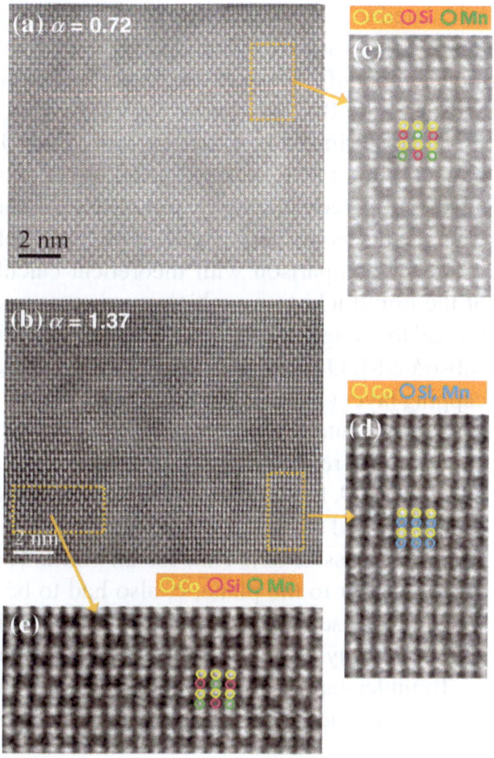

Fig. 18.4 High-resolution Z-contrast scanning TEM (STEM) images of **a** Mn-deficient $Co_2Mn_{0.72}Si_{0.88}$ film ($\alpha = 0.72$ in $Co_2Mn_\alpha Si_{0.88}$) and **b** Mn-rich $Co_2Mn_{1.37}Si_{0.88}$ film ($\alpha = 1.37$), both with the area of $\sim 16 \times 14\,nm^2$. **c** Enlarged image of a region of (**a**) showing the $L2_1$ structure. **d** and **e** Enlarged images of two different regions of (**b**) showing the B2 and $L2_1$ structures, respectively. Reprinted with permission from [12]. Copyright 2014, American Physical Society

The dominant $L2_1$ structure region observed for the Mn-deficient film having $\alpha = 0.72$ and the decreased region of the $L2_1$ structure along with the dominant region of the B2 structure for the Mn-rich film having $\alpha = 1.37$ are in good agreement with the findings from XRD. The decrease in the $L2_1$ area observed by Z-contrast STEM for the Mn-rich film ($\alpha = 1.37$) compared with the Mn-deficient film ($\alpha = 0.72$) agrees with the decreased XRD I(111)/I(022) for the film with $\alpha = 1.37$ compared with the film with $\alpha = 0.72$. These features are consistent with the lower Si occupation ratio at the nominal Si site in the proposed SSFU compositions for the Mn-rich film with $\alpha = 1.37$ than the Mn-deficient film with $\alpha = 0.72$.

In summary, the experimental findings regarding the structural properties observed for both the XRD intensity ratios and high-resolution STEM images support the proposed SSFU compositions.

18.2.4 Magnetic Properties

In this section, we describe how the experimental μ_s of nonstoichiometric $Co_2Mn_\alpha Si_\beta$ films varied with α. We will compare the experimental μ_s with the theoretically calculated total spin magnetic moments per formula unit, m_{spin}, and with the Slater-Pauling rule for half-metallic Co_2YZ alloys, $Z_t - 24$ (μ_B/f.u.) [29], where Z_t is the total number of valence electrons per formula unit provided by the SSFU composition and μ_B is the Bohr magneton.

The experimental μ_s was measured at 10 K. The saturation magnetization, M_s, of each film at 10 K was estimated by an extrapolation of the *M*-*H* curve measured using a superconducting quantum interference device magnetometer to $H = 0$ to subtract the contribution from the MgO substrate. Determining μ_s from the M_s value for comparison with theoretical calculations requires precise determination of the film thicknesses, which were measured using low-angle x-ray reflectivity. A typical low-angle x-ray reflectivity scan for a layer structure consisting of MgO(001) substrate/MgO buffer (10 nm)/$Co_2Mn_{1.32}Si_{0.88}$ (~30 nm)/MgO (2 nm) barrier/AlO_x capping layer (1 nm) is shown in Fig. 18.5 [12]. The solid and dotted curves represent the experimental and fitted curves as a function of angle 2θ. Clear oscillations in the experimental reflectivity scan observed up to $2\theta \sim 7°$ revealed the smooth interfaces and surface. A fit of the oscillatory x-ray reflectivity intensity for a 2θ range from 1.4 to 5.6° (dotted curve) using the Parratt formalism [49]provided the $Co_2Mn_{1.32}Si_{0.88}$ layer thickness of 32.34 nm for this film. The lattice constants in the plane, *a*, and perpendicular to the plane, *c*, also had to be determined for the films to obtain μ_s. *a* and *c* for each film were measured by making x-ray $\theta - 2\theta$ Bragg scans with a four-axis x-ray diffractometer with a scintillation counter.

To understand the origin of the experimentally obtained α dependence of μ_s of CMS films and TMR ratios of CMS MTJs, density-functional calculations based

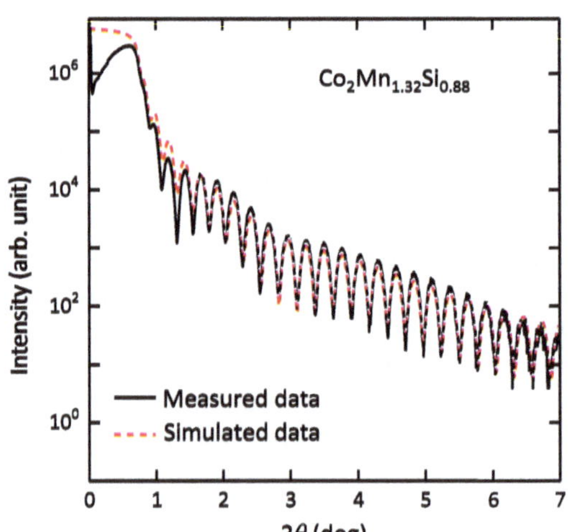

Fig. 18.5 A typical low-angle x-ray reflectivity scan for a layer structure consisting of MgO(001) substrate/MgO buffer (10 nm)/$Co_2Mn_{1.32}Si_{0.88}$ thin layer/MgO (2 nm) barrier/AlO_x capping layer (1 nm). The *solid* and *dotted curves* represent the experimental and fitted curves as a function of angle 2θ. A fit of the oscillatory x-ray reflectivity intensity was done for a 2θ range from 1.4° to 5.6°. Through fitting, the $Co_2Mn_{1.32}Si_{0.88}$ layer thickness was determined to be 32.34 nm. Reprinted with permission from [12]. Copyright 2014, American Physical Society

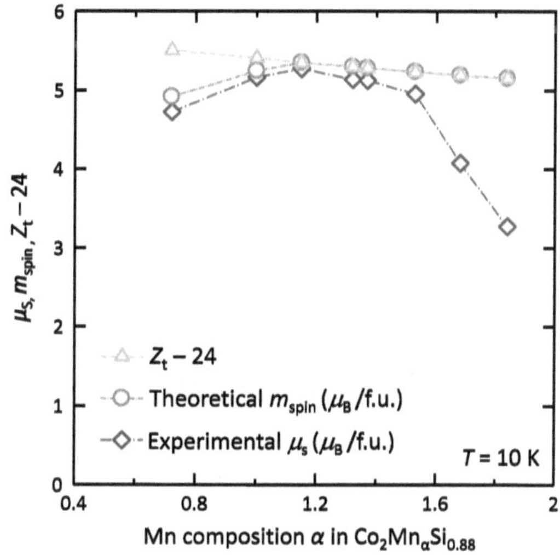

Fig. 18.6 Experimental μ_s values at 10 K and theoretical m_{spin} values obtained by KKR-CPA calculations as a function of α for $Co_2Mn_\alpha Si_{0.88}$ along with $Z_t - 24$, where Z_t is the total valence electron number/f.u. provided by the SSFU composition as a function of α for $Co_2Mn_\alpha Si_{0.88}$. Modified with permission from [12]. Copyright 2014, American Physical Society

on the Korringa-Kohn-Rostoker (KKR) method [50, 51] with coherent potential approximation (CPA) have been performed [12]. The KKR-CPA calculations were based on the SSFU composition model assuming antisite formation to accommodate nonstoichiometry, as described in Sect. 18.2.2.

Figure 18.6 plots the experimental μ_s values at 10 K for $Co_2Mn_\alpha Si_{0.88}$ films with α ranging from 0.72 (Mn-deficient CMS) to 1.84 (Mn-rich CMS) in comparison with the theoretical m_{spin} values obtained by the KKR-CPA calculations and with $Z_t - 24$. Let's first compare the α dependence of μ_s with that of $Z_t - 24$ in Fig. 18.6. For a strongly Mn-deficient composition ($\alpha = 0.72$), μ_s was markedly lower than $Z_t - 24$. However, μ_s increased with α increasing over $\alpha = 0.72$, growing closer to $Z_t - 24$, and was very close to $Z_t - 24$ for $\alpha = 1.15$–1.37. In contrast, the μ_s was appreciably lower than $Z_t - 24$ for $\alpha = 1.53$, and it decreased significantly with a further increase in α. We estimated critical value α_c over which μ_s decreased as $\alpha_c = 1.53$ for $\beta = 0.88$, midway between 1.37 and 1.68.

Let us now compare the experimental μ_s with the theoretically calculated m_{spin}. As shown in Fig. 18.6, the experimental μ_s is in good agreement with m_{spin} over a wide range of α from 0.72 to 1.37. However, μ_s deviated from and was lower than m_{spin} for $\alpha = 1.57$ and it decreased markedly from m_{spin} with a further increase in α, i.e., for $\alpha = 1.68$ and 1.84. In contrast, if we compare the α dependence of m_{spin} with that of $Z_t - 24$, the m_{spin} value, which is distinctly lower than for strongly Mn-deficient $\alpha = 0.72$, grows closer to the half-metallic value, $Z_t - 24$, with increasing α from 0.72. The m_{spin} values for an α range from 1.15 to 1.84 are in good agreement with the $Z_t - 24$ values. Note that the Co_{Mn} antisite according to the SSFU composition model is suppressed for this α range (Table 18.2).

Figure 18.7a, b show the α dependence of the Co and Mn spin magnetic moments, m_{spin}(Co) and m_{spin}(Mn), at the respective regular site and antisite deduced by the KKR-CPA calculations for $Co_2Mn_\alpha Si_{0.88}$ [12]. The KKR-CPA calculations show

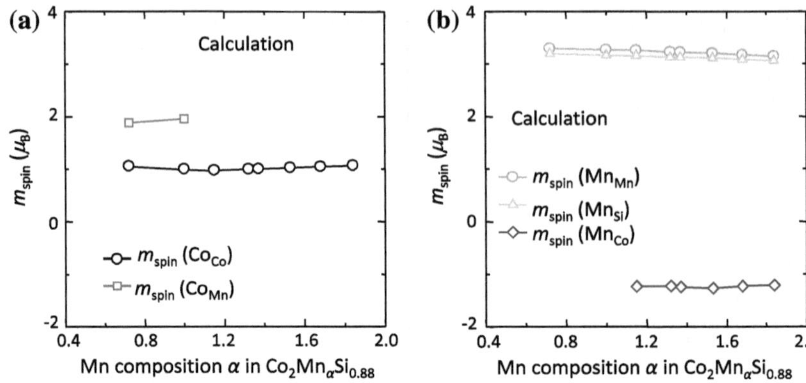

Fig. 18.7 Mn composition dependence of **a** the Co spin magnetic moment and **b** Mn spin magnetic moment at different sites deduced from the KKR-CPA calculations for $Co_2Mn_\alpha Si_{0.88}$. Reprinted with permission from [12]. Copyright 2014, American Physical Society

that the spin magnetic moment of Co at the nominal Mn site, which appears for Mn-deficient CMS (Mn deficiency generally corresponds to $\alpha < 2 - \beta$ for the expression $Co_2Mn_\alpha Si_\beta$), $m_{spin}(Co_{Mn})$, is much smaller than that of Mn at the nominal site, $m_{spin}(Mn_{Mn})$ ($m_{spin}(Co_{Mn}) = 1.89\mu_B$ in contrast to $m_{spin}(Mn_{Mn}) = 3.29\mu_B$ for $\alpha = 0.72$). This results in μ_s being lower than $Z_t - 24$ with decreasing $\alpha < 2 - \beta$. Thus, the loss of the half-metallicity in Mn-deficient CMS films, shown by μ_s being lower than $Z_t - 24$, originates from the Co_{Mn} antisite defects.

Finally, we discuss a possible origin of the observed significant drop in the experimental μ_s from the m_{spin} values for $\alpha > \alpha_c$ where the m_{spin} values were in good agreement with the half-metallic $Z_t - 24$ values. Note that the TMR ratios at 4.2 and 290 K for CMS MTJs also decreased with α increasing beyond a similar critical value, as will be shown in Sect. 18.2.5. Furthermore, the α_c value increased with decreasing $\beta < 1.0$ (larger Si deficiency) [12]. This suggests a correlation between α_c and the SSFU composition. The SSFU compositions were type III, $[Co_{2-x}Mn_x]Mn[Si_{1-y}Mn_y]$, for $\alpha \approx \alpha_c$. The critical value of x for the type III SSFU composition, x_c, corresponding to a given α_c, was found to be almost identical for different β values ranging from 1.0 to 0.88, i.e., $x_c \approx 0.17$. The coincident drops in μ_s and the TMR ratio for the critical Mn-rich composition, α_c, indicates the deviation of the SSFU composition from the half-metallic type III antisite-based SSFU compositions for Mn-rich CMS films with α being larger than α_c.

18.2.5 Tunneling Magnetoresistance Characteristics

We now describe the TMR characteristics of CMS MTJs with various Mn compositions α in $Co_2Mn_\alpha Si_\beta$ electrodes. Figure 18.8a, b show the TMR ratios at 4.2 and 290 K, respectively, for three series of CMS MTJs as a function of α, including CoFe-

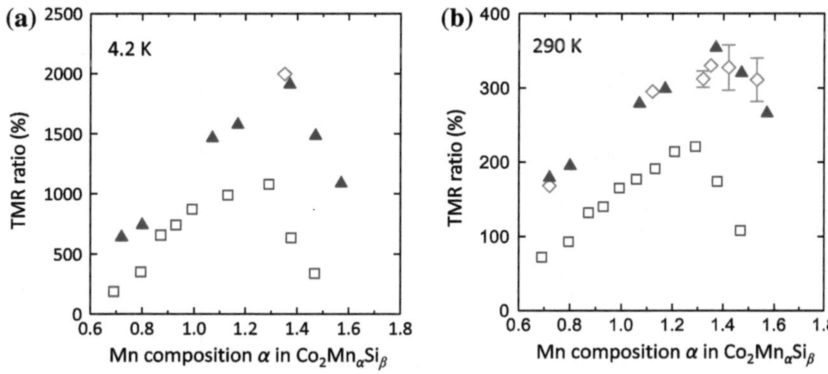

Fig. 18.8 Mn composition dependence of TMR ratios at **a** 4.2 K and **b** 290 K obtained for three series of $Co_2Mn_\alpha Si_\beta$/MgO/$Co_2Mn_\alpha Si_\beta$MTJs (CMS MTJs) as a function of α, including MgO-buffered MTJs with $Co_2Mn_\alpha Si_{1.0}$ electrodes [9] (*open rectangles*), CoFe-buffered MTJs with $Co_2Mn_\alpha Si_{0.96}$ electrodes [11] (*solid triangles*), and CoFe-buffered MTJs with $Co_2Mn_\alpha Si_{0.88}$ electrodes [12] (*open diamonds*). Modified with permission from [12]. Copyright 2014, American Physical Society

buffered CMS MTJs having $Co_2Mn_\alpha Si_{0.96}$ electrodes, CoFe-buffered CMS MTJs having $Co_2Mn_\alpha Si_{0.88}$ electrodes and MgO-buffered CMS MTJs having $Co_2Mn_\alpha Si_{1.0}$ electrodes [9–12]. As shown, the TMR ratios of the CoFe-buffered $\beta = 0.88$ and 0.96 CMS MTJs were markedly higher than those of the MgO-buffered $\beta = 1.0$ MTJs for all the α values ranging from a Mn-deficient composition to a Mn-rich composition. Note that CoFe-buffered CMS MTJs had a much larger misfit dislocation spacing of 6.3 ± 0.1 nm than that of 4.3 ± 0.1 nm for the MgO-buffered CMS MTJs, as described in Sect. 18.2.1. To clarify the significant increase in the TMR ratio for CoFe-buffered CMS, Fig. 18.9 plots how the TMR ratios at 4.2 K of these two kinds of MTJ depend on the average misfit dislocation spacing at the interfaces with a MgO barrier, where the highest TMR ratios at 4.2 K obtained as a function of α for the respective kinds of MTJ are plotted. As shown in Fig. 18.9, the CoFe-buffered CMS MTJs that had the much larger misfit dislocation spacing than the MgO-buffered CMS MTJs showed much higher TMR ratios [11]. This can be ascribed to the enhanced coherent tunneling contribution in the CoFe-buffered CMS MTJs arising from the increased area of epitaxial interfaces without a misfit dislocation because epitaxial interfaces are essential for conservation of the electron transverse momentum for tunneling.

We now describe the α dependence of the TMR ratio. Significantly, both MgO-buffered and CoFe-buffered CMS MTJs exhibited the distinct α dependence of the TMR ratios at 4.2 and 290 K regardless of the buffer layers, and showed almost identical dependence of the TMR ratio on α. In particular, it was found that the TMR ratios at 4.2 and 290 K increased significantly with α increasing from a Mn-deficient composition to a certain Mn-rich composition, α_c, beyond which it decreased. Indeed, the TMR ratio of the CoFe-buffered $\beta = 0.96$ CMS MTJs at 4.2 K, for example,

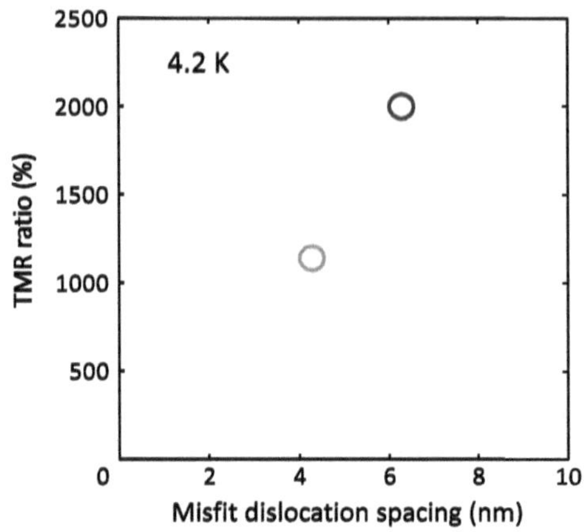

Fig. 18.9 Dependence of TMR ratios at 4.2 K on the average misfit dislocation spacing at the lower and upper interfaces with a MgO barrier for the CoFe-buffered $Co_2MnSi/MgO/Co_2MnSi$ MTJ (*blue open circle*) and MgO-buffered $Co_2MnSi/MgO/Co_2MnSi$ MTJ (*red open circle*), where the highest TMR ratios at 4.2 K obtained as a function of α for the respective kinds of MTJ are plotted. Modified with permission from [11]. Copyright 2012, American Institute of Physics (Color figure online)

increased significantly from 640 % for Mn-deficient $\alpha = 0.72$ to 1910 % for Mn-rich $\alpha = 1.37$. The TMR ratio at 290 K also notably increased with increasing α from 179 % for $\alpha = 0.72$ to 354 % for $\alpha = 1.37$.

Figure 18.10 shows typical TMR curves at 4.2 and 290 K for a CMS MTJ grown on a CoFe-buffered MgO(001) substrate with Mn-rich $Co_2Mn_{1.35}Si_{0.88}$. The MTJ exhibited clear exchange-biased TMR characteristics with high TMR ratios of 1995 % at 4.2 K and 330 % at 290 K. Table 18.3 summarizes typical TMR ratios at 4.2 and 290 K for an MTJ from each CoFe-buffered CMS MTJ series MTJ-A to MTJ-C, which had slightly different Si compositions β with respect to Co_2 in $Co_2Mn_\alpha Si_\beta$ electrodes. As summarized in Table 18.3, all three series of the CoFe-buffered CMS MTJs showed almost identical giant TMR ratios of up to 1995 % at 4.2 K and up to 354 % at 290 K.

We now discuss the origin of the Mn composition dependence of TMR ratios. Figure 18.11 plots two kinds of theoretical spin polarization, $P_{\text{th}}(spd)$ and $P_{\text{th}}(sp)$, obtained by the KKR-CPA calculations as a function of α for $Co_2Mn_\alpha Si_{1.0}$ with α ranging from 0.69 (Mn-deficient) to 1.44 (Mn-rich) [12]. The $P_{\text{th}}(spd)$ value was determined by considering all the contributions from s-, p-, and d-like electrons for the majority- and minority-spin density of states (DOS) at E_F, $D_M(spd)$ and $D_m(spd)$, respectively. It is defined as

$$P_{\text{th}}(spd) = \frac{D_M(spd) - D_m(spd)}{D_M(spd) + D_m(spd)}. \tag{18.1}$$

In contrast, the $P_{\text{th}}(sp)$ value was determined by considering only the contributions from s- and p-like electrons for the majority- and minority-spin DOS at E_F, $D_M(sp)$ and $D_m(sp)$, respectively. It is defined as

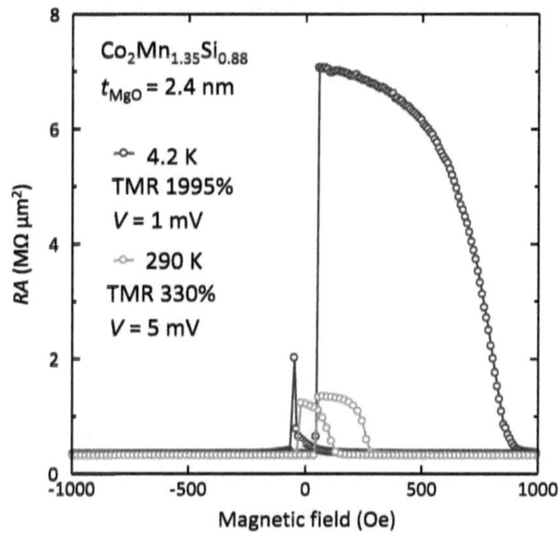

Fig. 18.10 Typical TMR curves at 4.2 and 290 K for a CMS MTJ consisting of (from the lower side) CMS lower electrode (3 nm)/MgO barrier (2.4 nm)/CMS upper electrode (3 nm) grown on a CoFe-buffered MgO(001) substrate with Mn-rich $Co_2Mn_{1.35}Si_{0.88}$ electrodes (MTJ-B). The magnetoresistance was measured with a magnetic field applied along the [1–10] axis of the CMS film using a dc four-probe method. The bias voltages were 1 mV at 4.2 K and 5 mV at 290 K. Reprinted with permission from [11]. Copyright 2012, American Institute of Physics

$$P_{\rm th}(sp) = \frac{D_{\rm M}(sp) - D_{\rm m}(sp)}{D_{\rm M}(sp) + D_{\rm m}(sp)}. \tag{18.2}$$

The values for both $P_{\rm th}(sp)$ and $P_{\rm th}(spd)$ for $Co_2Mn_\alpha Si_{1.0}$ with $\alpha \geq 1.13$ were almost completely half-metallic values as shown in Fig. 18.11: $P_{\rm th}(sp) = 0.995$ and $P_{\rm th}(spd) \geq 0.99$. The values for $P_{\rm th}(spd)(0.81)$ and $P_{\rm th}(sp)(0.96)$ for $\alpha = 1.0$ (stoichiometric Co_2MnSi) were lower than 1.0, the value for half-metals, due to the atomic sphere approximation in the KKR-CPA calculation, which assumes a spherically symmetric potential in each atomic sphere. First-principles calculations performed using a full potential method, such as the ultrasoft pseudopotential method

Table 18.3 Typical TMR ratios at 4.2 and 290 K for an MTJ from each MTJ series MTJ-A to MTJ-C having slightly different Si compositions β with respect to Co_2 in $Co_2Mn_\alpha Si_\beta$ electrodes, and the respective film compositions of $Co_2Mn_\alpha Si_\beta$ electrodes for each MTJ

MTJ series	Film composition of an MTJ from each MTJ series	Typical TMR ratios	
		at 4.2 K (%)	at 290 K (%)
MTJ-A	$Co_2Mn_{1.29}Si_{1.0}$	1804	344
MTJ-B	$Co_2Mn_{1.35}Si_{0.88}$	1995	330
MTJ-C	$Co_2Mn_{1.37}Si_{0.96}$	1910	354

Reprinted with permission from [11]. Copyright 2012, American Institute of Physics

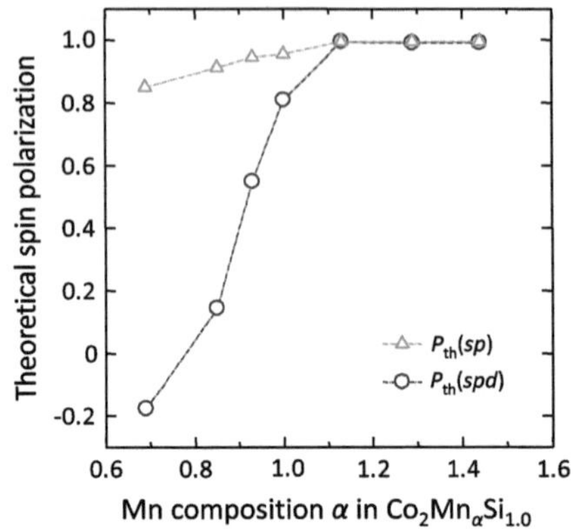

Fig. 18.11 Two kinds of theoretical spin polarization, $P_{\text{th}}(spd)$ (*open circles*) and $P_{\text{th}}(sp)$ (*open triangles*), obtained by the KKR-CPA calculations as a function of α for $Co_2Mn_\alpha Si_{1.0}$ with $\alpha = 0.69$ to 1.44. The values for $P_{\text{th}}(spd)$ (0.81) and $P_{\text{th}}(sp)$(0.96) for $\alpha = 1.0$ (stoichiometric Co_2MnSi) were lower than 1.0, the value for half-metals, due to the atomic sphere approximation in the KKR-CPA calculation, which assumes a spherically symmetric potential in each atomic sphere. First-principles calculations performed using a full potential method showed half-metallicity for stoichiometric Co_2MnSi [33, 52]. Reprinted with permission from [12]. Copyright 2014, American Physical Society

[33] or the full-potential linearized augmented plane wave method [52], showed half-metallicity for stoichiometric Co_2MnSi.

Note that the dependence of $P_{\text{th}}(spd)$ on α for the Mn-deficient region differed pronouncedly from that of $P_{\text{th}}(sp)$. $P_{\text{th}}(spd)$ decreased significantly with decreasing α for $\alpha < 1.0$ and became negative for a strongly Mn-deficient composition ($\alpha = 0.69$). If the tunneling current in CMS MTJs was dominantly determined by $P_{\text{th}}(spd)$, the TMR ratio for $\alpha < 1.0$ would substantially decrease with α, become almost zero one time for $P_{\text{th}}(spd) \sim 0$, and then increase again to a certain positive value. However, the α dependence of the TMR ratio showed a monotonic decrease with α for $\alpha < 1.0$, resulting in obvious disagreement with the α dependence of $P_{\text{th}}(spd)$. In contrast, $P_{\text{th}}(sp)$ for $\alpha < 1.0$ was positive and increased with increasing α from $P_{\text{th}}(sp) = 0.850$ for $\alpha = 0.69$ to 0.945 for $\alpha = 0.93$. Thus, the α dependence of $P_{\text{th}}(sp)$, in contrast to that of $P_{\text{th}}(spd)$, was found to qualitatively explain the α dependence of the TMR ratio except for $\alpha > \alpha_c$. This suggests that tunneling in Co_2MnSi/MgO-based MTJs is mainly due to itinerant s- and p-like electrons and that the contribution to tunneling from localized d-like electrons is negligible. In principle, the contribution from the delocalized $d(3z^2r^2)$-orbital component involved in the majority-spin Δ_1 symmetry should be included in the theoretical tunneling spin polarization. However, it can be neglected because the contributions from s and $p(z)$ orbitals are dominant. This consideration highlights the importance of clarifying the spin-resolved density-of-states at E_F that contribute to the tunneling spin polarization in MTJs.

Next, we discuss the origin of the decrease in the tunneling spin polarization for the Mn-deficient region. Figure 18.12a, b respectively show how the local density-of-

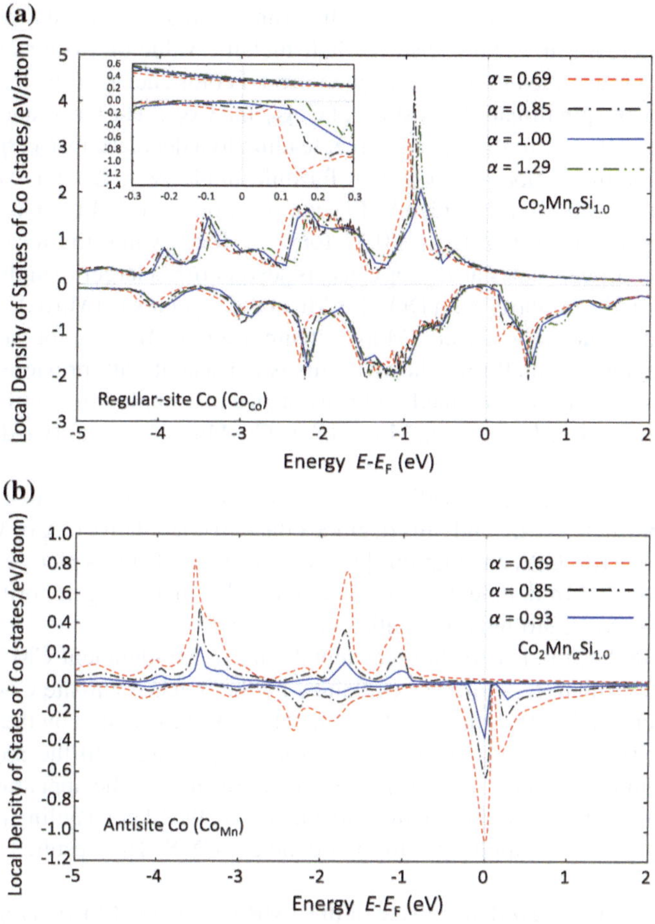

Fig. 18.12 Local density of states (LDOS) for **a** Co at the regular site (Co_{Co}) and **b** Co at the nominal Mn site (Co_{Mn}) for $Co_2Mn_\alpha Si_{1.0}$ for various α as a function of energy E relative to the Fermi level (E_F). *Inset* shows a magnification of the LDOS of the nominal site Co near E_F. Majority spin states are shown as positive LDOS and minority spin states as negative. Reprinted with permission from [12]. Copyright 2014, American Physical Society

states (LDOS) of the Co_{Co} and Co_{Mn} of $Co_2Mn_\alpha Si_{1.0}$ depended on α. Figure 18.12b shows the appearance of minority-spin in-gap states around E_F for the antisite Co (Co_{Mn}) for Mn-deficient $\alpha < 1.0$. As shown, the density of the minority-spin in-gap states around E_F for Co_{Mn} increased for the more strongly Mn-deficient $Co_2Mn_\alpha Si_{1.0}$. Furthermore, Co_{Mn} antisites for Mn-deficient Co_2MnSi affected the electronic state of Co atoms at the regular sites, resulting in the appearance of minority-spin states around E_F for Co_{Co}, although the LDOS around E_F for Co_{Co} was smaller than that for Co_{Mn}. These minority-spin in-gap states around E_F for both antisite Co_{Mn} and regular-site Co_{Co} were mainly composed of localized Co-$d(x^2 - y^2)$ orbital components along with a small amount of itinerant s- and $p(z)$-orbital components. Then, the decrease in $P_{th}(sp)$ from the half-metallic value of 1.0 for Mn-deficient $Co_2Mn_\alpha Si_{1.0}$ is ascribed to the s- and $p(z)$-orbital components of the localized density of minority-spin in-gap states at E_F of Co_{Mn} and Co_{Co}. The Co_{Mn} and Co_{Co} components affect $P_{th}(sp)$ to the same degree, leading to a decrease in $P_{th}(sp)$ from 1.0. This is because the number of Co_{Co} in the formula unit for $\alpha < 1.0$ (for example, 2 for $\alpha = 0.69$ in $Co_2Mn_\alpha Si_{1.0}$ of which SSFU composition is $Co_2[Mn_{0.75}Co_{0.17}Si_{0.08}]Si$) is much larger than that of Co_{Mn} (0.17 for $\alpha = 0.69$) while the local density of minority-spin in-gap states at E_F for Co_{Mn} is several times larger than that for Co_{Co}.

In contrast, the minority-spin DOS for Mn at the regular site (Mn_{Mn}), at the Si site (Mn_{Si} antisite), and at the Co site (Mn_{Co} antisite) for $Co_2Mn_\alpha Si_{1.0}$ for various values of α was negligibly small (not shown). This is consistent with previous theoretical results: half-metallicity is retained for the existence of a single Mn_{Co} antisite [37] and is robust against disorder between Mn and Si in $Co_2Mn_{1+x}Si_{1-x}$ and $Co_2Mn_{1-x}Si_{1+x}$ [39].

Picozzi et al. [37] theoretically predicted that a single Co_{Mn} antisite produces minority-spin states at E_F and thus degrades the half-metallicity of Co_2MnZ (Z = Si and Ge). The KKR-CPA calculations [12] demonstrated that this picture is generally valid for strongly Mn-deficient Co_2MnSi and revealed increasing minority-spin DOS around E_F for more strongly Mn-deficient $Co_2Mn_\alpha Si_{1.0}$.

In Sect. 18.2.4, the loss of the half-metallicity in Mn-deficient CMS, shown by μ_s being lower than $Z_t - 24$, was explained by the induced antisite Co_{Mn} of which m_{spin} is much smaller than that of Mn_{Mn}. It can also be explained by the appearance of minority-spin in-gap states around E_F for Co_{Mn} along with the appearance of minority-spin in-gap states around E_F for Co_{Co}. Because of the appearance of these minority-spin states around E_F in the half-metal gap, the Slater-Pauling rule for half-metallic Co_2YZ is not applicable for Mn-deficient CMS. This resulted in the lower μ_s than Z_t-24.

We now discuss the α dependence of the TMR ratio in the Mn-rich region defined by $\alpha > 2 - \beta$ for $Co_2Mn_\alpha Si_\beta$. Importantly, the TMR ratios at both 4.2 and 290 K of CMS MTJs with $Co_2Mn_\alpha Si_\beta$ electrodes increased even in the Mn-rich region up to $\alpha = \alpha_c$, where nominally no Co_{Mn} antisite is present in $Co_2Mn_\alpha Si_\beta$ for $\alpha \geq 2 - \beta$ according to the SSFU composition model. This result suggests that there is a slight difference between the Co_{Mn} antisite concentration of films prepared by sputtering under a nonequilibrium condition and the nominal concentration given by the SSFU composition model. It is then reasonable to speculate that the film of $\alpha \geq 2 - \beta$ still contained residual Co_{Mn} antisites and that the antisite concentration in the film was further reduced by an increase in α even in the Mn-rich region up to $\alpha = \alpha_c$, resulting in the further increase in the TMR ratio.

18.3 Spin Injection into Semiconductors Using Half-Metallic Co_2MnSi

The injection and detection of spin-polarized electrons using ferromagnet/semiconductor heterojunctions has attracted much interest for creating viable spintronic devices featuring nonvolatile, reconfigurable logic functions and ultralow power consumption. The spin injection from a ferromagnetic electrode of Fe or CoFe into GaAs, Si, or Ge has been observed optically in spin light emitting diodes (spin LEDs) [53–57] and also electrically in lateral spin-transport devices through nonlocal geometry [58–64]. Lou et al. demonstrated electrical spin injection in Fe/GaAs Schottky tunnel junctions through the observation of spin-valve signals and Hanle signals in a nonlocal configuration [58]. The observation of both signals provides direct evidence of spin injection and transport. Salis et al. investigated the temperature dependence of the spin signal in Fe/GaAs and reported achieving spin injection at room temperature [59]. Furthermore, Uemura et al. achieved spin injection in CoFe/n-GaAs Schottky tunnel junctions with an improved temperature dependence [64].

Along with spin injection and detection, understanding the spin interactions (such as spin-orbit and hyperfine interactions) in semiconductors is an important part of spintronics research. In particular, the interplay between electron spins and nuclear spins due to hyperfine interaction has been studied extensively for application to quantum information devices based on nuclear spins as well as for understanding the transport properties of electron spins in semiconductors [65–69]. The hyperfine interaction transfers the angular momentum from polarized electrons to nuclei, leading to an effective polarization of nuclear spins, referred to as dynamic nuclear polarization (DNP). In contrast, polarized nuclear spins affect electron spins as an effective magnetic field or Overhauser field. Thus, one can evaluate the degree of nuclear-spin polarization through the strength of the Overhauser field. Optical detection of the nuclear field has been demonstrated through oblique Hanle effect measurements for spin LEDs with an Fe electrode [70, 71]. More recently, electrical detection has been demonstrated through oblique Hanle effect measurements for lateral spin-transport devices with an Fe [72, 73] or (Ga,Mn)As electrode [74].

Highly spin-polarized ferromagnetic material is indispensable for creating highly spin-polarized states of both electrons and nuclei. Co-based Heusler alloys are one candidate for a highly polarized spin source. There have been several reports on spin injection through optical detection in spin LEDs with a Co-based Heusler alloy electrode, such as Co_2MnGe [20], $Co_{2.4}Mn_{1.6}Ga$ [21], or Co_2FeSi [22], and through electrical detection in lateral spin-transport devices with Co_2FeSi [24] or $Co_2FeAl_{0.5}Si_{0.5}$ [25]. In this section, we describe highly efficient spin injection into a GaAs channel by using a half-metallic CMS spin source. Clear electrical spin injection into GaAs was demonstrated through the observation of spin-valve signals and Hanle signals in a nonlocal geometry in the temperature range from 4.2 K to room temperature [23, 26]. The obtained spin injection efficiency was one order of magnitude higher than that with a CoFe spin source, indicating a high spin polarization of the CMS spin source [26]. An electrically induced DNP in GaAs was also demonstrated by using a spin injection technique [23] in which the nuclear spins of Ga and As atoms are dynamically polarized through the hyperfine interaction transferring the angular momentum from the electron spins electrically injected into GaAs to the nuclear spins. A nuclear magnetic field acting on the electron spins in the sample with a CMS spin source four times larger than that with a CoFe spin source was electrically detected. This result indicates that the CMS enables an efficient nuclear polarization of Ga and As atoms because the strength of the nuclear field is proportional to the nuclear polarization.

18.3.1 Experimental Methods in Spin Injection from Co_2MnSi into GaAs

Layer structures consisting of (from the substrate side) a 250-nm-thick undoped GaAs buffer layer, a 2.5-μm-thick n^--GaAs channel layer, a 15-nm-thick $n^- \rightarrow n^+$-GaAs transition layer, and a 15-nm-thick n^+-GaAs layer were grown by molecular beam epitaxy at 590 °C on semi-insulating GaAs(001) substrates. The doping concentration of the n^--GaAs channel was chosen to be 3×10^{16} cm^{-3} and the doping concentration of the n^+-GaAs was 5×10^{18} cm^{-3} to form a narrow Schottky barrier. The samples were then capped with an arsenic protective layer and transported in air to an ultrahigh vacuum chamber capable of magnetron sputtering. After the arsenic cap was removed by heating the samples to 300 °C, an ultrathin CoFe layer with a thickness of 1.1 nm or 1.3 nm and a 5-nm-thick CMS layer were deposited by magnetron sputtering at room temperature and successively annealed in situ at 400 °C. The CMS layer was deposited by cosputtering from a CMS target and a Mn target. The film composition of the CMS film was chosen to be Mn-rich $Co_2Mn_{1.30}Si_\beta$ ($\beta = 0.88$ or 0.84) to suppress harmful Co_{Mn} antisites [23, 26]. Last, a 5-nm-thick Ru cap layer was deposited using magnetron sputtering at room temperature (Fig. 18.13a).

Electron beam (EB) lithography and Ar ion-milling techniques were used to fabricate the lateral spin-transport devices shown in Fig. 18.13b. The size of the injector contact (contact-2) and detector contact (contact-3) were 0.5×10 μm and 1.0×10 μm, respectively, and the spacing (d) between them was 0.5 or 4.0 μm. We defined the x-axis along the channel direction, the y-axis along the longitudinal direction of the junction, and the z-axis along the out-of-plane direction. The sample was evaluated in a four-terminal nonlocal geometry in which the nonlocal voltage between contact-3 and contact-4 was measured under a constant current (I) supplied between contact-2 and contact-1 at 4.2 K as functions of the in-plane magnetic field for spin-valve effect measurements, the out-of-plane magnetic field for Hanle effect measurements, and the oblique magnetic field for oblique Hanle effect measurements. The bias voltage was defined with respect to the n-GaAs. As a reference, samples with a 5-nm-thick CoFe single layer were identically fabricated under the same conditions.

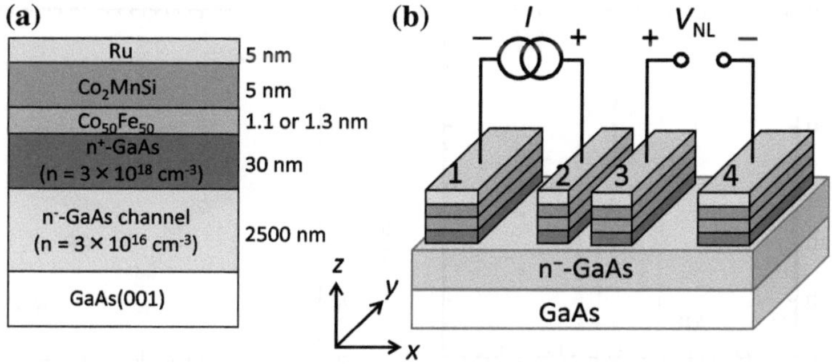

Fig. 18.13 Schematic diagram of **a** layer structure and **b** device structure for lateral spintransport devices. The film composition of the CMS film was chosen to be Mn-rich $Co_2Mn_{1.30}Si_{0.84}$ to suppress harmful Co_{Mn} antisites [26]. The ultrathin CoFe layer was inserted to improve the structural quality of the CMS layer. The sizes of the injector contact (contact-2) and detector contact (contact-3) of the lateral spin transport devices were $0.5 \times 10\,\mu m$ and $1.0 \times 10\,\mu m$, respectively, and the spacing (d) between them was 0.5 or 4.0 μm. The sample was evaluated in a four-terminal nonlocal geometry in which the nonlocal voltage between contact-3 and contact-4 was measured under a constant current (I) supplied between contact-2 and contact-1 at 4.2 K as functions of the in-plane magnetic field for spin-valve effect measurements, the out-of-plane magnetic field for Hanle effect measurements, and the oblique magnetic field for oblique Hanle effect measurements

18.3.2 Spin Injection Properties

Figure 18.14 shows a high-angle annular dark-field scanning transmission electron microscopy (HAADF STEM) image of a layer structure of Ru/CMS/CoFe(1.3 nm)/GaAs(001) [26]. An atomically flat and abrupt interface facing GaAs was formed. Figure 18.15a, b show a typical spin-valve signal and Hanle signal measured at 4.2 K for a CMS/CoFe (1.3 nm)/n-GaAs junction [26]. The junction was positively biased

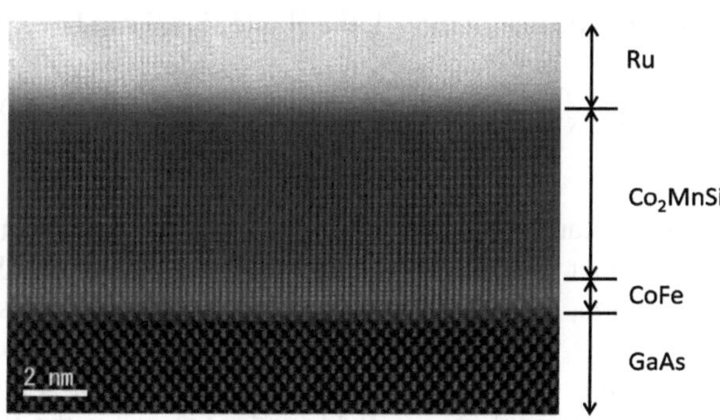

Fig. 18.14 High-angle annular dark-field scanning transmission electron microscopy (HAADF STEM) image of a layer structure of $Ru/Co_2MnSi/CoFe$(1.3 nm)/GaAs(001). An atomically flat and abrupt interface facing GaAs was formed. Reprinted with permission from [26]. Copyright 2014, American Institute of Physics

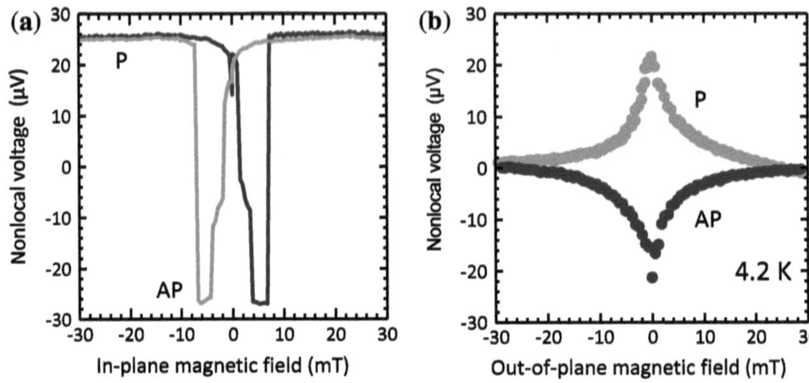

Fig. 18.15 **a** Typical spin-valve signal and **b** Hanle signal at 4.2 K in a nonlocal geometry for a Co_2MnSi/CoFe (1.3 nm)/*n*-GaAs junction. Clear spin-valve and Hanle signals were observed, indicating a spin injection from CMS into GaAs. Reprinted with permission from [26]. Copyright 2014, American Institute of Physics

with a supplied current $I = +10\,\mu A$, where electron spins were extracted from *n*-GaAs to CMS. An in-plane magnetic field (B_y) was applied along the longitudinal (y axis) direction of the junction. A clear spin-valve signal was observed due to switching between the parallel (P) and antiparallel (AP) states for the relative magnetization configurations between the injector and detector contacts at $B_y \cong +8$ and -8 mT. A nonlocal voltage change (ΔV_{NL}) of approximately 44 μV was obtained, where ΔV_{NL} is defined as ${V_{NL}}^{P} - {V_{NL}}^{AP}$ and where ${V_{NL}}^{P}$ and ${V_{NL}}^{AP}$ are nonlocal voltages for the P and AP configurations, respectively, between injector contact-2 and detector contact-3. Clear Hanle signals were also observed for both P and AP configurations, as shown in Fig. 18.15b. The observation of both the spin-valve signal and the Hanle signals is strong evidence of the injection, transport, and detection of spin-polarized electrons in the CMS/CoFe/*n*-GaAs lateral transport device. Based on the conventional spin diffusion model, the Hanle signal can be expressed by [75]

$$V_{NL}(B_z)/I = \pm \frac{P_{inj}P_{\det}}{2}\left(\rho \frac{l_{sf}}{S}\right)\left(\frac{2l_{sf}}{\tau_s}\right)\int_0^{\infty} \frac{1}{\sqrt{4\pi Dt}} \exp\left(-\frac{d^2}{4Dt}\right)\cos \Omega_L(B_z)t \exp\left(-\frac{t}{\tau_s}\right)dt, \tag{18.3}$$

where B_z is an out-of-plane magnetic field, $P_{inj(det)}$ is the spin polarization of the injector (detector) contact, ρ is the resistivity of the GaAs channel, S is the area of the channel cross-section, l_{sf} is the spin-diffusion length, d is the distance between contact-2 and contact-3, τ_s is the spin lifetime, I is the injection current, $D = l_{sf}^2/\tau_s$ is the diffusion constant, $\Omega_L(B_z) = g\mu_B B_z/\hbar$ is the Larmor frequency, g is an electron g factor in the channel ($g = -0.44$ for GaAs), μ_B is the Bohr magneton, and $\hbar$ is the reduced Planck's constant. The sign of $+(-)$ on the right-hand side of (18.3) corresponds to the P (AP) configuration. The observed Hanle curve can be fitted well with (18.3). The estimated values of τ_s and l_{sf} from fitting were 20 ns and 3 μm, respectively. These values are comparable to those observed for a CoFe

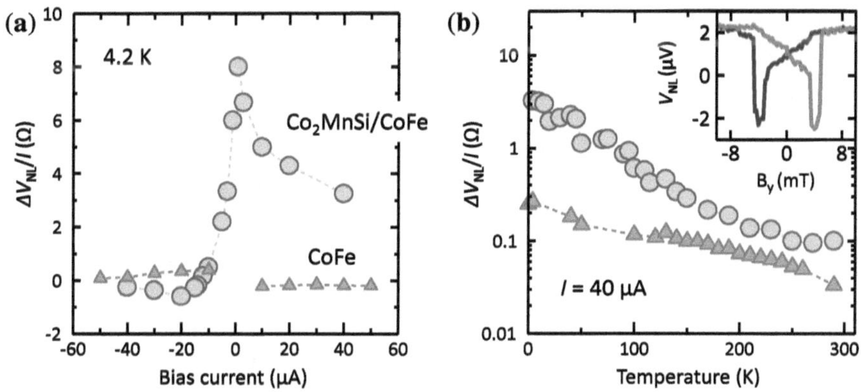

Fig. 18.16 **a** Bias-current dependence and **b** temperature dependence of $\Delta V_{NL}/I$ in spin-valve signals for a Co_2MnSi/CoFe (1.3 nm)/n-GaAs junction (circle) and a CoFe/n-GaAs junction (triangle). The inset of Fig. 18.16b shows a spin-valve signal for the Co_2MnSi/CoFe (1.3 nm)/n-GaAs junction at 290 K. The $\Delta V_{NL}/I$ values for the CMS/CoFe (1.3 nm)/n-GaAs junction were much higher than those of the CoFe/n-GaAs junction for almost the entire bias voltage range investigated and for the entire temperature range from 4.2 to 290 K, indicating that the CMS works as a highly polarized spin source. Reprinted with permission from [26]. Copyright 2014, American Institute of Physics

(5 nm)/n-GaAs sample with an identical channel structure [64]. The effective spin polarization defined by $(P_{inj}P_{det})^{1/2}$ was 0.52, a value almost 13 times larger than that for the CoFe/n-GaAs sample [26].

Here, we introduce $\Delta V_{NL}/I$ as a measure of spin injection efficiency because it is proportional to $P_{inj}P_{det}$ from (18.3). Figure 18.16a shows the bias-current dependence of $\Delta V_{NL}/I$ at 4.2 K for a CMS/CoFe (1.3 nm)/n-GaAs junction [26] and a CoFe (5 nm)/n-GaAs junction [64]. The $\Delta V_{NL}/I$ values for the CMS/CoFe (1.3 nm)/n-GaAs junction decreased as $|I|$ increased, and the sign of $\Delta V_{NL}/I$ changed at $I \approx -14\,\mu A$. This indicates that the magnitude and sign of $P_{inj}P_{det}$ strongly depends on the bias condition. Although there have been several experimental [23, 26, 58, 59, 64] and theoretical [76–78] investigations into whether the magnitude and sign of the spin polarization for a ferromagnet/GaAs heterojunction are affected by the bias condition, the origin is still an open question. Importantly, the $\Delta V_{NL}/I$ values for the CMS/CoFe (1.3 nm)/n-GaAs junction were much higher than those of the CoFe/n-GaAs junction for almost all of the bias region. In particular, the maximum value ($\sim 8\,\Omega$) was one order of magnitude higher than that of CoFe/n-GaAs. Thus, CMS works as a highly polarized spin source in the CMS/CoFe (1.3 nm)/n-GaAs junction even though a 1.3-nm-thick CoFe layer was inserted between the CMS and GaAs. Figure 18.16b shows $\Delta V_{NL}/I$ for the CMS/CoFe (1.3 nm)/n-GaAs junction and the CoFe (5 nm)/n-GaAs junction as a function of temperature (T) from 4.2 to 290 K. As shown in the inset of Fig. 18.16b, a clear spin-valve signal was observed even at 290 K for the CMS/CoFe (1.3 nm)/n-GaAs junction. Although the T dependence of $\Delta V_{NL}/I$ for the CMS/CoFe/n-GaAs junction is slightly stronger than that for the CoFe/n-GaAs junction, the value of $\Delta V_{NL}/I$ is still larger, even at 290 K. Table 18.4 summarizes

Table 18.4 Typical values of $\Delta V_{NL}/I$ reported in literature

Layer structure	d (μm)	$\Delta V_{NL}/I$
Co_2MnSi/CoFe/GaAs [26]	1.25	8.0 Ω (4.2 K)
		0.1 Ω (290 K)
CoFe/GaAs [64]	1.25	0.4 Ω (4.2 K)
		33 mΩ (290 K)
Fe/GaAs [59]	4.4	0.9 Ω (5 K)
GaMnAs/GaAs [74]	5	30 Ω (4.2 K)
Co_2FeSi/GaAs [24]	26	18 mΩ (22 K)
Fe/MgO/Si [61]	1.75	24 mΩ (8 K)
		1.0 mΩ (290 K)

d spacing between injector and detector

$\Delta V_{NL}/I$ reported in literature. It is noteworthy that the $\Delta V_{NL}/I$ of 0.1 Ω at 290 K obtained for the CMS/CoFe/*n*-GaAs junction is the highest yet reported for spin injections into GaAs or Si at room temperature.

18.3.3 Dynamic Nuclear Polarization

In this section, we describe the electrical detection of the Overhauser field through the observation of oblique Hanle signals. The steady-state Overhauser field ($\mathbf{B}_n$) induced by the DNP can be expressed by [79]

$$\mathbf{B_n} = f b_n \frac{\mathbf{B_{ob}} \cdot \mathbf{S}}{\mathbf{B_{ob}}^2 + \xi B_l^2} \mathbf{B_{ob}}, \tag{18.4}$$

where f ($\leq$1) is the leakage factor and b_n is the effective field due to the polarization of nuclear spins, which takes the negative value of -17 T in GaAs for the theoretical ideal case [80]. $\mathbf{S}$ is the average electron spin ($|\mathbf{S}| = 1/2$ corresponds to $P_{GaAs} = 100\,\%$), $\mathbf{B}_{ob}$ is the external magnetic field, B_l is the local dipolar field experienced by the nuclei, and ξ is a numerical coefficient on the order of unity, which depends on the nature of the spin-spin interactions [79, 80]. In the oblique Hanle effect measurement, $\mathbf{B}_{ob}$ is applied obliquely with respect to $\mathbf{S}$ so that electron spins make precession with $\mathbf{B}_n$ and $\mathbf{B}_{ob}$. Note that in the conventional Hanle effect measurement, in which $\mathbf{S}$ and $\mathbf{B}_{ob}$ are orthogonal, no nuclear field is generated because $\mathbf{B}_{ob} \cdot \mathbf{S} = 0$. Figure 18.17a shows a calculated nonlocal voltage using (18.3) and (18.4) as a function of B_{ob}. Here, we set $\mathbf{S} = S\mathbf{x}$ and $\mathbf{B}_{ob} = B_{ob}(\mathbf{x}\sin\theta + \mathbf{z}\cos\theta)$, where $\mathbf{x}$ and $\mathbf{z}$ are unit vectors along the x-axis and z-axis directions, respectively. Considering that electron spins experience the total magnetic field of $\mathbf{B}_{ob} + \mathbf{B}_{!n}$, and b_n is negative, the behavior of the nonlocal voltage shown in Fig. 18.17a can be explained as follows. At $B_{ob} > 0$, $\mathbf{B}_n$ and $\mathbf{B}_{ob}$ are antiparallel, and electron spins experience a smaller effective magnetic

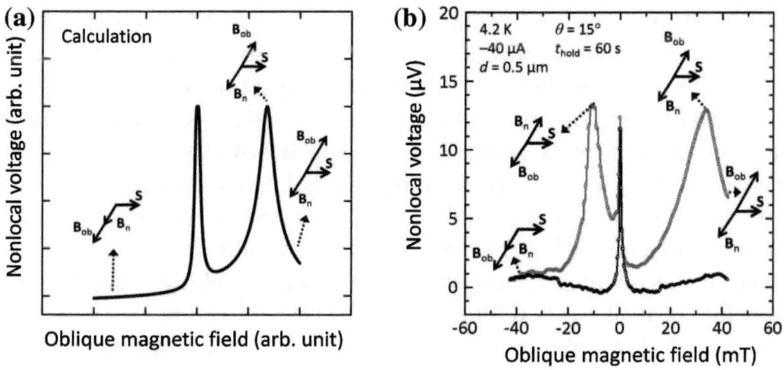

Fig. 18.17 **a** Calculated nonlocal voltage in the oblique Hanle effect using (18.3) and (18.4) as a function of oblique magnetic field (B_{ob}). The vector diagrams show the relative orientation of the average electron spin **S**, the oblique magnetic field $\mathbf{B}_{ob}$, and the nuclear field $\mathbf{B}_n$. **b** B_{ob} dependence of the nonlocal voltage for the CMS/CoFe(1.1 nm)/*n*-GaAs sample in the parallel magnetization configuration. The magnetic field was swept from +42 mT to −42 mT (red circles) and then swept back from −42 mT to +42 mT (black squares) with a sweep rate of 0.18 mT/s. Reprinted with permission from [23]. Copyright 2013, American Physical Society (Color figure online)

field than $|\mathbf{B}_{ob}|$. When $\mathbf{B}_n$ and $\mathbf{B}_{ob}$ cancel each other, electron spins get polarized and the nonlocal voltage shows a satellite peak. At $B_{ob} < 0$, on the other hand, no satellite peak appears, because $\mathbf{B}_n$ and $\mathbf{B}_{ob}$ are parallel and no cancellation occurs. Since it takes several hundred seconds or more for the nuclear field to reach the steady state [70–72], very slow sweeping of the external magnetic field is necessary for the steady-state measurement.

Figure 18.17b shows the B_{ob} dependence of the nonlocal voltage for a CMS/CoFe(1.1 nm)/n-GaAs sample in the parallel magnetization configuration. Since **S** is parallel to the x-axis direction in the CMS/CoFe(1.1 nm)/n-GaAs sample, $\mathbf{B}_{ob}$ was applied along the direction oblique by 15° from the z-axis in the $x - z$ plane so that electron spins were affected by $\mathbf{B}_{ob}$ and $\mathbf{B}_n$. The device was first initialized at $B_{ob} = +42$ mT for a hold time (t_{hold}) of 60 s at an injection current of $-40\,\mu$A, resulting in nuclear spins that became dynamically polarized. The magnetic field was then swept from +42 mT to −42 mT (negative sweep direction) and then swept back from −42 mT to +42 mT (positive sweep direction) with a sweep rate of 0.18 mT/s. This sweep rate was too fast for the nuclear field to reach the steady state. Compared to the steady-state signal, the observed transient oblique Hanle signal has two features: (1) an additional satellite peak observed at −10 mT in the negative sweep direction, and (2) no satellite peak observed in the positive sweep direction, indicating a clear hysteretic nature depending on the sweep direction.

To explain the observed oblique Hanle signal, we discuss the behavior of the nuclear spin in the transient state. In the negative sweep direction, the behavior of the nuclear field for $B_{ob} > 0$ is qualitatively similar to that for the steady-state nuclear field; i.e., the nuclear field is generated along the antiparallel direction to $\mathbf{B}_{ob}$ during an initial holding time at $B_{ob} = +42$ mT, and then $\mathbf{B}_{ob}$ and $\mathbf{B}_n$ cancel each other at

$B_{ob} = +33$ mT. In a similar way, the observation of the satellite peak at $B_{ob} = -10$ mT indicates that electron spins are repolarized due to the cancellation of $\mathbf{B}_{ob}$ and $\mathbf{B}_n$. However, this antiparallel state for $\mathbf{B}_n$ with respect to $\mathbf{B}_{ob}$ is the transient state because $\mathbf{B}_n$ and $\mathbf{B}_{ob}$ are in a parallel configuration at $B_{ob} < 0$ in the steady state, as indicated by (18.4). Then, $\mathbf{B}_n$ gradually goes to the steady state, and it almost reaches the steady state at $B_{ob} = -42$ mT, resulting in it being parallel to $\mathbf{B}_{ob}$. In the positive sweep direction, $\mathbf{B}_{ob}$ and $\mathbf{B}_n$ are parallel at $B_{ob} < 0$, and this parallel configuration is transiently kept just after $\mathbf{B}_{ob}$ is reversed from the negative direction to the positive one. Thus, no cancellation occurs between $\mathbf{B}_{ob}$ and $\mathbf{B}_n$, resulting in the disappearance of the satellite peak at $B_{ob} > 0$ for the positive sweep direction.

As discussed above, the transient oblique Hanle signals observed in the CMS/CoFe/n-GaAs sample can be qualitatively explained by adiabatic nuclear spin reversal; that is, the nuclear spins adiabatically rotate 180° when the magnetic field crosses zero [71, 73]. A similar adiabatic nuclear spin reversal has been reported in the electrical detection of spin-valve signals for lateral spin-transport devices with Fe/GaAs Schottky tunnel junctions [73], and in the optical detection of oblique Hanle signals for a spin LED with a MnSb ferromagnet electrode [71].

One can estimate the strength of the nuclear field from the oblique Hanle signal because $\mathbf{B}_{ob} + \mathbf{B}_n = 0$ is satisfied at the satellite peak position. Furthermore, one can estimate through transient analysis the time scale needed for the nuclear spins to reach a steady state. Figure 18.18 shows the t_{hold} dependence of the observed satellite peak position for $B_{ob} > 0$. The circles and triangles indicate the data for the CMS/CoFe(1.1 nm)/n-GaAs sample and those for a CoFe(5 nm)/n-GaAs sample, respectively. For both samples, the satellite peak position shows exponential

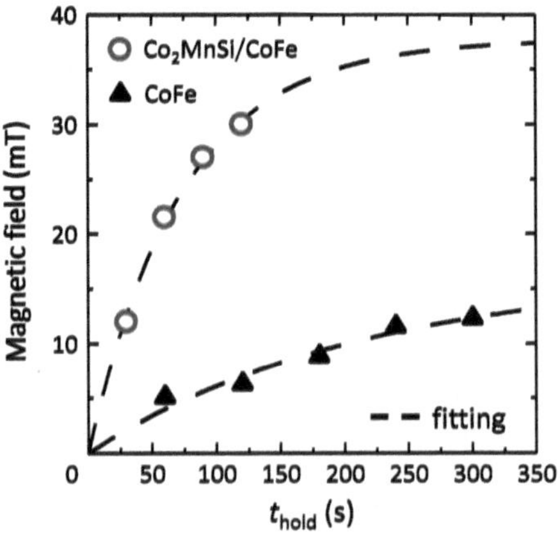

Fig. 18.18 t_{hold} dependence of the observed satellite peak position at $B_{ob}>0$ for a CMS/CoFe(1.1 nm)/*n*-GaAs sample and a CoFe/*n*-GaAs sample. The satellite peak position shows exponential dependence (*dash line*). The larger saturation field of 37 mT for the CMS/CoFe(1.1 nm)/*n*-GaAs sample than that of 16 mT for the CoFe/*n*-GaAs sample indicates a greater Overhauser magnetic field due to the higher spin polarization of the CMS. Reprinted with permission from [23]. Copyright 2013, American Physical Society

dependence. From the rate equation for the DNP, the time evolution of the nuclear field is given by [70, 71]

$$B_N(t) = B_N(\infty) \exp\left[-t\left(\frac{1}{T_P} + \frac{1}{T_1}\right)\right], \tag{18.5}$$

where $B_N(\infty)$ is the steady-state Overhauser field, which is given by (18.4), and T_P^{-1} and T_1^{-1} are the rate for nuclear polarization through DNP and that for nuclear spin relaxation through nuclear-lattice interaction, respectively. Importantly, the saturation value of the satellite peak position of 37 mT for the CMS/CoFe/n-GaAs sample is larger than that of 16 mT for the CoFe/n-GaAs sample, suggesting a greater Overhauser magnetic field due to the higher spin polarization of the CMS. The characteristic time, $(1/T_P + 1/T_1)^{-1}$, estimated from the exponential dependence of the satellite peak positions on t_{hold} is approximately 77 s for the CMS/CoFe/n-GaAs sample and 213 s for the CoFe/n-GaAs sample, respectively. These values are reasonable for the time scale needed for the nuclear spins to reach the steady state through the DNP [70, 71, 73]. The value of $|2\mathbf{S}|$, which corresponds to the spin polarization in the channel, estimated from the saturation value of the Overhauser field, was 5.4 % for the CMS/CoFe(1.1 nm)/n-GaAs sample and 2.3 % for the CoFe/n-GaAs sample. These values are comparable to the effective spin polarizations of $|P_{inj} \cdot P_{det}|^{1/2}$, estimated from the $|V_P|$ for both samples.

18.4 Conclusion

The effect of defects associated with nonstoichiometry on the half-metallicity of Co_2MnSi (CMS) was described. It was shown that the Mn composition (α) dependence of the saturation magnetization per formula unit (μ_s) of $Co_2Mn_\alpha Si_\beta$ thin films was in good agreement with the KKR-CPA calculations based on the antisite-based site-specific formula unit composition model. The decrease in μ_s for Mn-deficient CMS films from the half-metallic Slater-Pauling value, $Z_t - 24$, where Z_t is the valence electron number per formula unit, was explained by the smaller spin magnetic moment of the antisite Co_{Mn} in comparison with Mn at the nominal Mn site. It was also shown that the experimental α dependence of the TMR ratio of CMS/MgO/CMS MTJs (CMS MTJs) was qualitatively explained by the theoretical spin polarizations obtained from only the s- and p-orbital components. In particular, the decrease in the TMR ratio observed for CMS MTJs having Mn-deficient electrodes was ascribed to small s-and p-orbital components of the local density of minority-spin in-gap states at E_F that appeared for the antisite Co_{Mn} and regular-site Co_{Co}. It was found that the μ_s value was in good agreement with the half-metallic Slater-Pauling value when α was increased to a Mn-rich composition. It was also shown that the TMR ratios at 4.2 and 290 K systematically increased with increasing α for the range from a Mn-deficient composition to a certain Mn-rich composition. These behaviors of

Mn-rich CMS films and CMS MTJs with Mn-rich CMS electrodes were consistently explained by the suppression of harmful Co_{Mn} antisites. Given the concept of appropriately controlling the defects, giant TMR ratios of up to 1995 % at 4.2 K and up to 354 % at 290 K have been demonstrated for CMS MTJs with Mn-rich CMS electrodes featuring the enhanced half-metallicity. Furthermore, an efficient spin injection into GaAs and resultant efficient nuclear spin polarization have been demonstrated for lateral spintransport devices by applying Mn-rich CMS electrodes as a highly effective spin source.

In conclusion, it was shown that the suppression of Co_{Mn} antisites is essential to obtain the half-metallicity of Co_2MnSi and that detrimental Co_{Mn} antisite can be suppressed by preparing Co_2MnSi thin films with a Mn-rich composition. The giant TMR ratios of the CMS MTJs and the high electrical spin injection efficiency into semiconductor channels both obtained by applying half-metallic CMS electrodes as a spin source show the promise of half-metallic Co_2MnSi electrodes as a highly efficient spin source material for future spintronic devices.

References

1. R.A. de Groot, F.M. Mueller, P.G. van Engen, K.H.J. Buschow, Phys. Rev. Lett. **50**, 2024 (1983)
2. K. Inomata, S. Okamura, R. Goto, N. Tezuka, Jpn. J. Appl. Phys. **42**, L419 (2003)
3. S. Kämmerer, A. Thomas, A. Hütten, G. Reiss, Appl. Phys. Lett. **85**, 79 (2004)
4. T. Marukame, T. Kasahara, K.-I. Matsuda, T. Uemura, M. Yamamoto, Jpn. J. Appl. Phys. **44**, L521 (2005); T. Marukame, T. Ishikawa, K.-I. Matsuda, T. Uemura, M. Yamamoto. Appl. Phys. Lett. **88**, 262503 (2006)
5. M. Yamamoto, T. Marukame, T. Ishikawa, K.-I. Matsuda, T. Uemura, M. Arita, J. Phys. D: Appl. Phys. **39**, 824 (2006)
6. Y. Sakuraba, M. Hattori, M. Oogane, Y. Ando, H. Kato, A. Sakuma, T. Miyazaki, H. Kubota, Appl. Phys. Lett. **88**, 192508 (2006)
7. T. Ishikawa, T. Marukame, H. Kijima, K.-I. Matsuda, T. Uemura, M. Arita, M. Yamamoto, Appl. Phys. Lett. **89**, 192505 (2006)
8. T. Ishikawa, S. Hakamata, K.-I. Matsuda, T. Uemura, M. Yamamoto, J. Appl. Phys. **103**, 07A919 (2008)
9. T. Ishikawa, H.-X. Liu, T. Taira, K.-I. Matsuda, T. Uemura, M. Yamamoto, Appl. Phys. Lett. **95**, 232512 (2009)
10. M. Yamamoto, T. Ishikawa, T. Taira, G.-F. Li, K.-I. Matsuda, T. Uemura, J. Phys.: Condens. Matter **22**, 164212 (2010)
11. H.-X. Liu, Y. Honda, T. Taira, K.-I. Matsuda, M. Arita, T. Uemura, M. Yamamoto, Appl. Phys. Lett. **101**, 132418 (2012)
12. G.-F. Li, Y. Honda, H.-X. Liu, K.-I. Matsuda, M. Arita, T. Uemura, M. Yamamoto, Y. Miura, M. Shirai, T. Saito, F. Shi, P.M. Voyles, Phys. Rev. B **89**, 014428 (2014)
13. N. Tezuka, N. Ikeda, F. Mitsuhashi, S. Sugimoto, Appl. Phys. Lett. **94**, 162504 (2009)
14. W. Wang, E. Liu, M. Kodzuka, H. Sukegawa, M. Wojcik, E. Jedryka, G.H. Wu, K. Inomata, S. Mitani, K. Hono, Phys. Rev. B **81**, 140402(R) (2010)
15. K. Yakushiji, K. Saito, S. Mitani, K. Takanashi, Y.K. Takahashi, K. Hono, Appl. Phys. Lett. **88**, 222504 (2006)
16. T. Furubayashi, K. Kodama, H. Sukegawa, Y.K. Takahashi, K. Inomata, K. Hono, Appl. Phys. Lett. **93**, 122507 (2008)

17. Y. Sakuraba, K. Izumi, T. Iwase, S. Bosu, K. Saito, K. Takanashi, Y. Miura, K. Futatsukawa, K. Abe, M. Shirai, Phys. Rev. B **82**, 094444 (2010)
18. J. Sato, M. Oogane, H. Naganuma, Y. Ando, Appl. Phys. Express **4**, 113005 (2011)
19. M.J. Carey, S. Maat, S. Chandrashekariaih, J.A. Katine, W. Chen, B. York, J.R. Childress, J. Appl. Phys. **109**, 093912 (2011)
20. X.Y. Dong, C. Adelmann, J.Q. Xie, C.J. Palmstrøm, X. Lou, J. Strand, P.A. Crowell, J.-P. Barnes, A.K. Petford-Long, Appl. Phys. Lett. **86**, 102107 (2005)
21. M.C. Hickey, C.D. Damsgaard, I. Farrer, S.N. Holmes, A. Husmann, J.B. Hansen, C.S. Jacobsen, D.A. Ritchie, R.F. Lee, G.A.C. Jones, M. Pepper, Appl. Phys. Lett. **86**, 252106 (2005)
22. M. Ramsteiner, O. Brandt, T. Flissikowski, H.T. Grahn, M. Hashimoto, J. Herfort, H. Kostial, Phys. Rev. B **78**, 121303(R) (2008)
23. T. Akiho, J. Shan, H.-X. Liu, K.-I. Matsuda, M. Yamamoto, T. Uemura, Phys. Rev. B. **87**, 235205 (2013)
24. P. Bruski, Y. Manzke, R. Farshchi, O. Brandt, J. Herfort, M. Ramsteiner, Appl. Phys. Lett. **103**, 052406 (2013)
25. T. Saito, N. Tezuka, M. Matsuura, S. Sugimoto, Appl. Phys. Express **6**, 103006 (2013)
26. Y. Ebina, T. Akiho, H.-X. Liu, M. Yamamoto, T. Uemura, Appl. Phys. Lett. **104**, 172405 (2014)
27. S. Ishida, S. Fujii, S. Kashiwagi, S. Asano, J. Phys. Soc. Jpn. **64**, 2152 (1995)
28. S. Picozzi, A. Continenza, A.J. Freeman, Phys. Rev. B **66**, 094421 (2002)
29. I. Galanakis, P.H. Dederichs, N. Papanikolaou, Phys. Rev. B **66**, 174429 (2002)
30. P.J. Webster, J. Phys. Chem. Solids **32**, 1221 (1971)
31. W.H. Butler, X.-G. Zhang, T.C. Schulthess, J.M. MacLaren, Phys. Rev. B **63**, 054416 (2001)
32. J. Mathon, A. Umerski, Phys. Rev. B **63**, 220403(R) (2001)
33. Y. Miura, H. Uchida, Y. Oba, K. Nagao, M. Shirai, J. Phys.: Condens. Matter **19**, 365228 (2007)
34. S.S.P. Parkin, C. Kaiser, A. Panchula, P.M. Rice, B. Hughes, M. Samant, S.-H. Yang, Nat. Mater. **3**, 862 (2004)
35. S. Yuasa, T. Nagahama, A. Fukushima, Y. Suzuki, K. Ando, Nat. Mater. **3**, 868 (2004)
36. Y. Miura, K. Futatsukawa, S. Nakajima, K. Abe, M. Shirai, Phys. Rev. B **84**, 134432 (2011)
37. S. Picozzi, A. Continenza, A.J. Freeman, Phys. Rev. B **69**, 094423 (2004)
38. Y. Miura, K. Nagao, M. Shirai, Phys. Rev. B **69**, 144413 (2004)
39. I. Galanakis, K. Özdoğan, B. Aktaş, E. Şaşıoğlu, Appl. Phys. Lett. **89**, 042502 (2006)
40. B. Hülsen, M. Scheffler, P. Kratzer, Phys. Rev. B **79**, 094407 (2009)
41. H.-X. Liu, Y. Honda, K.-I. Matsuda, M. Arita, T. Uemura, M. Yamamoto, Jpn. J. Appl. Phys. **51**, 093004 (2012)
42. J.-P. Wüstenberg, R. Fetzer, M. Aeschlimann, M. Cinchetti, J. Minár, J. Braun, H. Ebert, T. Ishikawa, T. Uemura, M. Yamamoto, Phys. Rev. B **85**, 064407 (2012)
43. D. Asakura, T. Koide, S. Yamamoto, K. Tsuchiya, T. Shioya, K. Amemiya, V.R. Singh, T. Kataoka, Y. Yamazaki, Y. Sakamoto, A. Fujimori, T. Taira, M. Yamamoto, Phys. Rev. B **82**, 184419 (2010)
44. V.R. Singh, V.K. Verma, K. Ishigami, G. Shibata, T. Kadono, A. Fujimori, D. Asakura, T. Koide, Y. Miura, M. Shirai, G.-F. Li, T. Taira, M. Yamamoto, Phys. Rev. B **86**, 144412 (2012)
45. S. Ouardi, G.H. Fecher, S. Chadov, B. Balke, X. Kozina, C. Felser, T. Taira, M. Yamamoto, Appl. Phys. A: Mater. Sci. Process. **111**, 395 (2013)
46. X. Kozina, J. Karel, S. Ouardi, S. Chadov, G.H. Fecher, C. Felser, G. Stryganyuk, B. Balke, T. Ishikawa, T. Uemura, M. Yamamoto, E. Ikenaga, S. Ueda, K. Kobayashi, Phys. Rev. B **89**, 125116 (2014)
47. T. Saito, T. Katayama, T. Ishikawa, M. Yamamoto, D. Asakura, T. Koide, Appl. Phys. Lett. **91**, 262502 (2007)
48. T.M. Nakatani, Y.K. Takahashi, T. Ishikawa, M. Yamamoto, K. Hono, J. Magn. Magn. Mater. **322**, 357 (2010)
49. L.G. Parratt, Phys. Rev. **95**, 359 (1954)
50. W. Kohn, N. Rostoker, Phys. Rev. **94**, 1111 (1954)
51. H. Akai, P.H. Dederichs, Phys. Rev. B **47**, 8739 (1993)
52. Y. Miura, H. Uchida, Y. Oba, K. Abe, M. Shirai, Phys. Rev. B **78**, 064416 (2008)

53. H.J. Zhu, M. Ramsteiner, H. Kostial, M. Wassermeier, H.-P. Schönherr, K.H. Ploog, Phys. Rev. Lett. **87**, 016601 (2001)
54. X. Jiang, R. Wang, R.M. Shelby, R.M. Macfarlane, S.R. Bank, J.S. Harris, S.S.P. Parkin, Phys. Rev. Lett. **94**, 056601 (2005)
55. P. Renucci, V.G. Truong, H. Jaffrès, L. Lombez, P.H. Binh, T. Amand, J.M. George, X. Marie, Phys. Rev. B **82**, 195317 (2010)
56. M. Yasar, R. Mallory, A. Petrou, A.T. Hanbicki, G. Kioseoglou, C.H. Li, O.M.J. van 't Erve, B.T. Jonker, Appl. Phys. Lett. **94**, 032102 (2009)
57. B.D. Schultz, N. Marom, D. Naveh, X. Lou, C. Adelmann, J. Strand, P.A. Crowell, L. Kronik, C.J. Palmstrøm, Phys. Rev. B **80**, 201309(R) (2009)
58. X. Lou, C. Adelmann, S.A. Crooker, E.S. Garlid, J. Zhang, K.S.M. Reddy, S.D. Flexner, C.J. Palmstrøm, P.A. Crowell, Nat. Phys. **3**, 197 (2007)
59. G. Salis, A. Fuhrer, R.R. Schlittler, L. Gross, S.F. Alvarado, Phys. Rev. B **81**, 205323 (2010)
60. M. Ciorga, A. Einwanger, U. Wurstbauer, D. Schuh, W. Wegscheider, D. Weiss, Phys. Rev. B **79**, 165321 (2009)
61. T. Suzuki, T. Sasaki, T. Oikawa, M. Shiraishi, Y. Suzuki, K. Noguchi, Appl. Phys. Express **4**, 023003 (2011)
62. Y. Zhou, W. Han, L.-T. Chang, F. Xiu, M. Wang, M. Oehme, I.A. Fischer, J. Schulze, R.K. Kawakami, K.L. Wang, Phys. Rev. B **84**, 125323 (2011)
63. H. Kum, J. Heo, S. Jahangir, A. Banerjee, W. Guo, P. Bhattacharya, Appl. Phys. Lett. **100**, 182407 (2012)
64. T. Uemura, T. Akiho, M. Harada, K.-I. Matsuda, M. Yamamoto, Appl. Phys. Lett. **99**, 082108 (2011)
65. H. Sanada, Y. Kondo, S. Matsuzaka, K. Morita, C.Y. Hu, Y. Ohno, H. Ohno, Phys. Rev. Lett. **96**, 067602 (2006)
66. G. Yusa, K. Muraki, K. Takashina, K. Hashimoto, Y. Hirayama, Nature **434**, 1001 (2005)
67. R. Kaji, S. Adachi, H. Sasakura, S. Muto, Phys. Rev. B **77**, 115345 (2008)
68. P. Maletinsky, C.W. Lai, A. Badolato, A. Imamoglu, Phys. Rev. B **75**, 035409 (2007)
69. D. Gammon, Al L. Efros, T.A. Kennedy, M. Rosen, D.S. Katzer, D. Park, S.W. Brown, V.L. Korenev, I.A. Merkulov, Phys. Rev. Lett. **86**, 5176 (2001)
70. J. Strand, X. Lou, C. Adelmann, B.D. Schultz, A.F. Isakovic, C.J. Palmstrøm, P.A. Crowell, Phys. Rev. B **72**, 155308 (2005)
71. P. Van Dorpe, W. Van Roy, J. De Boeck, G. Borghs, Phys. Rev. B **72**, 035315 (2005)
72. M.K. Chan, Q.O. Hu, J. Zhang, T. Kondo, C.J. Palmstrøm, P.A. Crowell, Phys. Rev. B **80**, 161206(R) (2009)
73. G. Salis, A. Fuhrer, S.F. Alvarado, Phys. Rev. B **80**, 115332(R) (2009)
74. J. Shiogai, M. Ciorga, M. Utz, D. Schuh, T. Arakawa, M. Kohda, K. Kobayashi, T. Ono, W. Wegscheider, D. Weiss, J. Nitta, Appl. Phys. Lett. **101**, 212402 (2012)
75. J. Fabian, A. Matos-Abiaguea, C. Ertlera, P. Stano, I. Zutic, Acta Phys. Slovaca **57**, 565 (2007)
76. H. Dery, L.J. Sham, Phys. Rev. Lett. **98**, 046602 (2007)
77. A.N. Chantis, K.D. Belashchenko, D.L. Smith, E.Y. Tsymbal, M. van Schilfgaarde, R.C. Albers, Phys. Rev. Lett. **99**, 196603 (2007)
78. S. Honda, H. Itoh, J. Inoue, J. Phys. D: Appl. Phys. **43**, 135002 (2010)
79. V.G. Fleisher, I.A. Merkulov, in *Optical Orientation*, ed. by F. Meier, B.P. Zakharchenya (North-Holland, New York, 1984)
80. D. Paget, G. Lampel, B. Sapoval, V.I. Safarov, Phys. Rev. B **15**, 5780 (1977)

Chapter 19
Exchange Bias of Polycrystalline Heusler Alloy Thin Films

H. Endo, T. Nakayama, J. Sagar, G. Vallejo Fernandez, A. Hirohata and K. O'Grady

Abstract It is accepted that Heusler alloys have a great deal of potential for applications in spintronic devices due to their high magnetic moment and spin polarisation at the Fermi level. However for this potential to be realised it is necessary to create structures where one ferromagnetic layer is pinned allowing the other to rotate thus creating a GMR or TMR effect. One possible technique to pin one ferromagnetic layer is to exchange bias it in a similar manner to that used for CoFe ferromagnetic layers in devices such as read heads. In this work we review progress which has been made in attempting to exchange bias such films. It is found that it is possible to achieve a significant level of exchange bias using carefully controlled growth and annealing processes. A significant exchange bias can be achieved using the temperature limits normally associated with silicon device electronics. Hence one potential obstacle to the application of Heusler alloys has been successfully addressed.

19.1 Introduction

The potential of Heusler alloys for spintronic device applications is well established. Alloys such as Co_2FeSi have complex structures due to the mixture of high moment transition metal atoms with semi-conductor atoms. Indeed many Heusler alloys are ferromagnetic due to the inclusion of Fe, Co or Ni. Bulk alloys are relatively easy

H. Endo
Nihon University, Koriyama, Fukushima 963-8642, Japan

T. Nakayama
Nagaoka University of Technology, Nagaoka, Niigata 940-2188, Japan

J. Sagar · G.V. Fernandez · K. O'Grady (✉)
Department of Physics, University of York, York YO10 5DD, UK
e-mail: kevin.ogrady@york.ac.uk

A. Hirohata
Department of Electronics, University of York, York YO10 5DD, UK

A. Hirohata
PRESTO Japan Science and Technology Agency, Kawaguchi,
Saitama 332-0012, Japan

C. Felser and A. Hirohata (eds.), *Heusler Alloys*,
Springer Series in Materials Science 222, DOI 10.1007/978-3-319-21449-8_19

to prepare by arc melting or otherwise, and the degree of crystallinity obtained can be enhanced, e.g. by annealing, but often using temperatures as high as 700 °C [1]. Many of the alloys can be produced in single crystal thin film form by molecular beam epitaxial (MBE) growth [2] and they can also be produced by a range of sputtering or ion beam deposition techniques. It is this ability to produce single crystal or polycrystalline granular films that makes them suitable for potential applications in spintronic devices. Reviews of Heusler alloy thin films are at references [3, 4].

In the recent past there has been a surge in interest in these materials because theoretical calculations predict that they will exhibit 100 % spin polarisation at the Fermi level [1]. This implies that all electrons passing through the material will acquire full spin polarisation. This has advantages in spintronic devices over other techniques for the generation of spin polarised electrons such as the use of a ferromagnet or a spin tunnel junction. In principle no complex application of an applied field to magnetise a ferromagnet nor the challenging growth of e.g. an MgO spin tunnel barrier are required. Simply the deposition of the Heusler alloy would achieve the degree of spin polarisation required.

As a consequence there have been a number of recent reports of large magnetoresistance effects in devices based on Heusler alloys. However these reports are all associated with epitaxially grown films which would probably be impractical for the manufacture of large numbers of devices. Nonetheless values of the magnetoresistance coefficient (ΔR/R) as high as 354 % at room temperature have been obtained with a value of up to almost 2,000 % at 4.2 K. Hence the potential for Heusler alloys to achieve the required degree of spin polarisation is well established and the need to achieve applicable values using polycrystalline films is clear [5–8].

Figure 19.1 shows the theoretical calculation of the density of states for Co_2FeSi undertaken by Wurmehl et al. [9]. As can be seen, at the Fermi energy, which is the zero point on the ordinate, only one spin polarisation, in this case spin up, occurs.

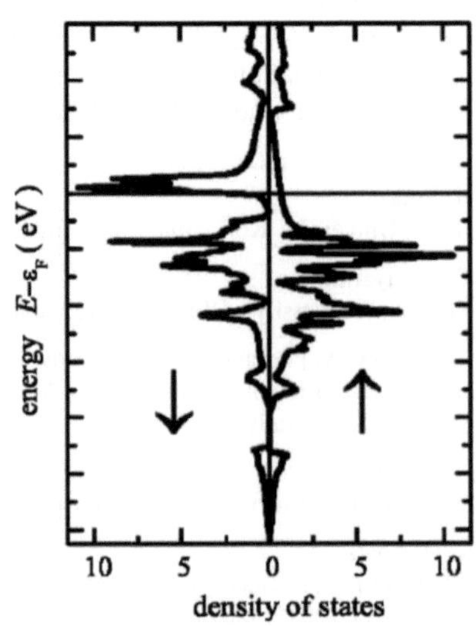

Fig. 19.1 Density of states of Co_2FeSi [9]

However the degree of spin polarisation predicted lies fairly close to the likely error in the calculation so there is a need for this result to be confirmed experimentally. Also in practice, spintronic devices will be produced by sputtering or ion beam deposition giving polycrystalline or granular films with a typical grain size of about 10 nm. Unlike the case for a single crystal film grown by MBE, this will give rise to grain boundary scattering of conduction electrons and also in the case of polycrystalline films, there will be variations in the orientation of the individual crystallites in the film. Hence while a given crystallite will 100 % spin polarise a conduction electron in a given direction, the neighbouring crystal will almost certainly not be oriented in the same direction and hence whilst it in turn will give 100 % spin polarisation, the direction of that polarisation will differ from that induced by the first grain. In practice it is possible to generate close to 100 % in plane texture in the films but still the direction of the axes within the plane will generally be random.

In the case of Co_2FeSi the alloy is ferromagnetic due to the presence of the ferromagnetic transition metals. In this case the distribution of easy axis directions within the plane will give a hysteresis loop that will not have 100 % loop squareness, i.e. the remanence to saturation ratio $(M_r/M_s) < 1$. This distribution of the easy axis directions even assuming that all the grains are thermally stable and not superparamagnetic, will give a spin polarisation of less than 100 % effectively moderated by the value of the squareness of the hysteresis loop. Assuming perfect thermal stability which will be dependent upon the grain volume and anisotropy of the Co_2FeSi, the maximum value of squareness and hence the degree of polarisation that can be achieved is equal to 0.82.

Co_2FeSi crystallises with a cubic structure which after annealing orders in the $L2_1$ state. Whilst there are no known measurements of the anisotropy of Co_2FeSi films it is reasonable to assume that the magnetocrystalline anisotropy (K_c) will be quite low and similar to that for other cubic materials such as fine grain Co which has a value of $K = K_c = 5 \times 10^5$ ergs/cc. However the grains are also likely to be irregular in shape and given the value of the saturation magnetisation (M_s) of the alloy ($M_s = 1{,}200$ emu/cc), significant shape anisotropy is expected to occur. The shape anisotropy constant K_s is given by

$$K_s = \frac{1}{2} {M_s}^2 (N_a - N_c) \tag{19.1}$$

where N_a and N_c are the shape demagnetising factors along the short and long axis of a grain respectively. For a particle elongation of the order of 15 % this would give rise to a shape anisotropy constant of the order of 2×10^6 ergs/cc. The well established criterion for superparamagnetic behaviour at which a ferromagnetic grain achieves thermal stability in zero field on a 100 s timescale is given by the Bean and Livingston criterion [10]

$$KV = 25kT \tag{19.2}$$

where V is the particle volume, k is Boltzmann's constant and T the temperature of measurement. For the values given above for an alloy such as Co_2FeSi, this would give rise to a critical diameter for superparamagnetic behaviour in this alloy of the order of 10 nm. However at the time of writing, to our knowledge, the actual critical size in such polycrystalline films has not been measured. Nonetheless with grain sizes in sputtered polycrystalline films generally lying in the range 5–15 nm, it would be expected that a significant fraction of the grains would be superparamagnetic thereby lowering the loop squareness of sputtered films and leading to a further reduction in the degree of spin polarisation achieved.

Hence if polycrystalline Heusler alloys are to be used in spintronic devices a technique for the enhancement of the anisotropy and a rectification of the lack of loop squareness is required. One obvious technique that would achieve both these requirements would be to exchange bias the films by growing them in contact with a suitable antiferromagnetic material.

19.2 Exchange Bias in Polycrystalline Films

Exchange bias was discovered in 1956 by Meiklejohn and Bean in oxidised Co/CoO nanoparticles suspended in mercury [11]. CoO is a cubic antiferromagnet with a Néel temperature of 19 °C. When the particles were field cooled to 77 K an anomalous shifted loop was observed which was explained as being due to the coupling between the ferromagnetic (F) Co and antiferromagnetic (AF) CoO on the surface of the particles. The field cooling results in an alignment of the surface layer spins of the AF with the F spins of the Co such that when the field is reversed a unidirectional interfacial anisotropy results giving rise to the loop shift. The original data of Meiklejohn and Bean is reproduced in Fig. 19.2 [11]. The phenomenon was largely of academic interest only until the development of the spin valve sensor for hard drive read heads in the 1990s. In such a sensor based on the giant magnetoresistive (GMR) or in the more recent tunnelling magnetoresistive (TMR) sensors, it is required that one of the F layers in the tri-layer stack have its magnetisation maintained in one direction known as the pinned layer, whilst the other is able to rotate under the influence of the stray field from the recording disc or tape, thereby creating the GMR effect. This can be achieved by using two F layers having very different coercivities but more commonly an AF alloy is used to pin one of the layers. Over the years for reasons of corrosion and thermal stability, the alloy $IrMn_3$ has become the AF of choice and is currently used in all magnetic recording sensors for both disc and tape applications. It is also used in a number of solid state magnetic memory devices such as those based on thermal reversal of one layer in a TMR stack commonly known as TA-MRAM [12].

A typical example of the resulting exchange bias stack is shown in Fig. 19.3 [13]. This measurement was made on a sheet film of IrMn grown on a thin layer of CoFe. As can be seen from Fig. 19.3 a very large loop shift in this case of 3.6 kOe can be generated and the intercept of the loop on the moment axis is to a first approximation, equal to the saturation magnetisation of the CoFe layer thereby giving a perfect degree of alignment of the grains in the F layer. Interestingly consideration of Fig. 19.3 shows

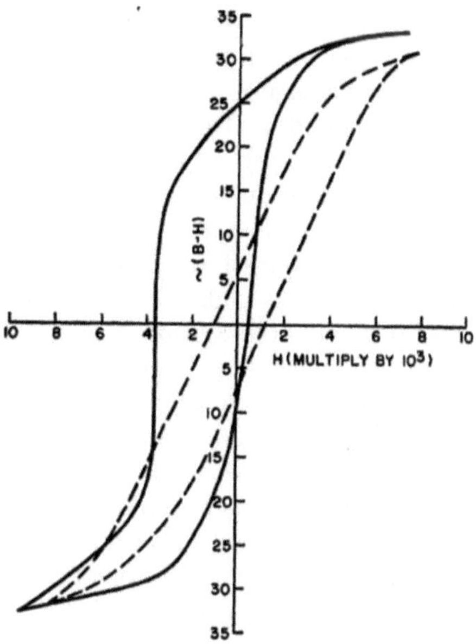

Fig. 19.2 Original data of Meiklejohn and Bean on Co/CoO nanoparticles [11]

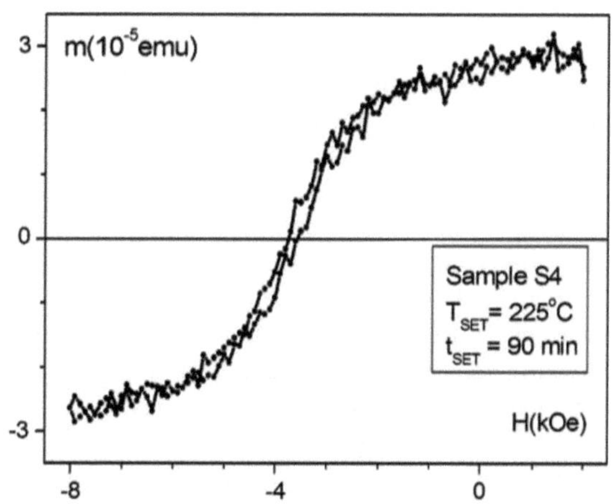

Fig. 19.3 Hysteresis loop for a typical TMR stack at room temperature [13]

that no matter how large a negative field is applied to this system it can never be demagnetised and the magnetisation will always return to the value of the saturation in the positive direction once the field is removed. When this result was published by our group in 2008 it represented a world record for a measured exchange bias system in the plane of the film [13].

To achieve this remarkable shift in the loop an AF with a very high anisotropy is required. To date the AF with the highest recorded anisotropy is IrMn [14] but higher values have been achieved perpendicular to the plane via the use of rare earth transition metal alloys which are synthetic antiferromagnets [15].

Until 2010 there was very limited understanding of those parameters which control exchange bias in polycrystalline films. It was, and still, is accepted that the actual direct mechanism of the coupling between the AF and F layers at the interface is that originally described by Meiklejohn and Bean. However there was little understanding of those thin film parameters such as the grain size, texture, etc. which determine the degree of thermal stability achievable. In 2010 our group published what is now regarded as a definitive model of the thermal stability of polycrystalline exchange bias films [16]. This work is a review article which summarises a large number of studies undertaken in the previous 10 years. The model shows that the thermal stability of a granular thin film exchange bias system is simply a grain volume effect. Hence the distribution of grain volumes gives rise to a distribution of temperatures at which individual grains become thermally unstable. This is known as the blocking temperature distribution where a thermally stable AF grain is described as being blocked. The stability of the AF grain system itself is modelled based on the well established Stoner-Wohlfarth theory of ferromagnetic fine particles [17]. In the Stoner-Wohlfarth model reversal occurs at $T = 0$ at a critical field which is resisted by the presence of an anisotropy field H_K. In our model of exchange bias a similar formalism is used giving an energy barrier of the form

$$\Delta E = K_{AF}V(1 - H^*/H_K^*) \tag{19.3}$$

Here K_{AF} is the anisotropy of the antiferromagnetic grains, H^* is the exchange field from the ferromagnet acting on the antiferromagnetic when the applied field is reversed along with the magnetisation of the F layer and H_K^* is a pseudo anisotropy field of the antiferromagnet which determines its thermal stability. Those parameters which determine the value of H_K^* have not yet been established but this critical field will clearly be related to the anisotropy of the AF grains.

In all real polycrystalline thin films a distribution of grain size and hence grain volumes will be present. This will give rise to the well established distribution of blocking temperatures. Figure 19.4 shows a measured grain diameter distribution for a system grown in our laboratories. In such polycrystalline films the structure is that of columnar growth with each grain in the F layer growing immediately on top of a grain in the AF layer which is deposited underneath it. However the grains in the F layer are exchange coupled together and hence the reversal process is via domain wall motion. For the case of the AF grains there is no intergranular coupling between them and the reversal follows the Stoner-Wohlfarth model for single domain particles. Of course there will be some indirect coupling between the AF grains via the F layer due to the exchange coupling both between the AF grains and the F grains and the coupling between individual F grains. However as will be seen this effect is relatively small.

In the case of IrMn a further complication arises because the Néel temperature of the alloy in bulk is of the order of 700 K and hence the structure cannot be aligned or 'set' above the Néel temperature without subsequent damage to the entire structure of the stack. The alignment process is therefore undertaken by a thermal activation process with the F layer saturated in say the positive field direction and the stack

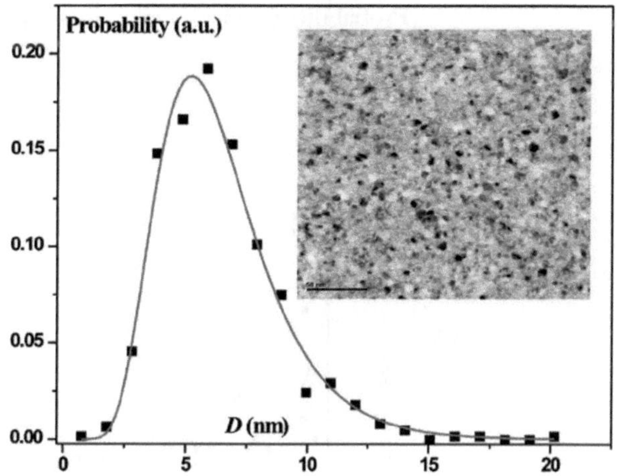

Fig. 19.4 Grain size distribution and TEM image for a sample grown using a HiTUS system [18]

annealed at a more modest temperature of about 300 °C for an hour or so. This in essence is a time dependent effect similar to the well established time dependence in all ferromagnetic materials. Nonetheless it is quite possible and indeed has been shown to be the case, that some of the grains in the AF layer will not be set at the temperatures which can be applied. This will limit the value of exchange bias achievable because certain F grains will now not be exchange biased because the AF grain upon which they have grown is itself not aligned. We denote the critical grain size above which AF grains cannot be set as V_{set}.

For the smaller grains in the size distribution there is also a strong probability that they may be too small to be thermally stable at the temperature of use. Essentially these grains behave in the same way as a superparamagnetic particle of a ferromagnetic material. Hence at a given temperature there will exist a critical grain volume in the AF layer (V_c) below which the orientation of the antiferromagnetic alignment of a grain will fluctuate under the influence of thermal energy and hence these grains will also not contribute to the resulting exchange bias. These two limits V_c and V_{set} are shown on the grain size distribution in Fig. 19.5.

Hence for a given system the value of exchange bias will be given by the area under the grain volume distribution between the two limits. That gives rise to an expression for the measured exchange bias of the form

$$H_{ex} = C^* \int_{V_c}^{V_{set}} f(V)DV \tag{19.4}$$

Here the parameter C^* represents the stiffness of the interfacial coupling between the AF and F layers. Again at the time of writing the parameters that control the value of C^* are not well defined but it has been shown that the nature of these interfaces is such that they behave independently of the degree of alignment in the AF grain from which they derive [16]. At the present time all read heads currently manufactured for

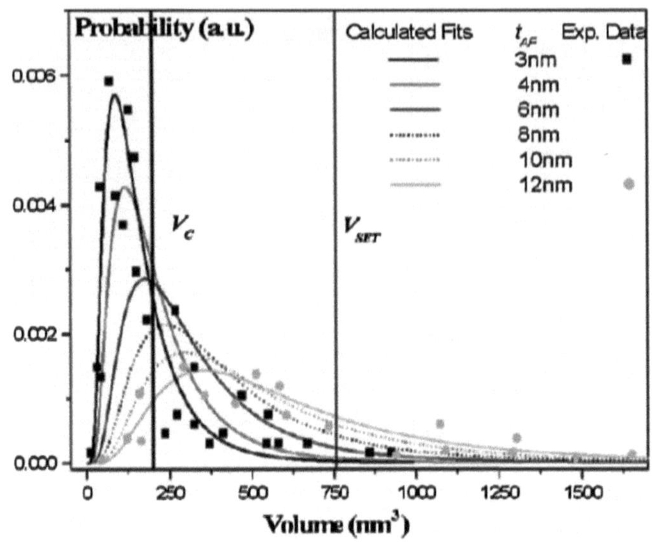

Fig. 19.5 Measured grain volume distributions for a set of samples with different AF thicknesses showing the two critical volumes responsible for exchange bias [19]

use in hard disc and tape drives have antiferromagnetic layers designed based upon the description of the effect provided above which is known as the York Model of Exchange Bias [16, 19].

Critical to the thermal stability of the AF grains is then the value of the anisotropy constant (K_{AF}). In a series of studies we have shown that the anisotropy of IrMn is in part controlled by the texture of the grains in the plane of the film. This in turn is controlled by the nature of the seed layer upon which they are grown. IrMn grows and crystallises readily in a FCC phase with the Mn atoms aligned along (111) planes. This then becomes the effective "easy axis" of the AF grains. Aley et al. showed that the optimum seed layer to induce (111) texture in the IrMn is an equi-atomic alloy NiCr preferably gown with a thickness of about 6 nm [20]. Figure 19.6 shows a grazing incidence X-ray diffraction scan for IrMn grown on NiCr, Ru and Cu seed layers. As can be seen from the figure, the IrMn (111) peak is present in the scans for IrMn grown on Ru and Cu but it is absent for the IrMn layer grown on NiCr seed layer. Given that these are very low angle XRD scans, the absence of the (111) peak indicates that the (111) plane is lying in the plane of the substrate. This will then optimise the anisotropy and hence the thermal stability of the IrMn layer.

Based on the York Model of Exchange Bias it is now possible to measure the effective anisotropy of an AF layer. If a sample of say IrMn has a layer of CoFe grown upon it and is then set, i.e. field cooled, from say 500 K back to a temperature where the AF stability is known to be established, then it is possible by heating the structure with the field in the negative direction progressively to higher temperatures to thereby reverse the AF grains in a systematic manner. Figure 19.7 shows a schematic diagram

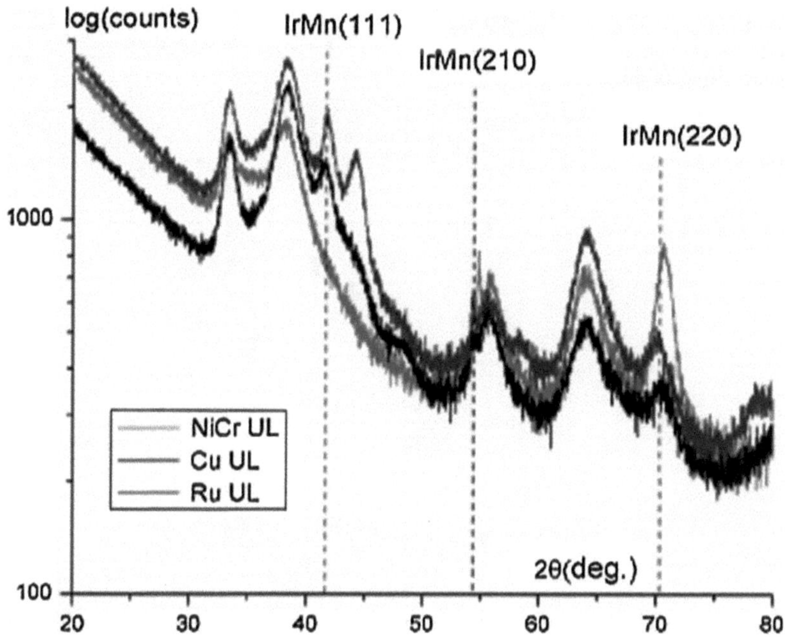

Fig. 19.6 Grazing incidence scans for a set of three IrMn/CoFe samples deposited on top of three different seed layer materials [20]

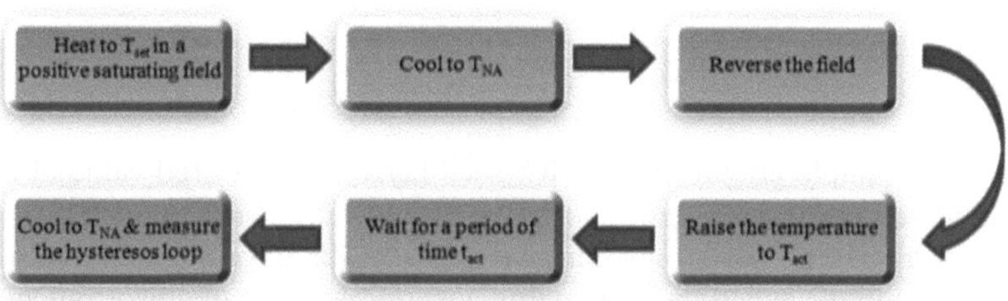

Fig. 19.7 Schematic diagram of the measurement protocols used

of the experimental procedure used. Note that the measurements are made only after cooling to a temperature T_{NA} where all AF grains are thermally stable. Figure 19.8 now shows the resulting variation of the loop shift H_{ex} with the temperature to which the structure was heated with the field reversed. We denote this temperature T_{ACT}. As can be seen there is a critical point at which the value of H_{ex} is equal to zero. At this temperature equal volumes of the AF are now oriented in opposite directions. This means that the grain size which was being activated at T_{ACT} is that of an AF grain having the median volume. This size can be determined easily using transmission electron microscopy (TEM) and hence given that the time of activation with the field reversed is equal to 30 min, K_{AF} can be determined according to

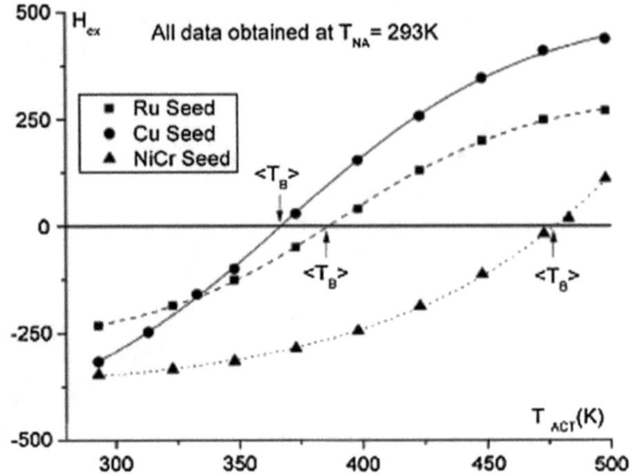

Fig. 19.8 Loop shift as a function of the activation temperature T_{ACT} for three IrMn/CoFe samples deposited on different seed layers [20]

$$K_{AF}(<T_B>) = \frac{\ln(1800 f_o) k T_B}{<V>} \tag{19.5}$$

where $<T_B>$ is the median temperature at which $H_{ex} = 0$, f_o is an attempt frequency generally taken to be $10^9\,s^{-1}$ [13] and $<V>$ is the medium grain volume. Of course given that IrMn is a cubic material and therefore this anisotropy K_{AF} is magnetocrystalline in origin, its value will be temperature dependent and varies according to

$$K_{AF}(T) = K_{AF}(0)\left[1 - \frac{T}{T_N}\right] \tag{19.6}$$

For the IrMn sample grown on the NiCr seed layer the measured value of the antiferromagnetic anisotropy is equal to 2.9×10^7 ergs/cc which to our knowledge is the largest value of an anisotropy constant ever measured for any thin film material. Reference to Fig. 19.8 also shows that this remarkable value of anisotropy gives rise to an average blocking temperature of 475 K compared to the value obtained for IrMn grown on a Cu seed layer of 330 K. This value for the sample grown on the NiCr seed layer is all the more remarkable because the median grain size in this particular film was only 4 nm.

Hence it has been shown that the correct growth of a suitable AF layer such as IrMn can provide an enormous increase in the anisotropy of the adjacent F layer which will be thermally stable at the operating temperature of most spintronic devices and furthermore generates a value of magnetisation in zero field equivalent to the saturation magnetisation. Hence the phenomenon of exchange bias can be used to induce the necessary properties in a polarising layer so as to provide the optimum spin polarisation.

19.3 Exchange Bias and Heusler Alloys

There are some reports of exchange bias being observed directly in Heusler alloys [21]. These reports relate to Heusler alloys with a typical composition X_2YZ with an FCC structure which undergo a martensitic transition. These materials exhibit shape memory properties but can also exhibit an exchange bias due to the multiple phases present. These effects are generally weak with shifted hysteresis loops of the order of 250 Oe and require very high fields used in the setting process. Obviously for modern applications in spintronic devices such effects are of limited interest because the magnitude of the loop shift is too small and the nature of the phase transitions required is such that the effects are unlikely to be observed in thin films.

The first extensive work on the generation of exchange bias in Heusler alloys in thin film form was undertaken by our group around the turn of the decade [22–25]. These works were all concerned with the coupling of the Heusler alloy Co_2FeSi to the standard alloy used for exchange bias in devices, IrMn. The samples were prepared by sputtering using a controlled ion beam deposition system where the plasma is generated in a side arm to the main chamber using a 3.2 kW rf antenna and subsequently steered onto a target using two solenoids and a DC bias voltage. This system produces a uniform plasma beam which covers the whole of the target giving **hi**gh **t**arget **u**tilisation **s**puttering and hence is known by the acronym HiTUS. The system allows the variation of the growth rate of the films via the variation of the DC bias voltage which is used to produce an initial seed layer generally using the alloy NiCr. Strong columnar growth then results controlling the properties of the IrMn film and the subsequent Heusler alloy grown in contact with it. It is in this way that the York Model of Exchange Bias described in the previous section was developed. The typical structures grown with their thicknesses is shown in Fig. 19.9.

The samples were grown in an in-plane field of 300 Oe and vacuum annealed in a field of 1 kOe along the same direction as that applied during the sputtering process. The annealing temperatures were varied from 220–400 °C both to transform the

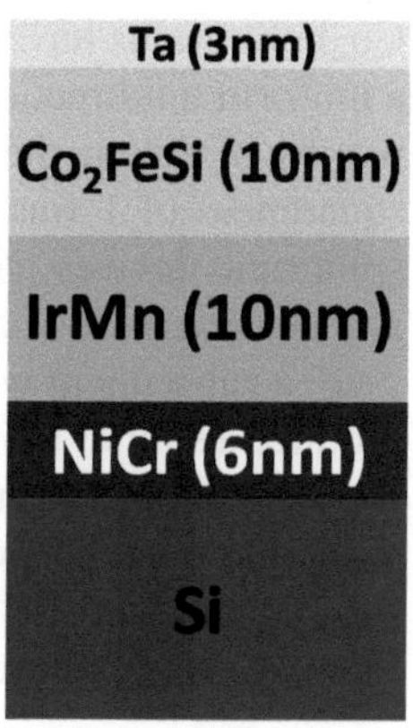

Fig. 19.9 Typical Heusler alloy based exchange biased structure structures grown

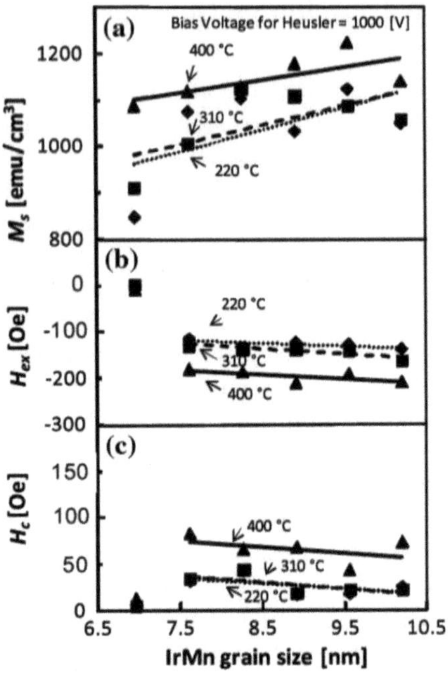

Fig. 19.10 H_{ex}, M_s and H_c values for IrMn/Co_2FeSi samples grown using different bias voltages [22]

Heusler alloy into its ordered $L2_1$ state and to set the IrMn AF layer at the same time. Significant values of exchange bias ranging between 100 and 250 Oe were observed with increasing grain size of the IrMn layer from 7.5 to 10.5 nm. These data together with the resulting change in the value of M_s and the half loop width H_c are shown in Fig. 19.10. This was the first report of the use of a classical exchange bias structure to shift the loop in a Heusler alloy. An example of a typical hysteresis loop for the sample with the largest value of exchange bias is shown in Fig. 19.11. These data show that using a classical exchange bias system it is possible to increase the loop squareness at the remanent point very close to the value of unity such that all the Co_2FeSi grains have their magnetic moments aligned in the same direction. Hence a uniform unidirectional spin polarisation would be expected from such films. Also worthy of note is the fact that the annealing temperature required to achieve a loop squareness of 1 was 400 °C which is a temperature suitable for the fabrication of spintronic devices without creating damage to the other layers in the stack notably any possible oxide barrier used to create a TMR based device [22].

In a subsequent work it was also shown that the use of a Cr seed layer could also produce a similar loop shift and that the strength of the interfacial exchange coupling was dependent upon the ordering of interfacial spin clusters which are known to form at the interface between the F and AF layers. This is an important result because it shows that control of the interfaces is critical to achieve an optimum exchange bias. However in this case much higher annealing temperatures of almost 600 °C were

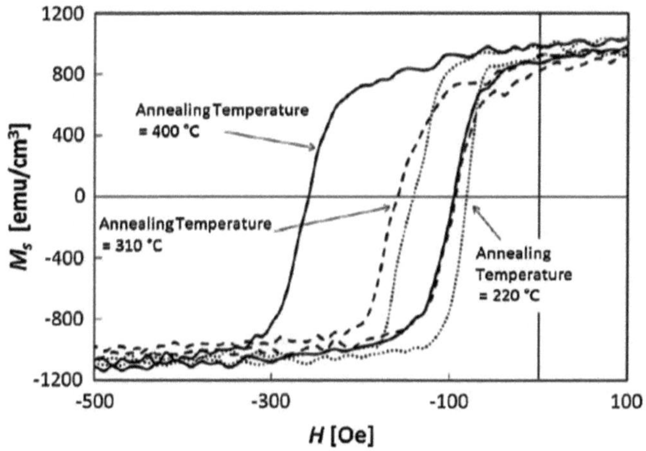

Fig. 19.11 Hysteresis loop for a sample at a bias voltage of 600 V (IrMn) and 1000 V (Co_2FeSi) for different annealing conditions [22]

required but that the resulting value of the saturation magnetisation of the Heusler alloy was close to the bulk value at 1,200 emu/cc [23].

In a further work [24] an attempt was made to use an insertion layer of Mn between the AF and F layer in the same system to enhance the interfacial coupling. Such a strategy was first proposed by Tsunoda et al. [26] and in a subsequent work [27] was shown to increase the moment on the interfacial spin clusters such that their alignment during the field annealing process was greater.

In the case of the Co_2FeSi Heusler alloy studied in these works it was found that the presence of an insertion layer of less than 2 nm thickness could produce a significant enhancement to the loop shift which was particularly marked when an annealing temperature of 500 °C was used even after a relatively short annealing time of 10 min. The origin of the enhancement of exchange bias due to a Mn insertion layer is still not yet fully understood. It has been shown [27] that the alignment of the interfacial spins is more readily achieved in the presence of a Mn insertion layer but there is also a strong probability that the well established substitution of Mn for Co at interfaces results in a Mn deficient IrMn composition at the interface thereby reducing the strength of the AF coupling within the IrMn layer. Given the marked effect of the use of a higher annealing temperature in this system it seems likely that there is an effect due to Mn diffusion into the Co_2FeSi layer and the presence of the insertion layer serves to replenish the level of Mn in the IrMn layer at the interface. However as has been seen in other systems [27], the effect of the Mn insertion layer is greatest when a layer 0.5 nm thick is used and the use of a thicker layer serves to reduce the resulting exchange bias. Hence a significant degree of interfacial diffusion of Mn does not occur. The variation of the resulting exchange bias for a range of annealing times and thickness of the Mn insertion layer is shown in Fig. 19.12. From this figure it is clear that the optimum interfacial Mn layer is between 1 and 2 atoms

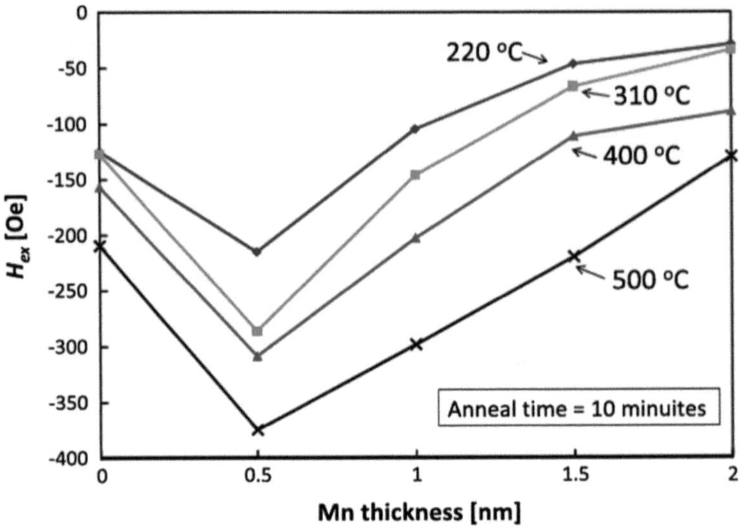

Fig. 19.12 Variation in the exchange bias for a range of annealing times and thicknesses of the Mn layer in an IrMn/Mn/Co_2FeSi system [24]

thick and the optimum annealing time is 500 °C for a short time of 10 min. Such a short annealing time is unlikely to have an adverse effect on other layers in a real device. Nonetheless the use of this strategy has resulted in an exchange bias of almost 400 Oe. Figure 19.13 shows the hysteresis loops for these systems and in the case of the use of a Mn insertion layer a loop squareness of almost exactly unity has been achieved.

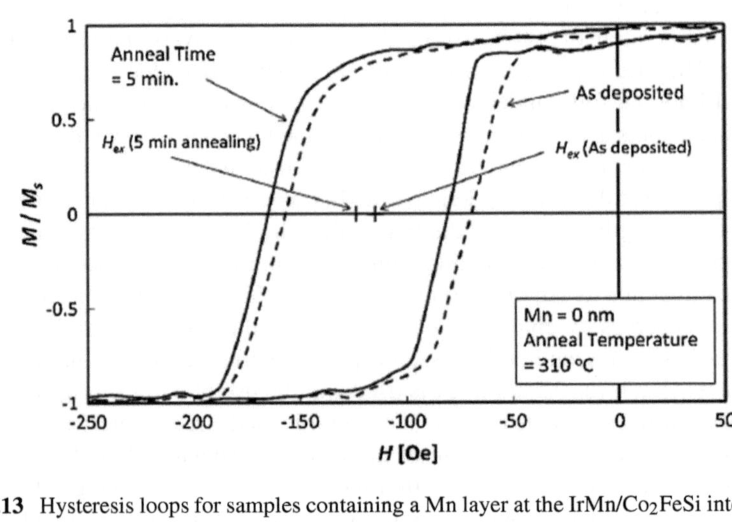

Fig. 19.13 Hysteresis loops for samples containing a Mn layer at the IrMn/Co_2FeSi interface [24]

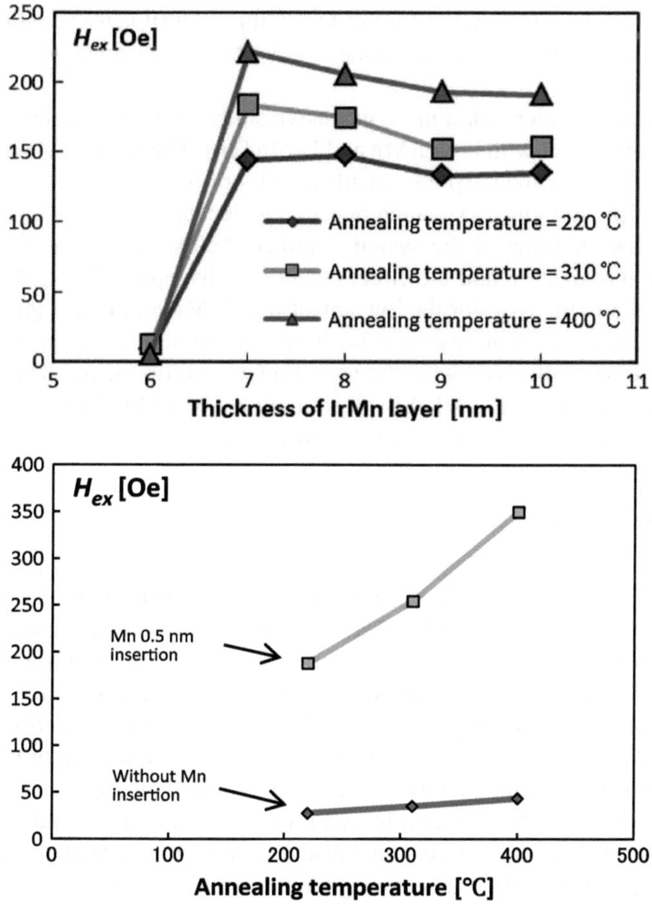

Fig. 19.14 Variation in the exchange bias as a function of the thickness of the AF layer for different annealing temperatures [25]

In a final work [25] a combined study of the thickness of the IrMn layer on the exchange bias in the presence of a 0.5 nm Mn insertion layer was undertaken. This work produced the classic result of a weak peak in the value of H_{ex} with the thickness of the IrMn layer. Based on the York Model of Exchange Bias this result is readily understood since the value of the final exchange bias is controlled by the fraction of thermally stable and set grains within the system. Grains of larger volume controlled by the film thickness will not be set at the normally accessible annealing temperatures. This result is shown in Fig. 19.14. Hence for thicker IrMn layers it is necessary to use higher annealing temperatures. This has implications for damage to other layers in the stack but as the layer thickness is increased the overall level of exchange bias achievable is somewhat greater. The critical parameter reported in this work was the dramatic effect of the 0.5 nm Mn insertion layer. An increase in the overall exchange bias from a few tens of Oe to almost 400 Oe was again reproduced showing that the control of the interface composition and exchange stiffness is the critical parameter.

Dubowik et al. undertook a more extensive study of three Heusler alloys which were grown in contact with both IrMn and FeMn [28]. The Heusler alloys used were Ni_2MnSn, Co_2MnSn and Co_2FeSi. In all cases loop shifts were observed indicating that it is possible to induce exchange bias in a wider range of Heusler alloys. However in the case of some of the systems studied the exchange bias exhibited very strong thermal relaxation and the effect essentially disappeared by 100 K. However this was generally the case for the low anisotropy FeMn antiferromagnet. Similarly to our observations with an insertion layer of Mn a significant improvement in the exchange bias was observed with the use of a Co insertion layer. This was particularly pronounced for the Co_2FeSi system coupled to IrMn. This again shows the importance of interface compositional control in such structures.

19.4 Conclusion

In this work the issue of the utilisation of Heusler alloys in thin film based spintronic devices has been discussed. It has been shown that significant magnetoresistance effects can be induced in Heusler alloy structures when they are grown using MBE techniques. This is because it is possible to achieve the necessary alignment of the crystallographic axes such that the spin polarisation across the thin film is uniform. However in the case of polycrystalline films where no such texture of the individual grains is possible, such magnetoresistive effects are likely to be significantly diminished. However it has been shown by work from our laboratory and more recently by others, that it is possible to use classical exchange bias structures based around alloys such as IrMn as the antiferromagnetic layer, to produce polycrystalline thin films of Heusler alloys where the loop squareness and thereby presumably the spin polarisation achievable should be close to 100 %. This is an important result for the possible commercialisation of Heusler alloys in spintronic devices. At the time of writing studies are under way to produce simple spintronic devices such as spin valve structures with an exchange biased Heusler alloy as the pinned layer and to evaluate their magnetoresistive performance.

Acknowledgments The authors acknowledge ongoing financial support from Seagate Technology (N. Ireland) and Western Digital Corp of Fremont CA.

References

1. T. Graf, C. Felser, S.S.P. Parkin, Prog. Solid State Chem. **39**, 1 (2011)
2. A. Hirohata, H. Kurchagashi, S. Okamura, M. Kikuchi, T. Masaki, T. Nozaki, N. Tezuka, K. Inomata, J. Appl. Phys. **97**, 103714 (2005)
3. A. Hirohata, J. Sagar, L. Lari, L.R. Fleet, V.K. Lazarov, Appl. Phys. A. **111**, 423 (2013)

4. S. Trudel, O. Gaier, J. Hamrle, B. Hillebrands, J. Phys. D: Appl. Phys. **43**, 193001 (2010)
5. J. Sato, M. Oogane, H. Naganuma, Y. Ando, Appl. Phys. Express **4**, 113005 (2011)
6. Y.K. Takahashi, A. Srinavasan, B. Varaprasad, A. Rajanikanth, N. Hase, T.M. Nakarani, S. Kasai, T. Furubayashi, K. Hono, Appl. Phys. Lett. **98**, 152501 (2011)
7. H. Liu, Y. Honda, T. Taira, K. Matsuda, M. Arita, T. Umura, M. Yamamoto, Appl. Phys. Lett. **101**, 132418 (2012)
8. Y. Sakuraba, M. Ueda, Y. Mium, K. Sato, S. Basu, K. Saito, M. Shirai, T.J. Konno, K. Takanashi, Appl. Phys. Lett. **101**, 252408 (2012)
9. S. Wurmehl, G.H. Fecher, H.C. Kandpal, V. Ksenofontov, C. Felser, H.-J. Lin, J. Morais, Phys. Rev. B **72**, 184434 (2005)
10. C.P. Bean, J.D. Livingston, Phys. Rev. Lett. **12**, 14 (1964)
11. W.H. Meiklejohn, C.P. Bean, Phys. Rev. **102**, 1413 (1956)
12. I.L. Prejbeanu, M. Kerekes, R.C. Sousa, H. Sibuet, O. Redon, B. Dieny, J.P. Nozieres, J. Phys.: Cond. Matt. **19**, 165218 (2007)
13. L.E. Fernandez-Olluton, K. O'Grady, S. Oh, M. Zhou, M. Pakala, IEEE Trans. Magn. **44**, 2824 (2008)
14. G. Vallejo-Fernandez, L.E. Fernandez-Outon, K. O'Grady, Appl. Phys. Lett. **91**, 212503 (2007)
15. S. Romer, M.A. Marioni, K. Thorwarth, N.R. Joshi, C.E. Corcitelli, H.J. Hug, S. Oezer, M. Parlinska-Wotjan, H. Rohrmann, Appl. Phys. Lett. **101**, 222404 (2012)
16. K. O'Grady, L.E. Fernandez-Outon, G. Vallejo-Fernandez, J. Magn. Magn. Mater. **322**, 883 (2010)
17. E.C. Stoner, E.P. Wohlfarth, Philos. Trans. R. Soc. Sec. **A240**, 599 (1948)
18. G. Vallejo-Fernandez, N.P. Aley, L.E. Fernandez-Outon, K. O'Grady, J. Appl. Phys. **104**, 033906 (2008)
19. G. Vallejo-Fernandez, L.E. Fernandez-Outon, K. O'Grady, J. Phys. D: Appl. Phys. **41**, 112001 (2008)
20. N.P. Aley, G. Vallejo-Fernandez, R. Kroeger, B. Lafferty, J. Agnew, Y. Lu, K. O'Grady, IEEE Trans Magn. **44**, 2820 (2008)
21. S. Giri, M. Patra, S. Majumdra, J. Phys. C. **23**, 073201 (2011)
22. H. Endo, A. Hirohata, J. Sagar, L.R. Fleet, T. Nakayama, K.O. Grady, J. Phys. D: Appl. Phys. **44**, 345003 (2011)
23. N.P. Aley, S. Takayama, A. Hirohata, K. O'Grady, IEEE Trans. Magn. **47**, 3490 (2011)
24. H. Endo, A. Hirohata, T. Nakayama, K. O'Grady, J. Phys. D: Appl. Phys. **44**, 145003 (2011)
25. H. Endo, A. Hirohata, J. Sagar, L.R. Fleet, T. Nakayama, K. O'Grady, IEEE Trans. Magn. **48**, 2896 (2012)
26. M. Tsunoda, S. Yoshitaki, Y. Ashizawa, D.Y. Kim, C. Mitsumata, M. Takahashi, Phys. Status Solidi **244**, 4470 (2007)
27. R. Carpenter, N.C. Cramp, K. O'Grady, IEEE Trans. Magn. **48**, 4351 (2012)
28. J. Dubowik, I. Goscianska, K. Zaleski, H. Glowinski, Y. Kudryavtsev, A. Ehresmann, J. Appl. Phys. **113**, 193907 (2013)

Part VII
New Properties

Chapter 20
Topological Insulators Within the Family of Heusler Materials

Stanislav Chadov and Claudia Felser

Abstract The first system, in which the quantized spin Hall effect has been demonstrated experimentally, was the zinc-blende binary HgTe semimetal. The aim of our contribution is to show that, by having the similar crystal structure but ternary chemical composition, Heusler compounds naturally represent much more diverse family of nontrivial semimetals, compared to the binaries. Here we will give an overview of these topologically nontrivial systems, by discussing the role of the spin-orbit coupling for the band inversion and the crystal symmetry mechanisms which lead to the semimetallic zero-gap state in these materials.

20.1 Zinc-Blende and Heusler Semiconductors

Successful demonstration of the nontrivial topological electronic structure in HgTe zinc-blende binary material [1–3] has triggered the subsequent revision of all members in the zinc-blende family, including the related structural groups. A tremendous number of systems was immediately processed using the first-principles band structure methods, since the structural information of these simple compounds has become available since long ago. Strong impact in this direction was given by theoretical studies [4, 5] which have searched for the nontrivial topological fingerprints within the family of Heusler materials. The idea behind is the obvious similarity between cubic Heusler and zinc-blende structures, which is demonstrated on Fig. 20.1a. Both zinc-blende and Heusler structures are based on the same fcc-type lattice, which provides four Wyckoff positions: 4*a* (0, 0, 0), 4*b* (1/2, 1/2, 1/2), 4*c* (1/4, 1/4, 1/4) and 4*d* (3/4, 3/4, 3/4). In case of the zinc-blende YZ binary only two Wyckoff positions are occupied, e.g. 4*a* or 4*b* by the main-group element Z, and 4*c* or 4*d*—by the

S. Chadov (✉) · C. Felser
Max-Planck-Institute Für Chemische Physik Fester Stoffe,
Nöthnitzer Str. 40, 01187 Dresden, Germany
e-mail: chadov@cpfs.mpg.de

C. Felser
e-mail: claudia.felser@cpfs.mpg.de

C. Felser and A. Hirohata (eds.), *Heusler Alloys*,
Springer Series in Materials Science 222, DOI 10.1007/978-3-319-21449-8_20

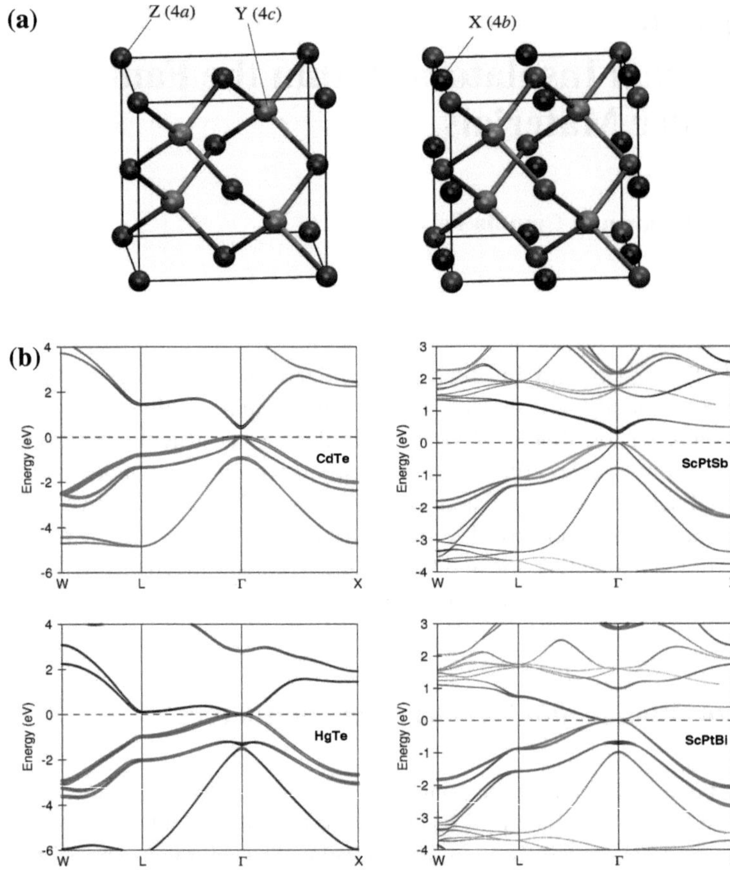

Fig. 20.1 **a** Similarity of the zinc-blende binary YZ (*left*) and Heusler ternary XYZ (*right*) crystal structures: *red spheres* represent the main-group element atoms (Z, 4*a* Wyckoff positions), *yellow*—transition metal elements (Y, 4*c* Wyckoff positions); *blue spheres* in Heusler structure represent the second type of transition metal (X, 4*b* Wyckoff positions), embedded in the zinc-blende YZ matrix. **b** The band structures of binary zinc-blende and ternary Heusler semiconductors reveal clear fingerprints: both CdTe and ScPtSb exhibit a direct gap at the Γ-point between the conduction s-like (*blue*) and the valence p-like (*red*) bands of Γ_8 and Γ_6 symmetries, respectively. On the other hand, the band structures of both heavier compounds, HgTe and ScPtBi, exhibit the band inversion caused by the strong spin-orbit coupling: Γ_6 (*blue*) is now situated *below* Γ_8 (*red*) which remains at the Fermi energy (*horizontal dashed line*) (Color figure online)

transition metal Y. The XYZ Heusler structure is obtained from the YZ zinc-blende by fulfilling it with one more transition metal atom X, which occupies 4*a* or 4*b* Wyckoff site. In principle, the remaining empty 4*c* or 4*d* Wyckoff position can be still filled up, by adding again one transition element, e.g. extra one Y atom. The corresponding chemical composition changes then from XYZ to XY_2Z. In order to distinguish between these two Heusler variants they are often referred as the "half-Heusler" (sometimes the "semi-Heusler") and the "full-Heusler" (or just "Heusler"). In "Strukturbericht" classification these structures are marked as $C1_b$ and $L2_1$, respectively. The important difference between them is the absence of the inversion symmetry in XYZ half-Heuslers, which takes place in the zinc-blende YZ structures as well.

The overwhelming manifold of half-Heusler and full-Heusler structures exhibit the so-called Slater-Pauling behavior [6, 7], which means that the total magnetization per atom m and the total number of valence electrons per atom n_v have a linear relation $m = n_v - 6$. Thus, the nonmagnetic systems within these families could be then easily preselected by picking out the compositions which give $n_v = 6$. Counting the total valence electron number per formula unit means that the nonmagnetic full-Heuslers will contain 24 electrons ($24 = 6 \times 4$), whereas the nonmagnetic half-Heuslers—18 electrons ($18 = 6 \times 3$). In the full-Heusler family the typical nonmagnetic representatives are e.g. Fe_2TiSi or Fe_2VAl; in the half-Heusler—NiTiSn or LuPtBi. The Slater-Pauling rule also holds for the non-stoichiometric compositions, by substantially expanding the scope of potential candidates. On the other hand, any Heusler system containing Mn will be magnetic, at least locally, disregarding the total number of its electrons: e.g. the 24-electron cubic $MnCo_{1.3}Ga_{1.7}$ [8], which obeys the Slater-Pauling rule by having the zero total magnetization, is the so-called "compensated ferrimagnet" which exhibits a break of the time reversal symmetry necessary for the nontrivial topology. In addition, most of the known Heusler semiconductors reside within the family of half-Heuslers which have a relatively large band gaps compared to the nonmagnetic full-Heuslers with 24 electrons, which contain more nonmagnetic metals than semiconductors. Due to the structural and chemical similarity between the half-Heusler and zinc-blende families, the later class obeys the same Slater-Pauling behavior, i.e. the zinc-blende semiconductors contain 18 valence electrons per formula unit, such as HgTe or CdTe.

In the band structure description, the difference between nontrivial and trivial representatives of zinc-blende semiconductors is known very well: e.g. the nontrivial HgTe exhibits the so-called *band inversion* compared to the trivial CdTe case (see Fig. 20.1b)—the eigenstate Γ_6 (marked as blue) with the negative parity is occupied in HgTe case, and unoccupied—in CdTe. Since all important physics in zinc-blende semiconducting systems occurs just at the Γ point of the Brillouin zone, the analysis is greatly simplified, despite that the systems are not inversion symmetric and the accelerated procedure of defining the parity of a total wave function [9, 10] cannot be applied. The sample topological and trivial band structure fingerprints are shown schematically in Fig. 20.2.

By projecting the electronic structure onto the on-site contributions, one can easily see that this Γ_6 eigenstate consists almost of the s-states of the transition metal element Y. In case of CdTe, two s electrons from the $5s$ shell of Cd completely fill the vacant $5p^4$ shell of Te (corresponding projection is marked by red color), which leads to the opening of a direct semiconducting band gap between these levels. Counting the parities of the occupied eigenstates one can conclude the trivial topological nature of CdTe [9]. Obviously, this conclusion should not depend on how large is the gap, for this reason, the band structure methods, based on the standard local density approximation (LDA) known to underestimate the gap widths in

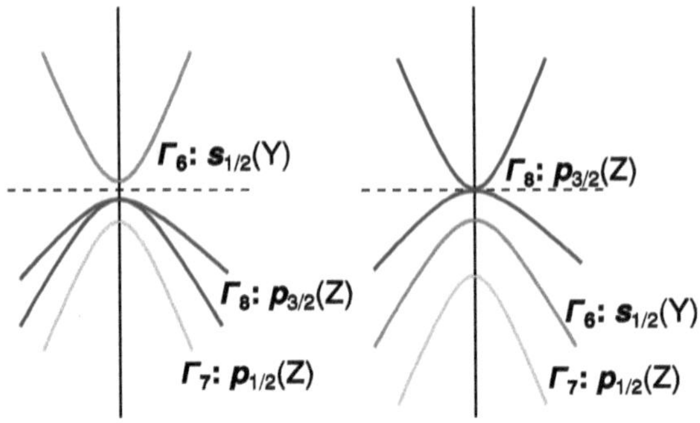

Fig. 20.2 Topologically trivial (*left*) and nontrivial (*right*) band structure fingerprints inherent both for the binary zinc-blende and ternary $C1_b$ (Heusler) semiconductors. The relevant electronic states in proximity of the Γ-point are colored as *blue* (Γ_6: negative parity, dominantly of $s_{1/2}$-character of the Y element), *red* (Γ_7: positive parity, dominantly of $p_{1/2}$-character of the Z element) and *dark-red* (Γ_8: positive parity, dominantly of $p_{3/2}$-character of the Z element) (Color figure online)

zinc-blende systems, give the correct topological classification of CdTe as far as the band gap opens at all. The situation with HgTe is different. The $6s^2$ (Γ_6) state of Hg lays deeper than $5s^2$ of Cd, and thus, is occupied. The $5p^4$ shell of Te remains in the same configuration, for this reason, being occupied partially it is confined at the Fermi energy. Strictly speaking, this is a sign of a metallicity, or, as it is often referred, the "semi-metallicity" (in order to stress that the Fermi level crossing occurs only at the single Γ point of the Brillouin zone). In the studies dealing with thermoelectrics, one can often encounter the term "gapless semiconductor", stressing that the system is still a semiconductor, but with a very small band gap. In fact, this band feature makes HgTe and similar systems good thermoelectrics: due to a metallic state at the Γ point they still exhibit rather high electrical conductivity, but the thermal conductivity mediated by phonons is rather low, due to the absence of the low-energy indirect (inelastic) transitions. Due to the spin-orbit coupling, the sixfold degenerated $5p$ state of Te is split into the lower energy twofold degenerated $p_{1/2}$ and higher energy fourfold degenerated $p_{3/2}$ subshells. The former one is fully occupied (it resides about -1.8 eV below the Fermi energy, see Fig. 20.1b), and it is the later one which is confined at the Fermi energy due to its half filling. Here it is important to stress that the spin-orbit coupling plays a minor role in setting on the band inversion, the primary factor is the sufficiently low energy of Hg $6s$ state compared to the $5p$ state of Te. Indeed, it is rather obvious by comparing these two systems, which contain the same Z element Te: as it follows from Fig. 20.1b, the $p_{3/2}$-$p_{1/2}$ energy distance (spin-orbit split Δ_{SO}) is almost the same in both cases and makes about 1 eV at the Γ point. In the following sample calculations (Fig. 20.3) we tune the strength of the spin-orbit coupling in the Te p shell of HgTe system by scaling coefficient λ: $0 < \lambda < 2$ ($\lambda = 1$ corresponds to the default value). As it follows, despite that the Δ_{SO} of Te strongly varies from 3.5 to 0 eV, the occupation and energy position of the s state of Hg almost does not change, which means that

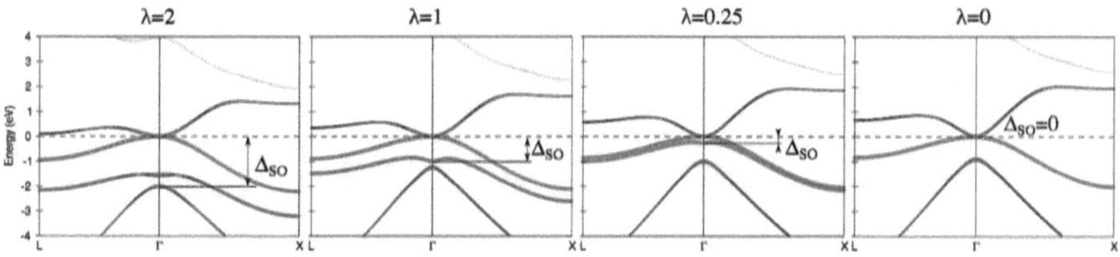

Fig. 20.3 Electronic structure of HgTe calculated by scaling down the amplitude of spin-orbit coupling (from $\lambda = 2$ to $\lambda = 0$) within the p shell of Te. *Red color* corresponds to the p-states of Te, *blue*—s states of Hg. Δ_{SO} marks the energy split between $p_{3/2}$ and $p_{1/2}$ subshells (Color figure online)

the system remains topologically nontrivial even in the nonrelativistic limit. Thus, the typical argument that the spin-orbit coupling is responsible for the band inversion is only partially true—it is not only the spin-orbit coupling which changes upon increasing the weight of a chemical element.

Similar properties are exhibited by half-Heusler semiconductors, with the only difference that in ternary XYZ compositions the zinc-blende substructure YZ does not provide the required 18 electrons, but this is improved by adding an extra transition metal X which restores the electronic balance. For example, in case of ScPtSb, the zinc-blende PtSb contains only 15 electrons, whereas Sc adds another three. Finally, comparing the band structures of the half-Heusler systems with zinc-blende, one easily distinguishes both trivial and nontrivial types in half-Heusler class. Figure 20.1b demonstrates this similarity, by comparing trivial CdTe with trivial ScPtSb and nontrivial HgTe with nontrivial ScPtBi. As it follows, the 6s state of Pt at Γ (Γ_6) appears to be relatively high in energy (with respect to the 5p shell of Sb) in ScPtSb than in ScPtBi (compared to the 6p shell of Bi). In case of ScPtSb $5p^3$ shell of Sb accepts one 6s electron from Pt and two 4s electrons from Sc; at the same time, the single 3d electron of Sc and nine 5d electrons of Pt fill up the Pt d-shell. For this reason, the bottom of the conduction band at Γ is formed by mixed s contributions from Pt and Sc. These properties where in detail discussed in [4, 5]. By going through the manifold of the half-Heusler materials, the authors have plotted the diagram which gives an idea how to search for the topologically nontrivial systems in this class (see Fig. 20.4). Trivial and nontrivial materials are distinguished by the sign of their band inversion $E_{\Gamma_6} - E_{\Gamma_8}$, positive and negative, respectively, plotted as a function of their lattice constant and the average nuclear number Z (averaged over the formula unit composition). Since the presented calculations where based on LDA (and what is also rather crude—spherical approximation to the atomic potentials), the whole diagram is somehow biased: e.g. the normal semiconducting gap in trivial materials is typically underestimated, for the same reasons, the position of the occupied s shell of the nontrivial materials is probably lower than it should be. This diagram was followed later by other authors [11–14] who have performed the similar calculations using different potential construction and different exchange correlation forms. However the general trend obtained in all computational studies remains the same—i.e. the probability to find a nontrivial system increases by increasing the lattice constant and

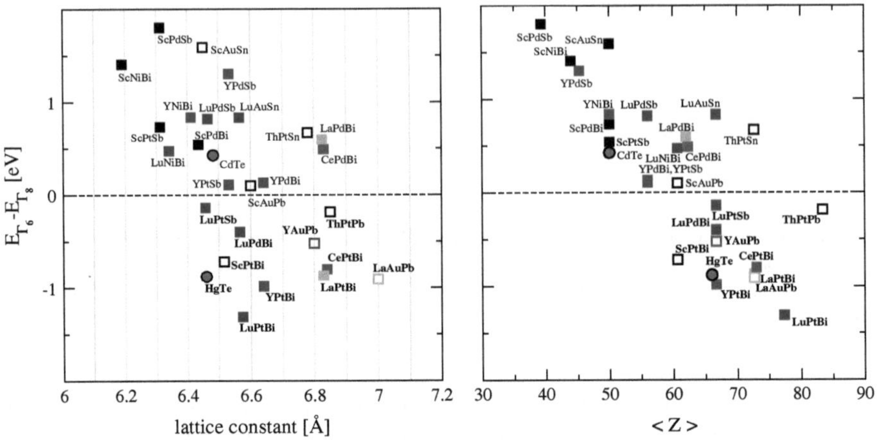

Fig. 20.4 The energy difference between Γ_6 and Γ_8 bands calculated as a function of lattice constant (*right*) and the mean nuclear number $\langle Z\rangle$ (*left*) [4]. Each YZ subgroup (YZ = NiSb, PdSb, PdBi, PtBi, AuSn, AuPb, PtPb) is marked by a certain color. Different colors of markers (*squares*, *circle* etc.) are assigned to a different stuffing elements (X = Sc, Y, La, Lu, Th). Compounds with $E_{\Gamma_6} - E_{\Gamma_8} > 0$ are topologically trivial, whereas those with $E_{\Gamma_6} - E_{\Gamma_8} < 0$ are the nontrivial candidates

the average $\langle Z\rangle$ number. The first trend reflects the fact that the energy position of the outer *s* shell of the transition metals X and Y lowers by reducing the crystal field split, the second trend—that this energy position is lower for the heavier transition metals. On the other hand, the spin-orbit coupling in the main group element *p* shell can play a direct role in setting on the topological order, if the *s* shell of transition metal atom is not particular deep in energy: in this case a large spin-orbit split can noticeably shift the $p_{3/2}$ subshell sufficiently high, above the *s* state. This mechanism becomes especially delicate in case of materials which are situated close to the topological phase transition—e.g. YPtSb or YPdBi. As it was shown in [4], a slight decrease or increase of the spin-orbit coupling in the *p*-shell drives the material into the nontrivial or trivial topological phase, respectively.

Later the spectral properties of topological half-Heusler materials where investigated in several photoemission studies. For example, the hard x-ray photoemission (HAXPES) on thin films of LuPdBi [14] (see Fig. 20.5) justifies the gapless state of the material and the possibility of the nontrivial topological order.

The diversity of ternary Heusler materials opens more possibilities to tune the band inversion due to larger number of combinations with 18 electron compositions. An obvious advantage of ternary compounds is their "multifunctionality". For instance, many of ternary zero-gap semiconductors (XAuPb, XPdBi, XPtSb and XPtBi) contain the rare-earth element X which can be used alternatively for managing the additional properties ranging from superconductivity (e.g. LaPtBi [15], YPtBi [16]) to antiferromagnetism (e.g. GdPtBi [17, 18]) and heavy-fermion behavior (e.g. YbPtBi [19]).

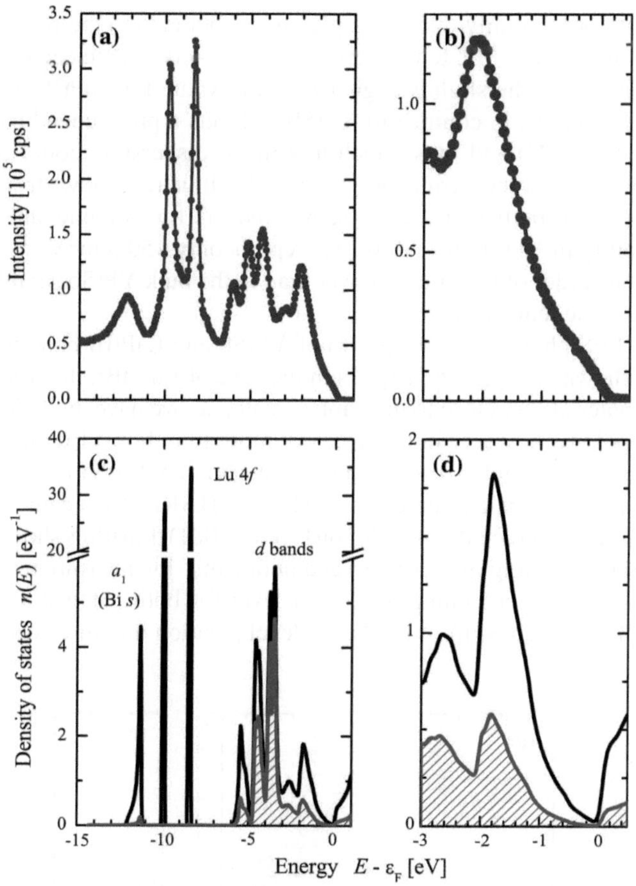

Fig. 20.5 Valence band of LuPdBi [14]. **a** and **b** The valence band spectra taken with a photon energy of about 8 keV. **c** and **d** The density of states, and the density localized at the Pd atoms is marked by the *shaded area*. **b** and **d** The valence band close to the Fermi energy on an enlarged scale

20.2 Closer to the Topological Phase Transition

The presented diagram (Fig. 20.4) may serve as a preliminary guide to design the materials with required topological order. In particular, it indicates how to adjust the systems situated closer to the border between topologically trivial and nontrivial regimes (as e.g. YPtSb, YPdBi or LuPtSb half-Heusler compounds, see Fig. 20.4). Such systems will possess Γ point centered Dirac cone as a bulk state which leads to unusual transport characteristics. For instance, recent experimental studies [20] on stoichiometric Heusler compounds close to topological/trivial transition regime revealed exceptionally high Hall mobility (of 4124 and 4275 [cm^{2}V^{-1}s^{-1}] for polycrystalline bulk samples of YPtSb and LaPtBi, respectively). These ultrahigh values of the mobility are not only due to "gaplessness" but also to the presence of linear dispersion of the bands close to the Fermi energy, where charge carriers behave like relativistic particles. The slightly higher mobility value for LaPtBi, which according to the calculations, in contrast to YPtSb, exhibits a pronounced band inversion ($E_{\Gamma_6} - E_{\Gamma_8} \approx -1$ eV) and thus, much less linear dispersion, could be due to the slightly higher DOS at the Fermi energy. This also illustrates how strongly sensitive are the transport properties of these systems. Indeed, for example, another study of the Hall mobility in YPtSb thin films [21] reports only 450 [$\mathrm{cm^2V^{-1}s^{-1}}$] at 300 K, which is by one order of magnitude lower than in the bulk YPtSb samples measured in [20] at the same temperature.

The HAXPES photoemission spectra of YPtSb and LaPtBi bulk polycrystalline samples are shown in Fig. 20.6 [22]. As in the case of LuPdBi, the basic conclusion is that both materials are close to the gapless state, i.e. we have an indication of their nontrivial topology, however in order to say more (e.g. how close is the system to the borderline), the angular-resolved photoemission (ARPES) is required. On the other hand the bulk-sensitive hard x-ray ARPES (HARPES) became possible only recently [23]. Another recent study [24] on LuPtSb (001) thin films shows by means of STM (scanning tunneling microscopy) and photoemission measurements a decrease in density of states approaching the Fermi level for both valence and conduction bands as well as a slight shift of the Fermi level position into the valence band. The

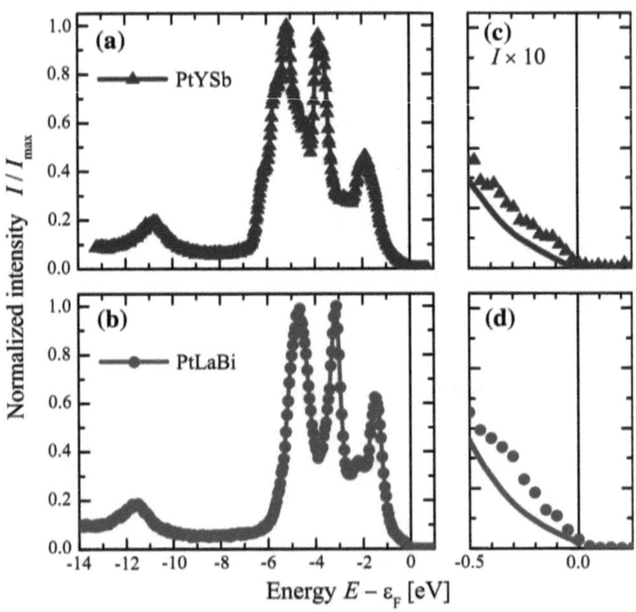

Fig. 20.6 Valence band spectra of YPtSb (**a**) and LaPtBi (**b**) [22]. Panels (**c**) and (**d**) show the corresponding regions close to the Fermi energy on an enlarged scale. The *lines* correspond to the calculated DOS normalized to the maximum. The spectra were taken at 5.9 keV energy

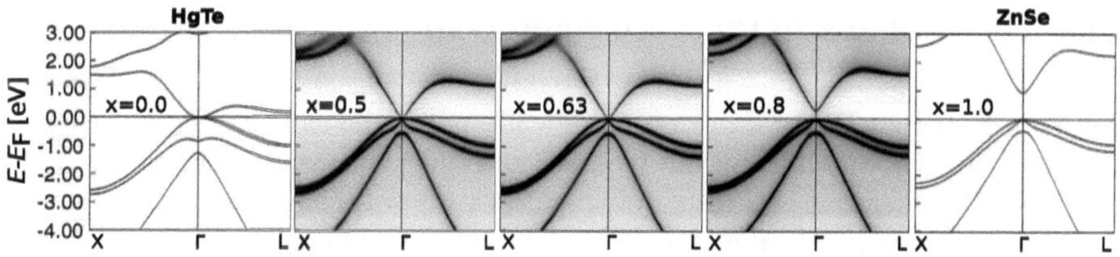

Fig. 20.7 Electronic structures calculated for the series of alloys $(Hg_{1-x}Zn_x)(Te_{1-x}Se_x)$, $x = 0, 0.5, 0.72, 0.8, 1$ [25]. Since the pure compounds HgTe ($x = 0$) and ZnSe ($x = 1$) belong to different topological classes, along this path a topological phase transition will take place. This phase transition at the concentration ($x \approx 0.63$) is indicated by the presence of the Dirac cone centered at the Fermi energy. All compositions where $x < 0.63$ are topologically nontrivial and characterized by zero gap, whereas $x > 0.63$ are trivial and exhibit a real band gap, which increases towards $x = 1$

Hall mobility is reported to be 550 [$cm^2V^{-1}s^{-1}$] at 300 K, which is close to the value for YPtSb thin films reported in [21]. It is expected that the carrier density can be reduced and the mobility increased with further optimization of the growth conditions.

The possibility of chemical adjustment towards the borderline behavior through the isovalent chemical substitution was theoretically demonstrated in [25] for the several zinc-blende systems using Coherent Potential Approximation technique (CPA) [26, 27] to emulate the random chemical disorder. Such topological phase transition is demonstrated on Fig. 20.7 within a sequence of electronic structures of the alloy series between nontrivial HgTe and trivial ZnSe, calculated using KKR method [28]. Topological transition occurs at approximately $x = 0.63$ amount of ZnSe. This indicates that only those particular systems will exhibit a Dirac cone at some critical concentration, which are located in the phase-space on the borderline between topologically trivial and non-trivial semiconductors. In all other cases this peculiar feature—the Dirac cone—will not appear in the band structure throughout the whole series of alloys, independent of concentration. Taking into account that these were the bulk studies, it indicates more general conclusion—that the surface of topological insulator first of all plays a role of the "transition regime" area, the same as observed in the bulk. The difference is that in the bulk the effective translational symmetry is preserved, so we observe two replica of the same Dirac cone, where the replica are superimposed with opposite spins on top of each other. On the surface due to the break of space-reversal symmetry only a single cone remains, exposing the adiabatic surface spin-current.

Due to the fact, that ternary Heusler semiconductors posses three elements instead of two in zinc-blende, such tuning can be realized by substituting each of them separately, which theoretically results into more combinations.

20.3 The Problem of the Zero Band Gap

Similarly to the HgTe, all known topologically nontrivial semiconducting Heusler systems with $C1_b$ structure in the bulk are zero-gap semiconductors (often called semimetals), rather than real insulating materials. As was mentioned above, the reason for that is the partial filling of the $p_{3/2}$ state and no additional intrinsic mechanism which further lifts its degeneracy. Thus, in addition to a pure spin-Hall effect, these systems will exhibit strong metallic longitudinal conductivity.

One possibility to turn these nontrivial materials into the insulators was proposed in [2]. The idea is to lift the undesired $p_{3/2}$ fourfold degeneracy by artificial symmetry break, i.e. by incorporating the system into the quantum-well structure which is not anymore cubic, and the split between in-plane and out-of-plane p-orbitals provides the desired band gap. Several theoretical studies later examined this possibility for the half-Heusler materials [4, 5, 12, 29] and showed basically the same result: the break of the cubic symmetry always opens the band gap in the bulk band structure. Of course, the strain has to be small enough (not more than 1–2 % of the lattice constant) in order to keep the rest of the band structure unchanged and to avoid the occurrence of the metallic state along other k-directions.

Another possibility to induce a non-conducting electronic structure is provided by the fact, that in contrast to the longitudinal conductivity, the Hall conductivity is much less sensitive to disorder. This property can be used to design the so-called Anderson-type topological insulators. In this case no band gap is created, but conductive electrons become localized by disorder, which converts the transport behavior of the material from metallic to semiconducting. At the same time, since no parity change occurs in the band structure, the system by definition becomes topological insulator of the Anderson type. So far such possibility was theoretically demonstrated in model studies [30–32] on HgTe-like system, by introducing electron localization generically.

20.4 Intrinsic Band Gap Mechanisms

Both zinc-blende and half-Heusler systems with nontrivial band structure are typically observed in case of the heavy anion atoms (e.g. Sn, Bi, Te). As we mentioned, the reason for that is the relative energy difference between the s shells of transition metals (cations) and the p shell of the main group element. On top of that, the role of the spin-orbit coupling within a p-shell is rather obsolete. Indeed, for such systems it is easy to show by a simple band structure calculation, that scaling down the spin-orbit coupling does not affect the non-trivial order (see Fig. 20.3). As it follows from Fig. 20.3, the relative energy position of the s-states (i.e., Γ_6) of Hg appears to be almost independent on the spin-orbit amplitude of the p-shell—the compound remains in the topological phase for all λ values.

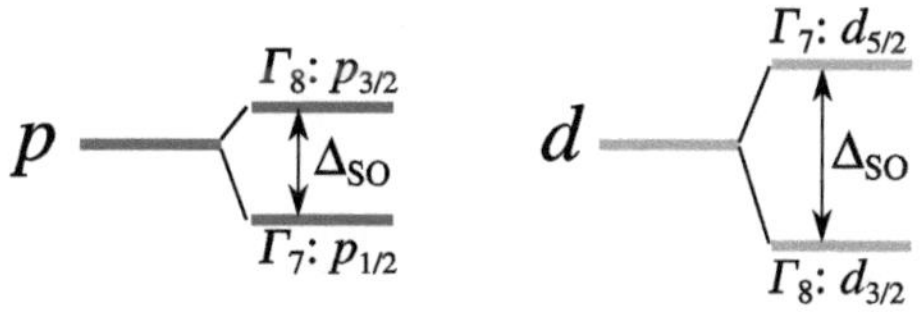

Fig. 20.8 Schematic plot of the spin-orbit split of t_2 states for angular momenta $L = 1$ and $L = 2$ in tetrahedral symmetry

An additional mechanism which influences the p-states in zinc-blende structures is the pd coupling. Its role was pointed out long ago (see e.g. [33, 34]). In the tetrahedral field the spin-orbit splits of the p and d shells are different in the sense of symmetry (see Fig. 20.8). In case of the p shell, the higher energy $p_{3/2}$ subshell belongs to the Γ_8 symmetry, whereas the lower energy $p_{1/2}$ —to the Γ_7. In case of the d shell, the higher energy subshell $d_{5/2}$ is of the Γ_7 symmetry (the same as the lower energy $p_{1/2}$), whereas the lower energy subshell $d_{3/2}$ is of the Γ_8 symmetry (the same as the higher energy $p_{3/2}$ subshell). If the pd-mixing is allowed, the energy split between the Γ_8 and Γ_7 p states can be approximately described as a combination of Δ^p_{SO} and Δ^d_{SO} spin-orbit splits of the p and d shells, respectively, weighted by the fraction Q_d of the d charge admixed into the p-shell [33]:

$$\Delta = E_{\Gamma_8} - E_{\Gamma_7} \approx \alpha\,(1 - Q_d)\,\Delta^p_{SO} + \beta Q_d \Delta^d_{SO}, \tag{20.1}$$

where α, β—are geometrical coefficients depending on the charge distribution of the state. In case of strong pd-coupling one finds a “negative” Δ, i.e., we observe the inversion between Γ_8 and Γ_7 states. If this occurs in the materials which contain four p electrons in the anion atom, the $p_{3/2}$ shell becomes fully occupied, whereas the $p_{1/2}$—fully empty. This opens a natural bulk band gap of the width Δ, which does not depend on the symmetry reduction as in case of tetragonal distortion. This situation can be straightforwardly demonstrated by band structure calculations, taking the familiar example of HgTe, by starting with nonrelativistic regime and increasing the spin-orbit split only in the 5d shell of Hg. As it follows from Fig. 20.9, the real band gap immediately opens. At the same time, within the whole range of λ values the 6s

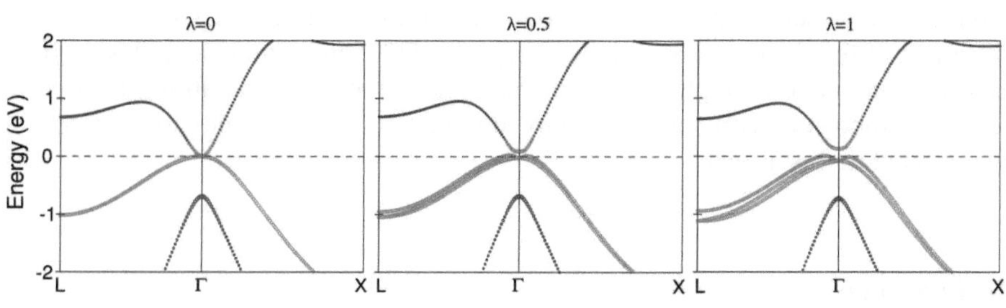

Fig. 20.9 Electronic structure of HgTe calculated by scaling the amplitude of spin-orbit coupling (from $\lambda = 0$ to $\lambda = 1$) within the d-shell of Hg. *Red color* corresponds to the p-states of Te, *blue*—s-states of Hg. For $\lambda > 0$ the band gap between Γ_7 and Γ_8 p-states is opened due to the pd-hybridization (Color figure online)

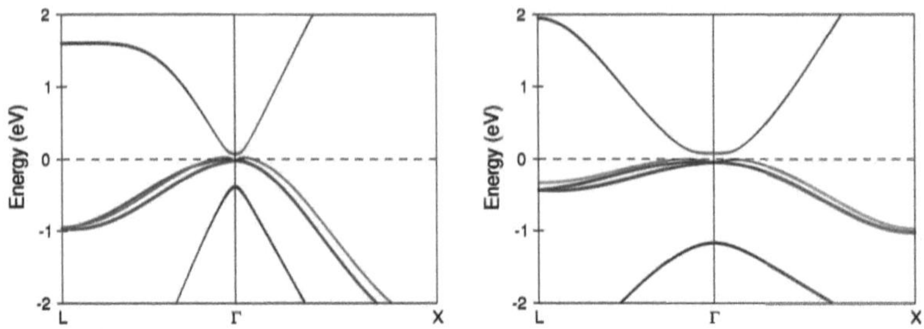

Fig. 20.10 Fully relativistic electronic structure of HgS (*left*) and HgO (*right*). HgS was calculated using the experimental lattice constant, which was also used for the hypothetical zinc-blende phase of HgO. *Red* and *dark-red* colors distinguish between $p_{1/2}$ (Γ_7) and $p_{3/2}$ (Γ_8) anionic states, respectively. *Blue marks* the $s_{1/2}$ (Γ_6) states of Hg (Color figure online)

state of Hg (Γ_6) remains occupied, thus the material stays topologically nontrivial. In real materials such situation occurs, e.g. for the zinc-blende phase of the HgS system [33, 34]. Theoretically HgS was recently proved to exhibit topologically nontrivial order [35]. Similar situation would happen even in a "less-relativistic" lighter material HgO, if it would exist in the zinc-blende form (see Fig. 20.10). Note, that both systems exhibit similar band structures as in case of HgTe with the spin-orbit coupling accounted only the for 5*d*-shell of Hg (Fig. 20.9, $\lambda > 0$). All this indicates that the spin-orbit interaction within the *p* shell of an anion for the zinc-blende semiconductors is not crucially important for the topological state, and moreover, is detrimental for the band gap.

It is interesting to admit, that except HgS, no more topological zinc-blende materials with nonzero gap have been found so far. The reason could be that in HgZ (Z = S, Se, Te) zinc-blende series all *p* electrons belong to the anion atom, whereas the valence shell of Hg does not contain *p* electrons. Indeed, for example, TlP zinc-blende phase, which theoretically was shown to exhibit a non-trivial phase (see, e.g. [25]), possesses all prerequisites similar to HgS, e.g. light anion (small spin-orbital split within the *p* shell) and heavy cation (strong spin-orbital split within the 5*d* shell), together with a large lattice constant ($a \approx 6.12$ Å). However, since Tl "donates" its own single *p* electron into the valence band, the spin-split of the resulting *p* shell would be overwhelmed by the strong spin-orbit coupling of Tl, which is stronger than the effect of the *d* admixture and thus keeps the TlP compound in a gapless regime. Heusler compounds, which allow to model more complicated situations due to their ternary composition compared to binary zinc-blende, may provide more interesting playground in this aspect.

References

1. B.A. Bernevig, T.L. Hughes, S.C. Zhang, Science **314**, 1757 (2006)
2. M. König, S. Wiedmann, C. Brüne, A. Roth, H. Buhmann, L. Molenkamp, X.L. Qi, S.C. Zhang, Science **318**, 766 (2007)

3. X. Dai, T.L. Hughes, X.L. Qi, Z. Fang, S.C. Zhang, Phys. Rev. B **77**, 125319 (2008)
4. S. Chadov, X.L. Qi, J. Kübler, G.H. Fecher, C. Felser, S.C. Zhang, Nat. Mater. **9**, 541 (2010)
5. H. Lin, L.A. Wray, Y. Xia, S. Xu, S. Jia, R.J. Cava, A. Bansil, M.Z. Hasan, Nat. Mater.**9**, 546 (2010)
6. J.C. Slater, Phys. Rev. **49**, 931 (1936)
7. L. Pauling, Phys. Rev. **54**, 899 (1938)
8. G.J. Li, E.K. Liu, Y.J. Zhang, Y. Du, H.W. Zhang, W.H. Wang, G.H. Wu, J. Appl. Phys. **113**, 103903 (2013)
9. L. Fu, C.L. Kane, E.J. Mele, Phys. Rev. Lett. **98**, 106803 (2007)
10. L. Fu, C.L. Kane, Phys. Rev. B **76**, 45302 (2007)
11. W. Feng, D. Xiao, Y. Zhang, Y. Yao, Phys. Rev. B **82**, 235121 (2010)
12. D. Xiao, Y. Yao, W. Feng, J. Wen, W. Zhu, X.Q. Chen, G.M. Stocks, Z. Zhang, Phys. Rev. Lett. **105**, 096404 (2010)
13. Z. Zhu, Y. Cheng, U. Schwingenschlögl, Phys. Rev. B **85**, 235401 (2012)
14. R. Shan, S. Ouardi, G.H. Fecher, L. Gao, A. Kellock, K.P. Roche, M.G. Samant, C.V. Barbosa, E. Ikenaga, C. Felser, S.S.P. Parkin, Appl. Phys. Lett. **102**, 172401 (2013)
15. G. Goll, M. Marza, A. Hamanna, T. Tomanica, K. Grubeb, T. Yoshinoc, T. Takabatake, Physica B **403**, 1065 (2008)
16. N.P. Butch, P. Syers, K. Kirshenbaum, A.P. Hope, J. Paglione, Phys. Rev. B **84**, 220504(R) (2011)
17. P.C. Canfield, J.D. Thompson, W.P. Beyermann, A. Lacerda, M.F. Hundley, E. Peterson, Z. Fisk, H.R. Ott, J. Appl. Phys. **70**, 5800 (1991)
18. R.A. Müller, N.R. Lee-Hone, L. Lapointe, D.H. Ryan, T. Pereg-Barnea, A.D. Bianchi, Y. Mozharivskyj, R. Flacau, Phys. Rev. B **90**, 041109(R) (2014)
19. Z. Fisk, P.C. Canfield, W.P. Beyermann, J.D. Thompson, M.F. Hundley, H.R. Ott, E. Felder, M.B. Maple, M.A.L. de la Torre, P. Visani, C.L. Seaman, Phys. Rev. Lett. **67**(23), 3310 (1991)
20. C. Shekhar, S. Ouardi, A.K. Nayak, G.H. Fecher, W. Schnelle, C. Felser, Phys. Rev. B **86**, 155314 (2012)
21. W. Wang, Y. Du, E. Liu, Z. Liu, G.H. Wu, J. Appl. Phys. **112**, 103910 (2012)
22. S. Ouardi, C. Shekhar, G.H. Fecher, X. Kozina, G. Stryganyuk, C. Felser, S. Ueda, K. Kobayashi, Appl. Phys. Lett. **98**, 211901 (2011)
23. A.X. Gray, C. Papp, S. Ueda, B. Balke, Y. Yamashita, L. Plucinski, J. Minár, J. Braun, E.R. Ylvisaker, C.M. Schneider, W.E. Pickett, H. Ebert, K. Kobayashi, C.S. Fadley, Nat. Mater. **10**, 759 (2011)
24. S.J. Patel, J.K. Kawasaki, J. Logan, B.D. Schultz, J. Adell, B. Thiagarajan, A. Mikkelsen, C.J. Palmstrøm, Appl. Phys. Lett. **104**, 201603 (2014)
25. S. Chadov, J. Kiss, J. Kübler, C. Felser, Phys. Status Solidi RRL **7**, 1 (2012)
26. P. Soven, Phys. Rev. **156**, 809 (1967)
27. W.H. Butler, Phys. Rev. B **31**, 3260 (1985)
28. H. Ebert, D. Ködderitzsch, J. Minár, Rep. Prog. Phys. **74**(9), 096501 (2011)
29. X.M. Zhang, W.H. Wang, E.K. Liu, G.D. Liu, Z.Y. Liu, G.H. Wu, Appl. Phys. Lett. **99**(7), 071901 (2011)
30. C.W. Groth, M. Wimmer, A.R. Akhmerov, J. Tworzydło, C.W.J. Beenakker, Phys. Rev. Lett. **103**, 196805 (2009)
31. H. Jiang, L. Wang, Q.F. Sun, X.C. Xie, Phys. Rev. B **80**, 165316 (2009)
32. J. Li, R.L. Chu, J.K. Jain, S.Q. Shen, Phys. Rev. Lett. **102**, 136806 (2009)
33. S.H. Wei, A. Zunger, Phys. Rev. B **37**, 8958 (1988)
34. P. Carrier, S.H. Wei, Phys. Rev. B **70**, 035212 (2004)
35. F. Virot, R. Hayn, M. Richter, J. van den Brink, Phys. Rev. Lett. **111**, 146803 (2013)

Index

C. Felser and A. Hirohata (eds.), *Heusler Alloys*,
Springer Series in Materials Science 222, DOI 10.1007/978-3-319-21449-8

D

E

F

G

H

T

MIX
Papier aus verantwortungsvollen Quellen
Paper from responsible sources
FSC® C105338

If you have any concerns about our products,
you can contact us on
ProductSafety@springernature.com

In case Publisher is established outside the EU,
the EU authorized representative is:
Springer Nature Customer Service Center GmbH
Europaplatz 3, 69115 Heidelberg, Germany

Printed by Libri Plureos GmbH
in Hamburg, Germany